The IMA Volumes in Mathematics and its Applications

Volume 85

Springer
New York
Berlin
Heidelberg
Barcelona
Budapest
Hong Kong
London
Milan
Paris
Santa Clara
Singapore
Tokyo

Institute for Mathematics and its Applications IMA

The **Institute for Mathematics and its Applications** was established by a grant from the National Science Foundation to the University of Minnesota in 1982. The IMA seeks to encourage the development and study of fresh mathematical concepts and questions of concern to the other sciences by bringing together mathematicians and scientists from diverse fields in an atmosphere that will stimulate discussion and collaboration.

The IMA Volumes are intended to involve the broader scientific community in this process.

Avner Friedman, Director

Robert Gulliver, Associate Director

* * * * * * * * * *

IMA ANNUAL PROGRAMS

1982–1983	Statistical and Continuum Approaches to Phase Transition
1983–1984	Mathematical Models for the Economics of Decentralized Resource Allocation
1984–1985	Continuum Physics and Partial Differential Equations
1985–1986	Stochastic Differential Equations and Their Applications
1986–1987	Scientific Computation
1987–1988	Applied Combinatorics
1988–1989	Nonlinear Waves
1989–1990	Dynamical Systems and Their Applications
1990–1991	Phase Transitions and Free Boundaries
1991–1992	Applied Linear Algebra
1992–1993	Control Theory and its Applications
1993–1994	Emerging Applications of Probability
1994–1995	Waves and Scattering
1995–1996	Mathematical Methods in Material Science
1996–1997	Mathematics of High Performance Computing
1997–1998	Emerging Applications of Dynamical Systems
1998–1999	Mathematics in Biology

Continued at the back

Stanislav A. Molchanov Wojbor A. Woyczynski
Editors

Stochastic Models in Geosystems

With 120 Illustrations

Springer

Stanislav A. Molchanov
Department of Mathematics
University of North Carolina
at Charlotte
Charlotte, NC 28223, USA

Wojbor A. Woyczynski
Center for Stochastic and Chaotic
Processes in Science and Technology
Case Western Reserve University
Cleveland, OH 44106, USA

Mathematics Subject Classifications (1991): 60G-XX, 60H-XX, 60H15, 76B15, 76B35, 76C15, 76M35, 76R50, 85A15, 86-02, 86A05, 86A10, 86A15

Library of Congress Cataloging-in-Publication Data
Stochastic models in geosystems / editors, Stanislav A. Molchanov,
Wojbor A. Woyczynski.
p. cm. – (The IMA volumes in mathematics and its
applications ; v. 85)
Papers from a workshop held at the University of Minnesota, May
1994.
Includes bibliographical references and index.
ISBN 0-387-94873-2 (hardcover : alk. paper)
1. Earth sciences – Mathematical models – Congresses. 2. Stochastic
processes – Congresses. I. Molchanov, S. A. (Stanislav A.)
II. Woyczynski, W. A. (Wojbor Andrzej), 1943- . III. Series.
QE33.2.M3S76 1997
550.1′5118 – dc20 96-38280

Printed on acid-free paper.

Production managed by Karina Gershkovich; manufacturing supervised by Johanna Tschebull.
Camera-ready copy prepared by the IMA.
Printed and bound by Braun-Brumfield, Inc., Ann Arbor, MI.
Printed in the United States of America.

9 8 7 6 5 4 3 2 1

ISBN 0-387-94873-2 Springer-Verlag New York Berlin Heidelberg SPIN 10550609

FOREWORD

This IMA Volume in Mathematics and its Applications

STOCHASTIC MODELS IN GEOSYSTEMS

is based on the proceedings of a workshop with the same title and was an integral part of the 1993–94 IMA program on "Emerging Applications of Probability." We would like to thank Stanislav A. Molchanov and Wojbor A. Woyczynski for their hard work in organizing this meeting and in editing the proceedings. We also take this opportunity to thank the National Science Foundation, the Office of Naval Research, the Army Research Office, and the National Security Agency, whose financial support made this workshop possible.

Avner Friedman

Willard Miller, Jr.

PREFACE

A workshop on Stochastic Models in Geosystems was held during the week of May 16, 1994 at the Institute for Mathematics and Its Applications at the University of Minnesota. It was part of the Special Year on Emerging Applications of Probability program put together by an organizing committee chaired by J. Michael Steele.

The invited speakers represented a broad interdisciplinary spectrum including mathematics, statistics, physics, geophysics, astrophysics, atmospheric physics, fluid mechanics, seismology, and oceanography. The common underlying theme was stochastic modeling of geophysical phenomena and papers appearing in this volume reflect a number of research directions that are currently pursued in these areas.

We have arranged the papers in alphabetical order. From the methodological mathematical viewpoint most of the contributions fall within the areas of wave propagation in random media, passive scalar transport in random velocity flows, dynamical systems with random forcing and self-similarity concepts, including multifractals.

The meeting was supported by IMA and by the Office of Naval Research. We would also like to thank IMA for its hospitality and competent support and, in particular, Avner Friedman and Willard Miller who assembled a very capable staff, and whose leadership created a welcoming and productive intellectual atmosphere at the Institute.

Stanislav A. Molchanov

Wojbor A. Woyczynski

CONTENTS

SEISMIC CODA WAVES: A STOCHASTIC PROCESS IN EARTH'S LITHOSPHERE

KEIITI AKI*

Abstract. Seismic coda waves are a natural wonder. Because they are formed by scattered waves from numerous heterogeneities in the lithosphere, nature does the averaging over a large volume of the earth and leads to beautiful simplicity such as the separability of seismic source, propagation path and recording site effects. In this review, we shall focus on the decay rate of coda amplitudes, called coda Q^{-1}, and present its significance as a geophysical parameter characterizing regional structures and earthquake processes in the lithosphere.

1. Introduction. When an earthquake or underground explosion occurs in the earth, seismic waves are propagated away from the source. After P waves, S waves and various surface waves are gone, the area around the seismic source is still vibrating. The amplitude of vibration is uniform in space, except for the local site effect, which tends to amplify the motion at soft soil sites as compared to hard rock sites. This residual vibration is called seismic coda wave, and decays very slowly with time. The rate of decay is the same independent of the locations of seismic source and recording station, as long as they are located in a given region. An example of coda waves is shown in Fig. 1.1. They are seismograms of a local earthquake near a seismic array in Norway constructed for monitoring the underground nuclear testing in USSR in early 1970's. The array aperature is about 100 km, and the epicentral distance is a few kilometers to the closest station, and more than 100 km to the farthest. In spite of the great difference in distance, which is of course reflected in arrival times and amplitudes of primary waves, the coda waves show a very similar amplitude and rate of decay among all stations. The signal is band-pass filtered around 4 Hz, and the coda lasts more than 200 seconds in this old stable part of the continent.

The closest phenomena to this coda waves is the residual sound in a room, first studied by W.C. Sabine (1922). If you shoot a gun in a room, the sound energy remains for a long time due to incoherent multiple reflections. This residual sound has a very stable, robust nature similar to seismic coda waves, independent of the locations where you shoot the gun or where you record the sound in the room.

The residual sound remains in the room because of multiple reflections at rigid wall, ceiling and floor of the room. Since we cannot hypothesize any room-like structure in the earth, we attribute seismic coda waves to back-scattering from numerous heterogeneities in the earth. We may consider seismic coda as waves trapped in a random medium. The seismic

* Southern California Earthquake Center, University of Southern California, Los Angeles, CA 90089.

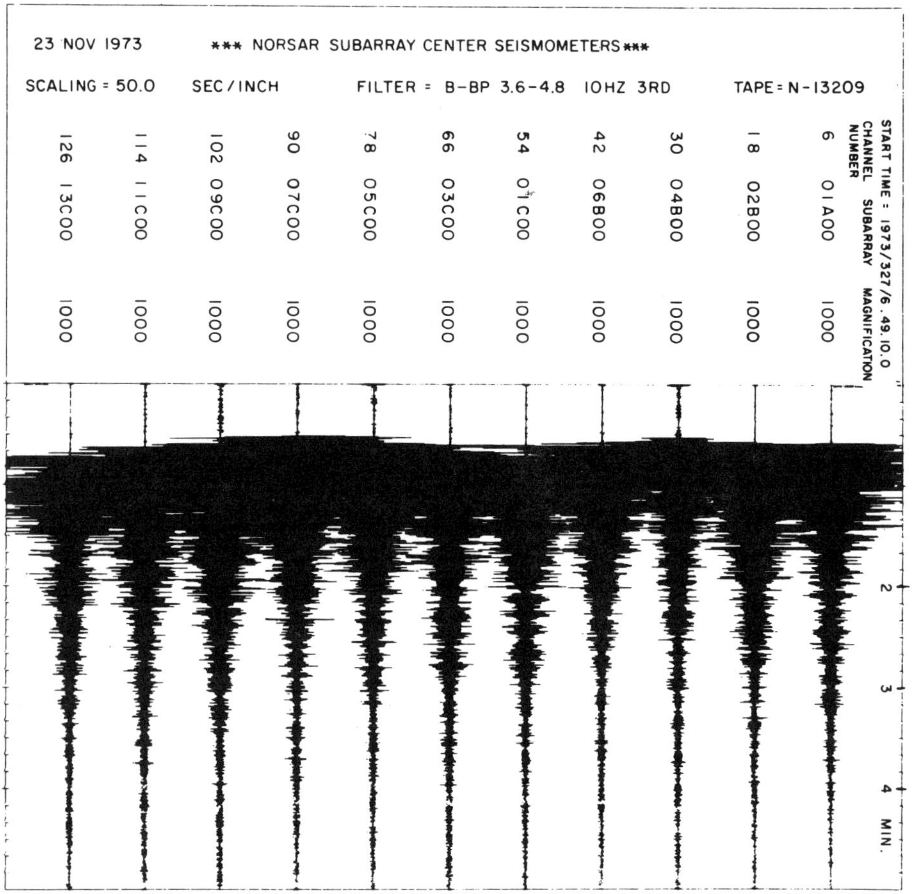

FIG. 1.1. *Short-period (band pass between 3.6 and 4.8 Hz) records of a local earthquake at Norsar near subarray 7C. The epicentral distance is a few kilometers to the closest station and more than* 100 *km to the farthest. The general level and decay rate of coda energy show no dependence on the epicentral distance.*

coda waves, however, do not represent the localization such as described by Papanicolaou (1994) in the present volume for waves trapped in plane random layers, because our observations on seismic coda can be explained by the equation for radiative energy propagation.

The seismic coda waves from a local earthquake can be best described by the time-dependent power spectrum $P(\omega|t)$, where ω is the angular frequency and t is the time measured from the origin time of the earthquake. $P(\omega|t)$ can be measured from the squared output of a band-pass filter centered at a frequency ω, or from the squared Fourier amplitude obtained from a time window centered at t. The most extraordinary property of $P(\omega|t)$ is the simple separability of the effects of seismic source, propagation path and recording site response, expressed by the following equation. The coda power spectrum $P_{ij}(\omega|t)$ observed at the ith station due to the jth earthquake can be written as

$$P_{ij}(\omega|t) = S_j(\omega)R_i(\omega)C(\omega|t) \tag{1.1}$$

for t greater than about $2t_\beta$, where t_β is the travel time of S waves from the jth earthquake to the ith station. Equation (1.1) means that $P_{ij}(\omega|t)$ can be written as a product of a term which depends only on the earthquake source, a term which depends only on the recording site and a term common to all the earthquakes and recording sites in a given region.

The above property of coda waves expressed by equation (1.1) was first recognized by Aki (1969) for aftershocks of the Parkfield, California earthquake of 1966. The condition that equation (1.1) holds for t greater than about $2t_\beta$ was found by the extensive study of coda waves in central Asia by Rautian and Khalturin (1978). Numerous investigators demonstrated the validity of equation (1.1) for earthquakes around the world, as summarized in a review article by Herraiz and Espinosa (1987). In general, (1.1) holds more accurately for a greater lapse time t and for higher frequencies (e.g., Su *et al.*, 1991). Coda waves are a powerful tool for seismologists because equation (1.1) offers a simple means to separate the effects of source, path and recording site. The equation has been used for a variety of practical applications, including the mapping of frequency-dependent site amplification factor (e.g., Su *et al.*, 1992), discrimination of quarry blasts from earthquakes (Su *et al.*, 1991), single station method for determining frequency-dependent attenuation coefficients (Aki, 1980), and normalizing the regional seismic network data to a common source and recording site condition (e.g., Mayeda *et al.*, 1992).

In the present paper, we shall not discuss applications to the seismic source or recording site response, but focus on the common decay function $C(\omega|t)$ on the right hand of equation (1.1). We shall first introduce coda Q to characterize $C(\omega|t)$ in the framework of single-scattering theory, and then summarize the current results on what the coda Q is in terms of scattering attenuation and intrinsic absorption. Then, we shall survey the spatial and temporal variation in coda Q^{-1} and describe how they are related to earthquake processes in the lithosphere.

2. Introducing coda Q (or coda Q^{-1}). The first attempt to predict the explicit form of $P(\omega|t)$ for a mathematical model of earthquake source and earth medium was made by Aki and Chouet (1975). Their models were based on the following assumptions.

1. Both primary and scattered waves are S waves.
2. Multiple scatterings are neglected.
3. Scatterers are distributed randomly with a uniform density.
4. Background elastic medium is uniform and unbounded.

The assumption (1.1) has been supported by various observations, such as the common site amplification (Tsujiura, 1978) and the common attenuation (Aki, 1980) between S waves and coda waves. It is also supported theoretically because the S to P conversion scattering due to a localized heterogeneity is an order of magnitude smaller than the P to S scattering as shown by Aki (1992) using the reciprocal theorem. Zeng (1993) have shown that the above difference in conversion scattering between P to S and S to P leads to the dominance of S waves in the coda.

Since the observed $P(\omega|t)$ is independent of the distance between the source and receiver, we can simplify the problem further by co-locating the source and receiver. Then, we find (Aki and Chouet, 1975, see also Aki, 1981, for more detailed derivation) that

$$P(\omega|t) = (\beta/2)g(\pi)|\phi_0(\omega|\beta t/2)|^2 \tag{2.1}$$

where β is the shear wave velocity, $g(\theta)$ is the directional scattering coefficient, and $\phi_0(\omega|r)$ is the Fourier transform of the primary waves at a distance r from the source. $g(\theta)$ is defined as 4π times the fractional loss of energy by scattering per unit travel distance of primary waves and per unit solid angle at the radiation direction θ measured from the direction of primary wave propagation.

Aki and Chouet (1975) adopted the following form for $|\phi_0(\omega|r)|$.

$$|\phi_0(\omega|r)| = |S(\omega)|r^{-1}\exp(-\omega r/(2\beta Q_c)), \tag{2.2}$$

where $|S(\omega)|$ is the source spectrum, r^{-1} represents the geometrical spreading, and Q_c is introduced to express the attenuation. Combining (2.1) and (2.2), and including the attenuation of scattered waves, we have

$$P(\omega|t) = \frac{2g(\pi)|S(\omega)|^2}{\beta t^2}\exp(-\omega t/Q_c) \tag{2.3}$$

Q_c is called "coda Q", and Q_c^{-1} is called "coda Q^{-1}".

The measurement of coda Q according to equation (2.3) is very simple. Coda Q^{-1} is the slope of straight line fitting the measured $ln(t^2P(\omega|t))$ vs. ωt. Since there is a weak but sometimes significant dependence of the slope on the time window for which the fit is made, it has become a necessary routine to specify the time window for each measured coda Q^{-1}.

Because of the simplicity of measurement of coda Q^{-1}, its geographical variation over a large area as well as its temporal variation over a long time can be studied relatively easily. Before presenting those results, however, we need to clarify what is the physical meaning of coda Q^{-1}.

3. Physical meaning of coda Q^{-1}. The physical meaning of coda Q^{-1} has been debated for almost twenty years. Within the context of the single scattering theory, coda Q^{-1} appears to represent an effective attenuation including both absorption and scattering loss. This idea prevailed for some time after Aki (1980) found a close agreement between coda Q^{-1} and Q^{-1} of S waves measured in the Kanto region, Japan. On the other hand, numerical experiments by Frankel and Clayton (1986), laboratory experiments by Matsunami (1991), and theoretical studies including multiple scattering effects (e.g., Shang and Gao, 1988) concluded that the coda Q^{-1} measured from the time window later than the mean free time (mean free path divided by wave velocity) should correspond only to the intrinsic absorption, and should not include the effect of scattering loss. The debates concerning this issue were summarized by Aki (1991).

In order to resolve the above issue, attempts have been made to separately determine the scattering loss and the intrinsic loss in regions where coda Q^{-1} has been measured. For this purpose, it is necessary to include multiple scattering in the theoretical model, either by the radiative energy transfer approach (Wu, 1985) or by the inclusion of several multiple-path contributions to the single-scattering model (Gao *et al.*, 1983a,b). Recently, Zeng *et al.* (1991) demonstrated that all these approaches can be derived as approximate solutions of the following integral equation for the seismic energy density $E(\boldsymbol{r},t)$ per unit volume at a location $\boldsymbol{r}$ and at time t due to an impulsive point source applied at $\boldsymbol{r}_0$ at $t=0$.

$$E(\boldsymbol{r},t)=E_o\left(t-\frac{|\boldsymbol{r}-\boldsymbol{r}_0|}{\beta}\right)\frac{e^{-\eta|\boldsymbol{r}-\boldsymbol{r}_0|}}{4\pi|\boldsymbol{r}-\boldsymbol{r}_0|^2}+\int_V \eta_s E\left(\xi,t-\frac{|\xi-\boldsymbol{r}|}{\beta}\right)\frac{e^{-\eta|\xi-\boldsymbol{r}|}}{4\pi|\xi-\boldsymbol{r}|^2}dV(\xi) \tag{3.1}$$

where symbols are defined as follows:

$E(\boldsymbol{r},t)$: seismic energy per unit volume at $\boldsymbol{r}$ and t

η: total attenuation coefficient: $\eta=\eta_s+\eta_i$ (energy decays with distance $|\boldsymbol{r}|$ as $\exp[-\eta|\boldsymbol{r}|]$)

η_i: intrinsic absorption coefficient

η_s: scattering attenuation coefficient

$Q_s^{-1}=\frac{\eta_s\beta}{\omega}$: scattering Q^{-1}

$Q_i^{-1}=\frac{\eta_i\beta}{\omega}$: absorption Q^{-1}

$B_o=\frac{\eta_s}{\eta}$: seismic albedo

$L_e=\frac{1}{\eta}$: extinction distance

$L=\frac{1}{\eta_s}$: mean free path

$\beta E_o(t)$: rate of energy radiated from a point source at time t.

The assumptions underlying equation (3.1) are less restrictive and more explicit than the assumptions used in deriving equations (2.1) and (2.3). The background medium is still uniform and unbounded, but scattering coefficients and absorption coefficients are explicitly specified, and all the multiple scatterings are included, although scattering is assumed to be isotropic. Equation (3.1) gives the seismic energy density as a function of distance and time in contrast to equations (2.1) and (2.3) which depend on time only. By comparing the predicted energy density in space and time with the observed, we can uniquely determined the scattering loss and the intrinsic absorption, separately.

An effective method using the Monte-Carlo solution of equation (3.1) was developed by Hoshiba *et al.* (1991) who calculated seismic energy integrated over three consecutive time-windows (e.g., 0 to 15, 15 to 30, 30 to 45 sec from the S arrival time) and plotted them against distance from the source. The method has been applied to the various parts of Japan (Hoshiba, 1993), Hawaii, Long Valley, California, central California (Mayeda *et al.*, 1992), and southern California (Jin *et al.*, 1994).

Fig. 3.1 shows typical examples of comparison between the predicted energy density and the observed. The observed energy density data come from many small earthquakes recorded at a single station GSC, southern California, and are normalized to a common source by the coda normalization method of Aki (1980). In spite of the simplified assumptions made in the prediction, the comparison with the observed is quite good, giving us some confidence in the estimated scattering and absorption coefficients.

Fig. 3.2 (Jin *et al.*, 1994) compares seismic albedo B_0, total attenuation coefficient $\eta(= Le^{-1})$ in km^{-1}, scattering attenuation coefficient η_s and intrinsic absorption coefficient η_i among Long Valley, central California, southern California, Hawaii, and Japan. Except for outlyers for Hawaii (at 1.5 and 3 Hz) which are probably due to inadequacy of model, the scattering and attenuation of seismic waves in the frequency range from 1 to 25 Hz are remarkably consistent in all these regions.

First, η_s tends to decrease with increasing frequency. This corresponds to the decrease of Q_s^{-1} faster than f^{-1} with increasing frequency f. In terms of the random medium model, this implies that the auto-correlation function may be more like Gaussian rather than exponential (Sato, 1982a,b; Wu, 1982; Frankel and Clayton, 1986). The Gaussian type medium shows much smoother variation than the exponential type below the correlation distance. There appears to be no difference between geothermal areas such as Hawaii and Long Valley and primarily non-geothermal active areas such as central and southern California.

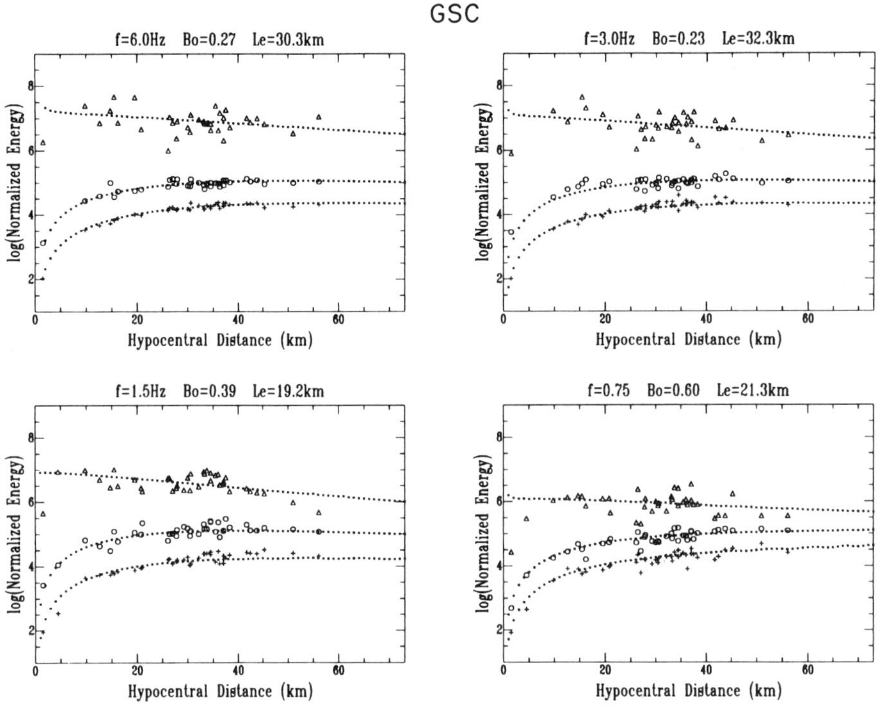

FIG. 3.1. *Observed coda energy in 4 frequency bands in three consecutive windows (triangle: 15 sec. from S-arrival, circle: 15 to 30 sec., and cross: 30 to 45 sec.) at station GSC in southern California normalized by the coda method. The dotted curves represent the prediction for best-fitting parameters. The center frequency f of the band, albedo* B_0, *and extinction distance* L_e *are shown above each figure.*

Fig. 3.2 also exhibits that the intrinsic absorption coefficient η_i shows a slight increase with increasing frequency. Because of the opposite frequency dependence of η_s and η_i, the intrinsic absorption dominates over the scattering at higher frequencies. Thus, the seismic albedo B_0 shown in Fig. 3.2 is less than 0.5 for all areas for frequencies higher than 5 Hz. For lower frequencies, we found large regional variation in B_0.

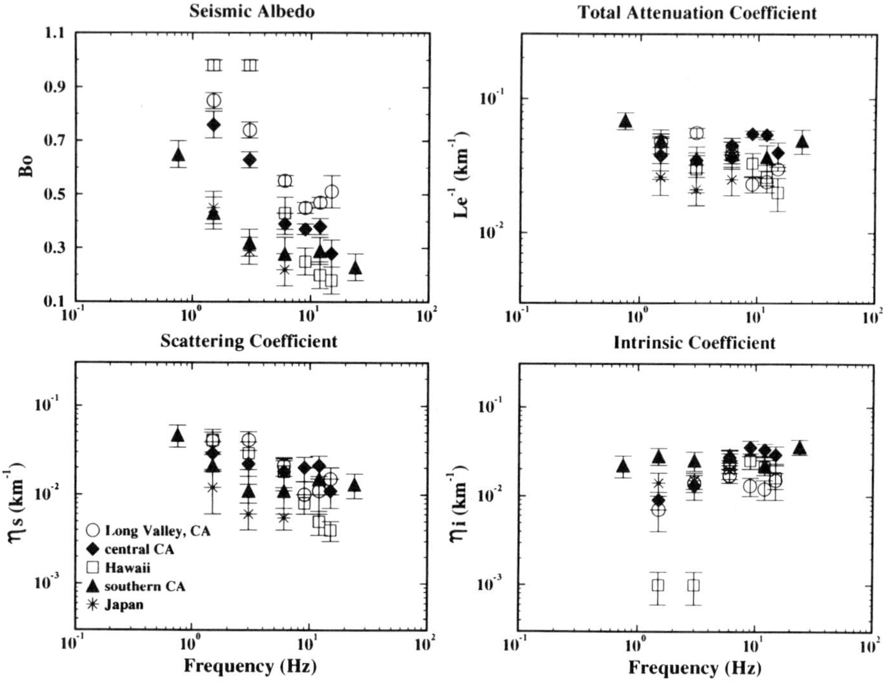

FIG. 3.2. *Albedo* B_0*, total attenuation coefficient* $\eta(L_e^{-1})$*, scattering attenuation coefficient* η_s *and intrinsic absorption coefficient* η_i *as a function of frequency for various regions.*

Now that we have several regions where intrinsic Q^{-1} and scattering Q^{-1} have been determined, we can compare the coda Q with them. Fig. 3.3 shows Q_i^{-1}, Q_s^{-1} and Q_t^{-1} $(= Q_i^{-1} + Q_s^{-1})$ as a function of coda Q^{-1} separately for Hawaii, southern California, central California, Long Valley and Japan. The coda Q^{-1} is determined for the lapse time interval 20 to 45s for southern California, 30 to 60s for Long Valley, Hawaii and central California, and 20 to 60s for Japan. In general, coda Q^{-1} lies between Q_i^{-1} and Q_t^{-1}. It is closer to Q_i^{-1} for Japan, and closer to Q_t^{-1} for all other regions. According to Gao and Aki (1994), who made numerical study of the departure of coda Q^{-1} from Q_i^{-1} for models of a scattering layer with a finite thickness, the above results may indicate that the thickness of scattering layer is greater than the mean free path under Japan, but comparable or smaller than the mean free path for the other regions.

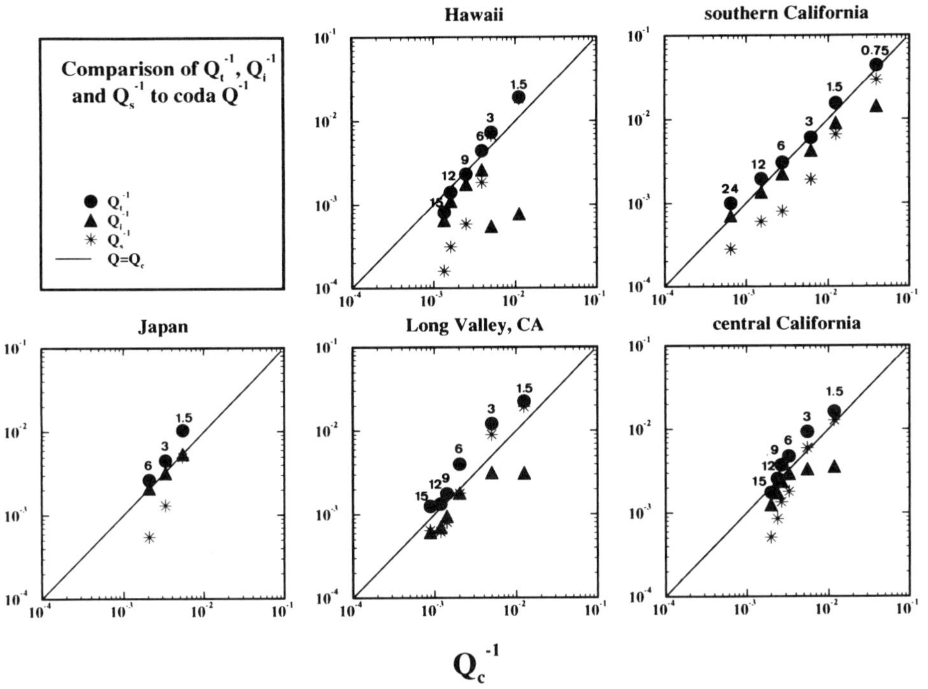

FIG. 3.3. *Comparison of coda Q^{-1} (Q_c^{-1}) with scattering $Q^{-1}(Q_s^{-1})$, intrinsic $Q^{-1}(Q_c^{-1})$ and total $Q^{-1}(Q_t^{-1})$ at various frequencies for five regions.*

Although, for a more complete understanding of coda Q, we need models with non-uniform scattering and absorption coefficients, Fig. 3.3 assures us empirically that coda Q^{-1} are bounded rather narrowly between intrinsic Q^{-1} and total Q^{-1}. With this understanding of coda Q^{-1}, we shall now proceed to the spectacular spatial and temporal correlation observed between coda Q^{-1} and seismicity.

4. Geographic variation in coda Q. The decay rate of coda waves shows a strong geographic variation. The example from Norway, an old stable region, given in Fig. 1.1 clearly suggests a very high Q because the duration of 200 seconds for 4 Hz means 800 cycles of vibration. On the other hand, the coda decays much more quickly in young active regions, such as Japan and California. for example, Singh and Herrmann (1983) found a systematic variation of coda Q at 1 Hz in the conterminous United States: more than 1000 in the central part decaying gradually to 200 in the western U.S.

The spatial resolution of the map of coda Q obtained by Singh and Herrmann (1983) was rather poor, because they had to use distant earthquakes to cover regions of low seismicity. As mentioned earlier, equation (1.1) holds for the lapse time greater than about twice the travel time for S waves. For a more distant earthquake, the coda part governed by equation

(1.1) starts later, making the region sampled by waves composing the coda greater and consequently losing the spatial resolution.

Peng (1989) made a systematic study of the spatial resolution of coda Q mapping as a function of the lapse time window selected for measuring coda Q. He used the digital data from the Southern California Seismic Network operated by Caltech and USGS, and calculated spatial auto-correlation function of coda Q^{-1} by the following procedure. Southern California is divided into meshes of size 0.2^o (longitude) by 0.2^o (latitude), and the average of coda Q^{-1} is calculated for each mesh using seismograms which have the mid-point of epicenter and station in the mesh. The average value for the ith mesh is designated as x_i. Then two circles of radius r and $r+20$ km are drawn with the center at the ith mesh, and the mean of coda Q^{-1} at mid-points located in the ring between the two circles is calculated and designated as $y_i(r)$. The auto-correlation coefficient $\rho(r)$ is computed by the following formula,

$$\rho(r) = \frac{\sum_{i=1}^{M}(x_i - \overline{x})(y_i(r) - \overline{y}(r))}{\sqrt{\sum_{i=1}^{M}(x_i - \overline{x})^2}\sqrt{\sum_{i=1}^{M}(y_i(r) - \overline{y}(r))^2}}$$

where, M is the total number of meshes, $\overline{x}$ is the mean of x_i, and $\overline{y}(r)$ is the mean of $y_i(r)$.$\rho(r)$ is calculated for coda Q^{-1} at four different frequencies $(1.5, 3, 6$ and 12 Hz) and three different lapse time windows, namely, 15 to 30 sec, 20 to 45 sec and 30 to 60 sec measured from the origin time. As shown in Fig. 4.1 through 4.3, the auto-correlation functions are similar among different frequencies, but depend clearly on the selected time window. The longer and later time window gives the slower decay in the auto-correlation with the distance separation. If we define the distance at which the correlation first comes close to zero as the "coherence distance", the average coherence distance is about 135 km for time window $30-60$ sec, 90 km for the window $20-45$ sec, and 45 km for the window $15-30$ sec.

The above observation offers a strong support to the assumption that coda waves are composed of S to S back-scattering waves, because the distance traveled by S waves with a typical crustal S wave velocity of 3.5 km/s in half the lapse time $60, 45$ and 30 sec are, respectively, $105, 79$ and 53 km, which are close to the corresponding coherence distance, namely, $135, 90$ and 45 km. In other words, the coda Q^{-1} measured from a time window represents the seismic attenuation property of the earth's crust averaged over the volume traversed by the singly back-scattering S waves.

Jin and Aki (1988) were able to construct a map of coda Q at 1 Hz for the mainland China with a high spatial resolution using earthquakes at short distances from each station as shown in Fig. 4.4. The coda Q (1 Hz) at individual stations estimated from these earthquakes for the time window from $2t_\beta$ to 100 sec. shown in Fig. 4.5, where the variation is smooth enough to draw contours of equal coda Q. The contour map of

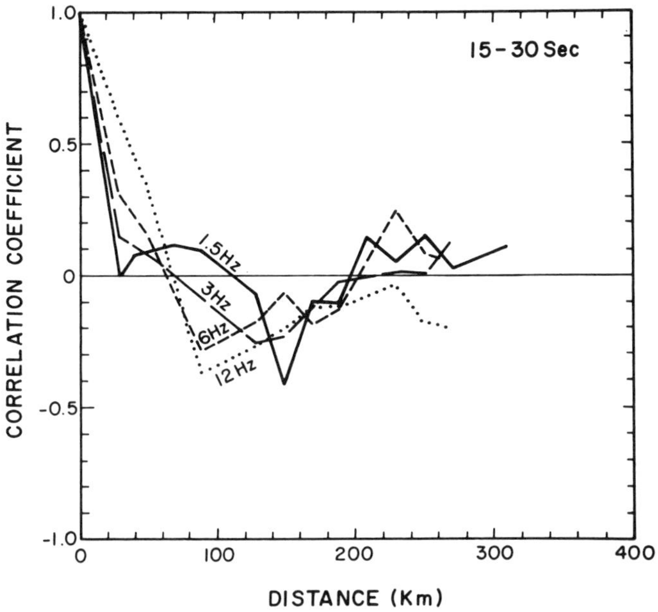

FIG. 4.1. *Spatial auto-correlation function of coda Q^{-1} at various frequencies measured from the time window* $15 - 30$ *sec. in southern California obtained by Peng (1989).*

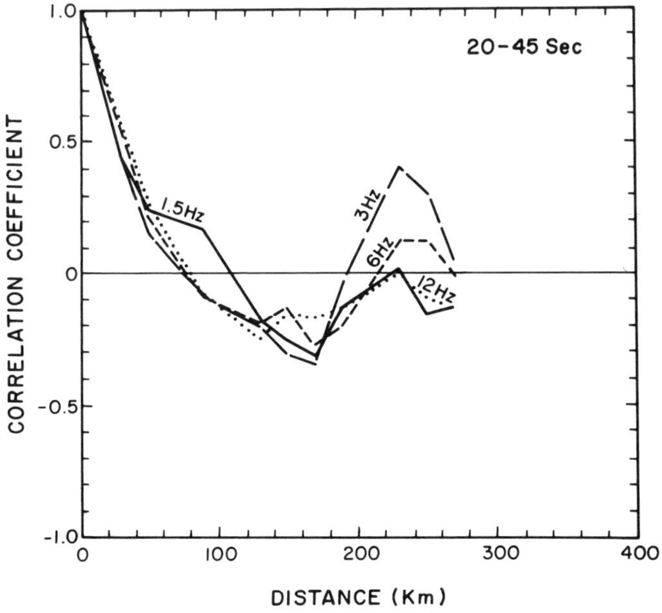

FIG. 4.2. *Spatial auto-correlation function of coda Q^{-1} at various frequencies measured from the time window* $20 - 45$ *sec. in southern California obtained by Peng (1989).*

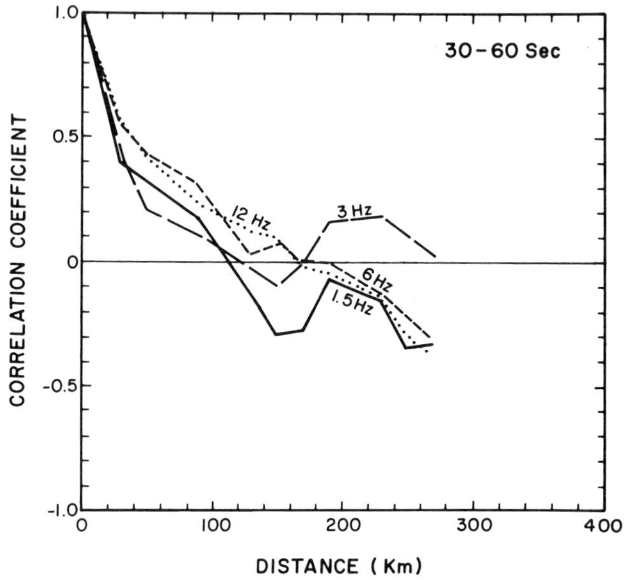

FIG. 4.3. *Spatial auto-correlation function of coda* Q^{-1} *at various frequencies measured from the time window* 30 − 60 *sec. in southern California obtained by Peng (1989).*

coda Q is compared with epicenters of major earthquakes with $M \geq 7$ in Fig. 4.6. A strong correlation was found between coda Q and seismicity. Seismically active regions, such as Tibet, western Yunnan and North China, corresponds to low coda Q region, and stable regions such as Ordos plateau, middle-eastern China, and the desert in southern Xinjiang have very high coda Q. The difference between the highest coda Q value and the lowest amounts to more than a factor of 20. Thus, mapping coda Q can be useful for the assessment of long-term seismic hazard.

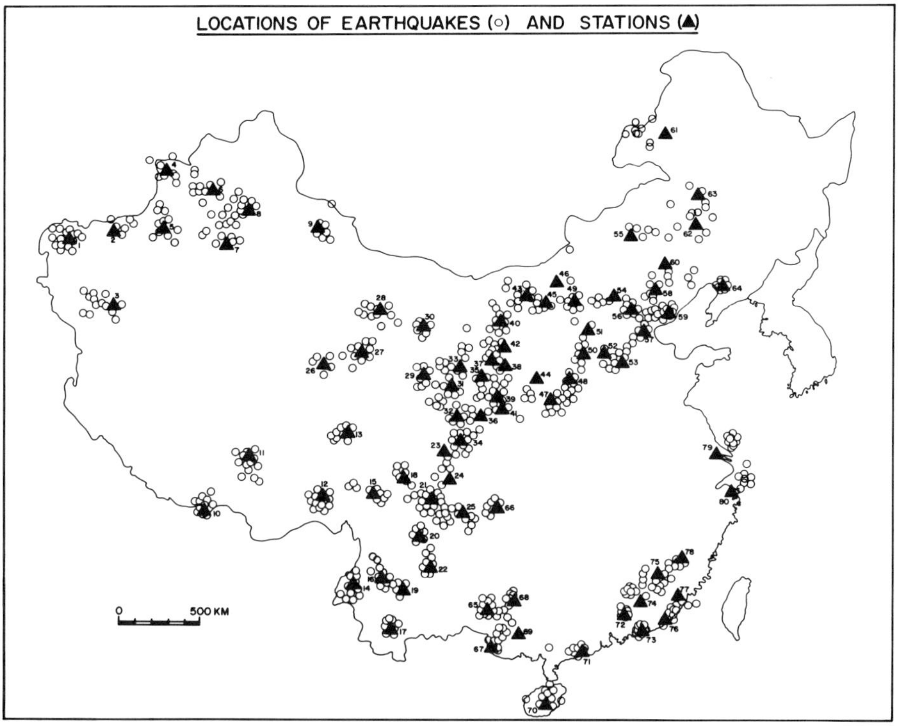

FIG. 4.4. *Distribution of earthquakes (circles) and stations (cross) used by Jin and Aki (1988) for coda Q measurements in China.*

Two different symbols are used to distinguish earthquakes that occurred before 1700 from those that occurred after 1700. As well known among Chinese seismologists, there has been a migration of epicenters from west to east during the past 300 years in North China. It is interesting to note that the coda Q values for the region active before 1700 is about twice as high as those for the region currently active. Jin and Aki (1988) suggests that the low coda Q region might also have migrated together with the high seismicity, referring to the Q values estimated by Chen and Nuttli (1984a,b) from intensity maps for past major earthquakes in the region. This leads to the most intriguing observation on coda Q, namely, its temporal change.

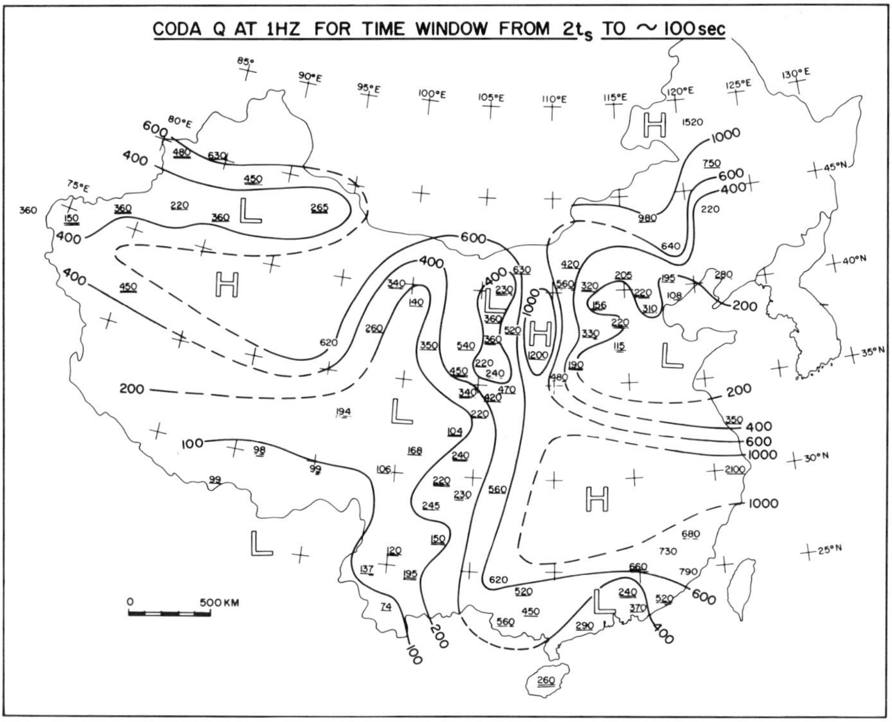

FIG. 4.5. *Measured coda Q values at* 1 *Hz with time window from* $2t_s$ *(*t_s *: S-wave arrival time from the origin time) to* 100 *sec. and the iso-Q lines, adapted from Jin and Aki (1988).*

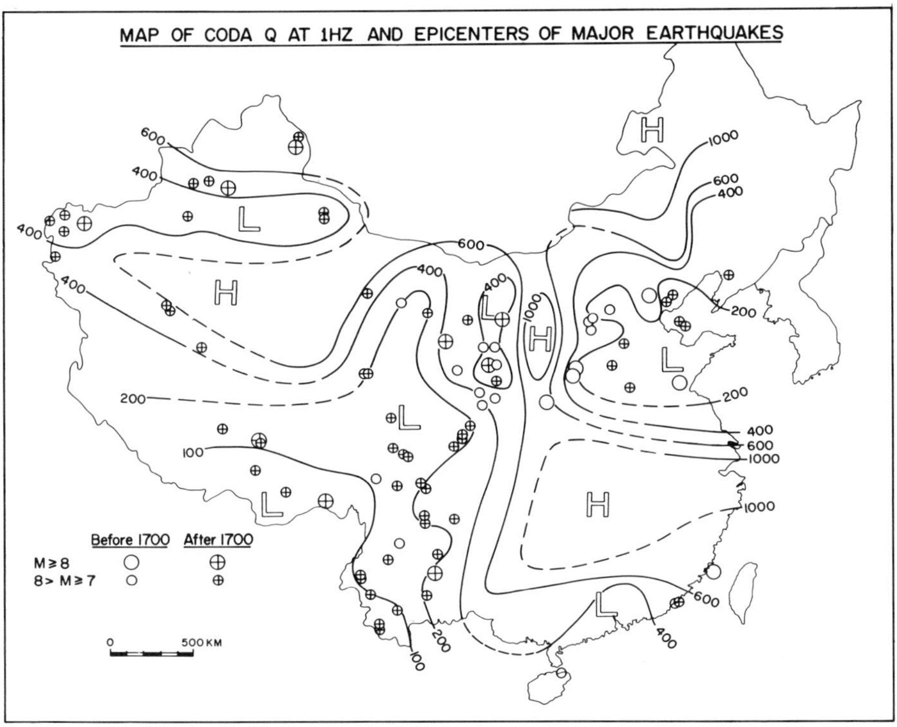

FIG. 4.6. *Map of coda Q at 1 Hz and epicenters of major earthquakes with* $M \geq 7$. *Different symbols are used for* $M \geq 8$ *and* $M < 8$, *and before and after* 1700, *adapted from Jin and Aki (1988). The high Q regions are devoid of earthquakes, and the low Q regions are full of them.*

5. Temporal change in coda Q. Chouet (1979) was the first to observe a significant temporal change in coda Q at Stone Canyon, California, which could not be attributed to changes in instrument response, or in the epicenter locations, focal depths, or magnitudes of earthquakes used for the measurement. The change was associated with neither the rainfall in the area nor with the occurrence of any particular earthquake, but showed a weak negative correlation with the temporal change in a seismicity parameter called "b-value" (Aki, 1985). The b-value is defined in the Gutenberg-Richter formula $\log N = a - bM$, where N is the frequency of earthquakes with magnitude equal or greater than M.

Numerous studies made since (see Sato (1988) for a critical review of early works) revealed that the temporal correlation between coda Q^{-1} and seismicity is not as simple as the spatial correlation described in the preceding section.

In a number of cases (Gusev and Lemzikov, 1984; Novelo-Casanova *et al.*, 1985; Jin and Aki, 1986; Sato, 1986; Faulkner, 1988; Su and Aki, 1990), coda Q^{-1} shows a peak during a period of $1-3$ years before the occurrence of a major earthquake. A similar precursory pattern showed up also before the 1989 Loma Prieta earthquake in central California, and the Landers earthquake in southern California (Jin and Aki, 1993). From the study of coda Q^{-1} over the period more than 50 years for both central and southern California, as shown in Fig. 5.1, Jin and Aki (1993) had to conclude that the coda Q^{-1} precursor is not reliable, because the precursory pattern, sometimes, is not followed by a major earthquake, and some major earthquakes were not preceded by the pattern.

A rather surprisingly consistent observation made by various studies is that coda Q^{-1} tends to take a minimum value during the period of high aftershock activity (Gusev and Lemzikov, 1984; Novelo-Casanova *et al.*, 1987; Faulkner, 1988). Furthermore, Tsukuda (1988) found in the epicentral area of the 1983 Misasa earthquake that a period of high coda Q^{-1} from 1977 to 1980 corresponds to a low rate of seismicity (quiescence). These observations suggest that the temporal change in coda Q^{-1} may not reflect change in seismic attenuation in the brittle part of the lithosphere, but may be related primarily to attenuation in the deeper ductile part.

Several convincing cases were made also for the temporal correlation between coda Q^{-1} and b-value. The result was at first puzzling because the correlation was negative in some cases (Aki, 1985; Jin and Aki, 1986; Robinson, 1987) and positive in other cases (Tsukuda, 1988; Jin and Aki, 1989). To resolve this puzzle, Jin and Aki (1989) proposed the creep model, in which creep fractures in the ductile part of the lithosphere are assumed to have a characteristic size in a given seismic region. The increased creep activity would then increase the seismic attenuation and at the same time produce stress concentration in the upper brittle part favoring the occurrence of earthquakes with magnitude M_c corresponding to the characteristic size of the creep fracture. Then, if M_c is in the lower end of the

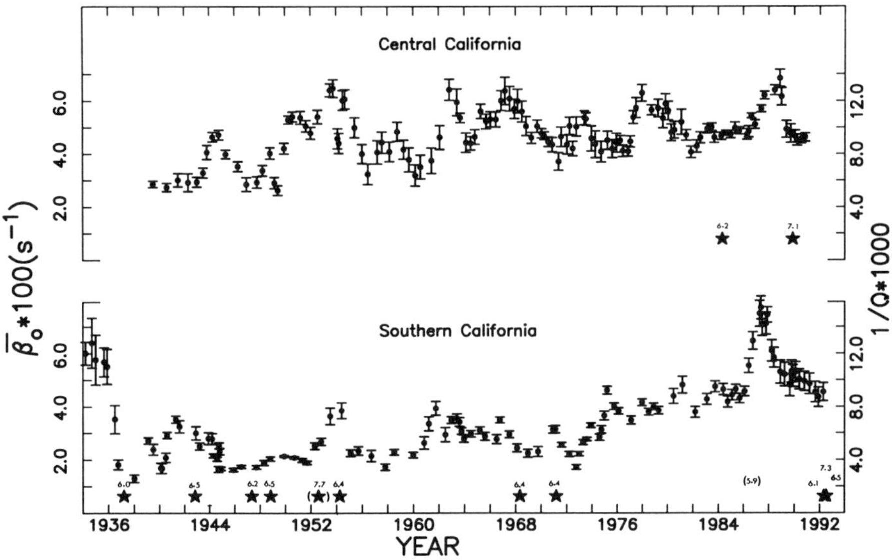

FIG. 5.1. *Mean values of $\pi f Q_c^{-1}$ plotted as a function of time for central and southern California (f is about 2 Hz). Stars indicate major earthquakes occurred within each study area. The number above each star is the magnitude.*

magnitude range from which the b-value is evaluated the b-value would show a positive correlation with coda Q^{-1}, and if M_c is in the upper end, the correlation would be negative.

The creep model is consistent with the observed behaviors of coda Q^{-1} during the periods of aftershocks and quiescence mentioned earlier. Another support for the deeper source of the coda Q^{-1} change comes from the observed coincidence between a large increase in coda Q^{-1} in southern California during 1986-87 (Peng, 1989; Jin and Aki, 1989), and the increase in electrical conductivity in the same region (Madden *et al.*, 1993) which is attributed to the lower crust.

If the creep model is correct, the strongest correlation should be found between coda Q^{-1} and the seismicity of earthquakes with M_c, and the correlation should always be positive. Indeed, Jin and Aki (1993) found a remarkable positive correlation between coda Q^{-1} and the fraction of earthquakes in the magnitude range $M_c \leq M \leq M_c + 0.5$ for both central and southern California. Fig. 5.2 shows the result for central California where the appropriate choice of M_c is 4.0. The correlation is highest (0.84) at the zero time lag and decays symmetrically with the time shift as shown in Fig. 5.3. A very similar result is obtained for southern California as shown in Fig. 5.4 where the appropriate choice of M_c is 3.0. The correlation is again the highest (0.81) at the zero time lag as shown in Fig. 5.5.

Thus, our current working hypothesis is that the temporal change in coda Q^{-1} reflects the degree of creep fractures in the ductile part of lithosphere. The ductile part of lithosphere is larger than the brittle part. The deformation in the ductile part is the source of stress in the brittle part. Although we found that the coda Q^{-1} precursor is not reliable, the study of spatial and temporal variation in coda Q^{-1} may be still promising for understanding the loading process that leads to earthquakes in the brittle part.

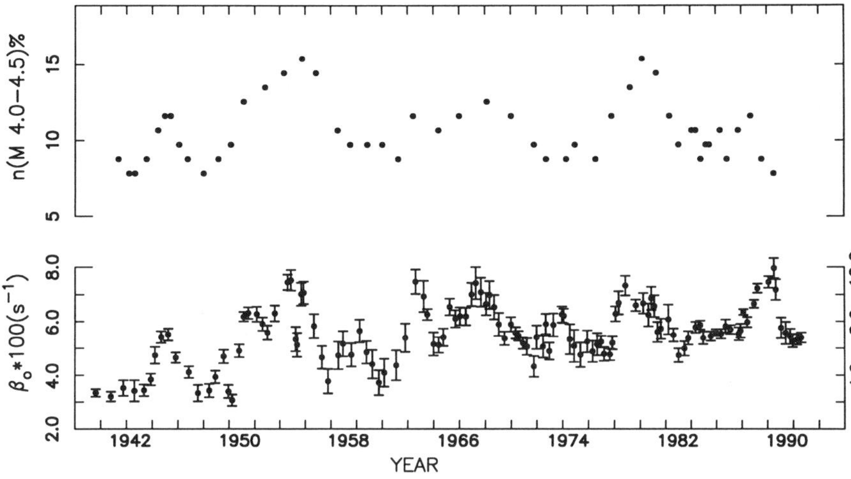

FIG. 5.2. *Comparison between temporal variations of* $\pi f Q_c^{-1}$ *(f is about 2Hz) and fractional frequency of earthquakes with magnitude* $4.0 \leq M \leq 4.5$, *for central California, from Jin and Aki (1993).*

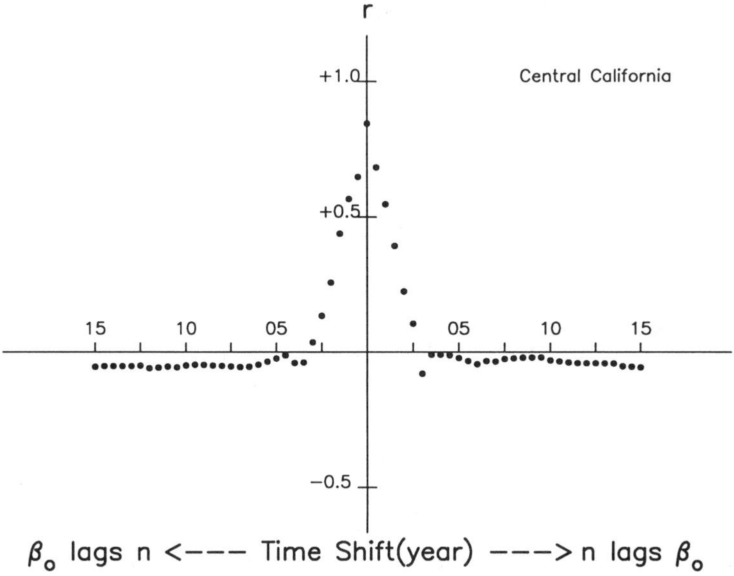

FIG. 5.3. *Cross-correlation function between the two time series shown in Fig. 5.2.*

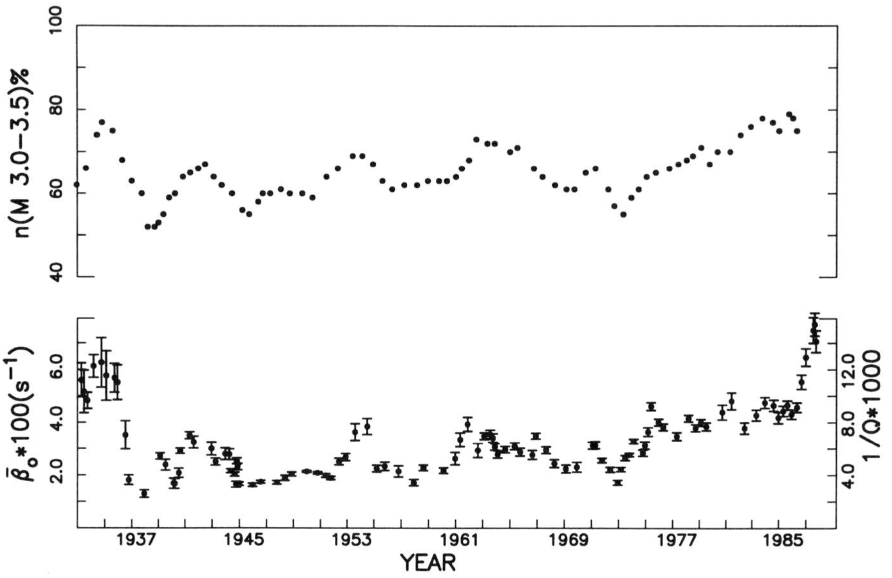

FIG. 5.4. *Comparison between temporal variations of* $\pi f Q_c^{-1}$ *(f is about 2 Hz) and fractional frequency of earthquakes with magnitude* $3.0 \leq M \leq 3.5$*, for southern California, from Jin and Aki (1993).*

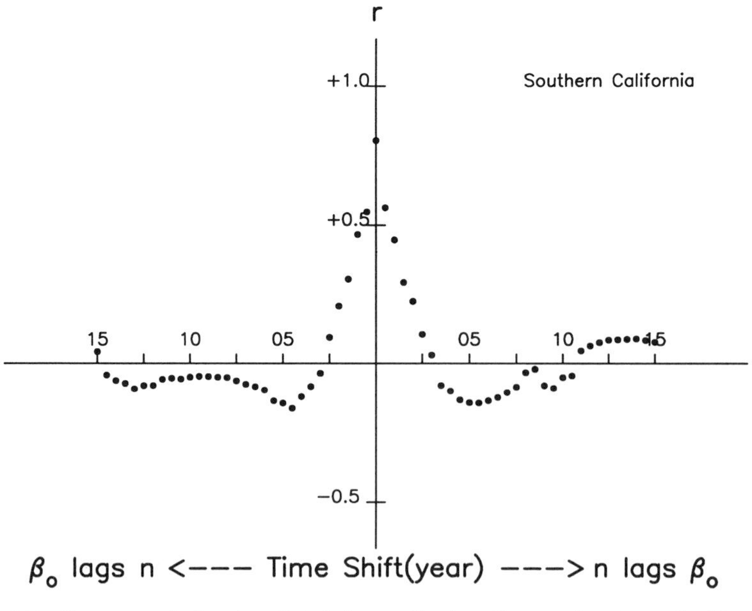

FIG. 5.5. *Cross-correlation function between the two time series shown in Fig. 5.4.*

6. Conclusions. Reviewing observational and theoretical studies of coda waves of local earthquakes made in the last quarter century, we reached the following conclusions.

(1) The coda waves are a powerful tool for seismologists because of the simple separability of the effects of source, path and recording site, and have been used for a variety of practical applications such as the mapping of frequency-dependent site amplification factor, discrimination of quarry blasts from earthquakes, single station method for determining frequency-dependent attenuation, and normalizing the regional seismic network data to a common source and receiver site condition.

(2) The definition of coda Q^{-1} in terms of the single-scattering model proved to be useful because of its simplicity. It has become, however, a necessary routine to specify the time window for each measured value of coda Q^{-1}.

(3) In order to clarify the physical meaning of coda Q^{-1}, attempts have been made to separately determine the scattering loss and the intrinsic absorption by the use of theories including multiple scattering. Although there is a need for further improving the interpretation theories, we found empirically that coda Q^{-1} are rather narrowly bounded between the intrinsic Q^{-1} and the total Q^{-1}.

(4) We found that the coda Q^{-1} measured from a time window represents the seismic attenuation property of the earth's crust averaged over the volume traversed by the singly back-scattered S waves. If the time window is longer and later, the resultant coda Q^{-1} map will have poorer spatial resolution.

(5) A map of coda Q^{-1} in the mainland China made with spatial resolution of about 200 km shows a strong correlation with the locations of large historic earthquakes ($M \geq 7$). The low Q regions are full of them, and the high Q regions are devoid of them. We found some evidence for the migration of a low Q zone associated with the high seismicity in North China during the past 300 years.

(6) A characteristic pattern of temporal coda Q^{-1} change sometimes occurs before a major earthquake, but we found that it is not a reliable precursor. Instead, the highest correlation (more than 0.8) with seismicity was found between coda Q^{-1} and the fraction of seismicity in a narrow magnitude range $M_c \leq M \leq M_c+0.5$, where M_c is characteristic to a given seismic region, for example, 3.0 for southern California and 4.0 for central California. This correlation was predicted by the creep model proposed by Jin and Aki (1989) for explaining the erratic (sometimes positive and sometimes negative) correlation between coda Q^{-1} and b-value. The creep model hypothesizes that the creep in the ductile part of lithosphere occurs over fractures with a unique size corresponding to M_c. High creep activity may increase coda Q^{-1} and enhance seismicity with magnitude around M_c because of stress concentration in the brittle part. This model is consistent with the observation that coda Q^{-1} shows a minimum value during the

period of high aftershock activity, and sometimes shows a peak during the quiescence. Another support for the deeper (ductile rather than brittle part of lithosphere) source of the coda Q^{-1} change comes from the observed coincidence between a large increase in coda Q^{-1} in southern California during 1986–87 and the simultaneous increase in electrical conductivity in the same region, which is attributed to the lower crust.

In summary, observations suggest that the temporal change in coda Q^{-1} may reflect the degree of creep fractures in the ductile part of lithosphere.

Acknowledgement

The author is greatful to Anshu Jin for her assistance in preparing figures. This work is supported by Department of Energy under grant DE-FG03-87ER 13807.

REFERENCES

[1] Aki, K., *Attenuation and scattering of short-period seismic waves in the lithosphere* in *Identification of Seismic Sources - Earthquake or Underground Explosion*, ed. E.S. Husebye and S. Mykkeltveit, D. Reidel Publishing Co., (1981), pp. 515–541.

[2] Aki, K., *Attenuation of shear waves in the lithosphere for frequencies from* 0.05 *to* 25 *Hz*, Phys. Earth Planet. Interiors, 21 (1980), pp. 50–60.

[3] Aki, K., *Scattering conversions P to S versus S to P*, Bull. Seis. Soc. Am., 82 (1992), 1969–1972.

[4] Aki, K., *Summary of discussions on coda waves at the Istanbul IASPEI meeting*, Phys. Earth Planet. Inter., 67 (1991), pp. 1–3.

[5] Aki, K., *Theory of earthquake prediction with special reference to monitoring of the quality factor of lithosphere by the coda method*, Earthq. Predict. Res., 3 (1985), pp. 219–230.

[6] Chen, P. and W.O. Nuttli, *Estimates of magnitude and short-period wave attenuation of Chinese earthquakes from Modified Mercali intensity data*, Bull. Seis. Soc. Am., 74 (1984a) pp. 957–968.

[7] Chen, P. and W.O. Nuttli, *Estimates of short-period Q values and seismic moments from coda waves for earthquakes of the Beijing and Yunnan regions of China*, Bull. Seis. Soc. Am., 74 (1984), pp. 1189–1208.

[8] Chouet, B., *Temporal variation in the attenuation of earthquake coda near Stone Canyon, Calif.*, Geophys. Res. Lett., 6 (1979), pp. 143–146.

[9] Faulkner, J., *Temporal variation of coda Q*, MS Thesis, University of Southern California, Los Angeles, (1988).

[10] Frankel, A. and R.W. Clayton, *Finite difference simulations of seismic scattering: implications for the propagation of short-period seismic waves in the crust and models of crust heterogeneity*, J. Geophys. Res., 91 (1986), pp. 6465–6489.

[11] Gao, L.S. and K. Aki, *Effect of finite thickness of scattering layer on coda Q of local earthquakes*, J. Geodynamics, 21 (1996), pp. 191–203.

[12] Gusev, A.A. and V.K. Lemzikov, *The anomalies of small earthquake coda wave characteristics before the three large earthquakes in the Kuril-Kamchatka zone (in Russian)*, Vulk. Seism., 4 (1984), pp. 76–90.

[13] Herraiz, M. and A.F. Espinosa, *Coda waves: a review*, PAGEOPH, 125 (1987), pp. 499–577.

[14] Hoshiba, M., H. Sato, and M. Fehler, *Numerical basis of the separation of scattering and intrinsic absorption from full seismogram envelope: A Monte-Carlo simulation of multiple isotropic scattering*, Pap. Meterol. Geophys., 42 (1991),

pp. 65–91.
[15] Hoshiba, M., *Separation of scattering attenuation and intrinsic absorption in Japan with the multiple lapse time window analysis from full seismogram envelope*, J. Geophys. Res., 98 (1993), pp. 15809–15824.
[16] Jin, A. and K. Aki, *Spatial and temporal correlation between coda Q and seismicity in China*, Bull. Seis. Soc. Am., 78 (1988), pp. 741–769.
[17] Jin, A. and K. Aki, *Spatial and temporal correlation between coda Q^{-1} and seismicity and its physical mechanism*, J. Geophys. Res., 94 (1989), pp. 14041–14059.
[18] Jin, A. and K. Aki, *Temporal changes in coda Q before the Tangshan earthquake of 1976 and the Haicheng earthquake of 1975*, J. Geophys. Res., 91 (1986), pp. 665–673.
[19] Jin, A. and K. Aki, *Temporal correlation between coda Q^{-1} and seismicity–evidence for a structural unit in the brittle-ductile transition zone*, J. Geodynamics, 17 (1993), pp. 95–120.
[20] Jin, A.K. Mayeda, D. Adams, and K Aki, *Separation of intrinsic and scattering attenuation in southern California using TERRAscope data*, J. Geophys. Res., 99 (1994), pp. 17835–17848.
[21] Madden, T.R., G.A. LaTorraca, and S.K. Park, *Electrical conductivity variations around the Palmdale section of the San Andreas fault zone*, J. Geophys. Res., 98 (1993), pp. 795–808.
[22] Matsunami, K. *Laboratory tests of excitation and attenuation of coda waves using 2D models of scattering media*, Phys. Earth Planet. Inter., 67 (1991), pp. 104–114.
[23] Mayeda, K., S. Koyanagi, M. Hoshiba, K. Aki, and Y. Zeng, *A comparative study of scattering, intrinsic and coda Q^{-1} for Hawaii*, Long Valley, and central California between 1.5 and 15 Hz, J. Geophys. Res., 97 (1992), pp. 6643–6659.
[24] Novelo-Casanova, D.A., E. Berg, Y. Hsu, and C.E. Helsley, *Time-space variation seismic S-wave coda attenuation (Q^{-1}) and magnitude distribution (b-values) for the Petatlan earthquake*, Geophys. Res. Lett., 12 (1985), pp. 789–792.
[25] Peng, J.Y., *Spatial and temporal variation of coda Q^{-1} in California*, Ph.D. thesis, University of Southern California, Los Angeles, (1989).
[26] Rautian, T.G. and V.I. Khalturin, *The use of coda for determination of the earthquake source spectrum*, Bull. Seis. Soc. Am., 68 (1978), pp. 923–948.
[27] Robinson, R., *Temporal variation in coda duration of local earthquakes in the Wellington region*, New Zealand, PAGEOPH, 125 (1987), pp. 579–596.
[28] Sabine, W.C., *Collected Papers on Acoustics*, Harvard University Press, Cambridge, Mass., (1922).
[29] Sato, H., A.M. Shomahmadov, V.I. Khalturin, and T.G. Rautian, *Temporal change in spectral coda Q associated with the $K = 13.3$ earthquake of 1983 near Garm, Tadjikistan region, in Soviet Central Asia* (in Japanese), (1987).
[30] Sato, H., *Amplitude attenuation of impulsive waves in random media based on travel time corrected mean wave formalism*, J. Acoust. Soc. Am., 71 (1982a), pp. 559–564.
[31] Sato, H., *Attenuation of S waves in the lithosphere due to scattering by its random velocity structure*, J. Geophys. Res., 87 (1982b), pp. 7779–7785.
[32] Sato, H. *Temporal change in attenuation intensity before and after Eastern Yamaashi earthquake of 1983, in central Japan*, J. Geophys. Res., 91 (1986), pp. 2049–2061.
[33] Sato, H., *Temporal change in scattering and attenuation associated with the earthquake occurrence - a review of recent studies on coda waves*, PAGEOPH, 126 (1988), pp. 465–498.
[34] Shang, T. and L.S. Gao, *Transportation theory of multiple scattering and its application to seismic coda waves of impulse source*, Sci. Sinica, Series V, 31 (1988), pp. 1503–1514.
[35] Singh, S.K. and R.B. Herrmann, *Regionalization of crustal coda Q in the conti-*

nental United States, J. Geophys. Res., 88 (1983), pp. 527–538.

[36] Su, F., and K. Aki, *Spatial and temporal variation in coda* Q^{-1} *associated with the North Palm Springs earthquake of 1986*, PAGEOPH, 133 (1990), pp. 23–52.

[37] Su, F., K. Aki, and N.N. Biswas, *Discriminating quarry blasts from earthquakes using coda waves*, Bull. Seis. Soc. Am., 81 (1991), pp. 162–178.

[38] Su, F., K. Aki, T. Teng, Y. Zeng, S. Koyanagi, and K. Mayeda, *The relation between site amplification factor and surficial geology in central California*, Bull. Seis. Soc. Am., 82 (1992), pp. 580–602.

[39] Tsukuda, T., *Coda Q before and after the 1983 Misasa earthquake of M6.2*, Tottori Pref., Japan, PAGEOPH, 128 (1988), pp. 261–280.

[40] Wu, R.S., *Attenuation of short period seismic waves due to scattering*, Geophys. Res. Lett., 9 (1982), pp. 9–12.

[41] Zeng, Y., F. Su, and K. Aki, *Scattering wave energy propagation in a medium with randomly distributed isotropic scatterers*, 1. Theory, J. Geophys. Res., 96 (1991), pp. 607–619.

[42] Zeng, Y., *Theory of scattered P and S waves energy in a random isotropic scattering medium*, Bull. Seis. Soc. Am., 83 (1993), pp. 1264-1277.

ONE DIMENSIONAL RANDOM WALK IN A RANDOM MEDIUM*

R.F. ANDERSON† AND S.A. MOLCHANOV†

1. Introduction. One of the fundamental problems in random medium (RM) theory is the relation between homogenization of the RM (i.e. the ability to describe such media at some scales as a homogeneous one) and localization. Homogenization describes transport processes in RM in terms of "effective" parameters, localization suppresses all forms of transport. For random walks in RM it is the relation between classical diffusive behavior of the random walk if time is large and the phenomenon of "trapping", which can produce subdiffusion asymptotics for the random walk. In the one-dimensional case, which is of course the simplest one in RM theory, under some restrictions on the random transition probabilities the homogenization theorems was proved by S. Kozlov [10] as an example of a more general theory. A general and elementary introduction to homogenization theory can be found in S. Molchanov [2]. There are deep relations between random walks in one dimensional RM and scattering of waves (say, seismic waves) in random layered media. See [3].

Ya Sinai [12] discovered a very general class of one-dimensional random walks which exhibited very strong trapping and as a result, very nonstandard transition properties. In the lectures [2] one can find an almost complete analysis of one-dimensional random walk in RM in the case of uniformly non-degenerate transition probabilities (the analog of uniform ellipiticity for diffusion processes). A different approach to the same problem is given in [13]. In this paper we'll study mainly a degenerate case, the so-called vent model, which is connected with the physics of (one-dimensional) semi-conductors (M. Rost, private communications). Again, some of the results of this paper one can find in [13] but in a different form. Our approach is essentially more probabilistic then in [13] and it gives additional "physical" information about the model. Other results (especially about "strong trapping" or Sinai's type behavior of the random walks) are new. We plan to continue our analysis for more general "quasi-one-dimensional graphs" elsewhere.

Let Z be the integers and let $\{p_x\}_{x\in Z}$be a strictly stationary and ergodic process defined on (Ω_m, μ) with $0 \le p_x \le 1$. Define $q_x = 1 - p_x$. For $\omega_m \in \Omega_m$ fixed, define the random walk $\{x_t\}_{t\geq 0}$ on (Ω, P_{ω_m}) with state

* This work was partially supported by ONR Grant N00014-95-1-0224.

† Department of Mathematics, University of North Carolina at Charlotte, Charlotte NC 28223.

space Z by

$$P_{\omega_m}(x_{t+1} = x+1 \mid x_t = x) = p_x(\omega_m) \qquad \text{and}$$
$$P_{\omega_m}(x_{t+1} = x-1 \mid x_t = x) = q_x(\omega_m)$$

Let

$$Q(d\omega, d\omega_m) = P_{\omega_m}(d\omega)\,\mu(d\omega_m)$$

be defined on $\Omega \times \Omega_m$ and

$$x_t(\omega, \omega_m) : t \geq 0$$

be the resulting process determined by $Q(d\omega, d\omega_m)$. We have constructed what is called random walk in a random medium. Note that there are two levels of randomness, the medium $\{p_x(\omega_m), q_x(\omega_m)\}_{x\in Z}$ and given the medium, the random walk.

REMARK 1.1. A quantity of importance in this study is

$$\xi_x = \ln \frac{p_x}{q_x}$$

Note that $\{\xi_x\}_{x\in Z}$ is also strictly stationary if, of course $q_x > 0$.

The goal of the theory is to determine when the medium can be summarized by effective parameters. The type of results we hope for are as follows:

RESULT 1. *SLLN. For almost all* ω_m,

$$P_{\omega_m}\left(\lim_{t\to\infty} \frac{x_t}{t} = a\right) = 1$$

where a is a constant independent of ω_m.

RESULT 2. *CLT. For almost all* ω_m ,

$$P_{\omega_m}\left(\frac{x_t - at}{\sigma\sqrt{t}} \leq x\right) \underset{t\to\infty}{\overset{L}{\to}} N(0,1)$$

where a is as above and σ is a constant independent of ω_m.

We refer to a as the Stokes drift and σ as the turbulent diffusivity and as a pair, effective parameters for the medium. As a further goal of the theory we seek formulas for a and σ .

2. Non-degenerate case. Suppose the following conditions are satisfied:

(1) For some $\varepsilon > 0$,

$$\varepsilon \leq p_x, q_x \leq 1 - \varepsilon \quad \text{(non-degeneracy)} .$$

(2) For some $\delta > 0$,

$$\alpha(\rho) = O\left(\frac{1}{\rho^{1+\delta}}\right) \quad as \quad \rho \to \infty \quad \text{(rapid decrease)} ,$$

where

$$\alpha(\rho) =$$
$$\sup(|\mu(A_1 A_2) - \mu(A_1)\mu(A_2)| : A_1 \in \sigma(\xi_x : x \le 0),\ A_2 \in \sigma(\xi_x : x \ge \rho))$$

is the strong mixing coefficient.

REMARK 2.1. Under the assumption of rapid decrease of $\alpha(\rho)$, if $R(\tau) = Cov(\xi_0 \xi_\tau)$and $\widehat{R}(\phi) : \phi \in [-\pi, \pi]$ is the corresponding spectral measure, then $\widehat{R}(\phi)$ has a density $\widehat{B}(\phi)$ with respect to Lebesque measure.

The following results one can find in Saint-Flour lectures [2].

2.1. Case when $\langle \xi_x \rangle = 0$.

THEOREM 2.1. *If $\widehat{B}(0) = 0$ ($\Longleftrightarrow Var\left(\sum_1^n \xi_x\right)$ is bounded by a constant independent of $n \Longleftrightarrow \xi_x = \eta_{x+1} - \eta_x$, $\{\eta_x\}$ stationary $\Longleftrightarrow \sum_\tau R(\tau) = 0$) then, if*

$$\left\langle e^{|\eta_x|} \right\rangle < \infty ,$$

we have SLLN and CLT (i.e. homogenization) with $a = 0$ and a known formula for σ.

THEOREM 2.2. *If $\widehat{B}(0) \neq 0$ ($\Longleftrightarrow Var\left(\sum_1^n \xi_x\right) = O(n) \Longleftrightarrow \sum_\tau R(\tau) \neq 0$) then Sinai's trapping occurs, that is for almost all $\omega_m \in \Omega_m$ there exist a random sequence $a_t(\omega_m)$, $t \ge 0$, such that for every $\delta > 0$,*

$$\lim_{t\to\infty} P_{\omega_m}\left(\left|\frac{x_t}{\ln^2 t} - a_t(\omega_m)\right| > \delta\right) = 0 .$$

2.2. Case when $\langle \xi_x \rangle \neq 0$.

THEOREM 2.3. *Suppose $\langle \xi_x \rangle = \alpha_0 < 0$ and for all δ, $0 < \delta < 3$,*

$$\left\langle \left(\sum_{x=1}^{n} \frac{p_x}{q_x}\right)^\delta \right\rangle < 1 .$$

Then, for suitable a and σ,

$$P_{\omega_m}\left(\frac{x_t - at}{\sigma\sqrt{t}} \le x\right) \xrightarrow[t\to\infty]{L} N(0,1) .$$

REMARK 2.2. $a = \langle (p_0 - q_0)\pi(\omega_m) \rangle$, where $\pi(\omega_m)$ is the invariant measure with respect to μ. σ has a more complicated but known expression.

2.3. Relation with the ruin problem. Define

$$u_n(x) = P_x(-1 \text{ is reached before } n), \quad -1 \le x \le n .$$

It is standard knowledge that $u_n(x)$ solves the problem

$$\begin{aligned} p_x u_n(x+1) + q_x u_n(x-1) &= u_n(x) , \\ u_n(-1) = 1, \quad u_n(n) &= 0 . \end{aligned} \tag{2.1}$$

Since $p_x + q_x = 1$, (2.1) has the form

$$u_n(x+1) - u_n(x) = \frac{q_x}{p_x}(u_n(x) - u_n(x-1)), \quad 0 \le x \le n-1 . \tag{2.2}$$

Thus,

$$\sum_{x=0}^{n-1} \frac{q_x}{p_x}(u_n(x) - u_n(x-1)) = u_n(n) - u_n(0) = -u_n(0) ,$$

or

$$u_n(0) = \sum_{x=0}^{n-1} \frac{q_x}{p_x}(u_n(x-1) - u_n(x)) . \tag{2.3}$$

This identity contains one of the boundary conditions. Using (2.2) again, for $0 \le x \le n-1$ we have

$$\begin{aligned} u_n(x) - u_n(x-1) &= \frac{q_{x-1}}{p_{x-1}}(u_n(x-1) - u_n(x-2)) \\ &= \frac{q_{x-1}}{p_{x-1}}\frac{q_{x-2}}{p_{x-2}}(u_n(x-2) - u_n(x-3)) \\ &= \cdots = \left(\prod_0^{x-1} \frac{q_y}{p_y}\right)(u_n(0) - 1) , \end{aligned}$$

or

$$u_n(x-1) - u_n(x) = \left(\prod_0^{x-1} \frac{q_y}{p_y}\right)(1 - u_n(0)) . \tag{2.4}$$

This contains the second boundary condition. Combining (2.3) and (2.4) yields

$$u_n(0) = \sum_{x=0}^{n-1} \frac{q_x}{p_x}\left(\prod_0^{x-1} \frac{q_y}{p_y}\right)(1 - u_n(0)) ,$$

that is

$$u_n(0) = \frac{\sum_{x=0}^{n-1}\left(\prod_0^x \frac{q_y}{p_y}\right)}{1+\sum_{x=0}^{n-1}\left(\prod_0^x \frac{q_y}{p_y}\right)} . \tag{2.5}$$

Next, from (2.2), in the same way as (2.3), we find

$$u_n(x) = u_n(0) + \sum_{k=0}^{x-1} \frac{q_k}{p_k}\left(\prod_0^{k-1} \frac{q_y}{p_y}\right)(u_n(0)-1) ,$$

which in combination with (2.5) results in the idenity

$$u_n(x) = \frac{\sum_{k=x}^{n-1}\prod_0^x \frac{q_y}{p_y}}{1+\sum_{x=0}^{n-1}\prod_0^x \frac{q_y}{p_y}} . \tag{2.6}$$

Letting $n \to \infty$ in (2.6) results in

$$u(x) = P_x(y > 0 \min X_y < 0) = \frac{\sum_{x=0}^{\infty}\left(\prod_0^x \frac{q_y}{p_y}\right)}{1+\sum_{x=0}^{\infty}\left(\prod_0^x \frac{q_y}{p_y}\right)} . \tag{2.7}$$

Next, define

$$\eta_x = \sum_{k=0}^{\infty}\prod_{y=x}^{k} \frac{q_y}{p_y} . \tag{2.8}$$

THEOREM 2.4. *Assume $\{p_x\}_{x\in Z}$ is stationary and ergodic and $\langle \ln\left(\frac{p_0}{q_0}\right)\rangle > 0$. Suppose $\left\langle \ln \frac{\eta_0}{1+\eta_0}\right\rangle$ exists. Then*

$$\lim_{x\to\infty} \frac{\ln u(x)}{x} = -\left\langle \ln\left(\frac{p_0}{q_0}\right)\right\rangle .$$

Proof. From (2.8) it is easy to see that

$$\eta_{x-1} = \frac{q_{x-1}}{p_{x-1}}(1+\eta_x) . \tag{2.9}$$

Moreover, combining (2.7),(2.8), and (2.9) yields

$$u(x) = \prod_{y=0}^{x} \frac{\eta_y}{1+\eta_y} .$$

Since $\{p_x\}_{x\in Z}$is stationary and ergodic it follows that $\{\eta_x\}_{x\in Z}$ and $\left\{\ln\left(\frac{\eta_x}{1+\eta_x}\right)\right\}_{x\in Z}$ are also stationary and ergotic. From (2.9) we find

$$\left\langle \ln\left(\frac{\eta_x}{1+\eta_x}\right)\right\rangle = \langle \ln(\eta_{x-1})\rangle - \langle \ln(1+\eta_x)\rangle = -\left\langle \ln\left(\frac{p_0}{q_0}\right)\right\rangle .$$

Hence, by ergodicity

$$\lim_{x\to\infty}\frac{\ln u(x)}{x} = \lim_{x\to\infty}\frac{\sum\limits_{y=0}^{x}\ln\left(\frac{\eta_y}{1+\eta_y}\right)}{x} = \left\langle \ln\left(\frac{\eta_0}{1+\eta_0}\right)\right\rangle = -\left\langle \ln\left(\frac{p_0}{q_0}\right)\right\rangle .$$

□

REMARK 2.3. The invariant measure $\pi(x,\omega_m)$ has a representation similar to (2.7), see [2].

3. Vent model. Suppose $\eta_k : \Omega_m \to I^+$, $k \geq 1$, are defined on (Ω_m, μ) as i.i.d. random variables. When needed, we will denote by η their generic copy. Assume that the η_k's do not take the value 0 with positive probability. Define

$$v_n = \sum_1^n \eta_k .$$

For $\omega_m \in \Omega_m$ fixed, define the random walk with reflection to the right at each point $v_n(\omega_m)$ ($v_n(\omega_m)$ are referred to as vent points) and jumping to the right with probability p and left with probability q at all other non-vent points. More precisely, x_t, $t \geq 0$, for fixed ω_m, will be a random walk living on the positive integers whose transition probabilities are for $x = 0$,

$$P_{\omega_m}(x_{t+1} = 1 \mid x_t = 0) = 1 ,$$

and for $x \geq 1$,

$$\begin{aligned} P_{\omega_m}(x_{t+1} = x+1 \mid x_t = x) &= 1, && \textit{if } x = v_n(\omega_m) \text{ for some } n ;\\ &= p, && \textit{if } x \neq v_n(\omega_m) \text{ for any } n ; \end{aligned}$$

and

$$\begin{aligned} P_{\omega_m}(x_{t+1} = x-1 \mid x_t = x) &= 0, && \textit{if } x = v_n(\omega_m) \text{ for some } n ;\\ &= q, && \textit{if } x \neq v_n(\omega_m) \text{ for any } n . \end{aligned}$$

Note, that we are in a very degenerate case in relation to that discussed in Section 2.

REMARK 3.1. Let y_t, $t \geq 0$, be the random walk with reflection at 0 and no other vents, that is

$$P(y_{t+1} = 1 \mid y_t = 0) = 1 ,$$

and for $x \geq 1$,

$$P(y_{t+1} = x+1 \mid y_t = x) = p \,,$$
$$P(y_{t+1} = x-1 \mid y_t = x) = q \,.$$

Define

$$t_n = \inf \{t : y_t = n\} \,.$$

It can be easily calculated and it is well known that

$$E_0[t_n] = \frac{2pq}{(q-p)^2}((\frac{q}{p})^n - 1) - \frac{n}{q-p} \,,$$

and

$$Var\,(t_n) = \frac{4p^2q^2}{(q-p)^4}(\frac{q}{p})^{2n} - \frac{10npq}{(q-p)^3}(\frac{q}{p})^n - \frac{2npq}{(q-p)^3} - \frac{2pq}{(q-p)^4}(2pq+1) \,.$$

Note that, if $q > p$ and n is large, then

$$E_0[t_n] \simeq \tfrac{2pq}{(q-p)^2}\left(\tfrac{q}{p}\right)^n$$

(3.1) and

$$Var\,(t_n) \simeq \tfrac{4p^2q^2}{(q-p)^4}\left(\tfrac{q}{p}\right)^{2n} \simeq (E_0[t_n])^2 \,.$$

If $q < p$ and n is large,

(3.2) $$E_0[t_n] \simeq \frac{1}{p-q}\,n \quad \text{and} \quad Var\,(t_n) \simeq \frac{2pq}{(p-q)^3}\,n \,.$$

For the case $p = q$, it is known that

$$E_0[t_n] \simeq c_{p,q}n^2 \quad \text{and} \quad Var\,(t_n) \simeq c_{p,q}n^4 \,.$$

Lastly, it is known that in the case $q > p$ (drift in the left direction),

(3.3) $$\frac{t_n}{E_0[t_n]} \xrightarrow[n\to\infty]{L} \text{exponential distribution with parameter 1.}$$

If $p = q$,

$$\frac{t_n}{n^2} \xrightarrow[n\to\infty]{L} \text{a distribution with characteristic function } \frac{1}{\cos h\,(\sqrt{z})},$$

and if $p > q$,

$$\frac{t_n}{n} \xrightarrow[n\to\infty]{} const$$

Returning to the discussion of the vent model, define

$$\tau_n(\omega, \omega_m) = t_{\eta_n(\omega_m)}(\omega)\ ,\ \ n \geq 1\ .$$

and

$$\tau(\omega, \omega_m) = t_{\eta(\omega_m)}(\omega)\ \ .$$

for a geneeric copy when needed.

REMARK 3.2. For ω_m fixed, τ_n, $n \geq 1$, are independent but not identically distributed.

REMARK 3.3. As random variables on (μ, Ω_m), $E_{\omega_m}[\tau_n]$, $n \geq 1$, and $Var_{\omega_m}(\tau_n)$, $n \geq 1$ are both i.i.d. families.

We will also use the notation

$$\langle \tau \rangle = \langle E_{\omega_m}[\tau] \rangle\ , \qquad T_n = \tau_1 + \tau_2 + \cdots + \tau_n, \tag{3.4}$$

and

$$\langle Var_{\omega_m}(\tau) \rangle = \langle E_{\omega_m}[(\tau - E_{\omega_m}[\tau])^2] \rangle$$

and

$$Var(E_{\varpi_m}[\tau]) = \langle (E_{\omega_m}[\tau] - \langle \tau \rangle)^2 \rangle\ .$$

Also, let

$$Q(d\omega, d\omega_m) = P(d\omega)\mu(d\omega_m).$$

THEOREM 3.1. *(SLLN) If* $\langle \eta \rangle < \infty$ *and* $\langle Var_{\omega_m}(\tau) \rangle < \infty$, *then*

$$P_{\omega_m}(\lim_{t \to \infty} \frac{x_t}{t} = \frac{\langle \eta \rangle}{\langle \tau \rangle}) = 1 \quad a.s.\ \omega_m.$$

REMARK 3.4. From above we see that the requirement $\langle Var_{\omega_m}(\tau) \rangle < \infty$ is equivelent to:

$$\langle (\frac{q}{p})^{2\eta} \rangle = \langle e^{2\eta \ln(q/p)} \rangle < \infty, \text{ if } \frac{q}{p} > 1\ ;$$
$$\langle \eta^4 \rangle < \infty, \text{ if } \frac{q}{p} = 1\ ;$$
$$\langle \eta^2 \rangle < \infty, \text{ if } \frac{q}{p} < 1.$$

Proof. The proof will be given in several steps, the first of which is a preliminary one.

LEMMA 3.1. *Suppose* Y_k, $k \geq 1$, *are i.i.d. with* $E[Y_k] < \infty$. *Then*

$$\lim_{n \to \infty} \sum_1^n \frac{Y_k}{k^2} < \infty\ a.s. \tag{3.5}$$

Proof. Let $0 < c < 1$. Now

$$\sum_1^\infty P(|Y_k| \geq ck^2) = \sum_1^\infty P(|Y| \geq ck^2) \leq \sum_1^\infty \frac{E[Y]}{ck^2} < \infty .$$

Thus, by the Borel-Cantelli Lemma,

$$P(|Y_k| \geq ck^2 \; i.o.) = 0 ,$$

and since

$$\sum_1^\infty \frac{1}{k^2} < \infty ,$$

(3.5) follows.

As remarked above, for fixed ω_m, τ_k, $k \geq 1$, are independent but not identically distributed. Since (also as remarked above) $Var_{\omega_m}(\tau_k)$, $k \geq 1$, form an i.i.d. sequence and $\langle Var_{\omega_m}(\tau_k)\rangle < \infty$, Lemma 3.1 implies, for $\omega_m \in A$, $\mu(A) = 1$, that

$$\lim_{n\to\infty} \sum_1^n \frac{Var_{\omega_m}(\tau_k)}{k^2} < \infty . \tag{3.6}$$

□

THEOREM 3.2. *(Kolmogorov) If $X_k : k \geq 1$ are independent and $\sum_1^\infty \frac{Var X_k}{k^2} < \infty$ then*

$$\lim_{n\to\infty} \frac{\sum_1^n X_k - E[X_k]}{n} = 0 \; a.s.$$

Using (3.6) and the Kolmorgorov's result

$$P_{\omega_m}\Big(\lim_{n\to\infty} \frac{\sum_1^n (\tau_k - E_{\omega_m}[\tau_k])}{n} = 0\Big) = 1 \; a.s. \; \omega_m . \tag{3.7}$$

Since $E_{\omega_m}[\tau_k]$, $k \geq 1$, are i.i.d., we have by the usual SLLN that

$$\lim_{n\to\infty} \frac{\sum_1^n E_{\omega_m}[\tau_k]}{n} = \langle\tau\rangle a.s. \; \omega_m . \tag{3.8}$$

Hence combining (3.7) and (3.8) we obtain

$$P_{\omega_m}\left(\lim_{n\to\infty}\frac{T_n}{n}=\langle\tau\rangle\right)=1\ a.s.\ \omega_m\ . \tag{3.9}$$

Let

$$\nu(t)=n \quad on\ (T_n\le t<T_{n+1})\ .$$

Our approach is based on the following identity:

$$\frac{x_t}{t}=\frac{x_t-x_{\nu(t)}}{t}+\frac{x_{\nu(t)}-\langle\eta\rangle\,\nu(t)}{t}+\frac{\nu(t)}{t}\langle\eta\rangle \tag{3.10}$$

The proof of our SLLN will be complete if we prove the following:

$$P_{\omega_m}\left(\lim_{t\to\infty}\frac{x_t-x_{\nu(t)}}{t}=0\right)=1\ a.s.\ \omega_m, \tag{3.11}$$

$$P_{\omega_m}\left(\lim_{t\to\infty}\frac{x_{\nu(t)}-\langle\eta\rangle\,\nu(t)}{t}=0\right)=1\ a.s.\ \omega_m, \tag{3.12}$$

and

$$P_{\omega_m}\left(\lim_{t\to\infty}\frac{\nu(t)}{t}=\frac{1}{\langle\tau\rangle}\right)=1\quad a.s.\ \omega_m. \tag{3.13}$$

Proof of (3.13): Recall that

$$\nu(t)=\sum_1^\infty nI_{(T_n\le t<T_{n+1})}.$$

Thus

$$\sum_1^\infty I_{(T_n\le t<T_{n+1})}\frac{n}{T_{n+1}}\le\frac{\nu(t)}{t}\le\sum_1^\infty I_{(T_n\le t<T_{n+1})}\frac{n}{T_n}\ ,$$

and since

$$P_{\omega_m}\left(\lim_{n\to\infty}T_n=\infty\right)=1\quad a.s.\ \omega_m\ ,$$

the proof is finished using (3.9).

Proof of (3.11): Note that

$$0\le\frac{x_t-x_{\nu(t)}}{t}\le\frac{\eta_{\nu(t)+1}}{t}=\frac{1}{t}\sum_0^\infty nI_{(\nu(t)=n)}\frac{\eta_{n+1}}{n}\ . \tag{3.14}$$

Because $\langle\eta\rangle < \infty$, for any $\varepsilon > 0$, $\mu(\eta_{n+1} \geq \varepsilon(n+1) \quad i.o.) = 0$. By (3.13) we can conclude

$$P_{\omega_m}\left(\lim_{t\to\infty} \nu(t) = \infty\right) = 1 \quad a.s.\ \omega_m\ ,$$

and so, for any $\varepsilon > 0$, again using (3.13),

$$\lim_{t\to\infty} \frac{1}{t}\sum_0^\infty nI_{(\nu(t)=n)} \leq \varepsilon \lim_{t\to\infty} \frac{\nu(t)}{t} = \varepsilon\frac{1}{\langle\tau\rangle}\ a.s.\ P_{\omega_m} \quad a.s.\ \omega_m\ .$$

Thus

$$P_{\omega_m}\left(\lim_{t\to\infty} \frac{\eta_{\nu(t)+1}}{t} = 0\right) = 1\ a.s.\ \omega_m\ ,$$

which implies (3.11) using (3.14).

Proof of (3.12): We see that

$$x_{\nu(t)} - \langle\eta\rangle\nu(t) = \sum_1^\infty nI_{(\nu(t)=n)}\left(\frac{x_n}{n} - \langle\eta\rangle\right)\ .$$

By the SLLN

$$\lim_{n\to\infty} \frac{x_n}{n} = \langle\eta\rangle a.s.\ \omega_m\ ,$$

and since

$$P_{\omega_m}(\lim_{t\to\infty} \nu(t) = \infty) = 1\ a.s.\ \omega_m\ ,$$

we see from (3.13) that

$$P_{\omega_m}\left(\lim_{t\to\infty} \frac{x_{\nu(t)} - \langle\eta\rangle\,\nu(t)}{t} = 0\right) = 1\ a.s.\ \omega_m\ .$$

THEOREM 3.3. *(CLT) Suppose for some* $\delta > 0$,

$$\langle E_{\omega_m}[(\tau - E_{\omega_m}[\tau])^{2+\delta}]\rangle < \infty\ .$$

Then

$$\lim_{t\to\infty} Q\left(\frac{x_t - \frac{\langle\eta\rangle}{\langle\tau\rangle}}{\sqrt{t}} \leq x\right) = \Phi_\sigma(x)\ ,$$

where

$$\sigma^2 = \frac{\langle\eta\rangle^2}{\langle\tau\rangle^3}(\langle Var_{\omega_m}(\tau)\rangle + Var E_{\omega_m}[\tau]) + 2\frac{\langle\eta\rangle}{\langle\tau\rangle^2}Cov(\eta, E_{\omega_m}[\tau]) + \frac{1}{\langle\tau\rangle}Var(\eta) \ .$$

The proof will be broken down into a series of steps.

Step 1. We state and prove the following Lemma.

LEMMA 3.2. *Suppose for some $\delta > 0$,*

$$\langle E_{\omega_m}[(\tau - E_{\omega_m}[\tau])^{2+\delta}]\rangle < \infty \ . \tag{3.15}$$

Then

$$\lim_{n\to\infty} Q\left(\frac{T_n - n\langle\tau\rangle}{\sqrt{n\langle Var_{\omega_m}(\tau)\rangle}} \le x\right) = \Phi_\sigma(x) \ , \tag{3.16}$$

where

$$\Phi_\sigma(x) \text{ is } N\left(0, \sigma^2\right) \text{ and } \sigma^2 = 1 + \frac{Var(E_{\omega_m}[\tau])}{\langle Var_{\omega_m}(\tau)\rangle} \ . \tag{3.17}$$

Recall the following well known result:

THEOREM 3.4. *(Lyapunov Condition for the CLT) If $S_n = \sum_1^n Y_k$ and $s_n = \sqrt{Var S_n}$, where Y_k, $k \ge 1$, are independent then if for some $\delta > 0$,*

$$\lim_{n\to\infty} \frac{1}{s_n^{2+\delta}} \sum_1^n E[(Y_k - \mu_{Y_k})^{2+\delta}] = 0 \tag{3.18}$$

then

$$\lim_{n\to\infty} \frac{\sum_1^n (Y_k - \mu_{Y_k})}{s_n^{1+\frac{\delta}{2}}} \stackrel{L}{=} N(0,1) \ .$$

Proof of Lemma 3.2: Under the asumption that $\langle E_{\omega_m}[(\tau - E_{\omega_m}[\tau])^{2+\delta}]\rangle < \infty$, for some $\delta > 0$, by the SLLN, $(E_{\omega_m}\left[(\tau_k - E_{\omega_m}[\tau_k])^{2+\delta}\right] : k \ge 1$, are i.i.d.)

$$\lim_{n\to\infty} \frac{\sum_1^n E_{\omega_m}\left[(\tau_k - E_{\omega_m}[\tau_k])^{2+\delta}\right]}{\left(\sum_1^n Var_{\omega_m}(\tau_k)\right)^{1+\frac{\delta}{2}}}$$

$$= \lim_{n\to\infty} \frac{\frac{1}{n}\sum_1^n E_{\omega_m}\left[(\tau_k - E_{\omega_m}[\tau_k])^{2+\delta}\right]}{\left(\frac{1}{n}\sum_1^n Var_{\omega_m}(\tau_k)\right)^{1+\frac{\delta}{2}}} \frac{n}{n^{1+\frac{\delta}{2}}}$$

$$= \frac{\left\langle E_{\omega_m}\left[(\tau - E_{\omega_m}[\tau])^{2+\delta}\right]\right\rangle}{(\langle Var_{\omega_m}(\tau)\rangle)^{1+\frac{\delta}{2}}} \lim_{n\to\infty} \frac{n}{n^{1+\frac{\delta}{2}}} = 0 \, .$$

So, by the Lyapunov condition (3.18),

$$\lim_{n\to\infty} \frac{\sum_1^n (\tau_k - E_{\omega_m}[\tau_k])}{\sqrt{\sum_1^n Var_{\omega_m}(\tau_k)}} \stackrel{L}{=} N(0,1) \; a.s. \; \omega_m \, . \tag{3.19}$$

Since $Var_{\omega_m}(\tau_k)$, $k > 1$, form an i.i.d. sequence,

$$\lim_{n\to\infty} \frac{1}{n}\sum_1^n Var_{\omega_m}(\tau_k) = \langle Var_{\omega_m}(\tau)\rangle$$

and we have (3.19)

$$\lim_{n\to\infty} \frac{\sum_1^n (\tau_k - E_{\omega_m}[\tau_k])}{\sqrt{n\,\langle Var_{\omega_m}(\tau)\rangle}} \stackrel{L}{=} N(0,1) \; a.s. \, \omega_m \, . \tag{3.20}$$

By the usual CLT, since $\{E_{\omega_m}[\tau_k]\}_{k\geq 1}$ form an i.i.d. sequence with

$$\langle E_{\omega_m}[\tau_k]\rangle = \langle\tau\rangle \text{ and } Var\left(E_{\omega_m}[\tau_k]\right) = Var\left(E_{\omega_m}[\tau]\right) \, ,$$

then with respect to $\mu\,(d\omega_m)$

$$\lim_{n\to\infty} \frac{\sum_1^n E_{\omega_m}[\tau_k] - n\langle\tau\rangle}{\sqrt{nVar(E_{\omega_m}[\tau])}} \stackrel{L}{=} N(0,1) \, . \tag{3.21}$$

Because the limiting distribution in the case of (3.20) does not depend on ω_m, and in the case of (3.21) depends only on ω_m, in the limit we will have independence and

$$\lim_{n\to\infty} Q\left(\frac{T_n - n\langle\tau\rangle}{\sqrt{n\langle Var_{\omega_m}(\tau)\rangle}} \leq x\right) = \Phi_\sigma(x) ,$$

where $\Phi_\sigma(x)$ is the normal distribution with mean 0 and variance

$$\sigma^2 = 1 + \frac{Var\left(E_{\omega_m}[\tau]\right)}{\langle Var_{\omega_m}(\tau)\rangle} . \tag{3.22}$$

REMARK 3.5. Note that in (3.20) the centering terms $E_{\omega_m}[\tau_k]$ are dependent upon the media. The consequence is that the CLT (Theorem 3.3 of this section) does not have the same character as that in Section 2. See Theorems 2.1 and 2.3.

Step 2. Define

$$\nu(t) = n \text{ on } (T_n \leq t < T_{n+1}) .$$

We follow a standard argument. By construction, we see that

$$Q(\nu(t) \geq n) = Q(T_n \leq t) .$$

As established in Step 1, for each $x \in R$,

$$\lim_{n\to\infty} Q(\frac{T_n - n\langle\tau\rangle}{\sqrt{n\langle Var_{\omega_m}(\tau)\rangle}} \leq x) = \Phi_\sigma(x) ,$$

where σ is given by (3.17). For x fixed, letting $n, t \to \infty$, in such a way that

$$\frac{t - n\langle\tau\rangle}{\sqrt{n\langle Var_{\omega_m}(\tau)\rangle}} \to x ,$$

we see that as $\to \infty$,

$$Q(\nu(t) \geq n) \to \Phi_\sigma(x) . \tag{3.23}$$

Note that

$$\frac{n - \frac{t}{\langle\tau\rangle}}{\sqrt{\frac{t\langle Var_{\omega_m}(\tau)\rangle}{\langle\tau\rangle^3}}} = \sqrt{\frac{t - n\langle\tau\rangle}{n\langle Var_{\omega_m}(\tau)\rangle}}\sqrt{\frac{n\langle\tau\rangle}{t}} .$$

If $n, t \to \infty$ in such a way that

$$\frac{t - n\langle\tau\rangle}{\sqrt{n\langle Var_{\omega_m}(\tau)\rangle}} \to x ,$$

then

$$1 - \frac{n\langle\tau\rangle}{t} = x\frac{n\langle Var_{\omega_m}(\tau)\rangle}{t} \to 0 \ ,$$

and thus

$$\frac{n - \dfrac{t}{\langle\tau\rangle}}{\sqrt{\dfrac{tVar_{\omega_m}(\tau)}{\langle\tau\rangle^3}}} \to -x \ . \tag{3.24}$$

Therefore, by (3.23) and (3.24),

$$Q\left(\frac{n - \dfrac{t}{\langle\tau\rangle}}{\sqrt{\dfrac{tVar_{\omega_m}(\tau)}{\langle\tau\rangle^3}}} \le -x\right) \to 1 - \Phi_\sigma(x) \ ,$$

where $\Phi_\sigma(x)$ is $N(0,\sigma^2)$ or, with respect to Q,

$$\lim_{t\to\infty} \frac{\nu(t) - \dfrac{t}{\langle\tau\rangle}}{\sqrt{\dfrac{t\,\langle Var_{\omega_m}(\tau)\rangle}{\langle\tau\rangle^3}}} \stackrel{L}{=} N(0,\sigma^2) \ , \tag{3.25}$$

and σ^2 given by (3.22).

Step 3. Note that

$$\frac{X_{\nu(t)} - \nu(t)\langle\eta\rangle}{\sqrt{t}} = \frac{\sum_1^\infty n^{\frac{1}{2}} I_{(\nu(t)=n)} \frac{\sum_1^n \eta_k - n\langle\eta\rangle}{n^{\frac{1}{2}}}}{\sqrt{t}}$$

and since

$$P_{\omega_m}\left(\lim_{t\to\infty}\frac{\nu(t)}{t} = \frac{1}{\langle\tau\rangle}\right) = 1 \ a.s. \ \omega_m \ .$$

With respect to Q

$$\frac{X_{\nu(t)} - \nu(t)\langle\eta\rangle}{\sqrt{t}} \stackrel{L}{\to} \Phi_{\sigma_1}(x) \text{ where } \sigma_1^2 = \frac{1}{\langle\tau\rangle}Var(\eta) \ . \tag{3.26}$$

Step 4. Since

$$\frac{X_t - \frac{\langle\eta\rangle}{\langle\tau\rangle}t}{\sqrt{t}} = \frac{X_t - X_{\nu(t)}}{\sqrt{t}} + \frac{X_{\nu(t)} - \langle\eta\rangle\,\nu(t)}{\sqrt{t}} + \langle\eta\rangle\,\frac{\nu(t) - \frac{t}{\langle\tau\rangle}}{\sqrt{t}} \, ,$$

and from (3.25) and (3.26) with respect to Q

$$\frac{X_{\nu(t)} - \langle\eta\rangle\nu(t)}{\sqrt{t}} + \langle\eta\rangle\frac{\nu(t) - \frac{t}{\langle\tau\rangle}}{\sqrt{t}} \to \Phi_\sigma(x) \, ,$$

where

$$\begin{aligned}\sigma^2 = \; & \frac{\langle\eta\rangle^2}{\langle\tau\rangle^3}\left(\langle Var_{\omega_m}(\tau)\rangle + Var\left(E_{\omega_m}[\tau]\right)\right) \\ & + 2\frac{\langle\eta\rangle}{\langle\tau\rangle^2} Cov\left(\eta, E_{\omega_m}[\tau]\right) + \frac{1}{\langle\tau\rangle} Var(\eta),\end{aligned}$$

to complete the proof we only need to show that

$$\lim_{t\to\infty} \frac{X_t - X_{\nu(t)}}{\sqrt{t}} = 0 \text{ in probability} .$$

THEOREM 3.5. *(Chow and Robbins [4]) Suppose Y_k, $k \geq 1$, are independent with $Y_k > 0$ for all k. Let $X_n = \sum_1^n Y_k$ and define*

$$\nu(t) = n \text{ on } X_n \leq t < X_{n+1} \, .$$

Suppose

$$\lim_{n\to\infty} \frac{1}{n} \sum_1^n E\left[Y_n\right] = \mu$$

and

$$\lim_{n\to\infty} \int_{(Y_n - E[Y_n] \geq n\varepsilon)} (Y_n - E[Y_n]) dP = 0 \, ,$$

for every $\varepsilon > 0$. Then

$$\lim_{t\to\infty} \frac{E\left[\nu(t)\right]}{t} = \frac{1}{\mu}.$$

For our use take $Y_k = \tau_k$, $k \geq 1$, and $X_n = T_n$. Since $E_{\omega_m}[\tau_k]$, $k \geq 1$, are i.i.d., by the SLLN

$$\lim_{n\to\infty} \frac{1}{n} \sum_1^n E_{\omega_m}[\tau_k] = \langle \tau \rangle a.s.\ \omega_m \ .$$

Also for any $\varepsilon > 0$,

$$\int_{(\tau_n - E_{\omega_m}[\tau] \geq n\varepsilon)} (\tau_n - E_{\omega_m}[\tau_n]) dP_{\omega_m} \leq \frac{1}{n\varepsilon} Var_{\omega_m}(\tau_n) \underset{n\to\infty}{\longrightarrow} 0 \quad a.s.\ \omega_m \ ,$$

since

$$\langle Var_{\omega_m}(\tau) \rangle < \infty \ .$$

Thus

$$\lim_{t\to\infty} \frac{E_{\omega_m}[\nu(t)]}{t} = \frac{1}{\langle \tau \rangle} \quad a.s. \quad \omega_m \ . \tag{3.27}$$

The proof of our CLT is finished by showing that

$$\lim_{t\to\infty} \frac{X_t - X_{\nu(t)}}{\sqrt{t}} = 0 \quad in \quad P_{\omega_m} - \text{probability} \quad a.s. \quad \omega_m \ .$$

To this end, note that

$$0 < X_t - X_{\nu(t)} \leq \eta_{\nu(t)+1} \ .$$

By Chebychev's inequality, for any $\varepsilon > 0$ and $\delta > 0$,

$$P_{\omega_m}(\eta_{\nu(t)+1} \geq \varepsilon\sqrt{t}) \leq \frac{E_{\omega_m}[\eta_{\nu(t)+1}^{2(1+\delta)}]}{t^{1+\delta}\varepsilon^{2(1+\delta)}} \ ,$$

so to complete the proof we will show

$$\lim_{t\to\infty} \frac{E_{\omega_m}[\eta_{\nu(t)+1}^{2(1+\delta)}]}{t^{1+\delta}} = 0 \ a.s.\ \omega_m \ . \tag{3.28}$$

Now

$$E_{\omega_m}[\eta_{\nu(t)+1}^{2(1+\delta)}] = \sum_0^\infty P_{\omega_m}(\nu(t) = n)\, \eta_{n+1}^{2(1+\delta)}(\omega_m)$$

$$= \sum_0^\infty n P_{\omega_m}(\nu(t) = n) \frac{\eta_{n+1}^{2(1+\delta)}(\omega_m)}{n} \ .$$

For any $\delta_1 > 0$,

$$\sum_1^\infty \mu\left(\eta^{2(1+\delta)} > n\right) \le \left\langle \eta^{2(1+\delta)(1+\delta_1)} \right\rangle \sum_1^\infty \frac{1}{n^{1+\delta_1}} < \infty .$$

Thus, by the Borel-Cantelli Lemma,

$$\lim_{t\to\infty} \frac{E_{\omega_m}[\eta^{2(1+\delta)}_{\nu(t)+1}]}{t^{1+\delta}}$$

is the same as

$$\lim_{t\to\infty} \frac{\sum_0^\infty n P_{\omega_m}(\nu(t) = n)\, I_{\left(\eta^{2(1+\delta)}_{n+1} \le n\right)} \dfrac{\eta^{2(1+\delta)}_{n+1}(\omega_m)}{n}}{t^{1+\delta}} . \tag{3.29}$$

Since

$$E_{\omega_m}[\nu(t)] = \sum_0^\infty n\, P_{\omega_m}(\nu(t) = n) ,$$

and from (3.27)

$$\lim_{t\to\infty} \frac{E_{\omega_m}[\nu(t)]}{t^{1+\delta}} = 0 \quad \text{a.s. } \omega_m$$

our proof is complete using (3.28) and (3.29).

4. Non-Gaussian limits. To go further in the discusssion of the types of possible limiting behavior we will make the assumption that the η's are geometric, that is

$$\mu(\eta_k = m) = (1-\beta)\beta^{m-1}, \quad m \ge 1.$$

We will focus our attention on limit laws for $\sum \tau_k$. See [13] for related results.

If $\left(\frac{q}{p}\right)^2 \beta < 1$ then we have a CLT and a SLLN since the condition established in the previous section are satisfied.

THEOREM 4.1. *(i) If $\left(\frac{q}{b}\right)\beta < 1$ but $\left(\frac{q}{p}\right)^2 \beta > 1$, let α be such that $\left(\frac{q}{p}\right)^\alpha \beta = 1$ (note that $1 < \alpha < 2$) and*

$$\lambda(n) = \frac{1}{\beta^{\{-\frac{\ln n}{\ln \beta}\}}}, \quad \textit{(Here } \{x\} = x - [x], \textit{ that is the fractional part of } x).$$

If we choose a subsequence $n' \to \infty$ so that

$$\left\{-\frac{\ln n'}{\ln \beta}\right\} \to \varepsilon_0 \text{ and } \lambda(n') \to \lambda_0 = \frac{1}{\beta^{\varepsilon_0}} ,$$

then

$$(4.1)\ \frac{\sum_{1}^{n'}(\tau_k - E_{\omega_m}[\tau_k])}{(n')^{\frac{1}{\alpha}}} \xrightarrow{L} \frac{2pq}{(q-p)^2}\left(\frac{p}{q}\right)^{\varepsilon_0}\sum_{1}^{\infty}\left(\frac{p}{q}\right)^{j-1}\sum_{1}^{\nu_j}(\xi_{j,l}-1)\ ,$$

where $\{\xi_{j,l}\}$ *are independent exponential random variables with parameter 1 and* ν_j, $j \geq 1$, *are independent Poisson random variables with parameters* $\frac{\lambda_0}{(1-\beta)\beta^j}$. *Moreover,*

$$(4.2)\qquad \frac{\sum_{1}^{n} E_{\omega_m}[\tau_k] - \langle\tau\rangle}{(n')^{\frac{1}{\alpha}}} \xrightarrow{L} \Phi_\alpha(x),$$

where $\Phi_\alpha(x)$ *is symmetric stable of index* α.

(ii) If $\left(\frac{q}{p}\right)\beta > 1$, *let* α *be such that* $\left(\frac{q}{p}\right)^\alpha \beta = 1$ *(note that* $0 < \alpha < 1$*) and*

$$\lambda(n) = \frac{1}{\beta^{\{-\frac{\ln n}{\ln \beta}\}}}, \quad (\textit{Here } \{x\} = x - [x],\ \textit{that is the fractional part of } x).$$

If we choose a subsequence $n' \to \infty$ *so that*

$$\left\{-\frac{\ln n'}{\ln \beta}\right\} \to \varepsilon_0 \text{ and } \lambda(n') \to \lambda_0 = \frac{1}{\beta^{\varepsilon_0}}\ ,$$

then

$$\frac{T_{n'} - n'\langle\tau\rangle}{(n')^{\frac{1}{\alpha}}} \xrightarrow{L} \frac{2pq}{(q-p)^2}\left(\frac{p}{q}\right)^{\varepsilon_0}\sum_{1}^{\infty}\left(\frac{p}{q}\right)^{j-1}\sum_{1}^{\nu_j}\xi_{j,l}$$

where $\{\xi_{j,l}\}$ *are independent exponential random variables with parameter 1, and* ν_j, $j \geq 1$, *are independent Poisson random variableswith parameters* $\frac{\lambda_0}{(1-\beta)\beta^j}$.

REMARK 4.1. The limit distributions in (4.1) and (4.2) are dependents as was the case in Theorem 3.3. The structure of the dependence is more complicated in this case and we do not address it. It involves the asymtotics of the maximal terms.

Proof of (ii). Let

$$M_n = max\{\ \eta_k\ :\ k = 1,\ 2, ...,\ n\}\ .$$

Define

$$\begin{aligned}
\nu_1(n) &= \#\{\eta_k' s : \eta_k = M_n \quad j \leq n\}\ , \\
\nu_2(n) &= \#\{\eta_k' s : \eta_k = M_n - 1 \quad j \leq n\}\ , \\
\nu_3(n) &= \#\{\eta_k' s : \eta_k = M_n - 2 \quad j \leq n\}\ , \\
&\vdots \\
\nu_{M_n}(n) &= \#\{\eta_k' s : \eta_k = 1 \quad j \leq n\}\,.
\end{aligned}$$

Let $m_n = \left[-\frac{\ln n}{\ln \beta}\right]$, and fix $a \in I$ and $l_0 \in I^+$. A direct calculation yields for $j_1, j_2, \cdots j_{l_0} \in I^+$, $j_1 \geq 1$, $m_n + a \geq 1$, and $j_1 + j_2 + \cdots + j_{l_0} \leq m_n + a$,

$$\mu\left(\nu_1(n) = j_1, \nu_2(n) = j_2, \cdots \nu_{l_0}(n) = j_{l_0} \mid M_n = m_n + a\right)$$

$$= \frac{n!}{j_1! j_2! \cdots j_{l_0}! \left(n - j_1 - j_2 - \cdots - j_{l_0}\right)!} \prod_{k=1}^{l_0} \left((1-\beta)\beta^{m_n+a-k}\right)^{j_k} \left(1 - \beta^{m_n+a-l_0}\right)^{n-j_1-j_2-\cdots-j_{l_0}} \times \frac{1}{\left(1-\beta^{m_n+a}\right)^n - \left(1-\beta^{m_n+a-1}\right)^n}$$

$$(4.3) \quad = \frac{1\left(1-\frac{1}{n}\right)\cdots\left(1 - \frac{n-j_1-j_2-\cdots-j_{l_0}+1}{n}\right)}{\left(1-\beta^{m_n+a-l_0}\right)^{j_1+j_2+\cdots+j_{l_0}}} \prod_{k=1}^{l_0} \frac{\left(\left(\frac{1-\beta}{\beta^k}\right)\beta^a n \beta^{m_n}\right)^{j_k}}{j_k!} \left(1 - \frac{\beta^a}{\beta^{l_0}}\beta^{m_n}\right)^n \times \frac{1}{\left(1 - \frac{n\beta^{m_n}\beta^a}{n}\right)^n - \left(1 - \frac{n\beta^{m_n}\beta^{a-1}}{n}\right)^n}$$

Define

$$\lambda(n) = \frac{1}{\beta^{\left\{-\frac{\ln n}{\ln \beta}\right\}}},$$

and note that

$$\lambda(n) = n\beta^{\left[-\frac{\ln n}{\ln \beta}\right]} = n\beta^{m_n}.$$

If $n' \to \infty$ in such a way that $\lambda(n') \to \lambda_0$, from (4.3) we find

$$\lim_{n' \to \infty} \mu\left(\nu_1(n') = j_1, \nu_2(n') = j_2, \cdots \nu_{l_0}(n') = j_{l_0} \mid M_{n'} = m_{n'} + a\right)$$

$$(4.4) \quad = \frac{e^{-\lambda_0\beta^a}}{e^{-\lambda_0\beta^a} - e^{-\lambda_0\beta^{a-1}}} \prod_{k=1}^{l_0} \left(\frac{\left(\lambda_0\left(\frac{1-\beta}{\beta^k}\right)\beta^a\right)^{j_k}}{j_k!} e^{-\left(\lambda_0\left(\frac{1-\beta}{\beta^k}\right)\beta^a\right)} \right),$$

where it must still be recalled that it is required that $j_1 \geq 1$.

Define

$$A_j(n) = \{k \leq n : \eta_k = M_n - j\}.$$

Now, for l_0 fixed

$$\frac{T_n}{n^{\frac{1}{\alpha}}} = \frac{\sum\limits_{1}^{n} \tau_k}{n^{\frac{1}{\alpha}}} = \frac{\sum\limits_{j=1}^{M_n} \sum\limits_{k \in A_j(n)} \tau_k}{n^{\frac{1}{\alpha}}}$$

$$= \frac{E_{\omega_m}\left[\tau_{M_n(\omega_m)}\right]}{n^{\frac{1}{\alpha}}} \sum_{j=1}^{M_n} \frac{E_{\omega_m}\left[\tau_{M_n-j(\omega_m)}\right]}{E_{\omega_m}\left[\tau_{M_n(\omega_m)}\right]} \sum_{k \in A_j(n)} \frac{\tau_k}{E_{\omega_m}\left[\tau_{M_n-j(\omega_m)}\right]}$$

$$(4.5) \quad \begin{aligned} &= \frac{E_{\omega_m}\left[\tau_{M_n(\omega_m)}\right]}{n^{\frac{1}{\alpha}}} \sum_{j=1}^{l_0} \sum_{k \in A_j(n)} \frac{E_{\omega_m}\left[\tau_k\right]}{E_{\omega_m}\left[\tau_{M_n(\omega_m)}\right]} \frac{\tau_k}{E_{\omega_m}\left[\tau_k\right]} \\ &+ \frac{E_{\omega_m}\left[\tau_{M_n(\omega_m)}\right]}{n^{\frac{1}{\alpha}}} \sum_{j=l_0+1}^{M_n} \sum_{k \in A_j(n)} \frac{E_{\omega_m}\left[\tau_k\right]}{E_{\omega_m}\left[\tau_{M_n(\omega_m)}\right]} \frac{\tau_k}{E_{\omega_m}\left[\tau_k\right]} . \end{aligned}$$

On the set $\{M_n = m_n + a\}$ and for n large enough, we see from (3.1) and (4.5) that

$$(4.6) \quad \frac{T_n}{n^{\frac{1}{\alpha}}} = \frac{2pq}{(q-p)^2} \frac{\left(\frac{q}{p}\right)^{m_n+a}}{n^{\frac{1}{\alpha}}} \sum_{j=1}^{l_0} \left(\frac{p}{q}\right)^{j-1} \sum_{k \in A_j(n)} \frac{\tau_k}{E_{\omega_m}\left[\tau_k\right]} + o\left(n^{\frac{1}{\alpha}}\right) .$$

Note that

$$n^{\frac{1}{\alpha}} = \left(\frac{q}{p}\right)^{-\frac{\ln n}{\ln \beta}} = \left(\frac{q}{p}\right)^{m_n} \left(\frac{q}{p}\right)^{\left\{-\frac{\ln n}{\ln \beta}\right\}}$$

and so

$$(4.7) \quad \frac{\left(\frac{q}{p}\right)^{m_n+a}}{n^{\frac{1}{\alpha}}} = \left(\frac{q}{p}\right)^{a} \left(\frac{p}{q}\right)^{\left\{-\frac{\ln n}{\ln \beta}\right\}} .$$

From (3.3) on $(M_n = m_n + a)$ for $k \in A_j(n)$, $1 \le j \le l_0$,

$$(4.8) \quad \frac{\tau_k}{E_{\omega_m}\left[\tau_k\right]} \underset{n\to\infty}{\overset{L}{\longrightarrow}} \text{exponential distribution with parameter } 1.$$

Thus, if $n' \to \infty$, so that

$$\left\{-\frac{\ln n'}{\ln \beta}\right\} \to \varepsilon_0 \text{ and } \lambda(n') \to \lambda_0 = \frac{1}{\beta^{\varepsilon_0}},$$

we have from (4.6), (4.7), and (4.8), that

$$
\begin{aligned}
&\langle \exp\left(-sI\left(n', l_0\right)\right) \mid M_{n'} = m_{n'} + a \rangle \underset{n' \to \infty}{\longrightarrow} \left(\frac{\exp(-\lambda_0 \beta^a)}{\exp(-\lambda_0\beta^a) - \exp(-\lambda_0 \beta^{a-1})}\right) \\
&\exp\left(- \frac{2pq}{(q-p)^2}\left(\frac{p}{q}\right)^{\varepsilon_0} \lambda_0 \left(\frac{1-\beta}{\beta}\right) \sum_{k=1}^{l_0} \frac{\left(\beta \frac{q}{p}\right)^{a-k+1} s}{1+ \frac{2pq}{(q-p)^2}\left(\frac{p}{q}\right)^{\varepsilon_0}\left(\frac{q}{p}\right)^{a-k+1} s}\right) \\
&\qquad - \left(\frac{\exp\left(-\lambda_0 \beta^{a-1}\right)}{\exp\left(-\lambda_0\beta^a\right) - \exp\left(-\lambda_0 \beta^{a-1}\right)}\right) \\
&\exp\left(- \frac{2pq}{(q-p)^2}\left(\frac{p}{q}\right)^{\varepsilon_0} \lambda_0 \left(\frac{1-\beta}{\beta}\right) \sum_{k=2}^{l_0} \frac{\left(\beta \frac{q}{p}\right)^{a-k+1} s}{1+ \frac{2pq}{(q-p)^2}\left(\frac{p}{q}\right)^{\varepsilon_0}\left(\frac{q}{p}\right)^{a-k+1} s}\right).
\end{aligned}
$$

(4.9)

Next, letting $l_0 \to \infty$, we obtain

$$
\begin{aligned}
&\left\langle \exp\left(- s\frac{T_{n'}}{(n')^{\frac{1}{\alpha}}}\right) \mid M_{n'} = m_{n'} + a \right\rangle \underset{n' \to \infty}{\longrightarrow} \left(\frac{\exp(-\lambda_0 \beta^a)}{\exp(-\lambda_0\beta^a) - \exp(-\lambda_0 \beta^{a-1})}\right) \\
&\exp\left(- \frac{2pq}{(q-p)^2}\left(\frac{p}{q}\right)^{\varepsilon_0} \lambda_0 \left(\frac{1-\beta}{\beta}\right) \sum_{k=1}^{\infty} \frac{\left(\beta \frac{q}{p}\right)^{a-k+1} s}{1+ \frac{2pq}{(q-p)^2}\left(\frac{p}{q}\right)^{\varepsilon_0}\left(\frac{q}{p}\right)^{a-k+1} s}\right) \\
&\qquad - \left(\frac{\exp(-\lambda_0 \beta^{a-1})}{\exp(-\lambda_0\beta^a) - \exp(-\lambda_0 \beta^{a-1})}\right) \\
&\exp\left(- \frac{2pq}{(q-p)^2}\left(\frac{p}{q}\right)^{\varepsilon_0} \lambda_0 \left(\frac{1-\beta}{\beta}\right) \sum_{k=2}^{\infty} \frac{\left(\beta \frac{q}{p}\right)^{a-k+1} s}{1+ \frac{2pq}{(q-p)^2}\left(\frac{p}{q}\right)^{\varepsilon_0}\left(\frac{q}{p}\right)^{a-k+1} s}\right) \\
&\qquad = \left(\frac{\exp(-\lambda_0 \beta^{a})}{\exp(-\lambda_0\beta^a) - \exp(-\lambda_0 \beta^{a-1})}\right) \\
&\exp\left(- \frac{2pq}{(q-p)^2}\left(\frac{p}{q}\right)^{\varepsilon_0} \lambda_0 \left(\frac{1-\beta}{\beta}\right) \sum_{k=0}^{\infty} \frac{\left(\beta \frac{q}{p}\right)^{a-k} s}{1+ \frac{2pq}{(q-p)^2}\left(\frac{p}{q}\right)^{\varepsilon_0}\left(\frac{q}{p}\right)^{a-k} s}\right)
\end{aligned}
$$

(4.10)

$$-\left(\frac{\exp\left(-\lambda_0\beta^{a-1}\right)}{\exp(-\lambda_0\beta^{a})-\exp(-\lambda_0\beta^{a-1})}\right)$$

$$\exp\left(-\tfrac{2pq}{(q-p)^2}\left(\tfrac{p}{q}\right)^{\varepsilon_0}\lambda_0\left(\tfrac{1-\beta}{\beta}\right)\sum_{k=0}^{\infty}\frac{\left(\beta\frac{d}{i}qp\right)^{a-1-k}s}{1+\dfrac{2pq}{(q-p)^2}\left(\dfrac{p}{q}\right)^{\varepsilon_0}\left(\dfrac{q}{p}\right)^{a-1-k}s}\right).$$

Finally, summing (4.10) over a's, we get

$$\left\langle \exp\left(-s\frac{T_{n'}}{(n')^{\frac{1}{\alpha}}}\right)\right\rangle \underset{n'\to\infty}{\longrightarrow}$$

$$\exp\left(-\tfrac{2pq}{(q-p)^2}\left(\frac{p}{q}\right)^{\varepsilon_0}\lambda_0\left(\tfrac{1-\beta}{\beta}\right)\sum_{j=-\infty}^{\infty}\frac{\left(\beta\dfrac{q}{p}\right)^{j}s}{1+\dfrac{2pq}{(q-p)^2}\left(\dfrac{p}{q}\right)^{\varepsilon_0}\left(\dfrac{q}{p}\right)^{j}s}\right).$$

This is in agreement with the limit distribution obtained in [13] .

Proof of (i). The analysis of the limiting behavior of (4.1) is accomplished in essentially the same way as in the case $0<\alpha<1$. For (4.2) we are in a more standard situation, namely $E_{\omega_m}[\tau_k]$, $k\geq 1$, are i.i.d. as random variables on (μ,Ω_m). Recall that

$$E_{\omega_m}[\tau_k]=\frac{2pq}{(q-p)^2}\left(\left(\frac{q}{p}\right)^{\eta_k}-1\right)-\frac{\eta_k}{q-p}.$$

Let $F(x)$ be the distribution function of $E_{\omega_m}[\tau_k]$ as a random variable on Ω_m. Then for x large and $n=\left[\frac{\ln x}{\ln\left(\frac{q}{p}\right)}\right]$, $\left(\left(\frac{q}{p}\right)^n\simeq x\ \right)$,

$$\begin{aligned}U(x)&=\int_{-x}^{x}y^2F(dy)\simeq c\sum_{k\leq n}\left(\tfrac{q}{p}\right)^{2k}(1-\beta)\beta^{k-1}\\&=c(1-\beta)\left(\frac{q}{p}\right)^2\left(\frac{\left(\left(\frac{q}{p}\right)^2\beta\right)^n-1}{\left(\frac{q}{p}\right)^2\beta-1}\right).\end{aligned}$$

Recall, that for this case α is determined by $\left(\frac{q}{p}\right)^{\alpha}\beta=1$. So

$$\left(\left(\frac{q}{p}\right)^2\beta\right)^n=x^{2-\alpha}.$$

Therefore

$$U(x)=x^{2-\alpha}L(x)$$

where

$$L(x) = \left(1 - \frac{1}{x^{2-\alpha}.}\right) \frac{(1-\beta)\left(\frac{q}{p}\right)^2}{\left(\frac{q}{p}\right)^2 \beta},$$

and it is easy to see that $L(x)$ is slowly varing. The result then is that the limiting distribution is the symmetric stable distribution of exponent α (see Feller [6]).

5. Sinai's type trapping behavior. Let $F(x)$ be a distribution function with density $p(x)$ with support on $[0, \infty)$, and require that $p(x) > 0$ for $x > 0$. Define

$$\rho_n = \int_{n-1}^{n} p(x) dx.$$

Suppose ξ_k, $k \geq 1$, are i.i.d. with distribution $F(x)$ and η_k, $k \geq 1$, are i.i.d. taking values in the positive integers with $P(\eta_k = n) = \rho_n : k \geq 1,\, n \geq 1$. Without loss of generality we can take our model so that

$$(\eta_k = n) = (n - 1 < \xi_k \leq n). \tag{5.1}$$

Define the random variables based on ξ_k, $k \geq 1$, by

$$Y_k = -\ln(1 - F(\xi_k)) : k \geq 1.$$

Also introduce the function

$$\phi(x) = -\ln(1 - F(x)).$$

By construction $Y_k : k \geq 1$ are i.i.d., exponentially distributed with parameter 1.

REMARK 5.1. One can easily verify that $\phi'(x) > 0$ and so $\phi(x)$ is strictly increasing and has an inverse

$$\psi(x) = \overline{F}^{-1}(e^{-x}).$$

Here $\overline{F}(x) = 1 - F(x)$.

5.1. Record values. Let $\xi_k : k \geq 1$ be as above. Define the times of the record values by

$$L_1 = 1 \qquad L_n = \inf\left\{j : j > L_{n-1},\, \xi_j > \xi_{L_{n-1}}\right\}.$$

Since $\phi(x)$ is strictly increasing, the record value times for ξ_k, $k \geq 1$ and Y_k, $k \geq 1$, are identical.

We will next consider the sequence of record values

$$\xi_{L_n},\ n \geq 1, \quad \text{and} \quad Y_{L_n},\ n \geq 1.$$

The following are well known results from record value theory for i.i.d. random variables:

A. There exist $Z_j,\ j \geq 1$, independent exponential random varibales with parameter 1 such that

$$Y_{L_n} = \sum_{j=1}^{n} Z_j .$$

B. Renyi (1962) proved for continuous random variables ($F(x)$ continuous) that

$$\frac{\ln L_n}{n} \to 1 \ a.s. \ ,$$

and

$$\frac{\ln L_n - n}{\sqrt{n}} \xrightarrow{L} N(0,1) \ .$$

C. Let $N_n = \#$ of record values be less than or equal to n. If $F(x)$ is continuous then

$$\lim_{n\to\infty} \frac{N_n}{\ln n} = 1 \ a.s. \ .$$

and

$$\frac{N_n - \ln n}{\sqrt{\ln n}} \xrightarrow{L} N(0,1) \ .$$

D. In the i.i.d. exponential case,

$$\frac{\ln Y_{L_n}}{n} \to 1 \ a.s. \ ,$$

and

$$\frac{\ln Y_{L_n} - n}{\sqrt{n}} \xrightarrow{L} N(0,1) \ .$$

Let $X_t = \#$ of record values of $Y_k,\ k \geq 1$, in the interval $[0,t]$. From A, we see that $X_t,\ t \geq 0$, is Poisson with parameter 1. Consider the interval $\Delta_n = \phi(n) - \phi(n-1) : n \geq 1$. Now

$$P(X_{\Delta_n} \geq 2) = 1 - e^{-\Delta_n}(1 + \Delta_n) \ ,$$

and so

$$\sum_{n=1}^{\infty} P\left(X_{\Delta_n} \geq 2\right) = \sum_{n=1}^{\infty} \left(1 - e^{-\Delta_n}\left(1 + \Delta_n\right)\right) = \sum_{n=1}^{\infty} \left(c\Delta_n^2 + o\left(\Delta_n^2\right)\right) .$$

Cases to be considered:
(i) $\overline{F}(x) \sim \frac{c}{x^\gamma}$,
(ii) $\overline{F}(x) \sim ce^{-x^\beta}$ $\quad 0 < \beta < 1$.
In case (i),

$$\phi(x) = -\ln\left(\frac{c}{x^\gamma}\right) \sim \gamma \ln x \ ,$$

so

$$\Delta_n = \phi(n) - \phi(n-1) = \gamma \ln\left(\frac{n}{n-1}\right) \simeq \gamma \frac{1}{n} .$$

Thus $\Delta_n \simeq \frac{c}{n}$, so that

$$\sum_{n=1}^{\infty} P\left(X_{\Delta_n} \geq 2\right) < \infty \ ,$$

and by the Borel-Cantelli Lemma,

$$P\left(X_{\Delta_n} \geq 2 \ i.o.\right) = 0 \ .$$

For case (ii), $\phi(x) = x^\beta$, so that

$$\Delta_n = \phi(n) - \phi(n-1) = n^\beta - (n-1)^\beta \leq \frac{1}{(n-1)^{1-\beta}} .$$

Thus

$$\sum_{n=1}^{\infty} P\left(X_{\Delta_n} \geq 2\right) \simeq \sum_{n=1}^{\infty} \left(\frac{1}{(n-1)^{1-\beta}}\right)^2 < \infty \ , \qquad if\ \beta < \frac{1}{2} .$$

Hence

$$P\left(X_{\Delta_n} \geq 2 \ i.o.\right) = 0 \ if\ \beta < \frac{1}{2} .$$

We now want to consider the record values for $\eta_k : k \geq 1$ where the η_k's are define by 5.1. Define

$$l_1 = 1 \ ,$$
$$l_n = \inf\left\{j : j > l_{n-1} \text{ and } \eta_j > \eta_{l_{n-1}}\right\} .$$

In the discrete case it is possible that record values are repeated. It is this question which we must address.

Define

$$\nu_n = \#(j : l_n < j < l_{n+1} \text{ such that } \eta_j = \eta_{l_n}) \ .$$

We want to be able to claim

$$P(\nu_n \geq 1 \ i.o.) = 0 \ .$$

Recall that

$$(\eta_k = n) = (n - 1 < \xi_k \leq n) \ .$$

Now,

$$\begin{aligned}&(\eta_j = \eta_{l_n} \text{ for some } j > l_n) \\ &= (m - 1 < \xi_{l_n} \leq m) \cap (m - 1 < \xi_j \leq m) \ .\end{aligned}$$

So,

$$(\nu_n \geq 1) = (m - 1 < \xi_{l_n} \leq m) \cap \left(\bigcup_{j=l_n+1}^{l_{n+1}-1} (m - 1 < \xi_j \leq m) \right) .$$

For both cases we are considering, for n large enough,

$$\xi_{l_{n+1}} - \xi_{l_n} > 1 \ ,$$

so that for

$$l_n + 1 \leq j \leq l_{n+1} - 1, \qquad \xi_j < \xi_{l_n} \ ,$$

the term which must be estimated is

$$\bigcup_{j=l_n+1}^{l_{n+1}-1} (\xi_{l_n} - \xi_j \leq 1) \ .$$

Building a plausible argument, recall that

(1) $L_n \approx e^n$,

(2) $\xi_{l_n} \approx \psi(n)$.

Therefore

$$\begin{aligned}P\left(\bigcup_{j=l_n+1}^{l_{n+1}-1} (\xi_{l_n} - \xi_j \leq 1) \right) &\leq \sum_{e^n}^{e^{n+1}} \int_{\psi(n)-1}^{\psi(n)} p(x)\, dx \\ &\simeq \sum_{e^n}^{e^{n+1}} p(\psi(n)) \simeq \left(e^{n+1} - e^n\right) p(\psi(n)) .\end{aligned}$$

For case (i), that is $\overline{F}(x) \simeq \frac{c}{x^\gamma}$, $\phi(x) \simeq \delta \ln n$, and $\psi(x) \simeq e^{\frac{x}{\delta}}$, we have

$$p(x) = -D_x \overline{F}(x) \simeq \frac{c}{x^{\delta+1}} ,$$

and so

$$p(\psi(n)) \simeq \frac{c}{e^{\frac{n}{\delta}(\delta+1)}}.$$

Hence

$$\left(e^{n+1} - e^n\right) p(\psi(n)) . \simeq e^n (e-1) \frac{c}{e^{\frac{n}{\delta}(\delta+1)}} . \simeq e^{-\frac{n}{\delta}}.$$

Therefore

$$\sum P(\nu_n \geq 1) \simeq \sum e^{-\frac{n}{\delta}} < \infty ,$$

and so

$$P(\nu_n \geq 1 \; i.o.) = 0.$$

For case (ii), that is $\overline{F}(x) \simeq e^{-x^\beta}$, $\phi(x) \simeq x^\beta$,and $\psi(x) \simeq x^{\frac{1}{\beta}}$

$$p(x) = -D_x \overline{F}(x) \simeq x^{\beta-1} e^{-x^\beta} ,$$

and so

$$p(\psi(n)) \simeq \left(n^{\frac{1}{\beta}}\right)^{\beta-1} \exp\left(-\left(n^{\frac{1}{\beta}}\right)^\beta\right) \simeq n^{\left(1-\frac{1}{\beta}\right)} e^{-n}.$$

Therefore

$$\sum P(\nu_n \geq 1) \simeq \sum n^{\left(1-\frac{1}{\beta}\right)} e^{-n} < \infty ,$$

so that, finally,

$$P(\nu_n \geq 1 \; i.o.) = 0 .$$

REMARK 5.2. For case (i), that is

$$\overline{F}(x) = \frac{c}{x^\gamma},$$

we have no power moments. To see this note that

$$\langle \tau^\alpha \rangle \simeq \sum \left(\frac{q}{p}\right)^{n\alpha} \rho_n,$$

where

$$\rho_n = \overline{F}(n-1) - \overline{F}(n) = \frac{c}{(n-1)^\gamma} - \frac{c}{n^\gamma} \simeq \frac{c\gamma}{n^{\gamma+1}}.$$

Thus

$$\langle \tau^\alpha \rangle \simeq \sum \left(\frac{q}{p}\right)^{n\alpha} \frac{c\gamma}{n^{\gamma+1}} = \infty,$$

for any α.

REMARK 5.3. For case(ii), that is

$$\overline{F}(x) = ce^{-x^\beta},$$

it can be cheeked that for

$$\alpha \ln \frac{q}{p} < \beta,$$

power moments exist.

5.2. Trapping behavior. We give a rough sketch of the ideas. Complete details will appear at a later time. Let

$$l_n = \text{ time of the nth record value of } \eta_k : k \geq 1.$$

Let $\tau_k : k \geq 1$ be the intervent times as before. Our goal is to show that

$$\lim_{n\to\infty} \frac{\sum\limits_{k\leq l_n} \tau_k}{\tau_{l_n}} = 1 \quad a.s. Q.$$

The meaning is this, $x_n : n \geq 1$ spends most of its time near the vent x_{L_n} between $1 \leq k \leq l_n$. That is $x_n : n \geq 1$ becomes trapped near the sequence $x_{l_n} : n \geq 1$.

Since

$$\frac{\tau_{l_n}}{\sum\limits_{k\leq l_n} \tau_k} \leq 1,$$

we need only to deal with a lower bound.

REMARK 5.4. For simple random walk with reflection at 0 and $q > p$, for n large enough,

$$\frac{t_n}{E[t_n]} \leq c \ln n.$$

To see this note that

$$P\left(\frac{t_n}{E[t_n]} \geq c \ln n\right) \leq \frac{E\left[\exp\left(\frac{t_n}{2E[t_n]}\right)\right]}{n^{\frac{c}{2}}}.$$

Since

$$\frac{t_n}{E[t_n]} \xrightarrow{L} \text{exponential with parameter } 1,$$

we would expect that

$$E\left[\exp\left(\frac{t_n}{2E\left[t_n\right]}\right)\right] \to \int_0^\infty e^{-\frac{x}{2}}dx = 2.$$

Strictly speaking, our argument is not correct since $\exp\left(\frac{x}{2}\right)$is not a bounded function. A correct argument can be constructed. Thus, if $c > 2$

$$\sum P\left(\frac{t_n}{E\left[t_n\right]} \geq c\ln n\right) < \infty,$$

and so, by Borel-Cantelli Lemma,

$$P\left(\frac{t_n}{E\left[t_n\right]} \geq c\ln n\ i.o.\right) = 0.$$

Let as before,

$$\nu_n = \#\left(j : l_n < j < l_{n+1} \text{ such that } \eta_j = \eta_{l_n}\right).$$

Suppose we know that

$$\mu\left(\nu_n \geq 1\ i.o.\right) = 0. \tag{5.2}$$

From the Remark 5.4 and 5.2 we see that for n large enough

$$\sum_{k<l_n} \tau_k \leq C\, l_n \ln \eta_{l_{n-1}} E_{\omega_m}\left[\tau_{l_{n-1}}\right] \quad a.s.\,Q.$$

Thus, for n large enough,

$$\frac{\frac{\tau_{l_n}}{E_{\omega_m}[\tau_{l_n}]}}{C\, l_n \ln \eta_{l_{n-1}} \cdot \frac{E_{\omega_m}\left[\tau_{l_{n-1}}\right]}{E_{\omega_m}[\tau_{l_n}]} + \frac{\tau_{l_n}}{E_{\omega_m}[\tau_{l_n}]}} \leq \frac{\tau_{l_n}}{\sum\limits_{k\leq l_n} \tau_k} \quad a.s.\,Q.$$

From a generalized version of Renyi's result which requires 5.2, we know that

$$\frac{\ln l_n}{n} \to 1\ a.s.\ \mu$$

or roughly speaking

$$l_n \simeq e^n.$$

Moreover, using the notation of Section 5.1,

$$\eta_{l_n} \simeq \overline{F}^{-1}\left(e^{-Y_{L_n}}\right),$$

and since

$$Y_{L_n} \simeq e^n,$$

we have

$$l_n \ln \eta_{l_{n-1}} \frac{E_{\omega_m}[\tau_{l_{n-1}}]}{E_{\omega_m}[\tau_{l_n}]}$$
$$\simeq e^n \ln\left(\overline{F}^{-1}\left(e^{-e^{n-1}}\right)\right)\left(\frac{p}{q}\right)^{\overline{F}^{-1}\left(e^{-e^n}\right)-\overline{F}^{-1}\left(e^{-e^{n-1}}\right)}.$$

We consider case $\overline{F}(x) = \frac{c}{x^\gamma}$. Since

$$\overline{F}^{-1}(x) = \frac{c}{x^{\frac{1}{\gamma}}},$$

and

$$e^n \ln\left(\overline{F}^{-1}\left(e^{-e^{n-1}}\right)\right)\left(\frac{p}{q}\right)^{\overline{F}^{-1}\left(e^{-e^n}\right)-\overline{F}^{-1}\left(e^{-e^{n-1}}\right)}$$
$$= e^n \ln\left(e^{e^{\frac{n-1}{\gamma}}}\right)\left(\frac{p}{q}\right)^{e^{e^{\frac{n}{\gamma}}}-e^{e^{\frac{n-1}{\gamma}}}}$$
$$= e^{n+\frac{n-1}{\gamma}}\left(\frac{p}{q}\right)^{e^{e^{\frac{n}{\gamma}}}\left(1-e^{e^{\frac{-1}{\gamma}}}\right)} \underset{n\to\infty}{\longrightarrow 0} \quad a.s.\, Q.$$

Therefore

$$\lim_{n\to\infty} \frac{\sum\limits_{k\leq l_n} \tau_k}{\tau_{l_n}} = 1 \quad a.s.\, Q.$$

REFERENCES

[1] Azov, D. Z., Bobrov, A. A., "The extreme terms of a sample and their role in the sum of independent variables", Theory of Probability and their Applications Vol. 5 (1960), pp. 377–396.

[2] Bakry,D., Gill,R., and Molchanov, S. Lectures in Probability, Theory Saint-Flour Summer School 1992, Lecture Notes in Mathematics No 1581, Springer-Verlag (1994).

[3] Burridge, R., Papanicolaou, G., White, B., "One-dimensional Wave Propagation in a Highly Discontinuous Medium," Wave Motion Vol. 10 (1988), pp.19–44.

[4] Chow, Y.S. and Robbins,H. "A renewal theorem for random variables which are dependent or non-identically distributed", Annals of Mathematical Statistics, Vol. 34 (1963), pp. 390–395.

[5] Darling D. A., "The role of the maximal term in the sum of independent random variables", Transactions of American Math. Society, Vol. 73 (1952), pp. 95–107.

[6] Feller, W., An Introduction to Probability Theory and its Applications Vol. 2, John Wiley and Sons, New York (1966).

[7] Glick, N., "Breaking records and breaking boards", American Math. Monthly Vol 85, (1978) pp. 2–26.

[8] Kesten, H., Kozlov, S., and Spitzer, F., "A limit law for random walk in a random environment", Compositio Mathematica, Vol. 30 (1975), pp. 145–168.

[9] Kesten, H., "The limit distribution of Sinai's random walk in random environment", Physica Vol. 138A, (1986), pp. 299–309.

[10] Kozlov, S.M., "The method of averaging and walks in inhomogeneous environments", Russian Math. Surveys, Vol. 45 (1985), pp. 73–145

[11] Resnick, S. I., "Record values and maxima" The Annals of Probability Vol. 1 (1973) pp. 650–662.

[12] Sinai, Ja., "The limiting behavior of a one-dimensional random walk in random medium", Theory of Probability and its Applications, Vol. 27, (1982), pp. 256–268.

[13] Solomon, F. "Random walks in a random environment", Annals of Probability, Vol. 3 (1975), pp. 1–31.

CASCADE OF SCALING GYROSCOPES: LIE STRUCTURE, UNIVERSAL MULTIFRACTALS AND SELF-ORGANIZED CRITICALITY IN TURBULENCE

Y. CHIGIRINSKAYA* AND D. SCHERTZER*

Abstract. Following V.I. Arnold and A.M. Obukhov, we consider the similarities between the Lie structure of the Navier-Stokes equations of hydrodynamic turbulence and the Euler equations of a gyroscope. We show that indeed a certain type of direct interaction yields a quite closer analogy than previously considered. Furthermore, the interactions built up dynamically on it yield a dynamical space-time cascade, the cascade of scaling gyroscopes, which should preserve most of the properties of the Navier-Stokes equations. We point out that it corresponds to a non-trivial tree-decomposition of the non-simple Lie structure of turbulence. We show how this cascade model can help to clarify fundamental questions of turbulence by investigating the possible multifractal universality and multifractal phase transitions to Self Organized Criticality.

Introduction. In this paper we consider a dynamical cascade process, the Scaling Gyroscopes Cascade (SGC), based on the Lie structure of the Navier-Stokes equations. This model is deduced from these equations by preserving only a certain type of direct interaction, the resulting indirect interactions being built dynamically along the tree-structure of the cascade. In its simplest form, this model is in fact one dimensional in space, however it does not suffer from the basic problem of the celebrated one dimensional caricature of the Navier-Stokes equation, i.e. the Burgers equation, since incompressibility is respected. Indeed, only the interactions which preserve incompressibility without requiring a pressure term will be selected.

The importance of the Lie structure of the Navier-Stokes equation has been underlined by Arnold (1966), Obukhov (1971), in particular in relation to the Euler equations of a rigid body motion. Starting with the same consideration we will concentrate on a somewhat opposite and more direct analogy. We will show that for the selected direct interactions the velocity field is a direct analogue of the angular momentum of a solid body rotation, whereas the vorticity is the analogue of the angular velocity. Therefore, the corresponding non-direct interactions can be understood as resulting from couplings of gyroscopes, yielding a (scaling) cascade of gyroscopes (see Fig. 1 for an illustration). One may note that in the following we will use the word "gyroscope" although the word "top" would have been more appropriate. However, the former is commonly used to designate the fundamental equation of motion which we will be using.

Cascades have a tree-structure since each eddy is a "mother" of two "daughter" eddies: the number of eddies of a given scale (required to define the velocity field) increases algebraically with the inverse of the scale, i.e. the scale ratio. Associated to it, we obtain a tree-decomposition on the

* Université P. et M. Curie, L.M.D., BP 99, 4 Pl. Jussieu F-75252 Paris Cedex 05, France.

FIG. 1. *Illustration of a scaling cascade of gyroscopes. We are not only using an apparently simple toy (who hasn't played with a top?) to explore complexity (non semi-simple Lie algebra as discussed later), but we are building up a rather explicit version of Richardson's famous poem which could be read as: big tops have little tops that feed on their momentum and little tops have smaller tops and so on to friction -in the molecular sense.*

semi-simple Lie algebra $so(3)$ of the non-simple Lie algebra of a subset of fluid turbulence interactions. The importance of achieving a decomposition of this type could be better appreciated by recalling that non-simplicity (i.e. the algebra admits a non-null radical) for turbulence is associated with scaling properties of the field as discussed by Schertzer and Lovejoy (1995), although it is important to note that it was done in the rather different framework of stochastic multiplicative cascade processes. Within this framework, different notions of multifractal intermittency have been developed: multifractal universality and transitions to Self Organized Criticality. We shall test the relevance of these notions by (comparison with) numerical simulations of the SGC. We will also find a surprisingly close agreement to various empirical studies on atmospheric turbulence.

1. Similarities between fundamental equations of solid and fluid mechanics.

1.1. Euler's theorems of rigid body motion. We recall that first Euler's theorem or Euler's equation for a rigid body (attached to a fixed point with no torque), often called the gyroscope equation, is merely the following:

$$\frac{d\underline{M}}{dt} = \underline{M} \wedge \underline{\Omega} \tag{1.1}$$

where $\underline{M}$ is its angular momentum and $\underline{\Omega}$ its rotation (both relative to the body frame); $\wedge$ is the vector product. The (quadratic) nonlinearity of the (apparently linear) equation results from the linear relationship between angular momentum and rotation via the (second order) moment of inertia tensor $\underline{\underline{I}}$ or its inverse ($\underline{\underline{J}} = \underline{\underline{I}}^{-1}$), both being symmetric:

$$\underline{M} = \underline{\underline{I}} \cdot \underline{\Omega} \; ; \quad \underline{\Omega} = \underline{\underline{J}} \cdot \underline{M} \; . \tag{1.2}$$

The equation of motion relative to the body frame (1.1) is equivalent to Newton's law of the conservation of angular momentum relative to space (M_s):

$$\frac{dM_s}{dt} = 0 \; . \tag{1.3}$$

This second Euler's theorem is in fact a particular case of Noether's theorem stating that there is an invariant associated with any equation of motion. We will discuss below the existence of two associated quadratic invariants. (1.1) corresponds to an adjoint action[1] on a Lie algebra a, which is a rather canonical evolution equation for the Lie algebras:

$$\frac{\partial X}{\partial t} = -ad_Y(X) \equiv [X, Y] \; ; \quad X, Y \in a \tag{1.4}$$

$[.,.]$ denoting the corresponding Lie bracket, which is a skew bilinear application on the Lie algebra and which satisfies the Jacobi identity:

$$\forall \;\; X, Y, Z \in a : \;\; [X,[Y,Z]] + [Y,[Z,X]] + [Z,[X,Y]] = 0 \; . \tag{1.5}$$

On a basis $\{E_i\}$ of the algebra, the Lie bracket is defined by the structure *constants* of the algebra C_{ij}^k:

$$[E_i, E_j] = C_{ij}^k \; E_k \; . \tag{1.6}$$

(1.1) is indeed equivalent to (1.4) when considering $so(3)$ which admits $\Re^3$ as a linear representation: the structure constants are the components of

---

[1] More precisely it corresponds to a co-adjoint equation, but on any finite Lie algebra one is allowed to identify systematically a dual space to original space. This explains the presence of the minus sign on the r.h.s. of (1.4).

the fundamental asymmetric tensor (the vector product ($\wedge$) corresponds to the Lie bracket). The first associated quadratic invariant to (1.1), is the square of the angular momentum (M^2), since:

$$\frac{d\underline{M}^2}{dt} = 2\ \underline{M} \cdot (\underline{M} \wedge \underline{\Omega}) = (\underline{M}, \underline{M}, \underline{\Omega}) \equiv 0 \tag{1.7}$$

(.,.,.) denoting the mixed product. On the other hand the kinetic energy of the body is defined as:

$$T = \frac{1}{2}\ \underline{M} \cdot \underline{\Omega} \equiv \frac{1}{2}\ \underline{M} \cdot (\underline{\underline{J}} \cdot \underline{M}) \tag{1.8}$$

and corresponds to a second quadratic invariant. It may be seen as a mere consequence that the equation of evolution of the angular velocity is dual to the one of angular momentum, since we have due to symmetry and time invariance of the inertia tensor:

$$\underline{\Omega} \cdot \frac{d\underline{M}}{dt} = \frac{d\underline{\Omega}}{dt} \cdot \underline{M} \tag{1.9}$$

therefore:

$$\frac{dT}{dt} = 2\ \underline{\Omega} \cdot \frac{d\underline{M}}{dt} \tag{1.10}$$

which is identically null, since:

$$\underline{\Omega} \cdot \frac{d\underline{M}}{dt} = \underline{\Omega} \cdot (\underline{M} \wedge \underline{\Omega}) = (\underline{\Omega}, \underline{M}, \underline{\Omega}) \equiv 0\ . \tag{1.11}$$

This invariance, as well as the related duality between angular velocity and angular momentum equations of evolution, corresponds more fundamentally to the fact that the motion is prescribed by a principle of least action: the motion describes a geodesic in the group of rotations for the kinetic energy, being a (left) invariant metric for the motion. This will be discussed further in section (2.2). One may note that there exists a rather straightforward extension of (1.1) to complex (three-dimensional) vectors with the Hermitian extension of the Euclidean scalar product (still denoted by a dot, whereas complex conjugates will be denoted by an overbar):

$$\frac{d\underline{M}}{dt} = \overline{\underline{M}} \wedge \overline{\underline{\Omega}}\ . \tag{1.12}$$

The Hermitian structure preserves the quadratic invariants $\underline{M}^2$ and T since the notion of mixed products is unchanged, (1.7) and (1.9)-(1.11) yield respectively:

$$\frac{d\underline{M}^2}{dt} = 2\Re\{(\overline{\underline{M}} \wedge \overline{\underline{\Omega}}) \cdot \underline{M}\} = 2\Re\{(\overline{\underline{M}}, \overline{\underline{\Omega}}, \overline{\underline{M}})\} \equiv 0\ ; \tag{1.13}$$

$$\frac{dT}{dt} = \Re\{\overline{\underline{M}} \wedge \overline{\underline{\Omega}}, \underline{\Omega}\} = \Re\{(\overline{\underline{M}}, \overline{\underline{\Omega}}, \overline{\underline{\Omega}})\} \equiv 0 \tag{1.14}$$

$\Re$ denoting the real part of complex variable.

1.2. The Lie structure of the Navier-Stokes equations. Consider the Navier-Stokes equations, for the velocity field $\underline{u}(\underline{x},t)$, written under Bernoulli's form (α being the kinematic pressure, e.g. for a constant fluid density[2] ρ_0: $\alpha = \frac{p}{\rho_0} + \frac{u^2}{2}$, p being the (static) pressure; ν is the fluid viscosity; $\underline{w}$ is the vorticity field):

$$(1.15)\qquad \begin{aligned} &\left(\tfrac{\partial}{\partial t} - \nu\Delta\right)\underline{u}(\underline{x},t) = \underline{u}(\underline{x},t) \wedge \underline{w}(\underline{x},t) - \underline{grad}(\alpha)\ ; \\ &\underline{w}(\underline{x},t) = \underline{curl}(\underline{u}(\underline{x},t)) \end{aligned}$$

with the associated incompressibility condition:

$$(1.16)\qquad div(\underline{u}(\underline{x},t)) = 0\ .$$

Taking the curl of Bernoulli's equation (1.15) one obtains the well known vorticity equation:

$$(1.17)\qquad \left(\frac{\partial}{\partial t} - \nu\Delta\right)\underline{w}(\underline{x},t) = [\underline{w}(\underline{x},t), \underline{u}(\underline{x},t)]$$

the Lie bracket being then defined as:

$$(1.18)\qquad [X,Y] = \underline{Y}\cdot\underline{\underline{grad}}(\underline{X}) - \underline{X}\cdot\underline{\underline{grad}}(\underline{Y})$$

due to the vector identity (for divergence free vectors):

$$(1.19)\qquad \underline{curl}(\underline{X}\wedge\underline{Y}) = \underline{Y}\cdot\underline{\underline{grad}}(\underline{X}) - \underline{X}\cdot\underline{\underline{grad}}(\underline{Y})\ .$$

The similarity pointed out by Arnold (1966) is between the vorticity equation (1.17) and Euler's gyroscope equation (1.1). However, the former is fundamentally much more involved than the latter because of the field nature of the velocity and vorticity, involving therefore an infinite-dimensional Lie algebra, e.g. introducing partial, instead of ordinary differentiations. On the other hand the Lie bracket is clearly of a different nature (even for a finite number of modes): the corresponding structure constants of the corresponding Lie algebra are rather different. In particular, one may note that the bracket is not dimensionless, therefore the structure constants are unfortunately scale-dependent. In the perspective of this similarity, the vorticity and the velocity are respectively the analogues of the angular momentum ($\underline{M}$) and of the rotation ($\underline{\Omega}$). The latter being simple three dimensional vectors, whereas the former are divergence-free vector fields (the space of divergence-free vector fields defined on a sub-set D of $\Re^3$ being noted SDiff D by Arnold). The field analogue of the inertial tensor, is the $\underline{curl}$, and although the latter is also linear it is not in general symmetric; as it can be inferred from the vector identity:

$$(1.20)\qquad div(\underline{X}\wedge\underline{Y}) = \underline{curl}(\underline{X})\cdot\underline{Y} - \underline{X}\cdot\underline{curl}(\underline{Y})\ .$$

One may note that the viscous term is analogous to a linear friction which could have been introduced in (1.1), (1.4).

---

[2] α is easily generalized to barotropic flows: $\frac{p}{\rho_0}$ is then replaced by $\int \frac{dp}{\rho(p)}$

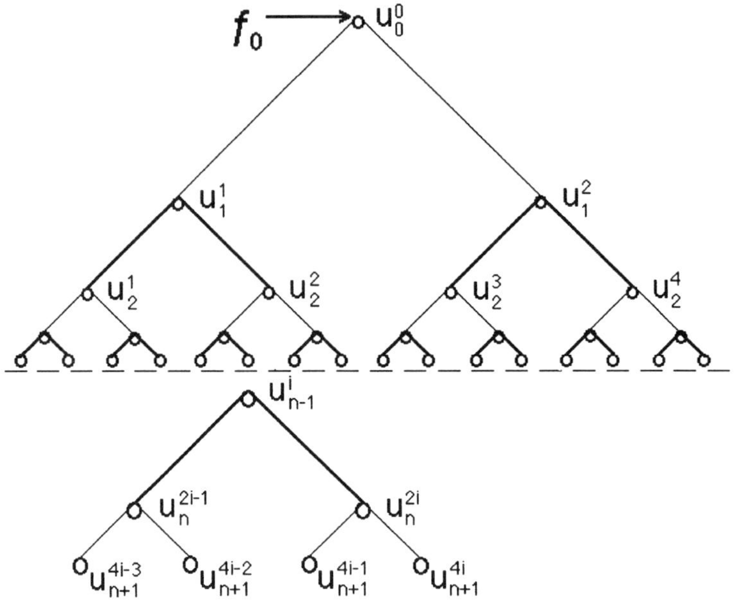

FIG. 2. *Scheme of the (spatio-temporal) multilevel cascade model. Each eddy in the cascade is a "daughter" of larger scale "mother" eddy and a "mother" of two smaller scale "daughter" eddies.*

However, we will consider a somewhat opposite and more direct analogy. Indeed, the incompressibility condition can be enforced on the advection term (it is the role of pressure to maintain it) with the help of the projector $\underline{\underline{P}}(\underline{\nabla})$ (resp. $\underline{\underline{\widehat{P}}}(\underline{k})$ in Fourier space) on divergence-free vector fields ($\underline{u} \in \mathrm{SDiff}\ D$):

$$P_{i,j}(\underline{\nabla}) = \delta_{i,j} - \nabla_i \nabla_j \Delta^{-1} \ ; \tag{1.21}$$

$$\widehat{P}_{i,j}(\underline{k}) = \delta_{i,j} - k_i k_j / k^2 \ . \tag{1.22}$$

Therefore Bernoulli's equations (1.15)-(1.16) yield an expression (either in physical space or Fourier) even more similar to (1.4) than the vorticity equation:

$$\left(\frac{\partial}{\partial t} - \nu\Delta\right) \underline{u}(\underline{x}, t) = \underline{\underline{P}}(\underline{\nabla}) \cdot \underline{u}(\underline{x}, t) \wedge \underline{w}(\underline{x}, t) \ ; \tag{1.23}$$

$$\left(\frac{\partial}{\partial t} + \nu k^2\right) \underline{\widehat{u}}(\underline{k}, t) = \underline{\underline{\widehat{P}}}(\underline{k}) \cdot \int_{\underline{p}+\underline{q}=\underline{k}} \underline{\widehat{u}}(\underline{p}, t) \wedge \underline{\widehat{w}}(\underline{q}, t) d^d \underline{p} \ . \tag{1.24}$$

Considering now that the velocity and vorticity are the (vector) fields analogous to the angular momentum ($\underline{M}$) and rotation ($\underline{\Omega}$) respectively, and that the field analogue of the inertial tensor, is now the $\underline{curl}$ inversion[3]:

$$(1.25) \qquad \underline{u}(\underline{x},t) = \underline{curl}^{-1}(\underline{w}(\underline{x},t)) = -\underline{\nabla}^{-2}[\underline{curl}(\underline{w}(\underline{x},t))] \ ;$$

$$(1.26) \qquad \underline{\widehat{u}}(\underline{k},t) = \frac{i\underline{k} \wedge \underline{\widehat{w}}(\underline{k},t)}{k^2} \ .$$

The dimensionless projector $\widehat{\underline{\underline{P}}}(\underline{k})$ corresponds to the velocity-vorticity vertex of interactions for a triad of wave vectors $(\underline{k}, \underline{p}, \underline{q})$ maintaining merely an orthogonality condition (corresponding to (1.16)):

$$(1.27) \qquad \underline{k} \cdot \underline{\widehat{u}}(\underline{k},t) = 0 \ .$$

It is obvious that for interactions for which the velocity-vorticity vertex reduces to the trivial vertex ($\underline{\underline{P}} = Identity$), the analogy between (1.23) (or (1.24)) and Euler's equation (1.1) is even closer: we may have the same structure constants for a limited number of interacting modes, in any case these constants are no longer scale-dependent. On the other hand, although the vector identity displayed in (1.20) shows that in general the *curl* and its inverse are not symmetric, this symmetry can be obtained for various conditions yielding a null l.h.s. of (1.20), as we will see bellow (section 2.2).

Summarizing the previous considerations, we may expect the existence of triads of (direct) interactions on which the Navier-Stokes equations reduce to Euler's gyroscope equation (1.1). Conversely, keeping only this interactions should yield a rather non-trivial tree-decomposition on the semi-simple $so(3)$ of the non-simple Lie algebra of a subset of fluid turbulence interactions. Let us recall that a non-simple algebra has a non-null radical, i.e. admits a sequence of ideals converging to an abelian ideal, whereas a semi-simple algebra does not admit any abelian ideal. The semi-simple algebra are all known due to the universal Cartan classification, e.g. the classical $so(n)$ and $su(n)$. On the contrary, the non-simple algebra escape to such a classification. The expected non-simplicity for turbulence is associated to scaling properties of the field (Schertzer and Lovejoy 1995). This underlines the importance of achieving a decomposition of this type.

The tree-structure is obtained by considering a cascade of triads composed by one "mother" and its two "daughter" eddies, the triad being labeled by the "mother" eddy, e.g. a subscript (n) indicating the level of cascade and a superscript (i) labeling the location of its center (we will render this labeling more explicit below). We expect to be able to rewrite

[3] one may also note that the energy flux is the analogue of the square of energy (see section 2.2), whereas in the framework of Arnold's analogy it is the enstrophy flux. The latter is only relevant (as invariant) for two-dimensional turbulence.

the corresponding interactions under a form similar to:

$$\frac{d\underline{M}_n^i}{dt} = \underline{M}_n^i \wedge \underline{\Omega}_n^i \; ; \quad \underline{\Omega}_n^i = \underline{\underline{J}}_n^i \cdot \underline{M}_n^i \tag{1.28}$$

where the $\underline{M}_n^i$ and $\underline{\Omega}_n^i$ define respectively the velocity and vorticity fields on the triad. The non-direct interactions will result from the fact that each eddy in the cascade being successively a "daughter" and a "mother", is therefore a common node for two successive triads in the tree (see Fig. 2 for illustration) and propagates indirect interactions between non-neighboring triads.

2. The Scaling Gyroscopes Cascade models.

2.1. Continuous Scaling Gyroscopes Cascade. This model is obtained considering interaction triads with trivial vorticity-velocity vertex interactions. Indeed, $(\underline{u} \wedge \underline{w})$ is divergence-free as soon as we consider some orthogonality constraints on nonlocal triad $(\underline{k}, \underline{p}, \underline{q})$. The nonlocality is not surprising since it allows one to reduce the convolution (in the Fourier space) to a simpler integral. The orthogonality constraint is rather obvious when considering eddy-viscosity interactions, i.e. the action of much smaller scale eddies on a given eddy, $\perp$ and $||$ denoting respectively the orthogonality and parallelism between vectors, we have:

$$\{|\underline{k}| \ll |\underline{p}| \approx |\underline{q}| \text{ and } \underline{p} \perp \underline{k}\} \;\Rightarrow\; (\widehat{\underline{u}}(\underline{p}) \wedge \widehat{\underline{w}}(\underline{q})) \perp \underline{k} \tag{2.1}$$

since :

$$(\underline{q} || \underline{p}) \;\Rightarrow\; (\widehat{\underline{u}}(\underline{p}) \wedge \widehat{\underline{w}}(\underline{q})) \, || \underline{p} \tag{2.2}$$

due to incompressibility of the velocity and vorticity fields:

$$\widehat{\underline{u}}(\underline{p}) \perp \underline{p} \text{ and } \widehat{\underline{w}}(\underline{q}) \perp \underline{q} \, . \tag{2.3}$$

On the other hand, considering interactions $|\underline{p}| \ll |\underline{k}| \approx |\underline{q}|$, the same property holds but without the same generality; since in order to obtain it we must further require $\widehat{\underline{u}}(\underline{p}) || \underline{k}$, however we will see below it is not as demanding as seen at first glance:

$$\begin{aligned} &\{|\underline{p}| \ll |\underline{k}| \approx |\underline{q}| \text{ and } \widehat{\underline{u}}(\underline{p}) || \underline{k}\} \;\Rightarrow\; (\widehat{\underline{u}}(\underline{p}) \wedge \widehat{\underline{w}}(\underline{q})) \perp \underline{k} \\ &\text{and } (\widehat{\underline{u}}(\underline{q}) \wedge \widehat{\underline{w}}(\underline{p})) \, || \underline{k} \end{aligned} \tag{2.4}$$

therefore the vorticity contributes only through high wave number ($|\underline{q}| \gg |\underline{p}|$).

We should also consider the dual interactions $|\underline{k}| \gg |\underline{p}| \approx |\underline{q}|$ since the Navier-Stokes equations are symmetric for any permutation of a triad $(\underline{k}, \underline{p}, \underline{q})$, e.g. there is a detailed conservation of energy for any triad (Leslie, 1973). However these interactions can be disregarded at first order since on

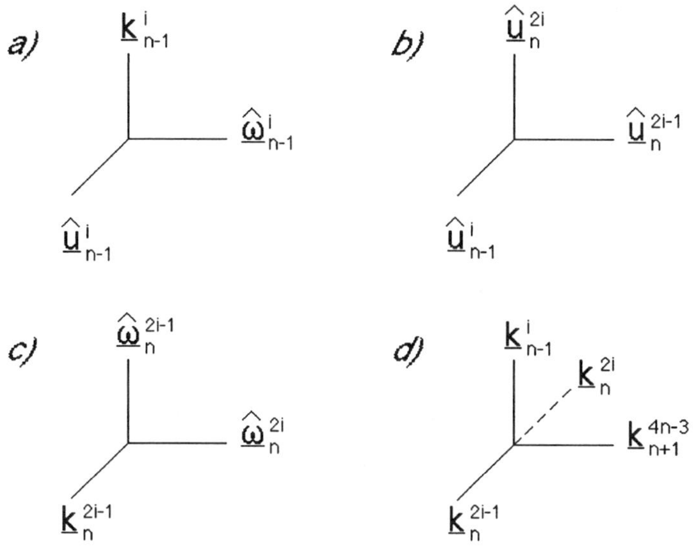

FIG. 3. *Illustration (in $\Re^3$) of orthogonal bases (in C^3) related to a triad of (direct) interactions between a "mother" and the two "daughter" eddies:*
a) wave-vector ($\underline{k}^i_{n-1}$), velocity ($\widehat{\underline{u}}^i_{n-1}$) and vorticity ($\widehat{\underline{w}}^i_{n-1}$) components for given eddy;
b) velocity components for given "mother" ($\widehat{\underline{u}}^i_{n-1}$) and its two "daughter" eddies ($\widehat{\underline{u}}^{2i-1}_n$, $\widehat{\underline{u}}^{2i}_n$);
c) "mother" wave-vector ($\underline{k}^{2i-1}_n$) and "daughter" eddies vorticity components ($\widehat{\underline{w}}^{2i-1}_n$, $\widehat{\underline{w}}^{2i}_n$);
d) wave-vectors of "mother" ($\underline{k}^i_{n-1}$), its two "daughter" ($\underline{k}^{2i-1}_n$, $\underline{k}^{2i}_n$) and one of its four "grand daughter" eddies ($\underline{k}^{2(2i-1)-1}_{n+1}$).

the one hand they correspond (at this order) to a mere advection of eddies of size $|\underline{k}|^{-1}$ by eddies of much larger sizes $|\underline{p}|^{-1}$ and $|\underline{q}|^{-1}$, and on the other hand the Navier-Stokes equations are of course Galilean invariant.

Therefore, for nonlocal orthogonal interactions the projector is no longer relevant at the first order and we have an equation similar to (1.28) (λ being the rather arbitrary nonlocalness parameter, and we temporarily do not explicit the orthogonality conditions discussed before and below):

$$(2.5)\qquad \begin{aligned} \left(\tfrac{\partial}{\partial t} + \nu k^2\right) \widehat{\underline{u}}(\underline{k}) \quad &= \quad \int_{|\underline{p}|\geq\lambda|\underline{k}|} \left(\widehat{\underline{u}}(\underline{p}) \wedge \overline{\widehat{\underline{w}}}(\underline{p})\right) d^d\underline{p} \\ &+ \quad \left(\int_{|\underline{p}|\leq\lambda^{-1}|\underline{k}|} \widehat{\underline{u}}(\underline{p}) d^d\underline{p}\right) \wedge \widehat{\underline{w}}(\underline{k}) \end{aligned}$$

where we used the complex conjugation symmetry which holds for the velocity field in Fourier space:

$$\forall \underline{k}: \quad \widehat{\underline{u}}(-\underline{k}) = \overline{\widehat{\underline{u}}}(\underline{k}) \tag{2.6}$$

due to reality of the corresponding field in real space:

$$\forall \underline{x}: \quad \overline{\underline{u}}(\underline{x}) = \underline{u}(\underline{x}) \ . \tag{2.7}$$

2.2. Discrete Scaling Gyroscopes Cascade. Now we can discretize the nonlocal orthogonal approximation derived above along a tree-structure (section 1.2). This is done by using already mentioned double index $\binom{i}{n}$ for eddies of the cascade: (n) indicating the level of the cascade (which is also the number of the corresponding steps in the cascade process) and the scale of the corresponding eddies ($l_n = L/\lambda^n$); (i) labeling its location (e.g. its center $\underline{x}_n^i$, the distance[4] betwen two neighbours centers being l_n). This indexes velocity field ($\widehat{\underline{u}}_n^i$) and vorticity field ($\widehat{\underline{w}}_n^i$) components , as well as their corresponding wave-vector ($\underline{k}_n^i$) and their directions ($\underline{e}_n^i$):

$$\widehat{\underline{u}}_n^i \equiv \widehat{\underline{u}}(\underline{k}_n^i) \ ; \quad \widehat{\underline{w}}_n^i = i\underline{k}_n^i \wedge \widehat{\underline{u}}_n^i \ ; \tag{2.8}$$

$$\underline{e}_n^i = \underline{k}_n^i / k_n \ ; \quad k_n = |\underline{k}_n^i| \ . \tag{2.9}$$

We will consider the tree-structure of interactions based on the fundamental triads of (direct) interactions $(\underline{k}_{n-1}^i, \underline{k}_n^{2i-1}, \underline{k}_n^{2i})$, between a "mother" and the two "daughter" eddies, therefore: $i = \overline{1, 2^n}$. Although these interactions are rather local in Fourier space (usually $\lambda = 2$), we will use the nonlocal approximations of the vertex and in light of the orthogonality conditions discussed before. (2.5) reduces then to:

$$\begin{aligned} \left(\tfrac{\partial}{\partial t} + \nu k_{n-1}^2\right) \widehat{\underline{u}}_{n-1}^i \quad = \quad & \widehat{\underline{u}}_n^{2i-1} \wedge \overline{\widehat{\underline{w}}}_n^{2i-1} \\ & + \ \widehat{\underline{u}}_n^{2i} \wedge \overline{\widehat{\underline{w}}}_n^{2i} \ + \ \widehat{\underline{u}}_{n-2}^{a(i)} \wedge \widehat{\underline{w}}_{n-1}^i \ ; \end{aligned} \tag{2.10}$$

$$a(j) = E\left(\frac{j+1}{2}\right) \tag{2.11}$$

$a(j)$ being the location index of its "ancestor", $E(x)$ being the integer part of any real number x:

$$E(x) \le x < E(x) + 1 \ . \tag{2.12}$$

---

[4] One may note that a more compact but less explicit notation could be achieved by considering the dyadic (in the present case, the binary) location of the eddies, its level corresponding to its (ultra metric) norm.

Each eddy of wave vector being at the same time "daughter" eddy and a "mother" eddy is a common component of two successive (direct) interacting triads:

$$\{\underline{k}_m^j\} = \{\underline{k}_m^j, \underline{k}_{m+1}^{2j-1}, \underline{k}_{m+1}^{2j}\} \cap \{\underline{k}_{m-1}^{a(j)-1}, \underline{k}_m^{2a(j)-1}, \underline{k}_m^{2a(j)}\} \; ; \tag{2.13}$$

As orthogonality plays a central role, one may first note that $(\underline{k}_{n-1}^i, \widehat{\underline{u}}_{n-1}^i, \widehat{\underline{w}}_{n-1}^i)$ form themselve as an orthogonal frame[5] (Fig. 3 *a)*) for any $({}^i_{n-1})$. On the other hand, (2.10) has a rather immediate connection to (1.28) if $(\widehat{\underline{u}}_{n-1}^i \, , \, \widehat{\underline{u}}_n^{2i-1} \, , \, \widehat{\underline{u}}_n^{2i})$ (Fig. 3 *b)*) and $(0 \, , \, \widehat{\underline{w}}_n^{2i} \, , \, \widehat{\underline{w}}_{n-1}^{2i-1})$ (Fig. 3 *c)*) are an orthogonal decomposition of $(\underline{M}_{n-1}^i)$ and $(\underline{\Omega}_{n-1}^i)$[6] respectively. This implies (Fig. 3 *d)*) that:

$$\underline{k}_n^{2i} = -\underline{k}_n^{2i-1} \, . \tag{2.14}$$

One may note that due to these different ortogonal proporties of a triad, the analogue of inertia tensor (i.e. $\underline{curl}^{-1}$) will be symmetric since l.h.s. of (1.20) will be null. (2.14) corresponds to the fact that not only $(\underline{e}_n^i \perp \underline{e}_{n-1}^i)$ but also:

$$\widehat{\underline{u}}(\underline{k}_{n-1}^i) = \widehat{u}_{n-1}^i \underline{e}_n^{2i-1} \; ; \quad \widehat{u}_{n-1}^i \in C \tag{2.15}$$

and:

$$\underline{e}_n^{2i} = -\underline{e}_n^{2i-1} \; ; \tag{2.16}$$

$$\widehat{\underline{u}}(\underline{k}_n^{2i-1}) = \widehat{u}_n^{2i-1} \underline{e}_{n+1}^{2(2i-1)-1} \; ; \quad \widehat{u}_n^{2i-1} \in C \; ; \tag{2.17}$$

$$\widehat{\underline{u}}(\underline{k}_n^{2i}) = \widehat{u}_n^{2i} \underline{e}_{n+1}^{2(2i)-1} \; ; \quad \widehat{u}_n^{2i} \in C \tag{2.18}$$

with:

$$\underline{e}_{n+1}^{2(2i-1)-1} = \underline{e}_{n-1}^i \wedge \underline{e}_n^{2i-1} \; ; \tag{2.19}$$

$$\underline{e}_{n+1}^{2(2i)-1} = \underline{e}_n^i \, . \tag{2.20}$$

The analogy for the triad of (direct) interactions $(\underline{k}_{n-1}^i, \underline{k}_n^{2i-1}, \underline{k}_n^{2i})$, with the gyroscope equation is now rather immediate in the frame work $(\underline{e}_{n-1}^i, \underline{e}_n^{2i-1}, \underline{e}_{n+1}^{4i-3})$ since the analogue of the momentum has the following

---

[5] The illustrations of Fig. 3 are drawn in $\Re^3$ instead of C^3.

[6] However one may consider a more degenerate case $(\widehat{\underline{w}}_{n-1}^{2i-1} - \widehat{\underline{w}}_n^{2i} \, , \, 0 \, , \, 0)$.

matrix representation (the introduced factor i being convenient to factor out the one of (1.26)):

$$\begin{bmatrix} \widehat{u}_n^{2i} \\ \widehat{u}_{n-1}^{i} \\ \widehat{u}_n^{2i-1} \end{bmatrix} = i\left[M_{n-1}^i\right] = i \begin{bmatrix} u_n^{2i} \\ u_{n-1}^{i} \\ u_n^{2i-1} \end{bmatrix} \tag{2.21}$$

and the analogue of angular velocity, keeping only the interacting components, corresponds to:

$$\begin{bmatrix} \widehat{\omega}_n^{2i-1} \\ 0 \\ \widehat{\omega}_n^{2i} \end{bmatrix} = \left[\Omega_{n-1}^i\right] = ik_n \begin{bmatrix} \widehat{u}_n^{2i-1} \\ 0 \\ \widehat{u}_n^{2i} \end{bmatrix} \tag{2.22}$$

and therefore the analogue ($\underline{\underline{J}}_{n-1}^i$) of the projection of inverse of the inertia tensor on the triad:

$$\underline{\Omega}_{n-1}^i = \underline{\underline{J}}_{n-1}^i \cdot \underline{M}_{n-1}^i \tag{2.23}$$

corresponds to:

$$\underline{\underline{J}}_{n-1}^i = k_n \underline{\underline{K}}\ ; \quad [K] = \begin{bmatrix} 0 & 0 & 1 \\ 0 & 0 & 0 \\ 1 & 0 & 0 \end{bmatrix} . \tag{2.24}$$

The presence of a zero eigenvalue for the analogue of the inverse of inertia tensor, and consequently for a component of the analogue of angular velocity, corresponds to properties of projections on the triad and not for the full system. This will have important consequences discussed below, and its is yielded by the fact that vorticity does not intervene in semi-nonlocal interaction with low wave number (2.4), but only with high wave-number as in the eddy viscosity interaction (2.1).

This achieves the tree-decomposition on $so(3)$. The equation of evolution of $\widehat{\underline{u}}_m^j$ corresponds therefore to the superposition and the coupling of two equations of gyroscope type, therefore to the following (in general complex) scalar equation of evolution for the velocity of the eddy of wave vector $\underline{k}_m^j$ obtained by considering explicitly the linear coupling between angular momentum and velocity (2.24)-(2.13), $a(j)$ defined in (2.11):

$$\begin{aligned} \left(\tfrac{d}{dt} + \nu k_m^2\right) u_m^j \quad &= \quad - \quad k_{m+1}\left[|u_{m+1}^{2j-1}|^2 - |u_{m+1}^{2j}|^2\right] \\ &\quad + \quad (-1)^{j+1} k_m u_m^j u_{m-1}^{a(j)} . \end{aligned} \tag{2.25}$$

Concerning the two quadratic invariants, the square of the angular momentum (M^2) corresponds to the turbulent energy:

$$\sum_{\substack{i \\ n-1}} {M_{n-1}^i}^2 = 2 \sum_{\substack{j \\ m}} |u_m^j|^2 \tag{2.26}$$

whereas the kinetic energy of the gyroscope is related to the helicity of the triad, more precisely to its k_{n-1}^i component:

$$T_{n-1}^i = \underline{M}_{n-1}^i \cdot \underline{\Omega}_{n-1}^i = u_n^{2i} \cdot \omega_n^{2i-1} + u_n^{2i-1} \cdot \omega_n^{2i} \,. \tag{2.27}$$

However, due to the zero eigenvalue of the analogue of the inertia tensor, this invariant is only local (to the triad). Indeed, the eddy-viscosity interactions involving the eddy of wave-vector k_{n-1}^i (as "daughter" eddy) break this conservation.

Finally the system can be rendered real by considering the parity symmetry, which is associated to statistical zero helicity for isotropic turbulence:

$$\underline{u}(-\underline{x}) = \underline{u}(\underline{x}) \;\Leftrightarrow\; \underline{\widehat{u}}(-\underline{k}) = -\overline{\underline{\widehat{u}}}(\underline{k}) \tag{2.28}$$

therefore due to complex conjugation symmetry (2.6):

$$u_n^i,\ \omega_n^i \in \Re\ ; \quad \underline{M}_n^i,\ \underline{\Omega}_n^i \in \Re^3 \tag{2.29}$$

which yields real vector equations of evolution (2.10), and correspondingly the real scalar equation of evolution for the velocity field (2.25).

One may note that under this form the geometric interpretation discussed in section 1, that the motion describes a geodesic in the group of rotations, is rather immediate since (2.25) has the structure of a geodesic equation:

$$\dot{u}_m^j = \sum_{\binom{i}{n},\binom{i'}{n'}} \Gamma^{\binom{j}{m}}_{\binom{i}{n}\binom{i'}{n'}} u_n^i u_{n'}^{i'} \tag{2.30}$$

where the Γ's are the Christophel symbols of the metric.

2.3. Severe limitations of over-simplification of the SGC: shell models. The spatial index can be eliminated and the model reduced to a temporal model, as soon as one observes that at each time there is a most active path on the tree connecting the largest structures to the smallest ones (with a unique eddy for each level) along which most of the energy transfer occurs.

This observation of the presence of intermittency in a similar model lead unfortunately to extreme over-simplifications, reducing the cascade of gyroscopes to the "one path model" (Obukhov and Dolzhansky, 1975).

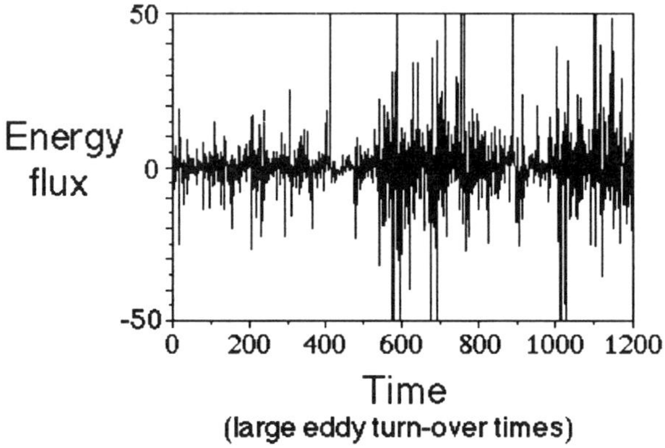

FIG. 4. *One of the time series of the energy flux on medium level (i.e. level 6), displaying rather strong intermittency.*

The corresponding approximation restricts the study to the temporal intermittency along a fixed path; whereas in the full model the path itself have a space-time variability. This tendency may explain why Obukhov, having pointed out that Euler's gyroscope equations have similarities to the Navier-Stokes equations, gave only a derivation for the one path model from the latter, and not for the full SGC model, contrary to what we have done in the previous sections. In fact this one path model may be obtained by some other direct phenomenological considerations (Gledzer, 1980) and is a predecessor of the "shell-models," which are obtained by averaging the flux energy over the different wave-vectors, i.e. averaging over spherical shells in Fourier space. For instance one may consider the following most energetic paths:

$$\left(\frac{d}{dt} + \nu k^2\right) u_n = k_n u_{n-1} u_n - k_{n+1} u_{n+1}^2 \; . \tag{2.31}$$

These models are only able to study the flow of energy through different scales (wave numbers) and ignore the increasing spatial degrees of freedom with resolution (i.e. with Reynolds number). Despite this fundamental deficiency, shell-models became extremely popular (see, e.g. Gledzer et al., 1981), whereas the original full model was forgotten. We will show that not only is this temporal approximation highly questionable and yields biased results, but totally unnecessary since the full SGC (i.e. space-time model) is as easy to handle as its gross approximation.

3. Multifractal analysis of the SGC.

3.1. Multifractal processes. Let us recall that the multifractal processes originated from the phenomenological assumption that in turbulence the successive cascade steps define the fraction of the flux transmitted to smaller scales and that a cascade from scale ratio λ to scale ratio $\lambda\lambda'$ is a rescaled version (by scale ratio λ) of a cascade from ratio 1 to λ'. Generally

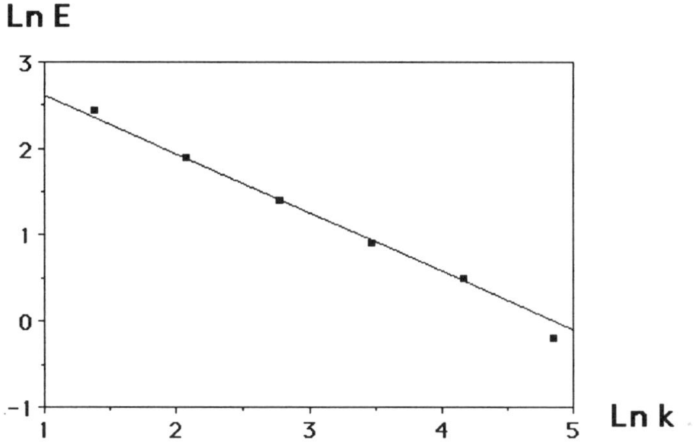

FIG. 5. *The Ln – Ln plot of the energy spectrum* $E_n \approx E(k_n) \cdot k_n$ *integrated over a wave number shell for levels* $n = \overline{2,7}$ *of the cascade. The observed slope (-2/3) is the analogue of the Kolmogorov-Obukhov value (-5/3).*

speaking, a stochastic multifractal process $\varepsilon(\underline{x}, t)$ is defined by an infinite hierarchy of singularities γ, i.e. at any scale resolution $\lambda = L/\ell$ (L being the outer scale, ℓ the scale of observation)—this process is scaling:

$$\varepsilon_\lambda \approx \lambda^\gamma \varepsilon_1 \tag{3.1}$$

$\gamma > 0$ being indeed the algebraic order of divergence of $\varepsilon_\lambda(\underline{x}, t)$; $\lambda \to \infty$. More precisely the frequency of occurrences of a given order of singularity is governed by the codimension/Cramer function $c(\gamma)$ (Schertzer and Lovejoy, 1987, 1989; Oono, 1989; Mandelbrot, 1991):

$$\Pr(\varepsilon_\lambda \geq \lambda^\gamma) \approx \lambda^{-c(\gamma)} \tag{3.2}$$

where "Pr" indicates "probability". It is equivalent (by the Mellon transform) to consider the scaling of the different orders (q) of moments with the associated scaling moment functions $K(q)$:

$$< (\varepsilon_\lambda)^q > \approx \lambda^{K(q)} \tag{3.3}$$

where the angle brackets indicate ensemble averages. In fact, $c(\gamma)$ and $K(q)$ are simply related by the Legendre transform (Parisi and Frisch, 1985).

3.2. Numerical simulations of the SGC. Contrary to the multiplicative processes, the SGC is fundamentally deterministic. Indeed, only the forcing, which must be introduced in order to obtain a quasi-equilibrium, could be stochastic. However, if the SGC has some universal features, it should be rather independent of the type of forcing used.

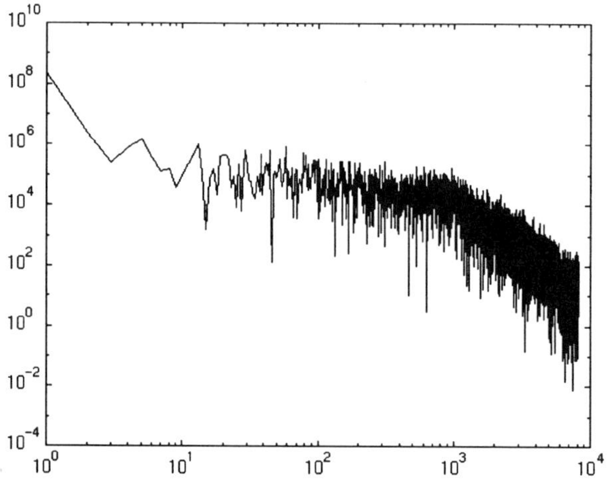

FIG. 6. *The Log – Log plot of the time series of the energy flux spectrum (data of Fig. 4) which is scaling over a large subrange. As expected, its slope is less steep (-0.8) than the slope of the vanishing intermittency (-1).*

We indeed checked that there does not seem to be fundamentally any difference between a caricatural deterministic forcing, such as a constant unit force only at the first level of the cascade, and the usual Gaussian forcing used rather universally for direct simulations of the Navier-Stokes equations. Indeed, the mean differences between both type of forcing intervene only at first time steps and large scales. As we are not relying on time and memory consuming simulations, contrary to direct numerical simulations, we take the deterministic forcing.

Indeed, due to the simplicity of the SGC, we performed long runs on work stations (e.g. 1024 large eddy turn-over times) for rather developed turbulence although we are using a somewhat elaborate time scheme, i.e. the fourth-order Runge-Kutta time scheme. The reason being that algebraic discretization of the scales yields important Reynolds numbers for a fairly low number of levels, in our simulations $Re \approx 10^5$ for 12 levels in the cascade.

We checked that the discrete energy spectrum integrated over a wave

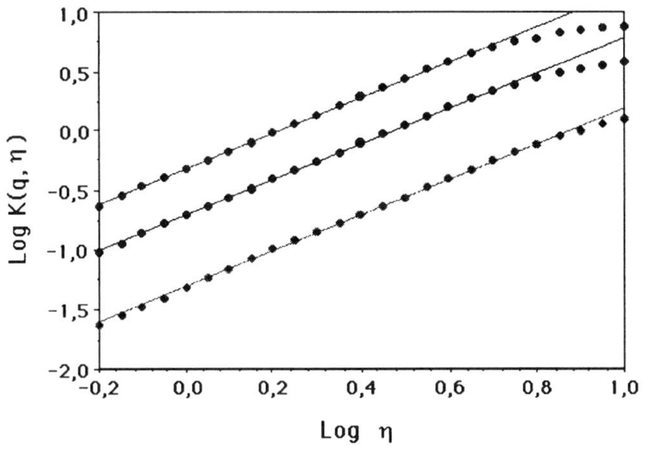

FIG. 7. *The DTM for $q = 0.8$, 1.5 and 2.0 (bottom to top) on the level 6 of the model. This yields the following estimates: $\alpha \approx 1.5 \pm 0.05$ (the slope of the Log $K(q, \eta)$ vs. Log η) and $C_1 \approx 0.25 \pm 0.05$ (from the intercept with the vertical axis).*

number shell $(k_n - k_{n-1})$ defined as:

$$E_n = \frac{1}{2^n} \sum_{i=1}^{2^n} |u_n^i|^2 \approx E(k_n) \cdot k_n \tag{3.4}$$

displayed a slope (-2/3) which is therefore the analogue of the Kolmogorov-Obukhov value (-5/3) (Fig. 5). This slope corresponds to the trivial scaling of the SGC when assuming a constant flux of energy, as it can be inferred from (2.25). However, as expected, time series of the corresponding flux of energy are highly intermittent, as displayed by Fig. 4. We will try to characterize the intermittency yielded by the SGC with the help of notions which have been developed in the framework of stochastic multifractal processes. One may already note that the spectrum of time series of energy flux (on the medium level, i.e. level 6) displays (Fig. 6) a rather large subrange with a slope of order -0.8, i.e. less steep than -1, latter corresponds to a vanishing intermittency exponent (e.g. Monin and Yaglom, 1972).

3.3. Multifractal universality. As there are only constraints of convexity on $c(\gamma)$ and $K(q)$ (the former being also increasing), an infinite number of parameters is generally necessary to specify them. The corresponding processes would be unmanageable either theoretically or empirically, unless only a few of the infinite number of parameters are physically relevant and determine the "universality" classes of these processes. The pioneering claim on lognormal universality in turbulent cascades (Yaglom, 1966), was shown to be questionable due to the singular small scale limit (Mandelbrot, 1974). More recently, there had been claims denying any universality for multifractals (Mandelbrot, 1989; Frisch, 1991; Gupta and Waymire, 1993). However, even if the singularity of the small scale limit does indeed prevent iterations of the process towards smaller scales from approaching a universal limit, this in no way contradicts the general idea of universality by considering other types of iterations (Schertzer and Lovejoy, 1987, 1994).

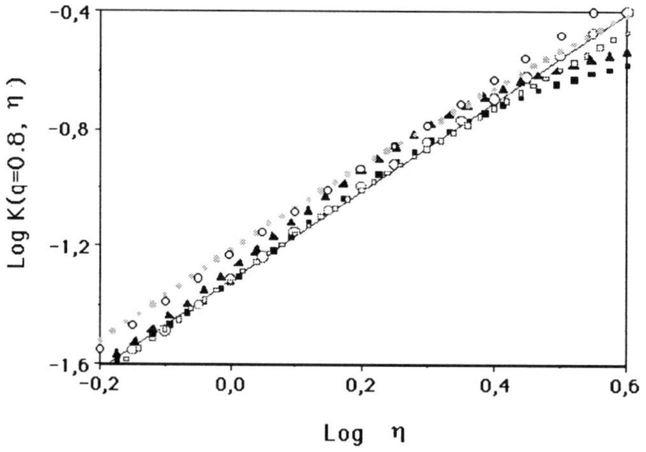

FIG. 8. *Curves of DTM (Log $K(q,\eta)$ vs. Log η) with the order of moment $q = 0.8$ for different levels of the SGC model.*

It was shown that two mechanisms (with possible combination) yield universality: (i) "nonlinear mixing" of these processes: multiplication of independent, identically distributed processes on the same scales; (ii) "scale densification" of the process: introducing more and more intermediate scales. In both cases, multiplying processes (ε) corresponds to adding generators (Γ) defined as:

$$\varepsilon_\Lambda \approx e^{\Gamma_\Lambda} \, . \tag{3.5}$$

Furthermore, two types of universality can be distinguished: "strong universality" when the generator is stable under renormalization, "weak universality" when the generator and its iterates are only loosely related; they no longer involve stability under rescaling and/or recentering. It seems reasonable that one must seek weak universality only when there is a failure of strong universality. The strong universal scaling functions $K(q)$ and $c(\gamma)$, corresponding to "Log-Lévy" statistics (where $\frac{1}{\alpha}+\frac{1}{\alpha'}=1$):

$$(3.6) \qquad c(\gamma)=C_1\left(\frac{\gamma}{C_1\alpha'}+\frac{1}{\alpha}\right)^{\alpha'}; \quad K(q)=\frac{C_1}{\alpha-1}\left(q^\alpha-q\right)$$

However, She and Leveque (1994) consider a weak universal multifractal process yielding "Log-Poisson" statistics (Dubrulle, 1994; Novikov, 1994; She and Waymire, 1995):

$$(3.7) \qquad \begin{aligned} &c(\gamma)=\left(1-\tfrac{\gamma^+-\gamma}{c\gamma^-}\left(1-\log\tfrac{\gamma^+-\gamma}{c\gamma^-}\right)\right)c\ ; \qquad \gamma\le\gamma^+\ ;\\ &c(\gamma)=\infty\ ; \qquad \gamma^+<\gamma\ ;\end{aligned}$$

$$(3.8) \qquad K(q)=q\gamma^+ +(\lambda_1^{-q\gamma^-}-1)c\equiv q\gamma^+-c+\left(1-\frac{\gamma^+}{c}\right)^q c$$

which turns out to be (Schertzer et al., 1995) the classical (and rather trivial) Poisson limit (using a smaller and smaller elementary step $\lambda_{1/N}=\lambda_1^{1/N}\to 1;\ N\to\infty$) of the α-model (Schertzer and Lovejoy, 1983, 1984; Bialas and Peschanski, 1986; Levich, and Tzvetkov, 1985). The latter model is the canonical (binomial) model generated by the (Bernoulli) two-state generator γ on elementary discrete step scale ratio λ_1:

$$(3.9) \qquad \begin{aligned} dP(\gamma) &= \lambda_1^{-c}\delta_{\gamma-\gamma^+}+(1-\lambda_1^{-c})\delta_{\gamma+\gamma^-}\ ; \quad \gamma^+,\ \gamma^-\ge 0\ ;\\ \lambda_1^{K(q)} &= \lambda_1^{q\gamma^+-c}+\lambda_1^{-q\gamma^-}(1-\lambda_1^{-c})\end{aligned}$$

Canonical means conservation on ensemble averages $\langle\varepsilon_\lambda\rangle=\langle\varepsilon_1\rangle$ for any λ; γ^+ is the upper bound of singularities; $c(\equiv c(\gamma^+))$ is its codimension and can be chosen rather arbitrarily; γ^- is the lower bound of singularities and is constrained so that the ensemble average $<(\lambda_1)^\gamma>=1$. The (monofractal) β-model is recovered for $\gamma^-=\infty$, $\gamma^+=c=C_1$. Assuming (non-fractal, D=1) filament-like structures for the highest order singularity and homogeneous eddy turn over times She and Leveque (1994) selected:

$$(3.10) \qquad C=2,\ \gamma^+=\frac{2}{3}\Rightarrow\lambda_1^{\gamma^-}=\frac{3}{2}\,.$$

The central limit theorem was used (Schertzer et al., 1991) to show that the (re-normalized) nonlinear mixing of (discrete) α-models leads to a (continuous) "lognormal" multifractal process (i.e. α=2).

We use the Double Trace Moment (DTM) analysis (Lavallee, 1991; Lavallee et al., 1993) in order to estimate the mean codimension C_1 and the Lévy index of multifractality α. The main idea of DTM is first to consider the normalized η powers of the field ε, then $\varepsilon_\lambda^{(\eta)}$ is defined as:

$$(3.11) \qquad \varepsilon_\lambda^{(\eta)}=\frac{(\varepsilon_\lambda)^\eta}{\langle(\varepsilon_\lambda)^\eta\rangle}$$

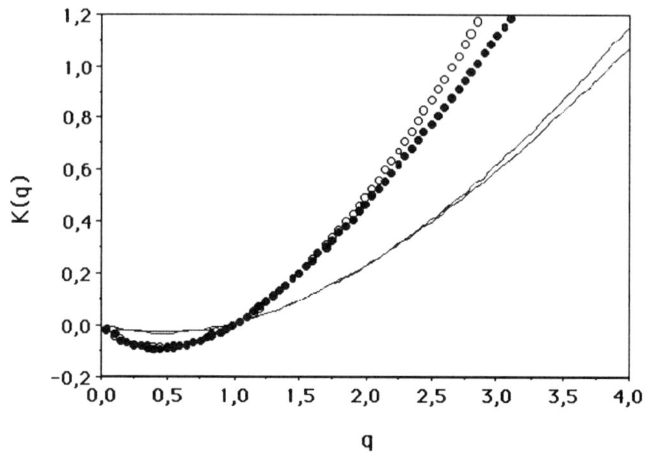

FIG. 9. *Estimate of the scaling function $K(q)$ obtained on the SGC runs, as well as the theoretical curves corresponding respectively to the strong universal multifractal corresponding to the SGC ($C_1 = 0.25$, $\alpha = 1.5$), to the weak universal multifractal (solid line) based on the She and Leveque choice of parameters, and its approximation by a corresponding strong universal multifractal (i.e. $C_1 = 0.11$ and $\alpha = 1.62$).*

with a moment scaling function $K(q,\eta)$, as a natural extension of (3.3):

$$\left\langle \left(\varepsilon_\lambda^{(\eta)}\right)\right\rangle \approx \lambda^{K(q,\eta)}\ ; \quad K(q,\eta) = K(q\eta) - qK(\eta)\ . \tag{3.12}$$

For universal multifractals, $K(q,\eta)$ has a particularly simple dependence on η:

$$K(q,\eta) = \eta^\alpha K(q)\ . \tag{3.13}$$

Fig. 7 and Fig. 8 display results of the DTM analysis of the time series of the energy flux respectively on level 6 of the model for different orders of moment q, and for different levels of the model with order of moment $q = 0.8$. This yields the following estimates: $C_1 \approx 0.25 \pm 0.05$, $\alpha \approx 1.5 \pm 0.05$, whereas the DTM technique applied to the Log-Poisson with the She and Leveque choice of parameters (3.10) yields $C_1 = 0.11$ and $\alpha = 1.62$. It is important to note that in the latter case the α is no longer an intrinsic parameter of the model, and its estimate should depend on the range of η used in the determination.

In order to better distinguish between the two types of universality, Fig. 9 displays the estimate of the scaling function $K(q)$ obtained on the SGC runs as well as the theoretical curves corresponding respectively to the strong universal multifractal corresponding to the SGC ($C_1 = 0.25$, $\alpha =$

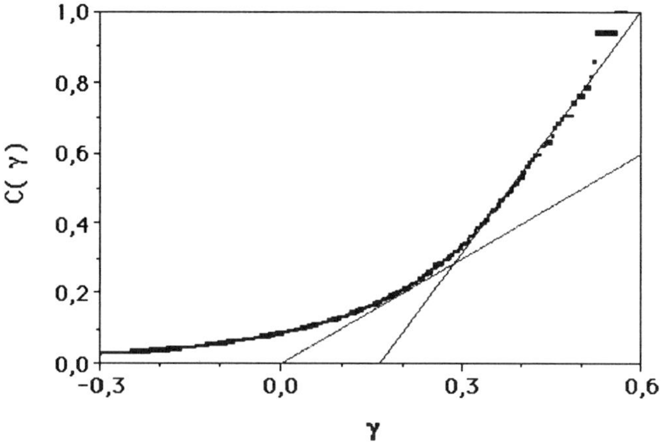

FIG. 10. *Estimate of the codimension of the flux (at level 6). It displays a clear linear asymptote with critical order* (q_D) *for the first-order phase transition to self organized criticality:* $q_D = 2.3 \pm 0.06$.

1.5), to the weak universal multifractal (Log-Poisson statistics) based on the She and Leveque choice of parameters (3.10), and its more or less corresponding strong universal multifractal (i.e. $C_1 = 0.11$ and $\alpha = 1.62$). The two latter curves diverge strongly from the former, mainly due to the difference of C_1 (roughly speaking a factor 2 between the two groups of curves), since it corresponds to the tangency of $K(1)$ at $q = 1$, i.e. the monofractal approximation of the "mean field" multifractal field. Therefore, the difference is rather of first order and is in favor of strong universality. However, not too surprisingly, the disagreement between weak universality and strong universality tends to increase with the order of moment, i.e. when we are exploring higher and higher, but also rarer and rarer, singularities which we will discuss in the next section.

3.4. Multifractal phase transitions and Self Organized Criticality. In a series of papers (Schertzer et al., 1993; Schertzer and Lovejoy, 1994, 1996) it was argued that not only do unbounded singularities pose interesting problems of observation and estimation, but are a requisite to the introduction, via first order multifractal phase transitions, of a non-classical Self-Organized Criticality (SOC), which is often desirable in order to explain the phenomenology of extreme events. Normal and Lévy ($\alpha > 1$) generators yield unbounded singularities, as is generally the case for canonical processes. On the contrary, micro-canonical conservation (i.e. per realization) of the flux of energy — e.g. the microcanonical version of the α-model called the "p-model" (Meneveau and Sreenivasan, 1987) — imposes $\gamma \leq D$

(the dimension of space). For $D = 1$, it is the celebrated inequality (expressed via $K(q)$) of Novikov (1971), who in fact imposed microcanonical conservation by considering the dissipation instead of the flux of energy. In the inertial range — especially in the limit of infinite Reynolds numbers — the relevance of the dissipation has been often questioned. Frisch (1991) argued for a physical bound to singularities due to the finite speed of sound, whereas Schertzer and Lovejoy (1996) considered both: the incompressible Navier-Stokes equations (without any characteristic velocity, infinite speeds of sound) and the physical issues of compressible turbulence involving compressibility effects. The corresponding hypersonic gradients are of course beyond the scope of the incompressible Navier-Stokes equations. The critical singularity of transition to SOC being denoted γ_D, the observed singularities SOC singularities $\gamma \geq \gamma_D$ are empirically bounded by γ_s, the maximum reachable singularity in the samples studied and have a codimension different from the theoretical one given by (3.6), since it should be linear in agreement with the algebraic fall-off of the probability distribution:

$$c(\gamma) = q_D\gamma - K(q_D) \; ; \quad \gamma \geq \gamma_D = K'(q_D) \tag{3.14}$$

The observed codimension for SOC singularities ($\gamma \geq \gamma_D$) follows the tangent instead of the theoretical parabola-like codimension. Consequently there is a divergence of higher order moments $q \geq_D$ for infinite samples. However, because of the finite size of empirical data sets, Legendre transform yields the corresponding spurious linear estimate of the scaling moment function:

$$K(q) = \gamma_s q - c(\gamma_s) \; ; \quad q \geq q_D \tag{3.15}$$

its slope γ_s depends on the number of the sample $N_s \approx \gamma^{D_s}$ (at the resolution λ), where the exponent D_s is the "sampling dimension" is given by:

$$c(\gamma_s) = D + D_s \; . \tag{3.16}$$

When the number of samples increases, $\gamma_s \rightarrow \infty$ which corresponds to divergence of higher moments. The α-model was developed to illustrate the generality of divergence of moments for multifractal fields, which is obtained as soon as the critical $q_D \approx (c - D)/(\gamma^+ - D)$ is greater than 1, which correspond to $c > \gamma^+ > D$, however the parameters (3.10) chosen by She and Leveque (1994) do not satisfy these conditions and therefore do not yield SOC.

This qualitative difference between weak and strong universality was already noted on Fig. 9, but is rendered more obvious by estimating the codimension of the flux (at level 6), since it displays (Fig. 10) a clear linear asymptote in correspondence to (3.14) and yielding a critical order

(q_D) for the first order phase transition to self-organized criticality: $q_D = 2.3 \pm 0.06$. This last value is very close agreement with empirical estimates for atmospheric turbulence in tropical conditions ($q_D \approx 2.3$, Chigirinskaya et al., 1994) as well as in mid latitude boundary layer ($q_D \approx 2.5$, Schmitt et al., 1992).

4. Discussions and perspectives. We started from rather speculative considerations on the abstract structure of the Navier- Stokes equations: its Lie structure. Nevertheless, these considerations lead us to derive a concrete dynamical model of cascade: the Scaling Gyroscopes Cascade. This model in turn can be used to investigate fundamental questions of turbulence. We specifically investigated the questions of multifractal universality and transitions to SOC in turbulence.

Numerical simulations of the SGC clearly support strong universality (the so-called Log-Lévy processes) rather than weak universality (Log-Poisson statistics). At the same time, although not independently, they display rather clear evidence of multifractal transition to SOC for extreme events. Due to the strong similarities of structure between the Navier-Stokes equations and the SGC, it is doubtful that the SGC could yield a result qualitatively different from the original equations. Furthermore, the corresponding numerical estimates of the universal exponents (C_1, α) and of the critical transition moment order q_D are extremely close to different estimates for atmospheric turbulence. We may conclude that presumably strong universality and Self-Organized Criticality are relevant for turbulence. We may also speculate that due to the relative simplicity of the SGC the universal exponents (C_1, α) and q_D may be analytically computable, at least in the framework of this model.

On the other hand, it would be obviously interesting to evaluate the sensitivity of these foreseen results to the variation of the structure constants of a semi-simple Lie algebra ($so(3)$ for the SGC with the components of the fundamental asymmetric tensor as the structure constants) on which the tree-structure of the cascade is built on. Indeed, this would involve other interactions than the ones preserved in the SGC models.

5. Acknowledgments. We heartily thank S. Lovejoy and A.M. Yaglom for enlightening discussions, A.E. Ordanovich, F. Schmitt, D. Marsan, D. Kwak and C. Naud for helpful comments. Partial supports by EEC-FI3P #CT93-0077 and (ARM) DOE #DE-FG03-90 contracts are gratefully acknowledged.

REFERENCES

Arnold, V.I., *Sur la géometrie différentielle des groupes de Lie de dimension infinie et ses applications à l'hydrodynamique des fluides parfaits*, Ann. Inst. Fourier (Grenoble), v.16, 1, pp. 319–361, 1966.

Bialas, A., Peschanski, R., Nucl. Phys. B, v. 273, p. 703, 1986.

Chigirinskaya, Y., Schertzer, D., Lovejoy, S., Lazarev, A., Ordanovich, A. *Nonlinear Processes in Geophysics*, v. 1, pp. 105–114, 1994.

Dubrulle, B., *Intermittency in Fully Developed Turbulence: Log-Poisson Statistics and Generalized Scale Covariance*. Phys.. Rev. Lett. , v. 73, 7, p.959–962, 1994.

Frisch, U., in *Turbulence and Stochastic Processes*, ed. by J.C.R. Hunt et al., The Royal Society, London, 1991.

Gledzer, E.B., Izv. Acad. Nauk USSR, Ser. MFG, 1, 1980.

Gledzer, E.B., Dolzhansky, E.V., Obukhov, A.M. *Systems of fluid mechanical type and their application*, Moscow, Nauka, p. 368 (in Russian), 1981.

Grossmann, S., Lohse, D., Europhys. Lett., v. 21, 2, pp.201–206, 1993.

Gupta, V., Waymire, E. C., J. Appl. Meteor., v. 32, p. 251, 1993.

Kolmogorov, A.N., Proc. Acad. Sci. USSR, Geochem. Sect. 30, pp. 299–303, 1941.

Lavallee, D., Ph.D. thesis, McGill University, Montreal, Canada, 1991.

Lavallee, D., Lovejoy, S., Schertzer, D., Ladoy, F., *Fractals in Geography*, Eds. De Cola, L., Lam, N., PTR, Prentice Hall, pp.158–192, 1993.

Leslie, D.C., *Developments in the theory of turbulence*, Clarendon Press, Oxford, p. 368, 1973.

Levich, E., Tzvetkov, E., Phys. Rep., v. 128, p. 1, 1985.

Mandelbrot, B., in *Fractals in the Natural Sciences* ed. by M. Fleischman et al., p. 3, Princeton University press, 1989.

Mandelbrot, B., *Intermittent turbulence in self-similar cascades: divergence of high moments and dimension of the carrier*, J. Fluid Mech. 62, 331–350, 1974.

Mandelbrot, B., Proc. R. Soc. London A, v. 434, p. 79, 1991.

Meneveau, C., Sreenivasan, K. R., Simple multifractal cascade model fro fully develop turbulence, Phys. Rev. Lett. 59, 13, 1424-1427, 1987.

Novikov, E.A., Appl. Math. Mech., v. 35, p. 231, 1971.

Novikov, E.A., *Infinitely divisible distributions in turbulence*, Phys. Rev. E 50, 5, R3303–R3305, 1994.

Obukhov, A.M., *On invariant characteristics of systems of fluid mechanical type*, Fluid Dynam. Trans., v.5, 2, p. 193–199, 1971.

Obukhov, A.M., Dolzhansky, E.V., *On simple models for simulation of nonlinear processes in convection and turbulence*, Geoph. Fluid. Dyn., v. 6, p. 195–209, 1975.

Oono, Y., Progr. theor. phys. Suppl., v. 99, p. 165, 1989.

Parisi, G., Frisch, U., in *Turbulence and predictability in geophysical fluid dynamics and climate dynamics*, ed. by M. Ghil et al., North Holland, pp. 84–88, 1985.

She, Z.S., Leveque, E., Phys. Rev. Lett., v. 72, p. 336, 1994.

She, Z.S., Waymire, E., Phys. Rev. Lett., v. 74, p. 262, 1995.

Schertzer, D., Lovejoy, S., in *Turbulence and chaotic phenomena in fluids*, Ed. T. Tatsumi, p. 505, North Holland, 1984.

Schertzer, D., Lovejoy, S., J. Geophys. Res., v. 92, p. 9692, 1987.

Schertzer, D., Lovejoy, S., in *Fractals, Physical origins and properties*, Ed. Pietronero, p. 49, Plenum Press, New York, 1989.

Schertzer, D., Lovejoy S., in *Fractals in the Natural and Applied Sciences*, Ed M.M. Novak, Elsevier Science B.V., pp. 325-339, 1994.

Schertzer, D., Lovejoy, S., *From scalar to Lie cascades: joint multifractal analysis of rain and clouds processes. -Space/Time Variability and Interdependence of Hydrological Processes*, R.A. Feddes ed., Cambridge University Press, pp. 153–173, 1995.

Schertzer, D., Lovejoy, S., *The Multifractal Transition Route to Self-Organized Criticality*, Physics Reports (to appear), 1996.

Schertzer, D., Lovejoy, S., Lavallee, D., Schmitt, F., in *Nonlinear Dynamics of Structures*, ed. by R. Z. Sagdeev et al., 213, World Scientific, 1991.

Schertzer, D., Lovejoy, S., Schmitt, F., *Structures in Turbulence and Multifractal Universality*, Small-Scale Structures in Three-Dimensional Hydro and Magnetohydrodynamic Turbulence, edited by M. Meneguzzi, A. Pouquet and PL Sulem, Springer-Verlag, Berlin, Lecture Notes in Physics, vol.462, pp.137-144, 1995.

Schmitt, F., Lavallee, D., Schertzer, D., Lovejoy, S., Phys. Rev. Lett., v.68, p.305, 1992.

Schmitt, F., Schertzer, D., Lovejoy, S., Brunet, Y., Fractals, v.1, 3, pp. 568–575, 1993.
Yaglom, A.M., Sov. Phys. Dokl., v. 2, p. 26, 1966.

A NON-LINEAR MODEL FOR FLUID PARCEL MOTIONS IN THE PRESENCE OF MANY LARGE AND MESO-SCALE VORTICES

L. JU. FRADKIN* AND A. R. OSBORNE†

Abstract. The trajectories of satellite tracked drifters that are assumed to represent fluid parcels motions in large and meso-scale oceanic flows have been shown previously i) to be physical fractals, ii) to behave as self-affine random walks, iii) to possess the scaling exponent approximately equal to the inverse of their fractal dimension, iv) to obey a superdiffusion law and v) to exhibit mild multiscaling (multifractality). A quasi-linear Torino model of drifter motion has proved successful in reproducing the first four of these features but not the fifth. In this paper a non-linear extension of this model taking into account vortex trapping and hopping is developed to remedy the situation.

1. Introduction. The problems of atmospheric and oceanic pollution as well as controled fusion have highlighted the need for models that can be used to make reliable predictions of spatial and temporal patterns of fluid parcel transport in two dimensions, since both the atmosphere and the oceans as well as Tokamak plasmas are known to be essentially two-dimensional. However, the ability to produce such models and assess their reliability is limited by lack of understanding of hydrodynamical and magnetohydrodynamical processes governing the interaction between small and large scales when the latter cannot be assumed well separated.

In order to get insight into scale interaction, in recent years much research effort has been directed at studying by numerical and analytical means "passive" tracer motion in prescribed Eulerian velocity fields. To give one example, advances in the area known as "chaotic advection" have demonstrated that large-scale random or deterministic flows can produce multi-scale tracer patterns leading to effective transport [1,2]. Most progress so far has been made for passive tracers in 2-D steady Eulerian flows in the presence of molecular diffusion or else in 2-D unsteady Eulerian flows which are time-periodic. Here we shall assume no molecular diffusion and concentrate on the unsteady large- and meso-scale Eulerian velocities that are a random superposition of waves with a constant linear phase speed. Such velocities have been proposed by the Torino group to reproduce some statistics of satellite-tracked drifter trajectories [3,4,5,6] (see also Fig. 1). Models of this type were first suggested by Davis [7] in an oceanographic context and by Horton [8] and Pettini *et al.* [9] for plasmas.

One of the specific questions addressed to in the field of chaotic advection is whether there exist different ranges of parameters where the mean square displacement (or other tracer statistics) scale in a simple manner.

---

* Now at the SEEIE, South Bank University, London SE1 OOA, U.K.
The work was carried out mainly when at the Department of Applied Mathematics and Theoretical Physics, University of Cambridge, U. K.
† Istituto di Fisica Generale, University of Torino, Italy.

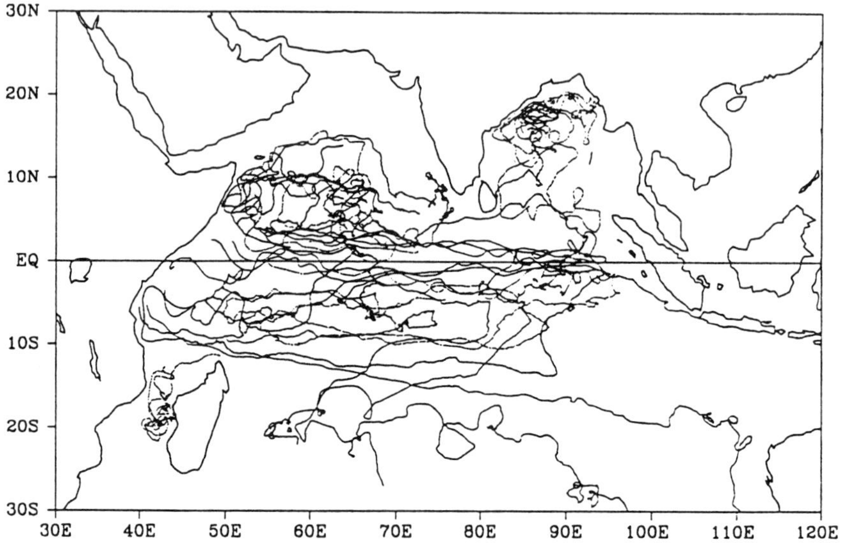

FIG. 1. *Several representative drifter trajectories.*

For the lack of a better word such ranges are referred to below as "regimes". Scaling itself is referred to as the Lagrangian asymptotic diffusion law. Predicting thresholds between different parameter regimes is an important related task. In many cases these thresholds cannot be established analytically and there is a need for extensive numerical experimentation.

2. Review of analytical methods for establishing diffusion laws. In the classical theory, where the Eulerian velocity is zero, the medium is assumed to be characterized by two scales that are well separated and are called micro- and macro-scopic respectively. An additional closure-type hypothesis is introduced saying that the Lagrangian tracer displacement behaves as a Gaussian random walk, and as a consequence, the classical asymptotic diffusion law turns out to be

$$< \mathbf{x}^2(t) >= 2\mathrm{D}t, \qquad \text{for } t \gg \tau_c, \tag{2.1}$$

where $\mathbf{x}(t)$ is the position of the tracer on the Lagrangian orbit at time t; τ_c is the correlation time along the orbit; and the angle brackets denote the average over an ensemble of realizations of the medium. The coefficient of proportionality D is called the (Lagrangian) diffusion coefficient.

Let us now move away from the classical situation and assume that a fluctuating Eulerian velocity is prescribed via a stream-function $\psi(\mathbf{x}, t)$ which has zero mean and a finite Fourier expansion

$$\psi(\mathbf{x}, t) = A \sum_{\boldsymbol{k}} \hat{\psi}_k(t) \cos\,(\boldsymbol{k} \cdot \mathbf{x} + \phi_{\boldsymbol{k}}). \tag{2.2a}$$

Here all the variables are non-dimensionalized; $\mathbf{x} = (x, y)$ is a position in space; t is time; $A > 0$ is the fluctuation amplitude; $\boldsymbol{k}$ is a wavevector with $k_{\text{min}} \leq k \leq k_{\text{max}}$, $\theta_{\text{min}} \leq \theta \leq \theta_{\text{max}}$ and (k, θ), polar coordinates in the wave-vector plane; $\hat{\psi}_k(t)$ is a real-valued *deterministic* function; and $\phi_{\boldsymbol{k}}$ are real random phases. Summation over $\boldsymbol{k}$ is a double summation over discrete values of k and θ. Then the Eulerian velocity is defined as

$$\text{(2.2b)} \qquad \mathbf{u}(\mathbf{x}, t) \equiv (\frac{\partial \psi}{\partial y}, -\frac{\partial \psi}{\partial x}) \equiv A \sum_{\boldsymbol{k}} \hat{\mathbf{u}}_k(t) \cos\,(\boldsymbol{k} \cdot \mathbf{x} + \phi_{\boldsymbol{k}}).$$

In order to study chaotic diffusion possible with (2.2) let us write the basic equation of tracer motion:

$$\text{(2.3)} \qquad \frac{d\mathbf{x}(t)}{dt} = \mathbf{u}(\mathbf{x}(t), t) \equiv \mathbf{v}(t),$$

where $\mathbf{v}(t)$ is the Lagrangian velocity. Then the tracer displacement is given by

$$\text{(2.4)} \qquad \mathbf{x}(t) = \int_{\text{o}}^{t} \mathbf{v}(t)\; dt,$$

where the integration is along the Lagrangian orbit and $\mathbf{x}(0) = 0$. In this case, the mean square displacement is

$$\text{(2.5)} \qquad < \mathbf{x}(t)^2 >= \int_{\text{o}}^{t} \int_{\text{o}}^{t} \rho_v(t', t'') dt' dt'',$$

where $\rho_v(t', t'') \equiv < \mathbf{v}(t') \cdot \mathbf{v}(t'') >$ is the Lagrangian velocity correlation function; and $< >$ is the ensemble average over different realizations of $\mathbf{u}$. This relationship has been derived by Taylor [10]. It does not automatically lead to a simple asymptotic diffusion law. By substituting (2.2) into (2.5) and assuming that

(i) the random phase assumption holds ($< \cos\,(\phi_{\boldsymbol{k}'} + \phi_{\boldsymbol{k}''}) >$, $< \sin\,(\phi_{\boldsymbol{k}'} + \phi_{\boldsymbol{k}''}) >= 0$),

(ii) the Eulerian velocity spectrum $\Phi_k(t', t'') \equiv \hat{\mathbf{u}}_k(t') \cdot \hat{\mathbf{u}}_k(t'')$ is stationary ($\Phi_k(t', t'') \equiv \Phi_k(\tau)$, with $\tau \equiv |t' - t''|$),

(iii) the increments of Lagrangian displacement are stationary as well ($\mathbf{x}(t'') - \mathbf{x}(t') \equiv \Delta\mathbf{x}(\tau)$), and

(iv) the sum in wave vector space may be approximated by an integral, expression (2.5) may be re-cast [11] as

$$< \mathbf{x}(t)^2 >= A^2 \Delta k^{-2} \Delta\theta \int_{\text{o}}^{t} dt' \int_{\text{o}}^{t'} d\tau \int_{k_{\text{min}}}^{k_{\text{max}}} k\; \Phi(k, \tau) < \cos\,[\boldsymbol{k} \cdot \Delta\mathbf{x}(\tau)] > dk,$$

(2.6)

where $\Phi(k,\tau) \equiv \Phi_k(\tau)$ and $\Delta\theta \equiv \theta_{\max} - \theta_{\min}$. In general, this does not lead to a simple asymptotic diffusion law either, since in general it is not known how $< \cos\,[\boldsymbol{k} \cdot \Delta\mathbf{x}(\tau)] >$ depends on k and τ. Three non-linear regimes have been studied analytically so far:

a) the quasi-linear regime (better known as the first-order smoothing or mixing length approximation), $\lambda \ll 1$;

b) the weakly non-linear regime, $1 \leq \lambda \leq \lambda_o$; and

c) the strongly non-linear regime, $\lambda \geq \lambda_1 \gg 1$,

where the non-linearity parameter $\lambda \equiv A/c$ and c is a characteristic speed of ψ-variation (e.g. see (3.1) below). λ_o and λ_1 are two thresholds which do not necessarily coincide.

In the first, quasi-linear, regime the RHS of (2.6) simplifies by noting that to the zeroth order, $\Delta\mathbf{x}(\tau) = 0$. Therefore [12], in this regime we obtain

$$< \cos \boldsymbol{k} \cdot \Delta\mathbf{x}(\tau) >= 1. \tag{2.7}$$

In the weakly non-linear regime, the most recent calculations have been conducted using a version of the renormalization-group technique [13]. The main additional assumptions in that work are

(i) on each scale 2π/k, the Péclet number Pe_k is smaller than unity. This implies [13] that the classical Péclet number which is defined with reference to the scale $2\pi/k_{\max}$ and is akin to the non-linearity parameter λ cannot be arbitrarily large (a point which is often overlooked); and

(ii) on each scale $2\pi/k$, the correlation time τ_{ck} is smaller than characteristic time of variation of the corresponding mean field. This implies that on each scale convection is negligible compared to diffusion.

It has been shown in Reference [13] that these assumptions allow one to say that the scales are widely separated in very precise statistical sense. Furthermore, they are equivalent to the closure-type hypothesis that the Lagrangian displacements possess a Gaussian distribution on larger scales. The resulting approximation to the ensemble average of the characteristic function $< \cos \boldsymbol{k} \cdot \Delta\mathbf{x}(\tau) >$ turns out to be

$$< \cos \boldsymbol{k} \cdot \Delta\mathbf{x}(\tau) >= \exp\,(-k^2 \mathrm{D}_k \tau) \tag{2.8}$$

Here D_k is the Lagrangian diffusion coefficient which varies from scale to scale.

In the strongly non-linear regime, the tracer motion which is Hamiltonian in physical space (see (2b) and (2.3)) is constrained by the existence of adiabatic invariants, which turn out to be the enclosed areas where fluid parcels may get trapped for a very long time before hopping out.

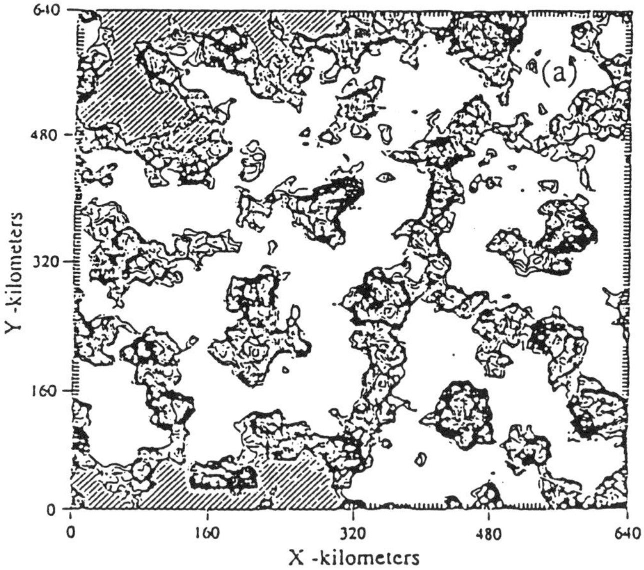

FIG. 2. *Momentary stream-function contours (reproduced from Ref. [3])* .

Such techniques as the adiabatic invariant approximation [14, 15], Manhattan model [16] and statistical topography [17] have been used to describe chaotic transport in special cases possible in this limit. They all rely on the fact that in the strongly non-linear regime the Lagrangian orbits circle the instantaneous ψ-contours many times before the contour plot undergoes a perceptible transformation.

3. The Torino model for fluid parcel motions. We now turn to the stream-function studied by Osborne and Caponio [3]. It consists of a random superposition of waves with a constant linear phase speed c and the Hurst exponent H,

$$(3.1)\quad \psi(\mathbf{x},t) = A\sum_{\boldsymbol{k}} k^{-H-1}\cos(\boldsymbol{k}\cdot\mathbf{x} - kct + \phi_{\boldsymbol{k}}), \qquad \tfrac{1}{2} < H < 1$$

where $H = \gamma/2 - 1$, and γ is the parameter used in Reference [3]. At any given moment in time the corresponding contour plot looks like an "islands-in-the-sea" picture (Fig. 2). The total area of the "islands", i.e. the nests of closed contours, is almost the same as the area in between. As time progresses the islands, that is vortices, shimmer around their original positions in a random manner.

As we have already mentioned, the many vortices model (3.1) has been developed by the Torino group to reproduce some features of satellite-tracked drifter trajectories which in their turn are believed to be similar to orbits followed by fluid parcels [18]. The corresponding quasi-linear orbits are computed by substituting (3.1) into (2.2b), neglecting $\boldsymbol{k} \cdot \mathbf{x}$ and then integrating (2.1). For an appropriate choice of H, the orbits have the same fractal dimension D, the same property of behaving as statistical self-affine walks and follow the same superdiffusion law as the drifter trajectories do.

Let us discuss these similarities in greater detail. Starting with the fractality, we emphasize that the concept is used here in the "physical" sense only, that is it applies to a limited range of scales. In the case of the satellite-tracked drifters the time scales in question are in between 1 and 10 days and the spatial scales vary from 60 to 150 km. The quasi-linear orbits simulated with (3.1) are physical fractals as well but on a larger range of scales, from 6 to 800 km. For an appropriate choice of $H(\approx 0.7)$, the fractal dimension D turns out to be the same for both real and simulated orbits. D is estimated using either the box counting or the Grassberger & Procaccia correlation dimension [18].

Turning to the scaling properties, the drifter trajectories have been shown to be physical self-affine random walks with the scaling exponent that may be approximated by the inverse of their fractal dimension [18]. The quasi-linear displacements are random (Fig. 3) and can be shown to have a scaling exponent equal to H. This means that they also behave as statistically self-affine random walks and satisfy the relationship

$$D \approx H^{-1}. \tag{3.2}$$

Osborne and Caponio [3] explain this reciprocity relationship between D and H as follows: in the quasi-linear approximation, due to isotropicity, the x and y motions are statistically independent and fractional functions of t. According to Mandelbrot [22] this means that they are fractional Brownian trails and their fractal dimension $D = \min(2, H^{-1})$. For $H > 1/2$, this implies relationship (3.2).

Let us consider the resulting diffusion laws. Analysis of drifter trajectories based on the time averages indicates [19] that they possess the superdiffusion exponent of about $2D^{-1}$. As to the Torino quasi-linear model, the corresponding Eulerian velocity spectrum is

$$\Phi(k, \tau) = A^2 k^{-2H} \cos\,(kc\tau), \tag{3.3}$$

and taking the ensemble average, the Lagrangian mean square displacement may be shown to be

$$< \mathbf{x}(t)^2 >= A^2 \Delta k^{-2} \Delta\theta \int_{0}^{t} dt' \int_{0}^{t'} d\tau \int_{k_{\min}}^{k_{\max}} k^{-2H+1} \cos(kc\tau) dk. \tag{3.4}$$

Let us restrict our attention to the intermediate times,

$$k_{\max}^{-1} \ll c\tau \ll k_{\min}^{-1}, \tag{3.5}$$

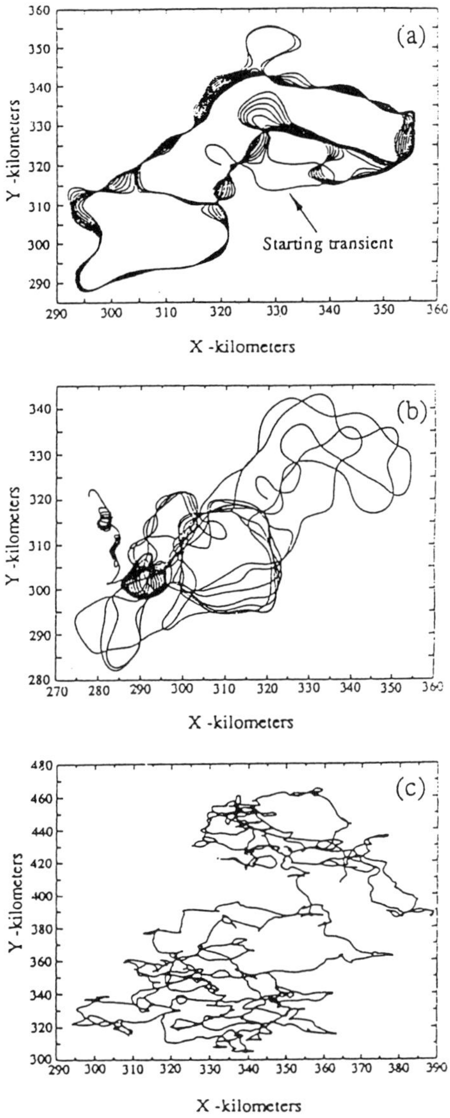

FIG. 3. *Particle trajectories in a) the strongly non-linear regime, b) intermediate non-linear regime, c) quasi-linear regime (reproduced from Ref. [3]).*

and assume that the contribution of small times may be neglected. Then noting that $0 < 2 - 2H < 1$, we can introduce a new variable $k' = kc\tau$ (cf. Ref.[21]) to deduce from (3.4) that for the intermediate times (3.5) the quasi-linear trajectories also obey the superdiffusion law

$$< \mathbf{x}(t)^2 > \sim \lambda^2 (ct)^{2H} \text{ if } \lambda \ll 1. \tag{3.6}$$

Thus, assuming that the time and ensemble average coincide (a reasonable assumption when the latter is taken only over the orbits exhibiting the same dynamical behavior), the quasi-linear and drifter trajectories disperse in the same manner. A different proof of (3.6) may be found e.g. in Reference [15].

To summarize the material presented so far, using the Torino model with a very small non-linearity parameter λ is consistent with the fractal and self-affine nature of drifter trajectories, reciprocal relationship (3.2) and superdiffusion law (3.6). However, the Torino group has also observed that unlike the quasi-linear orbits the real drifter trajectories exhibit mild multiscaling. They refer to it as mild multifractality. We will explain below why a change in name is desirable. Further tests have revealed that the trajectories simulated using intermediate values of λ exhibit mild multiscaling as well, whenever the trajectories simulated using large values of λ do not. All this has been established in a standard way by calculating the generalized correlation dimensions D_q and plotting D_q against q [3, 18]. Such calculations are based on the trajectory points equally spaced in *time*. If the *spatial* distribution of these points is inhomogeneous, then D_q depends on q in a non-linear fashion [25]. Since infinitely many scaling exponents D_q's and not just one D are required to describe an inhomogeneous fractal orbit such orbits are said to exhibit multi-fractality. The spatial inhomogeneity may also be characterized by covering the trajectory with balls of radius ℓ and then repeating the operation with smaller and smaller ℓ to check how the pointwise probability for a trajectory point to belong to a ball of a particular radius scales with this radius when the latter is small (*ibid.*) Note that non-fractal inhomogenous orbits including spirals may also produce non-linear dependence of the generalized correlation dimensions on q, and this is why we prefer to say that this non-linearity establishes multiscaling rather than multifractality.

Why would some trajectories behave as inhomogeneous fractals whenever others are homogeneous? The visual inspection of drifter and simulated trajectories (see Figs. 1 and 3 respectively) leads to the conclusion that when the non-linearity parameter is small the simulated trajectories slide from one instantaneous ψ-contour to another and when it is large they circle contours many times over before the latter transform. The trajectories which behave as inhomogeneous fractals, whether real or simulated, appear to consist of vortex trappings and hoppings instead. This allows us to describe them as two interwoven homogeneous fractals. A difference in the corresponding fractal dimensions will cause the mixture to be inhomogeneous.

It is not surprising that observed multiscaling is very mild. Indeed, as we have shown above, in the quasi-linear regime we have relationship (3.2). On the other hand, according to Isichenko [17] in the strongly non-linear regime we have

$$D = (3 + 10H)/7. \tag{3.7}$$

When $H \approx 0.7$, the value identified by data analysis and used in all simulations, the two RHS's are very close.

Thus, the presence of multiscaling (however mild) suggests that the regime can not be modeled as quasi-linear. This means that the superdiffusion exponent has to be re-calculated. The Torino group has established numerically [3] that the superdiffusion exponent does not vary with the non-linearity parameter by much. At the first glance this appears to be surprising because for large λ the argument of cosine in (2.6) may not be neglected any more. Let us investigate this situation further. The weakly-nonlinear approximation described above cannot apply, because in this regime the tracer displacements still behave in a Gaussian manner on larger scales. On the other hand, if after averaging over an ensemble of Eulerian velocity realizations the trappings/hoppings picture is still preserved, it is conceivable that the Lagrangian displacements may be described using a non-Gaussian closure-type hypothesis.

Indeed, let us assume that non-linearity is not negligible and that there exists *an intermediate range* of k and τ defined by conditions (3.5) and (A.12) and (A.15) is such that the Fourier transform of the probability density, $P(\Delta\mathbf{x}, \tau)$ (see Appendix A for the definition), behaves in the following manner

$$P(\boldsymbol{k}, \tau) \sim k^{-\alpha}\tau^{-\beta}, \qquad 0 < \alpha,\ \beta < 1, \tag{3.8}$$

(see Appendix). Note that at this stage both k and τ are non-dimensional, let us say normalised by k_{min} and $\tau_{\text{min}} = 1/ck_{\text{max}}$ respectively. Then the approximations (2.7) and (2.8) may be replaced by

$$\begin{aligned} < \cos\left[\boldsymbol{k}\cdot\Delta\mathbf{x}(\tau)\right] > \ &\equiv \int\int_{-\infty}^{\infty} \cos\left[\boldsymbol{k}\cdot\Delta\mathbf{x}(\tau)\right](P(\Delta\mathbf{x},\tau)d\Delta\mathbf{x} \\ &\sim k^{-\alpha}\tau^{-\beta}. \end{aligned} \tag{3.9}$$

The exponents α and β may be called spatial and temporal fractal dimensions respectively, similarly to the exponents α and β introduced in Afanasyev et al. [3]. By assuming that $0 < 2 - 2H - \alpha < 1$, introducing variable $k' = kc\tau$ and substituting (3.9) into (2.6) we find that the classical diffusion law (2.1) may be replaced by

$$< \mathbf{x}^2(t) > \sim t^{2H+\alpha-\beta}. \tag{3.10}$$

Using our formula (A.13) and introducing variables $u' = ut$, $q' = qt^{\beta/\alpha}$ and $\mathbf{x}' = \mathbf{x}t^{-\beta/\alpha}$ we can see that for any self-affine trajectory generated by our random walk, the anomalous diffusion law in the intermediate time range specified above is

$$
\begin{aligned}
< \mathbf{x}^2(t) > &\equiv \int_{-\infty}^{\infty} \mathbf{x}^2(t)P(t,\mathbf{x})d\mathbf{x} \\
&\sim t^{2\beta/\alpha}. \qquad (3.11)
\end{aligned}
$$

Comparing (3.11) and (3.10) it follows that we have the relation

$$2H + \alpha - \beta = 2\beta/\alpha. \qquad (3.12)$$

Moreover, comparing Shlesinger [24] and Afanasyev *et al.* [23] we have

$$\beta/\alpha = D^{-1} \qquad (3.13)$$

or

$$2H + \alpha - \beta = 2D^{-1}. \qquad (3.14)$$

This is possible if $\alpha - \beta$ is small, and thus (3.2) is a good approximation to (3.14). To reiterate, the fact that the superdiffusion exponent hardly changes with λ has been observed by Osborne and Caponio [3] in their numerical work. The results pertaining to *very strong* non-linearity and described in Reference [17] have been mentioned above. Our model appears to be the first possible analytical explanation of how this might be possible for intermediate non-linearities as well. Of course, it remains to be seen whether the intermediate non-linear regimes indeed exist, where α and β are well-defined, $\alpha - \beta \ll 1$ and $\alpha/\beta \approx D$. Computer experiments are in progress to establish this as well as to pinpoint thresholds between different types of non-linear regimes. The preliminary results show that the *intermediate* non-linear regime amenable to a description based on a singular Levy distribution may lie within the $1.585 \leq \lambda \leq 20$ range.

4. Conclusions. To conclude, we have shown that the large and meso-scale Torino model with intermediate values of the non-linearity parameter λ may well account for the same features of drifter trajectories as the quasi-linear model does and reproduce mild multiscaling as well. Note that the fractal trajectories we consider in this paper are characterised by just one pair of spatial and temporal fractal dimensions α and β and exhibit classical multifractality, that is behave as inhomogeneous fractals. Note that Afanasyev *et al.* [23] use a different definition of multifractality, describing orbits which are characterised by *a set* of α's and β's.

Three points of general interest are worth stressing: in addition to models discussed in Aref [26] the non-linear Torino model provides us with yet another example of more or less the same fractal dimension corresponding to two different dynamical behaviors: a) fast sliding from contour to contour in the quasi-linear regime and b) vortex trapping and hopping in the intermediate range. Moreover, it shows that in the latter case the corresponding Lagrangian orbits will be multiscaling. Finally, we believe that the statistical orbit description based on combination of vortex trappings and hoppings may find a wider application than is detailed in this paper.

Acknowledgements.

In the course of this work the first author was sponsored under the Culham Laboratory, UKEA agreement EMR 478 M.

Appendix A.

Originally the Montroll-Weiss equation has been derived to describe a discrete random walk on square lattice, with a step describing hopping from one *square* to another. We propose to use the same equation to describe particle motion in the field of many randomly moving vortices as a random walk, with a step involving hopping from one *vortex* to another.

Let us then derive the Montroll-Weiss equation following Ref.[13]. Let us designate a particular moment in time as $t = 0$. Let us assume that the vortices move randomly each within its own portion of space, without invading that of another. (This appears to be a good assumption when dealing with the Torino model. If it is dropped a more complex formalism results). Let $\mathbf{x}$ be a vector determining the position in space of the center of mass of such portion in space associated with a particular vortex and thus label this vortex. Let $P_j(\mathbf{x})$ be the probability density for finding this particle in the vortex $\mathbf{x}$ at the jth vortex hopping. Then, assuming vortices and the associated portions of space are very small, so that there is a very small difference between particle position and $\mathbf{x}$, we have

$$P_j(\mathbf{x}) = \int_{-\infty}^{\infty} W(\mathbf{x}-\mathbf{x}')\ P_{j-1}(\mathbf{x}')d\mathbf{x}', \tag{A.1}$$

where $W(\mathbf{x}-\mathbf{x}')$ is a stationary transition probability. Let $P_j(\mathbf{q})$ and $W(\mathbf{q})$ be the Fourier transforms of $P_j(\mathbf{x})$ and $W(\mathbf{x})$ respectively. Assuming that $P_0(\mathbf{q}) = 1$, so that initially the particle is at the vortex near the origin of the position-space, it follows from (A.1) that

$$P_j(\mathbf{q}) = [W(\mathbf{q})]^j. \tag{A.2}$$

Replacing the discrete time parameter j with its continuous counterpart t, the inverse Fourier transform of (A.2) leads to

$$P(\mathbf{x},t) = \int_{-\infty}^{\infty} \exp[-i\mathbf{q}\mathbf{x} + t\ \ln W(\mathbf{q})]\ d\mathbf{q}. \tag{A.3}$$

Note that mean squared Lagrangian displacement is

$$< \mathbf{x}^2 >= \int_{-\infty}^{\infty} \mathbf{x}^2 \; P(\mathbf{x},t) d\mathbf{x}. \tag{A.4}$$

Now, let $Q(\mathbf{x},t)$ be the probability density for finding a particle in vortex $\mathbf{x}$ at time t immediately after hopping. Then

$$P(\mathbf{x},t) = \int_{o}^{t} \phi(t-\tau') \; Q(\mathbf{x},\tau) d\tau, \tag{A.5}$$

where $\phi(t)$ is the probability density for the particle staying in vortex $\mathbf{x}$ for a period of length t. This probability density may be expressed via the probability density $\psi(t)$ for the time interval between two consequtive hoppings being t:

$$\phi(t) = \int_{t}^{\infty} \psi(t') \; dt'. \tag{A.6}$$

Furthermore, let $\psi_j(t)$ be the probability density for the particle making the jth hopping at time t. Then

$$\psi_j(t) = \int_{o}^{t} \psi(t-\tau)\psi_{j-1}(\tau) \; d\tau, \tag{A.7}$$

with the initial condition $\psi_1(t) \equiv \psi(t)$. Let $\psi_j(u)$ be the Laplace transform of $\psi_j(t)$. Then similarly to (A.2)

$$\psi_j(u) = [\psi(u)]^j. \tag{A.8}$$

It then follows that

$$Q(\mathbf{x},t) = \sum_{j=o}^{\infty} \psi_j(t) P_j(\mathbf{x}). \tag{A.9}$$

All the above equations combine to give the Montroll-Weiss equation

$$\begin{aligned} P(\mathbf{x},t) &= \frac{1}{2\pi i} \int_{c-i\infty}^{c+i\infty} du \, \frac{1-\psi(u)}{u} e^{ut} \\ &\cdot \frac{1}{(2\pi)^s} \int_{-\infty}^{\infty} dq \frac{e^{-i\mathbf{q}\mathbf{x}}}{1-\psi(u)W(\mathbf{q})}, \end{aligned} \tag{A.10}$$

where s is dimension of the state space.

In addition to (3.5) let as assume that we are dealing with singular Levy distributions,

$$\text{(A.11)}\qquad \begin{aligned} W(\mathbf{q}) &= \begin{cases} 1 - Aq^{\alpha} & \text{if } q < |q|_{\max}, \\ 0 & \text{otherwise}, \end{cases} \\ \psi(u) &= \begin{cases} 1 - Bu^{\beta} & \text{if } |u| < u_{\max}, \\ 0 & \text{otherwise}, \end{cases} \end{aligned}$$

where $q \equiv |\mathbf{q}|$ and $0 < \alpha, \beta < 1$ (assuring positivity of probabilities $W(\mathbf{q})$ and $\psi(t)$). Let us again restrict our attention to scales that are not too large, so that

$$\text{(A.12)}\qquad \begin{aligned} u_{\max} t_{\min} &\equiv u_{\max}/ck_{\max} \gg 1, \\ q_{\max} x_{\min} &\equiv 2\pi q_{\max}/k_{\max} \gg 1. \end{aligned}$$

Then substituting (A.11) into (A.10) the Montroll-Weiss equation reduces to

$$\text{(A.13)}\qquad P(\mathbf{x}, t) \approx \frac{1}{2\pi i} \int_{c-i\infty}^{c+i\infty} du \, \frac{1}{(2\pi)^s} \int_{-\infty}^{\infty} dq \frac{Bu^{\beta-1}}{Bu^{\beta} + A|q|^{\alpha}} e^{ut - i\mathbf{qx}}.$$

Finally, the Fourier transform gives

$$\text{(A.14)}\qquad P(\boldsymbol{k}, t) = \int_{c-i\infty}^{c+i\infty} du \, \frac{Bu^{\beta-1}}{Bu^{\beta} + Ak^{\alpha}} e^{ut}.$$

If the scales are not too small, so that

$$\text{(A.15)}\qquad Bu_{\max} k_{\min}^{-\alpha/\beta} \ll A,$$

then yet another change of variable, $u' = ut$, leads to (3.8).

Note that conditions (A.12), (A.15) and (3.5) give precise meaning to the concept of "intermediate scales."

REFERENCES

[1] H. Aref J. Fluid Mech. **143**, 1 (1984).
[2] J. Ottino *Kinematics of Mixing: Stretching, Chaos and Transport* (Cambridge University Press, Cambridge, 1989).
[3] A. R. Osborne and R. Caponio Phys. Rev. Lett. **64**, 1733 (1990).
[4] M. G. Brown and K. B. Smith J. Phys. Ocean. **20**, 139 (1991); Phys. Fluids A **3(5)** 1186 (1991).
[5] B. G. Sanderson, A. Goulding and A. Okubo Tellus **42A**, 550 (1990); B. G. Sanderson and D. A. Booth Tellus **43A**, 334 (1991).
[6] P. H. LeBlond Private communication (1990).

[7] R. E. Davis J. Mater. Res. **41**, 163 (1983).
[8] W. Horton 1990 Phys. Rep **192**(1), 1 (1990).
[9] M. Pettini *et al.* Phys. Rev. A **38**(1), 344 (1989).
[10] G. I. Taylor, Proc. London Math. Soc., **A20**, 196 (1921)
[11] J. H. Misguish, R. Balescu, H. L. Pécseli, T. Mikkelsen, S. F. Larsen, and Qiu Xiaoming, *Plasma Physics and Controlled Fusion*, **29** (7), 825 (1987)
[12] J. A. Krommes in Handbook of Plasma Physics, Vol. 2, edited by M. N. Rosenbluth and R. Z. Sagdeev (North Holland Physics Publishing, Amsterdam, 1983).
[13] L. Fradkin Plasma Phys. and Controll. Fusion **33**(6), 685 (1991).
[14] D. F. Escande Phys. Rep. **121**(3), 166 (1984).
[15] A. N. Neishtadt, D. K. Chaikovskii, and A. A. Chernikov Sov. Phys. JETP **72**(3), 423 (1991).
[16] J.-P. Bouchard *et al.* Phys. Rev. Lett. **64**, 2503 (1990).
[17] M. B. Isichenko Rev. Mod. Phys. **64**(4), 961 (1992).
[18] A. R. Osborne, A. D. Kirwan, Jr., A. Provenzale, and L. Bergamasco Tellus **41B**, 416 (1989).
[19] A. Provenzale, A. R. Osborne, A. D. Kirwan, Jr., and L. Bergamasco in *Proceeding of the Enrico Fermi School: Nonlinear Topics in Ocean Physics* edited by A. R. Osborne (North-Holland, Amsterdam, 1981).
[20] A. R. Osborne, R. Caponio, and A. Pastorello Mod. Phys. Lett. A **6**(2) 83 (1991).
[21] I. S. Gradshteyn, and I.M. Ryzhik 1980 *Table of Integrals, Series and Products*, 3.761(9) (Academic Press, New York, 1980).
[22] B. B. Mandelbrot *The Fractal Geometry of Nature* (Freeman, San Francisco, 1982).
[23] V. V. Afanasyev, R. Z. Sagdeev, and G. M. Zaslavsky Chaos **1**(2), 143 (1991).
[24] M. F. Shlesinger Physica D **38**, 304 (1989).
[25] G. Paladin, and A. Vulpiani Phys. Rep. **156**(4), 147 (1987).
[26] H. Aref Phil. Trans. Soc. Lond. A **333**, 273 (1990).

SCALE-DEPENDENT OCEAN WAVE TURBULENCE

ROMAN E. GLAZMAN*

Contents:

1. Introduction. Wave turbulence is a common feature of nonlinear wave motions observed when external forcing acts during a long period of time, resulting in developed spectral cascades of energy, momentum and, possibly, other conserved integrals. In the ocean, wave turbulence occurs on scales from capillary ripples and to those of baroclinic inertia-gravity and Rossby waves. In general, oceanic wave motions are characterized by rather complicated dispersion laws containing characteristic scales such as, for instance, the Rossby radius of deformation. The resulting absence of scale invariance makes many problems of wave turbulence intractable by standard, small-perturbation-based techniques. As a result, present theoretical understanding has been limited to short- and long-wave asymptotic regimes (Zakharov et al., 1992). Another, more fundamental limitation of the small perturbation theories is the assumption that the wave amplitude be small in relation to the wavelength. Thus, rare (and highly intermittent) events of appearance of the strongly nonlinear wavelets are disregarded at the outset. A number of laboratory and field measurements reveal rather peculiar wave spectra which cannot be explained by scale-invariant and/or weak-turbulence theories. The peculiarities include multiple breaks of power laws and saturation of the otherwise monotonous dependence of wave spectra on external forcing (Jähne and Riemer, 1990; Hwang et al, 1993; Hara et al., 1994; LeTraon et al., 1990). Furthermore, field observations show occurrence of breaking waves in which the nonlinearity is locally very high. In the case of wind-generated surface gravity waves, these are observed as whitecaps. Observation of breaking events in the baroclinic inertia-gravity waves requires measurements at a depth of the ocean thermocline – hundreds meters below the surface. Breaking waves are manifested as spots of small-scale turbulence resulting from overturning of internal waves at the interface between two layers of different densities. These rare events may coexist quite nicely with the generally low energy level in the wave field. However, their effect on the overall (i.e.,

* Jet Propulsion Laboratory, California Institute of Technology, Pasadena, CA 91109.

averaged over a large time and area) spectrum can be rather important. These features find a remarkably simple explanation in the framework of a recently developed heuristic approach called "multiwave-interaction theory" (Glazman, 1992-95). In what follows we review the basic ideas and results, compare theoretical predictions with experimental data for cases of capillary-gravity and baroclinic inertia-gravity waves, and discuss possible avenues for further development of the theory.

2. Observations of ocean wave turbulence. Until recently, the standard example of ocean wave turbulence has been that of deep-water surface gravity waves generated by wind. Characteristic wavelengths of these waves range from 1m to 200m. Early field observations, conducted mostly in near-shore regions, showed the "equilibrium" range of the power spectrum as dominated by

$$S(\omega) = \beta g^2 \omega^{-5} \tag{2.1}$$

where β is a constant. This is known as the Phillips' saturated spectrum (Phillips, 1958). The corresponding 2-d wavenumber spectrum is

$$F(k) = Bk^{-4} \tag{2.2}$$

where $B \approx \beta/2$. These spectra represent the regime of strong wave turbulence: the equilibrium is reached due to the breaking of steep wave crests, that is to a highly non-local energy transfer to small scales. The wavelength of breaking waves is within the range described by (2.1)-(2.2). These spectra used to be regarded as universal (Pierson and Moskovitz, 1964; Pierson, 1991). However, later observations – at longer wind fetches – revealed an extended range of frequencies dominated by

$$S(\omega) = \alpha g U \omega^{-4} \tag{2.3}$$

where U is the wind speed (at 10m height) and α is a constant (Toba, 1973; Forristall, 1981; Kahma, 1981; Donelan et al., 1985). This range corresponds to a direct cascade of energy through the spectrum. However, the cascade is not necessarily conservative (Phillips, 1985). In the wavenumber space, spectrum (2.3) takes the form

$$F(k) = Ag^{-1/2} U k^{-7/2} \tag{2.4}$$

where $A \approx \alpha/2$. In terms of the energy flux (per unit surface area, per unit mass of water), Q, through the spectrum, (2.4) can be written as $F(k) = A' g^{-1/2} Q^{1/3} k^{-7/2}$ where A' is the "Kolmogorov constant." Thus, the prediction of weak turbulence theory (Zakharov and Filonenko, 1966) is confirmed.

Zakharov and Zaslavskii (1982) pointed to a possibility of an even flatter spectrum based on the conservation of wave action P, in an inverse spectral cascade:

$$F(k) = A_p' P^{1/3} k^{-10/3} \tag{2.5}$$

This spectrum occurs at yet lower frequencies – below the "generation range." An experimental observation of this spectrum by Grose et al. (1972) never received much attention in the oceanographic literature.

A detailed analysis of the gravity wave spectrum for a broad range of wave development stages presented in (Glazman, 1994) was based on buoy observations. In particular, it showed that the exponent p in $F(k) \sim k^{-p}$ slowly grows as the wavenumber increases away from the spectral peak. Therefore, different regimes of energy and action fluxes dominate in different subranges of the spectrum.

Another example of wave turbulence is given by capillary-gravity (CG) waves (Jähne and Riemer, 1990; Hwang, et al., 1993; Hara, et al., 1994). The CG spectra exhibit pronounced "breaks" at certain scales pointing to an important role played by the intrinsic scale $(\sigma/g)^{1/2}$ of the problem. Here σ is the surface tension coefficient divided by the water density and g is the acceleration due to gravity. The CG wave spectra measured at different wind speeds by Hwang et al. (1993) are illustrated in Figure 1(b).

Manifestation of wave turbulence in long baroclinic waves (called inertia-gravity (IG) waves) was recently discovered by re-interpreting 1-d wavenumber spectra of sea surface height (SSH) spatial variations on scales of 10 to 1000 km (Glazman, 1995(b)). Examples of 1-d SSH spectra based on satellite altimeter measurements are reported by Gordon and Baker (1980), Fu (1983), Gaspar and Wunsch (1989), Le Traon et al., (1990; 1994) and others. A typical spectrum is illustrated in Figure 2. These spectra are dramatically different from what one would observe if SSH variations were dominated by the 2-dimensional eddy turbulence. Indeed, according to Kraichnan's (1967) prediction, the kinetic energy spectrum of 2-d turbulence is given by k^{-3} or $k^{-5/3}$ laws for the energy and enstrophy cascades, respectively. Assuming geostrophy, these laws translate respectively into k^{-5} or $k^{-11/3}$ spectra for SSH oscillations along altimeter ground tracks. The altimeter-observed SSH spectra are much flatter. Besides, they exhibit spectral breaks pointing to an important role played by the intrinsic spatial scale, the Rossby radius of deformation, characterizing IG wave turbulence.

3. Kinetic equations of weak-turbulence theory. Assuming wave fields to be near Gaussian, a closed form equation for second statistical moments can be derived by means of small-perturbation techniques (Zakharov et al., 1992). For a decay dispersion law in the case of resonant 3-wave interactions, the kinetic equation is

$$\frac{\partial N(\boldsymbol{k},t)}{\partial t} = \pi \int [|V_{k12}|^2 f_{k12} \delta(\boldsymbol{k} - \boldsymbol{k}_1 - \boldsymbol{k}_2) \delta(\omega_k - \omega_1 - \omega_2)$$
$$+2|V_{1k2}|^2 f_{1k2} \delta(\boldsymbol{k}_1 - \boldsymbol{k} - \boldsymbol{k}_2) \delta(\omega_1 - \omega_k - \omega_2)] d\boldsymbol{k}_1 d\boldsymbol{k}_2 + \gamma(k) N(\boldsymbol{k},t), \tag{3.1}$$

where $N(\boldsymbol{k},t) = F(\boldsymbol{k},t)/\omega(k)$ is the spectral density of wave action, $F(\boldsymbol{k},t)$ is the spectral density of wave energy, V_{k12} is the interaction coefficient for

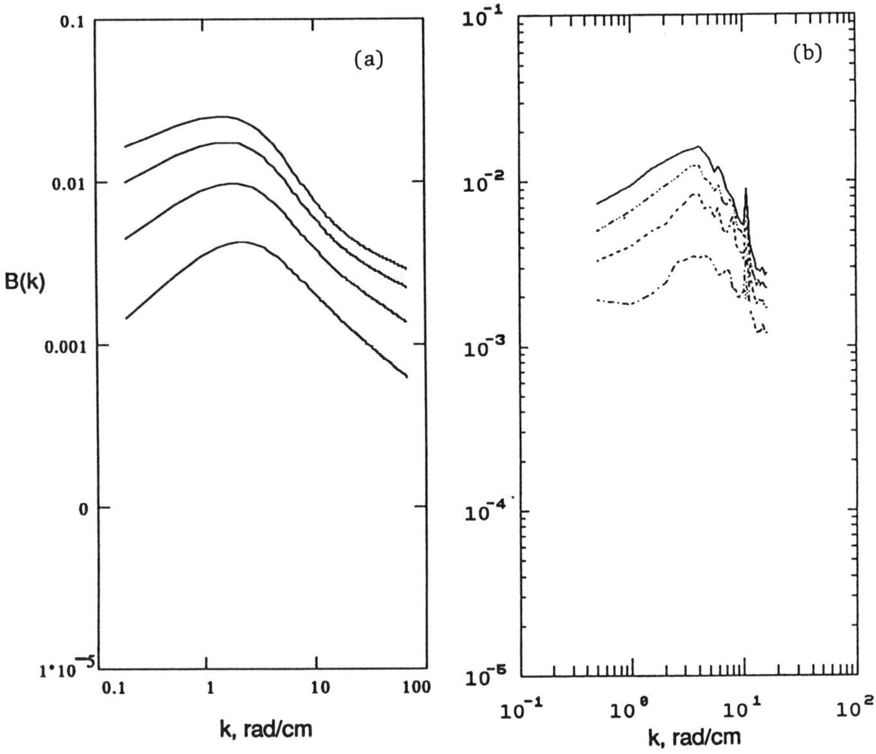

FIG. 1. *Theoretical and experimental spectra of "surface curvature" $B(k) = k^4 F_\eta(k)$ for wind speed values, $U(m/sec)$: 5.7, 7.0, 8.5 and 9.9 - increasing upward. (a) is based on equations (4.18),(4.19) and on an empirical formula for ν as a function of wind: $\nu(U) = 0.2 + 0.6U$ where U is in m/sec. The energy flux is determined as $Q = cU^3$ where the meaning and value of constant c are given in (Glazman, 1995(a)). (b) is a subset of measurements reported by Hwang et al. (1993). (Reproduced by courtesy of the authors).*

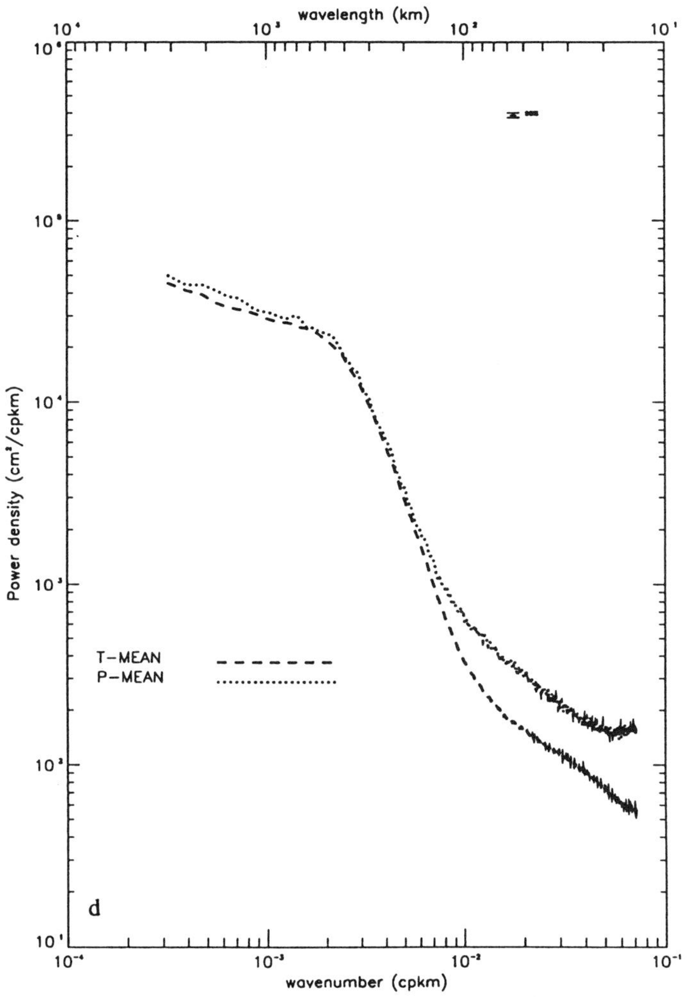

FIG. 2. *1-dimensional spectra of surface height variations observed by Topex altimeter along satellite "ground tracks" for mid-latitude regions. (Reproduced from (Le Traon et al., 1994) by courtesy of the authors).*

wave triads, and $\gamma(k)$ is the growth (decay) rate due to external forcing (dissipation). The molecular viscosity corresponds to $\gamma(k) = -2\nu k^2$. A conservative cascade occurs when all external sources and sinks are outside the inertial range: $\gamma(k) \equiv 0$. Furthermore,

$$f_{k12} = N_1 N_2 - N_k(N_1 + N_2), \quad N_l = N(\boldsymbol{k}_l, t). \tag{3.2}$$

For a non-decay dispersion law–when 3-wave interactions are non-resonant–the kinetic equation (after eliminating non-resonant terms by an appropriate canonical transformation) has the form (Zakharov et al., 1992):

$$\frac{\partial N(\boldsymbol{k},t)}{\partial t} = \frac{\pi}{2}\int [|T_{k123}|^2 f_{k123}\delta(\boldsymbol{k}+\boldsymbol{k}_1-\boldsymbol{k}_2-\boldsymbol{k}_3)\delta(\omega_k +$$

$$+\omega_1 - \omega_2 - \omega_3) d\boldsymbol{k}_1 d\boldsymbol{k}_2 d\boldsymbol{k}_3 + \gamma(k)N(\boldsymbol{k},t), \text{ where} \tag{3.3}$$

$$f_{k123} = N_2 N_3 (N_1 + N_k) - N_1 N_k (N_2 + N_3), \tag{3.4}$$

and T_{k123} is the interaction coefficient for resonant tetrads. Explicit expressions for V_{k12} and T_{k123} for various wave processes are given, for instance, by Zakharov (1984) and Zakharov et al. (1992). Taking into account the physical significance of the collision integrals in (3.1) and (3.3), these equations can be written in a more instructive form (Phillips, 1977):

$$\frac{\partial N(\boldsymbol{k},t)}{\partial t} + \nabla_k \cdot \boldsymbol{P}^{(n)} = \gamma(k) N(\boldsymbol{k},t), \tag{3.5}$$

where $\nabla_k \cdot \boldsymbol{P}^{(n)}$ is the divergence in the wave-vector space of the action flux due to n-wave interactions. In the next section this form is generalized to account for a higher number of resonantly interacting Fourier components.

In the simplest, scale-invariant case, the dispersion law and the interaction coefficients are homogeneous functions of their arguments:

$$V(\lambda\boldsymbol{k}, \lambda\boldsymbol{k}_1, \lambda\boldsymbol{k}_2) = \lambda^m V(\boldsymbol{k}, \boldsymbol{k}_1, \boldsymbol{k}_2), \tag{3.6}$$

$$T(\lambda\boldsymbol{k}, \lambda\boldsymbol{k}_1, \lambda\boldsymbol{k}_2, \lambda\boldsymbol{k}_3) = \lambda^m T(\boldsymbol{k}, \boldsymbol{k}_1, \boldsymbol{k}_2, \boldsymbol{k}_3),$$

$$\omega(k) \propto k^s. \tag{3.7}$$

For instance, for capillary and gravity waves on deep water we have (Zakharov, 1984)

$$m = 9/4, \quad s = 3/2 \quad \text{and} \tag{3.8a}$$

$$m = 3, \quad s = 1/2, \tag{3.8b}$$

respectively. Assuming the nonlinear interactions to be local in the wave-number space, equations (3.1), (3.3), (3.6)-(3.8) yield scaling relationships for the characteristic interaction time (the "turnover time"):

$$t^{-1} \approx N k^{2(m+1)} \omega^{-1} \quad \text{for 3-wave interactions;} \tag{3.9a}$$

(3.9b) $$t^{-1} \approx N^2 k^{2(m+2)} \omega^{-1}, \quad \text{for 4-wave interactions.}$$

The relevant small parameter in the perturbation expansion is $\epsilon = ak$, where a is the characteristic wave amplitude related to $N(k)$ by $N d\boldsymbol{k} \propto a^2/\omega$ – in the case of gravity waves and $N d\boldsymbol{k} \propto (ak)^2/\omega$ – in the case of capillary waves.

It is easy to check that equations (3.6)-(3.8) allow expressing the turnover time in terms of ϵ:

(3.10a) $$t^{-1} \approx \omega\epsilon^2,$$

(3.10b) $$t^{-1} \approx \omega\epsilon^4.$$

This form remains valid for a broad class of nonlinear wave systems which are not necessarily scale-invariant and whose degree of nonlinearity is measured by a more general quantity

(3.11) $$\epsilon = u/c,$$

where u and c are characteristic particle and phase velocities, respectively, at scale k. For deep-water waves (3.11) reduces to $\epsilon = ak$. For shallow-water waves, (3.10) and (3.11) are employed directly (Glazman, 1995(b)).

The 1-d form of (3.5) is obtained either by transforming to the frequency space (Zakharov, 1984), or by multiplying by k and integrating over the polar angle in the 2-d wave-vector space. This yields

(3.12) $$\partial\overline{N}/\partial t + \partial P/\partial k = \gamma\overline{N}, \text{ where}$$
$$\overline{N}(k,t) = \int_{-\pi}^{\pi} N(k,\theta,t)k d\theta, \quad \text{and} \quad P(k,t) = \hat{i}\int_{-\pi}^{\pi} \boldsymbol{P}(k,\theta,t)k d\theta.$$

Similar expressions can be derived for the wave energy and momentum fluxes. The steady-state inertial cascades ($\gamma(k) = 0$) of wave action and energy are given by

(3.13) $$P(k) = P_0, \quad \text{and} \quad Q(k) = Q_0,$$

where $Q(k) = \omega(k)P(k)$ is the energy flux.

Weak-turbulence theory is most useful for scale-invariant systems and purely inertial cascades–when solutions of (3.13) are given by power laws. Departures from (3.6) and (3.7) make the task of solving the collision integral quite formidable. Furthermore, the quasi-Gaussian assumption underlying the kinetic equation, even for the lowest degree of wave nonlinearity, disregards possible intermittency in the wave field. Finally, the small-perturbation approach does not allow one to explore effects of higher-order wave-wave interactions

$$\omega_0 \pm \omega_1 \pm ... \pm \omega_m = 0, \quad \text{and} \quad \boldsymbol{k}_0 \pm \boldsymbol{k}_1 \pm ... \pm \boldsymbol{k}_m = 0 \tag{3.14}$$

in which $m > 4$. These interactions become highly important at least in some localized regions of the wave field, and they may lead to wave breaking. Due to extreme mathematical difficulties of accounting for scale-dependence and high-order nonlinearity, a heuristic approach may have considerable advantages over the formal perturbation theory.

4. Multiwave interaction approach. The present approach (Glazman, 1992,1993, 1995) is based on a set of scaling relationships which become most obvious if introduced in the language of the previous section.

Let us consider an inertial energy cascade. The external source is assumed to act at lower frequencies – outside our inertial spectral subrange. The rate Q_0 of energy input from an external source, assumed to be known, equals the rate of energy transfer down the spectrum, hence the dissipation rate at high wavenumbers. Following the standard scaling procedure (e.g., Frisch et al., 1978; Larraza et al. 1990), Q is related to the characteristic time of nonlinear interactions between Fourier components t_j, and to the energy E_j, transferred from a cascade step j to step $(j+1)$:

$$Q = E_j / t_j. \tag{4.1}$$

To be more specific, one may introduce characteristic wavenumber scales k_j and k_{j+1} marking the (tentative) boundaries of individual cascade steps. The net transfer of energy from longer waves (with characteristic wavelength $2\pi/k_j$) to shorter waves (with characteristic wavelength $2\pi/k_{j+1}$) is similar to the production of smaller eddies by unstable larger eddies in the 3-d turbulence. Following this analogy, one can introduce a constant ratio r for the cascade process:

$$k_{j+1} = r k_j, \tag{4.2}$$

where $r > 1$. A specific value of r is not required for our subsequent development. Apparently, the energy transferred during time t_j, in a single step of the cascade, is related to the spectral density $F(k)$ of the wave energy by the formula

$$E_j = \int_{k_j}^{k_{j+1}} \int_{-\pi}^{\pi} G(k, \theta) k d\theta dk = \int_{k_j}^{k_{j+1}} F(k) k dk, \tag{4.3}$$

where $G(k, \Theta)$ is the 2-d "angular" spectral density and $F(k)$ is the 2-d energy spectrum after integration over all wave propagation directions Θ. The limited width $(k_{j+1} - k_j)$ of a cascade step allows one to introduce characteristic scales for all dynamical quantities at each step j.

Provided t_j can be expressed as

$$t_j = t_j(k_j, E_j), \tag{4.4}$$

equation (4.1) serves in place of the kinetic equation (3.13) to determine the energy at each step of the cascade:

$$E_j = E_j(Q, k_j). \tag{4.5}$$

The continuous spectrum $F(k)$ is then found by differentiating (4.3) over k_j and using (4.2):

$$\partial E_j / \partial k_j = F(k_j r) k_j r^2 - F(k_j) k_j .$$

If the spectrum falls off sufficiently fast with an increasing wavenumber, the first term in the r.h.s. becomes negligible compared to the second term, and one can write

$$F(k_j) \approx -\frac{1}{k_j}\frac{\partial E_j}{\partial k_j}. \tag{4.6}$$

For the special case of $F(k) \propto k^{-p}$, this approximation is valid if

$$r^{-p+2} << 1. \tag{4.7}$$

Otherwise, it can be checked *a posteriori*. Equation (4.7) replaces a more rigorous criterion (derived in weak turbulence theory) for the wave-wave interactions to be local (e.g., Zakharov et al., 1992).

In a weakly nonlinear case, the turnover time can be formally obtained by scaling the collision integral - e.g. (Phillips, 1985). However, we shall introduce this timescale in a less formal and more general fashion. To this end let us notice that the nonlinearity of most wave processes is measured by (3.11) where $c = \omega/k$. Respectively, the lowest-order nonlinear terms in deterministic equations of motion (for properly normalized and partially-time-averaged Fourier amplitudes $a(\boldsymbol{k}, t) \propto \epsilon$) are of order ϵ^2 and the subsequent terms are of order ϵ^3, ϵ^4, etc. However, since the kinetic equation is derived for second statistical moments (i.e., for the wave action spectral density, $N(\boldsymbol{k}) = F(\boldsymbol{k})/\omega$), the equations of statistical theory are developed in powers of ϵ^2: Each additional Fourier component accounted for in the interaction integral adds terms ϵ^2 times as great as the preceding term. Suppose, we could derive a general, closed-form equation for $N(k)$, for an arbitrary number of resonantly interacting wave components. Symbolically, this equation can be written as

$$\frac{\partial N(\boldsymbol{k};t)}{\partial t} + \nabla_k \cdot \boldsymbol{P}^{(3)} + \nabla_k \cdot \boldsymbol{P}^{(4)} + \nabla_k \cdot \boldsymbol{P}^{(5)} + ... = \gamma(\boldsymbol{k}) N(\boldsymbol{k}, t), \tag{4.8}$$

where "partial" collision integrals $\nabla_k \cdot \boldsymbol{P}^{(n)}$ account for n-wave interactions (of which resonant interactions are most important). $\gamma(k)N(\boldsymbol{k}, t)$ represents the external source (or sink, or both) where $\gamma(k)$ is an increment (decrement) of wave growth (attenuation). The collision integrals scale as $\epsilon^{2(n-1)}$. Weakly nonlinear waves (i.e., $\epsilon << 1$) permit neglecting all collision integrals except the first one. Indeed, if $\epsilon \approx 0.1$, the first interaction term in (4.8) is 10^2 times as large as the subsequent terms. However, this is not the case if the nonlinearity is stronger. For a weak inequality $\epsilon < 1$, we would have to retain a series of terms (up to $n \approx 6$ for the case of $\epsilon \approx 0.5$) in order to maintain the same accuracy as in our example with $\epsilon \approx 0.1$. The number of "effective" terms to be retained is thus a function of the degree of wave nonlinearity ϵ. When $\epsilon \longrightarrow 1$, interactions of all orders become of comparable importance. This case of strong wave turbulence (i.e., $n \longrightarrow \infty$) results in the "saturated" spectra. Larraza et al. (1990) showed that the Phillips spectrum $F(k) \sim k^{-4}$ for deep-water gravity waves is just one such example.

The appropriate characteristic time of resonant wave-wave interactions should be taken as the slowest among all individual turnover times associated with partial fluxes. This translates into the highest value of ν among all "effective" collision integrals. Therefore, the appropriate turnover time, found by a scaling of the terms in (4.8), is

$$t_j^{-1} \approx \omega \epsilon^{2(\nu-2)}. \tag{4.9}$$

Expressing ϵ in terms of the relevant parameters, namely k, ω and wave amplitude $a(k)$, or energy E_j at a given step of the cascade, equations (4.4)-(4.6) yield the spectrum of the wave energy. While it is intuitively clear that ν should be an increasing function of the external energy input, its determination is an open issue. Some empirical and semi-empirical models have been proposed (Glazman, 1992, 1993, 1995(a)), and one of them is employed in the next section.

4.1. Capillary-gravity wave turbulence. We will first consider capillary-gravity waves on deep water for which the ratio of water particle to wave phase velocity is

$$\epsilon = ak. \tag{4.10}$$

The equipartition between the kinetic and potential energies allows one to express the total wave energy as twice the potential energy:

$$\mathcal{E} = 2\mathcal{E}_p = \rho g \int \eta^2 d\boldsymbol{x} + 2\rho\sigma \int (\sqrt{1 + |\nabla\eta|^2} - 1) d\boldsymbol{x}. \tag{4.11}$$

Here, $\eta = \eta(\boldsymbol{x}, t)$ is the elevation of the fluid surface above the zero-mean level, g is the acceleration due to gravity, and σ is the coefficient of surface tension divided by fluid density ρ. Using the ensemble-averaged form of (4.11) we note that

$$\langle \eta^2 \rangle = \int_0^\infty F_\eta(k) k dk, \qquad \text{and} \qquad \langle (\nabla \eta^2) \rangle = \int_0^\infty k^2 F_\eta(k) k dk,$$

where $F_\eta(k)$ is the (2-dimensional) spectral density of surface height oscillations averaged over the polar angle Θ. Assuming $|\nabla\eta|^2 << 1$ (which is well justified under natural sea conditions), the surface density of the wave energy is given by

$$E = \rho g \int_0^\infty F_\eta(k)\left[1 + \frac{\sigma}{g}k^2\right] kdk. \tag{4.12}$$

Obviously, the spectra of wave energy and surface height are related by equality

$$F(k) = \rho g F_\eta(k)\left[1 + \frac{\sigma}{g}k^2\right]. \tag{4.13}$$

Replacing the semi-infinite integration range by a narrow spectral window one can express E_j in terms of the characteristic wave amplitude a_j for a given step in the cascade:

$$E_j = \rho g \int_{k_j}^{k_{j+1}} F_\eta(k)\left[1 + \frac{\sigma}{g}k^2\right] kdk \approx \rho g a_j^2 (1 + \frac{\sigma}{g}k_j^2). \tag{4.14}$$

This equation immediately yields the expression for $\epsilon = \epsilon(E_j, k_j)$ which can be used in (4.9). The dispersion law for CG waves is

$$\omega^2 = gk + \sigma k^3. \tag{4.15}$$

Let us introduce the non-dimensional wavenumber

$$K = kL, \tag{4.16}$$

where the intrinsic scale of the problem is

$$L = (\sigma/g)^{1/2}. \tag{4.17}$$

After a little algebra, eqs. (4.4)-(4.6) and (4.13) yield the power spectrum of surface height spatial variations:

$$F_\eta(k) = B \cdot K^{-(4\nu - 11/2)/(\nu-1)}\left(1 + K^2\right)^{-3/2(\nu-1)} \cdot \Lambda(K,\nu), \tag{4.18}$$

where

$$B = \alpha'(Q/\rho w^3)^{1/(\nu-1)}L^4,\ \alpha' = \alpha\frac{(2\nu - 7/2)}{(\nu - 1)},\ w = (\sigma g)^{1/4}, \tag{4.19a}$$

and

$$\Lambda(K,\nu) = 1 - \frac{2\nu - 5}{2\nu - 7/2}\frac{K^2}{1 + K^2}. \tag{4.19b}$$

The numerical constant α is a non-dimensional "Kolmogorov constant" of proportionality. In the limits of long and short waves, this spectrum yields Zakharov-Filonenko (1966,1967) power laws for gravity and capillary waves, respectively.

Using additional expressions for Q and ν as functions of wind speed, equations (4.18)-(4.19) describe the (rather complicated) shape of the CG wave spectrum and its dependence on external factors. Beside wind forcing, these factors include the magnitude of the spectrum at the lower-wavenumber boundary of the inertial subrange. By comparing (4.18)-(4.19) with experimental data, we find that a fixed value of ν, such as $\nu = 3$ or $\nu = 4$, leads to a drastic disagreement with observations (Glazman, 1995(a)). If, however, this effective number is allowed to increase with an increasing wind, the agreement becomes quite reasonable. Figure 1(a) illustrates the predicted spectrum in terms of the "saturation function" $B(k) = k^4 F\eta(k)$. We used $\nu = c_0 + c_1 U$ where c_0 and c_1 are empirical coefficients found earlier (Glazman, 1995(a)).

4.2. Inertia-gravity (IG) wave turbulence. Nonlinear shallow-water equations for a rotating fluid have the form

$$\begin{aligned} &\left(\tfrac{\partial}{\partial t} + \boldsymbol{U}\cdot\nabla\right)\boldsymbol{U} + f\boldsymbol{k}\times\boldsymbol{U} = -g\nabla\eta, \\ &\tfrac{\partial\eta}{\partial t} + \nabla\cdot((H+\eta)\boldsymbol{U}) = 0, \end{aligned} \tag{4.20}$$

where $\boldsymbol{U}$ is the horizontal velocity vector averaged over the layer depth H, and $\boldsymbol{k}$ is the unit vector along the Earth rotation axis. For simplicity, the Coriolis parameter is assumed to be constant (f-plane approximation) and the gravity force $\boldsymbol{g}$ parallel to $\boldsymbol{k}$. (In the rest of this paper, symbol $\boldsymbol{k}$ is employed for a different purpose: it designates the wavenumber vector. We hope this will not cause any confusion.) The nonlinear terms in (4.20) become especially important in the case of baroclinic waves. Therefore, we shall treat H as the thermocline depth and g as the reduced gravity - implying a 1.5 layer model in which the density of the upper layer is slightly lower than that of the (semi-infinite) lower layer. The amplitude $\eta(\boldsymbol{x}, t)$ of the density interface oscillation may constitute an appreciable fraction of the thermocline depth. The ocean surface plays only a passive role: its response to the oscillations of $\eta(x,t)$ is very weak and linear and with an opposite sign (e.g., Le Blond and Mysak, 1978). However, since the statistics of the density interface oscillations are identical (up to a constant of proportionality) to those of the SSH variations, we shall view the spectrum of $\eta(x,t)$ as the SSH spectrum.

The dispersion relationship of the corresponding linear theory is

$$\omega^2 = f^2 + C_0^2 k^2, \qquad \text{where} \qquad C_0 = \sqrt{gH}. \tag{4.21}$$

This equation forbids 3-wave resonance. Therefore, the lowest order resonance occurs in wave tetrads (Falkovich and Medvedev, 1992). The intrinsic scale of the problem (the Rossby radius of deformation) is

$$R = C_0/f. \tag{4.22}$$

For the 1st mode baroclinic waves, this number is typically between 30 and 80 km.

At high wavenumbers, equations (4.20) describe non-dispersive waves, while the low-wavenumber limit corresponds to inertial ("gyroscopic") waves. The kinetic and potential energies of IG waves (per unit volume, per unit mass of water) are

$$EK = \frac{\langle |\boldsymbol{U}|^2 \rangle}{2}, \quad EP = \frac{g\langle \eta^2 \rangle}{2H}. \tag{4.23}$$

For a narrow frequency band between k_j and k_{j+1}, these energies can be related to the characteristic scales of the wave amplitude a_j, and wave number k_j. The energy ratio increases with an increasing wavelength (e.g., [Gill, 1982]). The linearized theory yields

$$\frac{EK_j}{EP_j} \approx 1 + \frac{2}{(kR)^2}. \tag{4.24}$$

Physically, the absence of energy equi-partition is due to the fact that the orbits of water particles are not strictly vertical (as would be the case for pure gravity waves). Their inclination is greater, as the relative importance of the Coriolis force increases. Since the total energy is $E = EK + EP$, it is useful to express both components in terms of E_j:

$$EK_j \approx \frac{E_j}{2}\left(1 + \frac{1}{1+(kR)^2}\right), \tag{4.25}$$

$$EP_j \approx \frac{E_j}{2}\frac{(kR)^2}{1+(kR)^2}. \tag{4.26}$$

In view of (4.25), the characteristic particle velocity at scale k is given by

$$u^2(k) \approx E_j\left(1 + \frac{1}{1+(kR)^2}\right). \tag{4.27}$$

Based on (4.21), the characteristic phase speed is

$$c^2(k) \approx C_0^2\left(1 + \frac{1}{(kR)^2}\right). \tag{4.28}$$

Now, we can express the interaction time, t_j, in terms of $E_j, k, \omega(k)$ and C_0:

$$t_j^{-1} \approx C_0 R^{-1}(1+K^2)^{1/2}\left[\frac{E_j}{C_0^2}\left(1 - \frac{1}{(1+K^2)^2}\right)\right]^{\nu-2}, \tag{4.29}$$

$$\text{where} \quad K = kR \tag{4.30}$$

is the non-dimensional wavenumber. It is also convenient to non-dimensionalize other quantities:

$$\tilde{Q}_0 = Q_0(R/C_0^3), \quad \tilde{E}_j = E_j/C_0^2, \quad \tilde{F}(\tilde{k}) = F(k)/(C_0 R)^2. \tag{4.31}$$

With t_j^{-1} given by (4.29), equation (4.1) can be solved for E_j. In the non-dimensional form, the result is

$$\tilde{E}_j \approx \tilde{Q}_0^{1/(\nu-1)} z^{-1/2(\nu-1)} (1 - 1/z^2)^{-(\nu-2)/(\nu-1)} \tag{4.32}$$

where we introduced

$$z = 1 + K^2. \tag{4.33}$$

According to (4.21), this variable has a simple interpretation: $z = (\omega(k)/f)^2$. Equation (4.6) becomes

$$\tilde{F}(K) \approx -2 \frac{\partial \tilde{E}_j}{\partial z}\bigg|_{z=1+K^2}. \tag{4.34}$$

This yields the 2-d spectrum of the total wave energy:

$$\tilde{F}(k) = \alpha \frac{\tilde{Q}_0^{1/(\nu-1)}}{(\nu-1)} z^{(\nu-7/2)/(\nu-1)} (z^2-1)^{-(2\nu-3)/(\nu-1)} (z^2 + 4\nu - 9). \tag{4.35}$$

The ("Kolmogorov") constant α enables us to replace sign "$\approx$" with "$=$". The surface height spectrum (i.e., the potential energy spectrum) is found based on (4.26):

$$\tilde{F}_\eta(K) = \alpha \frac{\tilde{Q}_0^{1/(\nu-1)}}{2(\nu-1)} \frac{(z-1)(z^2+4\nu-9)}{(z^2-1)^{(2\nu-3)/(\nu-1)} z^{5/2(\nu-1)}}. \tag{4.36}$$

Since the angular dependence in our 2-d spectra is neglected, the corresponding 1-d spectrum is simply $\tilde{F}_\zeta(K)K$. The plot of this spectrum is shown in Figure 3 for several values of ν. In the high wavenumber limit, the spectrum behaves as K^{-s} where $s = 1+1/(\nu-1)$. Apparently, the regime of $\nu \to \infty$ corresponds to a surface which is discontinuous in the mean square. Physically, this means that short waves form bores ("shocks") and break, and the energy cascade becomes non-local. The main oceanographic implication of this internal wave breaking process is the production of small-scale turbulence in the ocean thermocline and increased vertical mixing. Being relevant for the direct energy cascade, (4.35) applies to wavenumbers above the generation range. At lower wavenumbers, one must consider the inverse cascade. For weakly non-linear waves (when $\nu = 4$), the kinetic equation admits an additional physically meaningful solution corresponding to the inverse cascade of wave action (Falkovich and Medvedev, 1992). Since it is the wave action that is conserved in the cascade, equation (4.1) is replaced by

$$P_0 = N_j/t_j, \tag{4.37}$$

where $N_j = \int_{k_n}^{k_{n+1}} F(k)\omega^{-1}kdk$. In the same fashion as before, we ultimately arrive at the formula

$$\tilde{F}(K) = \beta\frac{2\tilde{P}_0^{1/3}}{3}z^{1/3}(z^2-1)^{-5/3}. \tag{4.38}$$

The spectrum of surface height variations becomes

$$\tilde{F}_\eta(K) = \beta\frac{\tilde{P}_0^{1/3}}{3}z^{-2/3}(z^2-1)^{-5/3}(z-1). \tag{4.39}$$

The "Kolmogorov" constant β is different from that appearing in (4.35). Obviously, these two constants are related because (4.38) must merge with (4.35) at common wavenumbers. Figure 3 shows that the spectra merge quite smoothly (at $K \approx 1.2$) for all values of ν. In the long wave limit, (4.39) tends to $k^{-4/3}$ which describes purely inertial oscillations. Spectrum (4.39) is confirmed by satellite-altimeter observations (Wunsch and Stammer, 1995) on scales much greater than the Rossby radius of deformation.

Plausible sources of baroclinic wave energy include tidal forcing (when barotropic tides interact with topographic features of ocean basins), fluctuations of wind stress and atmospheric pressure, and various types of instability of ocean currents, eddies, etc. In view of the current poor quantitative knowledge of these sources, it is difficult to suggest any specific dependence of Q_0 and P_0 on external factors. Comparison of theoretical (4.36),(4.39) and observed spectra, Figures 2 and 3, points to a rather high degree of nonlinearity typical for "short" ($KR >> 1$) baroclinic waves.

5. Discussion. The heuristic approach described in section 4 enables one to analyze rather realistic wave processes occurring in oceans, atmosphere, and other nonlinear media. At the present time, the most urgent task is to gain a quantitative understanding of the effective number ν of interacting wave harmonics as a function of the external factors.

The present approach can be applied to other cases of wave turbulence, such as Rossby waves, deep-water internal waves, etc. The fact that spectra of scale-dependent wave turbulence exhibit pronounced breaks at certain wavenumbers which are related to the intrinsic scales of the media, and are also strongly dependent on the actual degree of the wave nonlinearity, points to a possible use of the theory for extracting deep-ocean parameters from observed spectra of sea surface height variations. For instance, SSH spectra based on satellite altimeter measurements contain information on the ocean stratification and deep-water processes in the thermocline because the baroclinic Rossby radius is a function of the density gradient

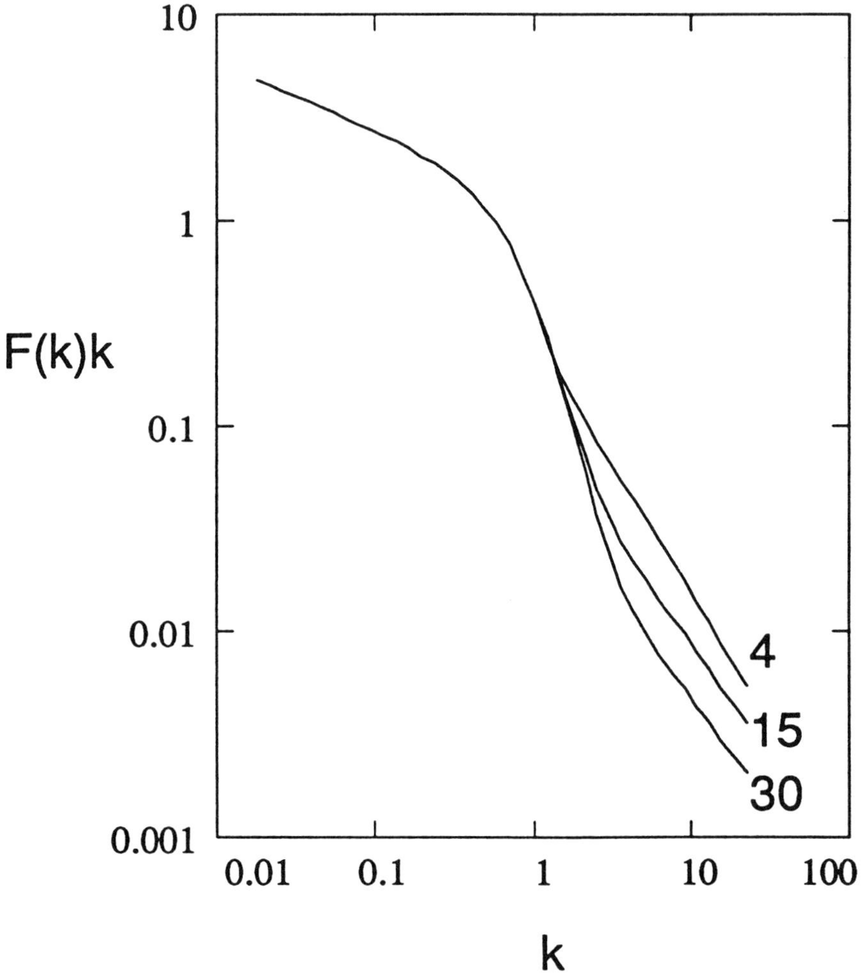

FIG. 3. *Non-dimensional spectra of surface height variations,* $K\tilde{F}_\eta(K)$. *Equation (4.39) is used for* $K \leq 1.2$. *Equation (4.36) is used for* $K \geq 1.2$, *for several values of the effective number of resonantly interacting Fourier harmonics,* ν, *as designated at the curves.*

while the degree of the wave nonlinearity controls intensity of internal wave breaking, hence vertical mixing in the thermocline.

Acknowledgments. This work was performed at the Jet Propulsion Laboratory, California Institute of Technology, under contract with the National Aeronautics and Space Administration.

REFERENCES

[1] Donelan, M.A., J. Hamilton, and W.H. Hui, *Directional spectra of wind- generated waves*, Phil. Trans. R. Soc. London, A 315, 509-562 (1985).

[2] Falkovich, G.E. and S.B. Medvedev, *Kolmogorov-like spectrum for turbulence of inertial-gravity waves*, Europhys. Let. 19(4), 279-284, (1992).

[3] Forristall, Z., *Measurements of a saturated range in ocean wave spectra*, J. Geophys. Res., 86, 8075-8084 (1981).

[4] Frisch, U., Sulem, P.-L. & Nelkin, M., *A simple dynamical model of intermittent fully developed turbulence*, J. Fluid Mech., 87, 719-736 (1978).

[5] Fu, L.-L., *On the wavenumber spectrum of oceanic mesoscale variability observed by the Seasat altimeter*, J. Geophys. Res., 88(C7), 4331-4341 (1983).

[6] Gaspar, P. and C. Wunsch, *Estimates from altimeter data of baroclinic Rossby waves in the Northwestern Atlantic Ocean*, Jour. Phys. Oceanogr., 19, 1821-1844 (1989).

[7] Gill, A.E., *Atmosphere-Ocean Dynamics*, Academic Press, New York, 662 pp. (1982).

[8] Glazman, R.E., *Multiwave interaction theory for wind-generated surface gravity waves*, Journ. Fluid Mech., vol. 243, 623-635 (1992).

[9] Glazman, R.E., *A cascade model of wave turbulence with applications to surface gravity and capillary waves*, Fractals, 1(3), 513-520 (1993).

[10] Glazman, R.E., *Surface gravity waves at equilibrium with a steady wind*, J. Geophys. Res. 99(C3), 5249-5262 (1994).

[11] Glazman, R.E., 1995(a). *A simple theory of capillary-gravity wave turbulence*, J. Fluid Mech., vol. 293, 25-34, (1995a).

[12] Glazman, R.E., *Spectra of baroclinic inertia-gravity wave turbulence*, Journ. Phys. Oceanogr., 26(7), 1256-1265, (1996).

[13] Gordon, A.L. and T.N. Baker, *Ocean transients as observed by Geos-3 coincident orbits*, J. Geophys. Res., 85(C1), 502-506, (1980).

[14] Grose, P.L., K.L. Warsh and M. Garstang, *Dispersion relations and wave shapes*, J. Geophys. Res., 3902-3906 (1972).

[15] Hara, T., E.J. Bock, and D. Lyzenga, *In situ measurements of capillary-gravity wave spectra using a scanning laser slope gauge and microwave radars*, J. Geophys. Res., 99(C6), 12,593-12,602 (1994).

[16] Hwang, P.A., Trizna, and J. Wu, *Spatial measurements of short wind waves using a scanning slope sensor*, Dyn. Atm. and Oceans 20, 1-23, (1993).

[17] Kahma, K.K., *A study of the growth of the wave spectrum with fetch*, J. Phys. Oceanogr., 11, 1503-1515 (1981).

[18] Kraichnan, R.H., *Inertial ranges in two-dimensional turbulence*, Phys. Fluids, 10(7), 1417-1423, (1967).

[19] Jähne, B. and K.S. Riemer, *Two-dimensional wave number spectra of small- scale water surface waves*, Journ. Geophys. Res., 95(C7), 11,531-11,546 (1990).

[20] Larraza, A., S.L. Garrett, and S. Putterman, *Dispersion relations for gravity waves in a deep fluid: Second sound in a stormy sea*, Phys. Rev. A 41(6), 3144-3155 (1990).

[21] LeBlond, P.H. and L.A. Mysak, *Waves In The Ocean*, Elsivier, New York, 602 pp. (1978).

[22] Le Traon, P.Y., M.C. Rouquet and Boissier, *Spatial scales of mesoscale variability*

in the north Atlantic as deduced from Geosat data, J. Geophys. Res., 95(C11), 20,267-20,285 (1990).

[23] Le Traon, P.Y., J. Stum, J. Dorandeu, and P. Gaspar, *Global statistical analysis of Topex and Poseidon data*, J. Geophys. Res., 99(C12), 24,619-24,631 (1994).

[24] Phillips, O.M., *The equilibrium range in the spectrum of wind-generated waves*, J. Fluid Mech., 4, 426-434 (1958).

[25] Phillips, O.M., *The Dynamics of the Upper Ocean*, Cambridge Univ. Press, New York, 336 pp. (1977).

[26] Phillips, O. M., *Spectral and statistical properties of the equilibrium range in wind-generated gravity waves*, J.Fluid Mech., 156, 505-531 (1985).

[27] Pierson,W.J. and L. Moskowitz, *A proposed spectral model for fully developed wind seas based on the similarity theory of S.A. Kitaigorodskii*, J. Geophys. Res., 69, 5181-5190 (1964).

[28] Pierson,W.J., *Comment*, J. Geophys. Res., 96(C3), 4973-4977 (1991).

[29] Toba, Y., *Local balance in the air-sea boundary processes, III, On the spectrum of wind waves*, J. Oceanogr. Soc. Jpn, 29, 209-220 (1973).

[30] Wunsch, C. and D. Stammer, *The global frequency-wavenumber spectrum of oceanic variability estimated from Topex/Poseidon altimeter measurements*, J. Geophys. Res. (Oceans). In press (1995).

[31] Zakharov, V.E., *Kolmogorov spectra in weak turbulence problems*, Handbook of Plasma Physics, Eds. M.N.Rosenbluth and R.Z. Sagdeev. Chapter 5.1. Elsevier Science Publishers, 4-36 (1984).

[32] Zakharov, V.E. and N.N. Filonenko, *The energy spectrum for stochastic oscillation of a fluid's surface*, Doklady Academii Nauk S.S.S.R., 170(6), 1292-1295, (in Russian) (1966).

[33] Zakharov, V.E. and N.N. Filonenko, *Weak turbulence of capillary waves*, J. Appl. Math. Tech. Physics, No. 4, 506-515 (1967).

[34] Zakharov, V.E. and M.M. Zaslavskii, *The kinetic equation and Kolmogorov spectra in the weak turbulence theory of wind waves*, Izvestiya, Atmospheric and Oceanic Physics (English Translation), 18(9), 747-753 (1982).

[35] Zakharov, V.E., V.S. L'vov and G. Falkovich, *Kolmogorov Spectra of Turbulence I: Wave Turbulence*, Springer Verlag, Berlin, 264 pp. (1992).

A SURVEY OF CASCADES WITH APPLICATIONS FROM GEOSCIENCES

VIJAY K. GUPTA* AND ED WAYMIRE†

Abstract. Our basic objective in this paper is to provide a mostly expository introduction to the salient mathematical features of cascades with some indication of their role in applications of interest in the geosciences. References to more comprehensive treatments are provided along the way.

Key words. Cascade, martingale, precipitation, turbulence, tree

AMS(MOS) subject classifications. primary 60J80,86A05; secondary 60J85, 60K05

1. Introduction and cascade preliminaries. In this paper we consider a class of mass or energy distributions on a compact space X represented by a class of Borel measures possessing a natural multiplicative structure, referred to as *cascade structure*. In subsequent sections we then describe some specific examples of both deterministic and random cascades of interest in the geosciences. To explain ideas the discussion will be for the case of cascades on the *d-cube* $X = [0,1]^d$, and mostly for $d = 1$. Our basic objective in this paper is to provide a mostly expository introduction to the salient mathematical features of cascades with some indication of their role in applications of interest in the geosciences; additional applications along these lines are also given in Gupta and Waymire (1995). References to more comprehensive treatments are provided along the way.

Cascade measures on the d-cube are a special class of mass distributions on $X = [0,1]^d$ defined recursively in terms of a natural hierarchical sequence of scales of resolution of the cube. For simplicity of the exposition let us first take $d = 1$ and consider the hierarchy of scales associated with the successive binary partitions of the unit interval $X = [0,1]$ at the nth scale given by the 2^n subintervals naturally coded-up as

$$\Delta(t_1, \ldots, t_n) = \left[\sum_{i=1}^{n} t_i 2^{-i}, \sum_{i=1}^{n} t_i 2^{-i} + 2^{-n}\right) \tag{1.1}$$

for $t_i \in \{0,1\}$. Ternary, quartary, and more generally b-ary partitions, for a parameter $b = 2,3,4,\ldots$, are obtained similarly and can be coded-up in terms of b-ary expansions $\sum_{i\geq 1} t_i b^{-i}, t_i \in \{0,1,\ldots,b-1\}$ of the partition points. Similarly one may consider a b-ary partition of the d-cube and code up the cells in the case $d \geq 2$. For simplicity, return

* Center for the Study of Earth From Space/CIRES, Department of Geological Sciences, University of Colorado, Boulder, Colorado 80309.

† Departments of Mathematics and Statistics, Oregon State University, Corvallis, OR 97331.

to $d = 1$ and the binary partition of $X = [0,1]$. The defining parameters of the model are a denumerable collection of non-negative weights $W_0, W_1, W_{00}, W_{01}, W_{10}, \ldots, W_{t_1 t_2 \ldots t_n}, \ldots$ on a probability space $(\Omega, \mathcal{F}, P)$, referred to as *cascade generators.* Given the generators one distributes mass at resolution n according to the measure defined by

$$\mu_n(A) = \int_A Q_n(x)dx \tag{1.2}$$

where the density $Q_n(x)$ is piecewise constant having having 2^n pieces of the multiplicative form

$$Q_n(x) = W_{t_1 \ldots t_n} W_{t_1 \ldots t_{n-1}} \cdots W_{t_1 t_2} W_{t_1} \tag{1.3}$$

for $x \in \Delta(t_1 t_2 \ldots t_n)$, respectively. Now the cascade measure is defined by the fine scale limiting measure $\mu_\infty = \lim_{n\to\infty} \mu_n$ when the limit a.s. exists in the sense of vague convergence. While such limits may exist, in general it is not the case that the limit measure will have a density, i.e. the limit measures are typically thinly supported singular measures. In particular for each x the density $Q_n(x)$ will typically die off to zero or explode to infinity as $n \to \infty$.

In order to obtain fine scale limits of the type described above one must impose certain *conservation* conditions on the cascade generators. A natural condition introduced by Kahane (1987) is that of *positive T-martingale.* In particular one requires that the sequence $\{Q_n\}$ of random non-negative Borel (jointly) measurable functions on $X \times \Omega$ is, for each fixed $x \in X$, a martingale sequence $\{Q_n(x)\}$ with respect to some adapted filtration $\mathcal{F}_n$. The T in "T-martingale" refers to the extrapolation of the usual martingale property to that of a process indexed by an underlying metric space T; here T = X. In this case one observes that for each non-negative bounded continuous function f on X, the sequence $\int_X f(x)Q_n(x)dx$ forms a non-negative martingale and therefore one obtains a limiting positive linear functional from the martingale convergence theorem. That this defines a cascade measure μ_∞ from the Riesz representation theorem and, in fact, such that a.s. $\int_X f(x)Q_n(x)dx \to \int_X f(x)\mu_\infty(dx)$ for all bounded Borel functions was established by Kahane (1987) as a consequence of a remarkable T-martingale decomposition theorem; see Waymire and Williams (1995a,1995b) for discussion and further applications of this decomposition.

The first systematic treatment of cascades of this type is that of Kahane and Peyriere (1976) in connection with a special class of models from statistical turbulence considered by Kolmogorov (1941, 1962), Yaglom (1966), and Mandelbrot (1974). For these models the generators W_γ are i.i.d. non-negative mean one random variables and the T-martingale property is easily verified. A natural extension to dependent cascades is that in which the generator sequences along paths are identically distributed and conditionally independent given the generators up to the last common ancestor.

This class may also be easily checked to be a T-martingale, however the methods used for the i.i.d. generators in Kahane and Peyriere (1976) are not applicable in this generality. A general approach for analyzing such dependent cascades is given in Waymire and Williams (1994,1995a,b), with a special focus on Markov and exchangeable dependence.

2. Examples. The *deterministic binomial cascade* on $X = [0,1]$ is obtained by a $(b = 2)$ binary partition with generators of the cascade density given by $W_0 = 2m_0, W_1 = 2m_1$ and subject to the mass conservation condition

$$m_0 + m_1 = 1. \tag{2.1}$$

In the definition of the mass density note that the length scale is $\frac{1}{2}$ at the $n = 1$ level of resolution. Mass is then redistributed multiplicatively in the same proportions so that the generators for the iteration are simply of the form:

$$W_{t_1 t_2 \cdots t_n} = 2m_{t_n}, t_i \in \{0,1\}. \tag{2.2}$$

That is, the piecewise constant density at the n-th resolution is given by

$$Q_n(x) = W_{t_1 \ldots t_n} W_{t_1 \ldots t_{n-1}} \cdots W_{t_1 t_2} W_{t_1} = 2^n m_{t_1} m_{t_2} \cdots m_{t_n}, x \in \Delta(t_1 t_2 \ldots t_n) \tag{2.3}$$

and the mass of the n-th level cell $\Delta(t_1 t_2 \ldots t_n)$ is

$$\mu_n(\Delta(t_1 t_2 \ldots t_n)) = m_{t_1} m_{t_2} \cdots m_{t_n}, \tag{2.4}$$

where each $m_{t_i} = m_0$ or m_1 depending on $t_i \in \{0,1\}$. In the case of this example one has the following rather special equation

$$\mu_\infty(\Delta(t_1 \ldots t_n)) = \mu_n(\Delta(t_1 \ldots t_n)). \tag{2.5}$$

The special nature of this equation will become more clear from the considerations below. As noted above observe that, unless $m_0 = m_1 = \frac{1}{2}$, either $Q_n(x) \to 0$ or $Q_n(x) \to \infty$ as $n \to \infty$ for each fixed x since there are only two parameter values m_0, m_1. However if $m_0 = m_1 = \frac{1}{2}$ then $Q_n(x) \equiv 1$ and the resulting measure is simply Lebesgue measure.

The more general multinomial cascade is obtained exactly in the same way using a b-ary partition and cascade parameters $m_0, \ldots, m_{b-1}$ for the generators subject to the conservation condition $m_0 + \cdots + m_{b-1} = 1$.

The randomized version along the following lines defines the *random multinomial cascade* . In particular let us again consider the binary partition with two parameters m_0, m_1 as above but now, at each stage we independently randomly select from these two values for a generator value with equal probabilities. That is

$$W_{t_1 t_2 \cdots t_n} = \begin{cases} 2m_0, & \text{with probability } \frac{1}{2} \\ 2m_1, & \text{with probability } \frac{1}{2}. \end{cases} \tag{2.6}$$

All else is the same as above, though now the values of the cascade measure are randomly varying. Note also that the martingale conservation condition implies a *conservation of averages.* That is, the expected values

$$EW_{t_1t_2\cdots t_n} = 2m_0 \times \frac{1}{2} + 2m_1 \times \frac{1}{2} = 1 \tag{2.7}$$

are conserved.

For our third example consider b-ary partitions of X = [0,1]. A simple but important choice for the generator distribution of a random cascade is that of independent and identically distributed generators distributed as

$$W = \begin{cases} b^{\beta} & \text{with probability } p = b^{-\beta} \\ 0 & \text{with probability } 1 - b^{-\beta}. \end{cases} \tag{2.8}$$

Note that "Y_n := number of non-zero values at the nth generation," $n = 1, 2, \ldots$, is a Bienayme-Galton-Watson branching process having Binomial (b,p) offspring distribution with expected number $bp = b^{1-\beta}$. Observing that

$$\mu_n(X) = Y_n \times (b^{\beta})^n \times b^{-n} = \frac{Y_n}{(b^{1-\beta})^n} \tag{2.9}$$

one sees from the martingale convergence theorem that the critical value of β for non-degeneracy is given by $\beta_c = 1$. In the case of this example one has

$$\mu_\infty(\Delta(t_1 \ldots t_n)) = Z_\infty(t_1 \ldots t_n)\mu_n(\Delta(t_1 \ldots t_n)) \tag{2.10}$$

where the prefactor $Z_\infty(t_1 \ldots t_n)$ is statistically independent of the nth level mass $\mu_n(\Delta(t_1 \ldots t_n))$ and is distributed as the total mass $\mu_\infty(X)$. Observe that the case of the deterministic cascade in (2.5) is made special by the fact that the corresponding prefactor is identically one. Other interesting critical phenomena associated with this particular cascade model arise in connection with the intermittancy structure of the limit mass. In particular Chayes, Chayes, Durrett (1978) have computed the critical probability at which nontrivial connected components of mass can occur in the 2-dimensional Beta model.

More generally, as noted in the introduction, one may generate non-negative values according to any independent and identically distributed choice of the generators, eg. logNormal, Gamma, logPoisson, etc., so long as a suitable average conservation condition is satisfied (else the limits will not exist). In fact, as also noted above, one may replace the i.i.d. generators by identically distributed discrete parameter stochastic processes along paths which are conditionally independent given the process up to the last common ancestor of the tree coordinates. An interesting and somewhat unexpected symmetry breaking critical phenomena occurs here which may be described as follows. Suppose that the generator processes are ergodic

Markov chains. Then one might expect that survival of this dependent cascade may coincide with that of i.i.d. generators distributed according to the ergodic invariant distribution, when in fact this turns out to be the case if and only if the Markov chain satisfies a certain time-reversibility condition; see Waymire and Williams (1995b,c) for details.

3. Fine scale structure and scaling exponents. In this section we will briefly consider the fine scale structure of random cascades. We summarize here two standard approaches to the computation of fine scale structure in terms of certain *multifractal* scaling exponents.

As noted earlier the typical behavior is for the cascade density to either die off or diverge in the fine scale limit. So while there is no value in trying to identify a limiting density, one may investigate the rate at which density dies off or explodes as a function of scale. For example, consider the deterministic binomial cascade discussed above. One has for the leftmost interval at the length scale resolution $l_n = 2^{-n}$

$$\mu_\infty[0, l_n) = \mu_\infty(\Delta(00\cdots 0)) = m_0^n = l_n^{\alpha_0} \tag{3.1}$$

and for the rightmost subinterval

$$\mu_\infty[1 - l_n, 1] = \mu_\infty(\Delta(11\cdots 1)) = m_1^n = l_n^{\alpha_1} \tag{3.2}$$

where

$$\alpha_i = -\log_2 m_i, i = 0, 1. \tag{3.3}$$

Similarly, if $\Delta(t_1 \ldots t_n), n = 1, 2, \ldots$ is a nested sequence of intervals shrinking to a point x then in the fine scale limit $l_n = 2^{-n} \to 0$ as $n \to \infty$

$$\mu_\infty(\Delta(t_1 \ldots t_n)) \sim l_n^{\alpha(x)} \tag{3.4}$$

in the sense that

$$\alpha(x) = \lim_{l_n \to 0} \frac{\log \mu_\infty(\Delta(t_1 \ldots t_n))}{\log l_n}. \tag{3.5}$$

The parameter $\alpha(x)$ is a *local Holder exponent* or *order of the singularity* at the location x.

Since the Holder exponents measure the local strengths of the singularities in the limit cascade, it is of interest to determine "how much" of the space is occupied by singularities of a given order α. A convenient measure of the "size" of the region of growth of order α is in terms of another scaling exponent $f(\alpha)$ giving the Hausdorff dimension of that region. This exponent function is often referred to as the *singularity spectrum.* In the case of the binomial cascade one has (eg see Falconer 1990, Evertsz and Mandelbrot 1992)

$$f(\alpha) = -\left\{\frac{A - \alpha}{A - a}\log_2\left(\frac{A - \alpha}{A - a}\right) + \frac{\alpha - a}{A - a}\log_2\left(\frac{\alpha - a}{A - a}\right)\right\}, \tag{3.6}$$

where A, a are the largest and smallest Holder exponents.

Another related computation of the fine scale structure of cascades may be made in terms of the behavior of a *cascade partition function* defined in analogy with statistical mechanics, or *cascade structure function* in analogy with statistical turbulence, as follows. For each $q \in \mathbf{R}$ define

$$S_{l_n}(q) = \sum_{(t_1,\ldots,t_n)} \mu_\infty^q(\Delta(t_1 \ldots t_n)) \tag{3.7}$$

at the scale of resolution $l_n = 2^{-n}$. In the case of the Binomial cascade one easily has

$$S_{l_n}(q) = (m_0^q + m_1^q)^n = l_n^{-\log_2(m_o^q + m_1^q)} = l_n^{\tau(q)} \tag{3.8}$$

where, in general,

$$\tau(q) = -\lim_{l_n \to 0} \frac{\log S_{l_n}(q)}{\log l_n}. \tag{3.9}$$

That is, for the Binomial cascade,

$$\tau(q) = \log_2(m_0^q + m_1^q). \tag{3.10}$$

In this case one may check that $f(\alpha)$ and $\tau(q)$ are related as *Legendre transform pairs.*

The problem of extending these computations to random cascades was considered in Holley and Waymire (1992) where it was shown that the corresponding relationships are valid under certain boundedness assumptions on the generators. While one may easily check these relationships to be valid for the expected values without any assumptions on the generators, and this calculation is routinely carried out in the literature, the extension to samplepathwise results is delicate. The importance of this problem for data analysis will be made clear in section 4.

As a last fine scale structure consideration we wish to describe a dimension disintegration formula. The *dimension spectrum* of $\mu \in M^+(T)$ is the Lebesgue-Stieltjes measure ν on $[0, \infty)$ defined by the distribution function

$$F(\infty) := \mu(T), \quad F(\alpha) := \sup\{\mu(B) : B \in \mathcal{B}(T), \dim B \leq \alpha\}, \quad 0 \leq \alpha \leq \infty. \tag{3.11}$$

THEOREM 3.1. *(Cutler, Kahane-Katznelson). Let $\mu \in M^+(T)$ with dimension spectrum ν. If ν is discrete then μ has a unique disintegration*

$$\mu(\cdot) = \int \mu_\alpha(\cdot)\nu(d\alpha),$$

where μ_α is unidimensional of dimension α. In general, the disintegration formula holds in the weak sense.

If the dimension spectrum of $0 \neq \mu \in M^+(T)$ is Dirac point mass at $\beta \in [0, \infty]$ then μ is said to be *unidimensional of dimension* β. In spite of a nontrivial multifractal singularity spectrum described above and obtained in Holley and Waymire (1992) for iid generators, it is also the case that the cascade is unidimensional in the sense just defined. So let us consider a simple example to see how a nontrivial disintegration may arise. The example we consider is a simple mixture of Beta models as follows. We start with a Markov chain on $\{0, b^{\beta_1}, b^{\beta_2}, 1\}$ with transition probabilities given by $q_{0,0} = 1, q_{b^{\beta_i}, b^{\beta_i}} = b^{-\beta_i}, q_{b^{\beta_i},0} = 1 - b^{-\beta_i}, q_{1,b^{\beta_i}} = p, i = 1, 2, q_{1,1} = 1 - 2p$, for $0 < \beta_i < 1, 0 < p < \frac{1}{2}$. Now construct a cascade whose generators along any path comprise a Markov chain with this transition law with $W_\emptyset = 1$ and such that along two distinct paths the Markov chains are conditionally independent given the generators up to the last common ancestor. In this case the Markov cascade μ_∞ has the spectral disintegration given by

$$\mu_\infty = \mu_\infty^{(1)} \nu(\{1 - \beta_1\}) + \mu_\infty^{(2)} \nu(\{1 - \beta_2\}), \tag{3.12}$$

where $\mu_\infty^{(i)}$ is an (iid generated) β_i−cascade and the dimension spectrum ν is a (harmonic) hitting probability for the Markov chain; see Waymire and Williams (1995b,c) for details.

Having provided some background to the rich mathematical structure of this class of random measures, let us now turn to some illustrations of their occurrence in natural phenomena of interest in the geosciences.

4. Examples from geosciences. In this section some naturally occuring cascade examples are described. No attempt is made to be complete. This is merely to acquaint the reader with some interesting examples from the geosciences and related references.

4.1. Statistical turbulence. As already noted, the notion of a random cascade has its origins in Kolmogorov's statistical theory of turbulence. The idea is that a large scale stirring motion will produce eddies which contain some random factor of the total energy, which in turn produce new eddies at lower scales ad infinitum, resulting in a redistribution of the large scale energy to lower scales. This view makes it possible to make statistical assumptions about the distributions of velocities and energies associated with the turbulent flow over a range of scales identified by Kolmogorov as the *inertial range*. While the Navier-Stokes equations are typically assumed in this theory, very little direct use is made of these equations in formulating the statistical model. Rather, one makes a self-similarity scaling assumption about the distribution of the velocity differences and/or energy dissipation rates and then derives the form of the exponents by a dimension analysis. More specifically, in this application $\epsilon(r) = \mu_\infty(\Delta_r(x))$ represents nth-level approximation to the energy dissipation rate per unit volume in a region of radius $r = b^{-n}$, say, at x at a fixed time. In Kolmogorov's classic 1941 paper this random field was assumed to be constant in r. However, a scaling assumption and dimension analysis on the velocity differences $u(x+r) - u(x)$ was used to write

$$E\{u(x+r) - u(x)\}^2 = (\text{const}) E\epsilon_r^{\frac{2}{3}} |r|^{\frac{2}{3}}, \tag{4.1}$$

which may be extrapolated to higher moments h (and equality in distribution) as

$$E\{u(x+r) - u(x)\}^h = (\text{const}) E\epsilon_r^{\frac{h}{3}} |r|^{\frac{h}{3}}. \tag{4.2}$$

In response to a remark of Landau on the intermittancy of turbulence, Kolmogorov (1962) modified his earlier assumption on the constancy of energy dissipation rates and allowed the energy dissipation to vary as well. As a result (4.2) was changed to

$$E\{u(x+r) - u(x)\}^h = \text{const.}\, |r|^{\xi_h} \tag{4.3}$$

and

$$E\epsilon_r^h = \text{const.}\, |r|^{\tau_h}. \tag{4.4}$$

In this way multiscaling exponents are introduced along with the relationship

$$\xi_h = \tau_{\frac{h}{3}} + \frac{h}{3}. \tag{4.5}$$

In Kolmogorov's 1941 paper the constancy of energy dissipation rendered $\xi_h = \frac{h}{3}$. However, in the 1962 modification it was assumed that the energy dissipation rate was logNormally distributed. This leads to quadratic in h for τ_h and, consequently, ξ_h. However, modern data analysis refutes these exponents and hence this form of the distribution; see Mandelbrot (1974), Benzi etal(1984). Recently some scaling properties were uncovered by She and Leveque (1994) from which a logPoisson distribution of the generators could be determined; the latter result was obtained by Dubrulle (1994), and independently by She and Waymire (1995). Further physical arguments have led to consideration of continuous cascades in place of discrete cascades described above.

4.2. Rainfall. During the 1950's and 1960's a central theme of rainfall research in hydrology was to fit parameters of various time series models to point rainfall measurements on time scales ranging from hourly, daily, monthly and yearly. Examples of time series models include mth order wet/dry Markov chain models, renewal sequences, moving average models; see Katz (1985), Waymire and Gupta (1981). As attention turned to more physically based approaches to take into account observed clustering in space and time, this theme was further explored into the following decades using compound Poisson and compound Neyman-Scott time series models; see Rodriguez-Iturbe (1986). The lesson learned was that up to second order, moment characteristics could be "reasonably well reproduced by a variety of models at a given scale". Better understanding of the temporal evolution would require the physical structure of storm events which is furnished by spatial observations.

The hierarchical structure of rainfall fields takes the form of clusters of high intensity rain cells embedded in clusters of lower intensity regions, called small mesoscale areas (SMSA), which are in turn embedded in rainbands of identifiably lower intensity, called large mesoscale areas (LMSA), embedded in a still larger scale synoptic rain area of lower rainrate. This structure is supported by radar and raingauge observations of the type analyzed by Austin and Houze (1972); see Gupta and Waymire (1993). While this structure is the supposed consequence of combined effects of vertical and horizontal motions, the precise dynamics of rainfall formation are not available. As a rule of thumb, the (possibly artificial) scales of these regions decrease by successive factors of $\frac{1}{10}$ from the synoptic scale through LMSA, SMSA, and down to a cell, while the corresponding rainrates nearly double at each level until the scale of a cell where this rule generally breaks down; supercells are possible where the rainrate may be larger than the SMSA by several orders of magnitude.

One of the earliest studies of the spatial and temporal variability of rainfall was that of LeCam (1961) based on spatial cluster point processes and random measures of the type also occuring in the study of the clustering of galaxies, earthquake aftershock sequences, population growth, etc. A substantial body of research has evolved over the past decade involving various problems associated with this approach to describing rainfall. Among these is a reasonably accurate computation of spatial/temporal correlation structure down to the scales of cells; see Zawadzki (1973), Waymire, Gupta, Rodriguez-Iturbe (1984), Rodriguez-Iturbe, Cox, Eagleson (1986), Phelan and Goodall (1990), Bell (1987), Smith and Karr (1985), Kedem and Chiu (1987). However, the lower scale high intensity regions have not been adequately represented in this approach.

Considerations of scaling properties of rain data have led to random cascade models of the type described in section 2; see Lovejoy and Shertzer (1990), Tessier, Lovejoy, Schertzer (1993), Gupta and Waymire (1990,1993). The basis for such models in the context of rainfall will be explained next.

One of the fascinating symmetries observed in nature is that of self-similarity. The simplest form of this symmetry in a statistical context is that of simple scaling of the probability distributions. A random field $\{R(x)\}$ is said to be simple scaling if

$$\{R(\lambda x)\} = \{\lambda^{\theta} R(x)\} \tag{4.6}$$

where equality is in the sense of joint distributions. The *scaling exponent* θ is a real number parameter. A self-similar theory was explored for rainfall by analogy to turbulence by Lovejoy and Mandelbrot (1985). Note that this was also the early hypothesis of Kolmogorov (1941) in his formulation of statistical turbulence theory as discussed at equation (4.2) above.

One approach to testing this hypothesis is to compute moments of order h as a function of scale λ. One gets the log-log relation

$$\log ER^h(\lambda x) = h\theta \log \lambda + c_h. \tag{4.7}$$

In particular simple scaling translates into two properties:
i. log-log linearity between a specified moment and length scale.
ii. a linear change in slope $s(h) = \theta h$ of the line as a function of moment order.
Familiar examples of simple scaling behavior include Gaussian white noise, fractional Brownian noise, increments of Brownian motion and Levy stable processes.

In the application of this test to spatial rainfall data one finds a very interesting result. Namely, property (i) is preserved but the slope function $s(h)$ in (ii) is nonlinear; see Gupta and Waymire (1990). This is also the situation in statistical turbulence theory as discussed in the previous section. Moreover, this is precisely the structure one may compute for a cascade distribution;see Holley and Waymire (1992). While both turbulence and rainfall share general features of extreme variability and intermittancy, the physics of rainfall is much more elusive by comparison to turbulence. The vindication of Kolmogorov's statistical theory of turbulence from the equations of fluid dynamics is a major unsolved mathematical problem, eg. see Constantin (1995), which at this stage has not even been properly formulated in the context of rain.

Let $\Delta_\lambda(i), i = 1, 2, \ldots$ denote a partition of a region X of d-dimensional space into cells at the length scale λ. Then one may easily compute

$$\log_b E\left[\sum_i \mu_\infty^h(\Delta_\lambda(i))\right] = -d\chi_b(h)\log(\lambda) + \log E\mu_\infty^h(X), \tag{4.8}$$

where

$$\chi_b(h) = \log_{b^d} EW^h - (h-1) \tag{4.9}$$

and, using the classical Cramer-Chernoff theorem for large deviations (eg. see Durrett, 1991),

$$\frac{\log_b Prob[\mu_\infty(\Delta_\lambda) > \lambda^{d\alpha}]}{\log \lambda} \to -d\chi_b^*(1-\alpha), \quad \lambda \to 0 \tag{4.10}$$

where

$$\chi_b^*(a) = \sup_h [ah - \chi_b(h)] \tag{4.11}$$

is the Legendre transform of $\chi_b(h)$. Note from (4.8) that the structure function exponent for random cascades is given by $-d\chi_b(h)$. The real problem from the point of view of applications is that in order to apply these formulae to sample realizations one needs to be able to "drop the expectations". However because of long-range spatial correlations the cascade fields are nonergodic and one cannot simply replace expected values by spatial averages. Since data typically consists of a single spatial sample realization, this problem is quite important. Nonetheless it was partially solved by Holley and Waymire(1992) for a large class of cascades; namely those with suitably bounded generators. In particular, the Beta model, for example, can be tested within this framework. In fact the Beta model provides an extremely simple yet qualitatively accurate representation of the rainfall cluster and intermittancy structure; see Over and Gupta (1994).

Another set of calculations which provide a role for cascades is that of the so-called *fractional wetted area.* Here one is interested in the probability $p(\lambda) = P(R(\lambda) > 0)$ as a function of scale λ. The general formula of the form (2.10) makes it possible to easily calculate this probability; eg. see Holley and Waymire (1991). In particular one obtains an exact log-log linearity of the form

$$\log_b p(\lambda) = c_1 \log \lambda + c_2$$

where $c_1 = -2\log_b(1 - P(W = 0))$. A discussion of this formula in the context of rainfall data can be found in Gupta and Waymire (1993) and in Over and Gupta (1994). The reader is invited to consider the same calculation for a d-dimensional white noise rainfield. It is interesting that one finds approximate log-log linearity in the limit as $\lambda \to 0$ but with a slope fixed at the spatial dimension d.

Acknowledgements. This research was partially supported by grants from NSF and NASA.

REFERENCES

[1] Austin, P. M., and R. A.Houze, *Analysis of the structure of precipitation patterns in New England* , J. Appl. Met., 11, 926-935 (1972).

[2] Bell, T. L., *A space-time stochastic model of rainfall for satellite remote sensing studies*, J. Geophys. Res., 92(D8), 9631-9643 (1987).

[3] Benzi, R., S. Ciliberto, C. Baudet, F.Massaioli, R.Tripiccione, and S. Succi, Phys. Rev. E. **48**, 29 (1993).

[4] Chayes, J, L. Chayes, and R. Durrett, *Connectivity of Mandelbrot's percolation process*, Prob. Theor. Related Fields, 77, 307-324 (1988).

[5] Constantin, P., *Some results and open problems regarding incompressible fluids*, Notices of the AMS, **43**(6),658-663 (1995).

[6] Dubrulle,B., *Intermittancy in fully developed turbulence: log Poisson statistics and generalized scale invariance*, Phy. Rev. Letters, 73(7), 959-962 (1994).

[7] Durrett, R., *Probability: Theory and Examples*, Wadsworth & Brooks/Cole, CA (1991).

[8] Evertsz, C.J.G., and B.B. Mandelbrot, In: H.O. Peitgen, H. Jergens,and D. Saupe, *Chaos and Fractals: New Frontiers of Science, Appendix B on Multifractal Measures*, Springer Verlag (1992).

[9] Falconer, K., *Fractal Geometry Mathematical Foundations and Applications*, John Wiley and Sons Inc., New York, New York (1990).

[10] Feder, J., *Fractals*, Plenum Press, New York, New York (1988).

[11] Feigenbaum, M., M.H. Jensen, and I. Procaccia, *Time ordering and thermodynamics of strange sets: Theory and experimental tests*, Phy. Rev. Letters, 57(13), 1503-1506 (1986).

[12] Gupta, V. K., and E. Waymire, *A Stochastic Kinematic Study of Subsynoptic Space-Time Rainfall*, Water Resour. Res., 15(3), 637-644 (1979).

[13] Gupta, V. K., and E. Waymire, *Multiscaling Properties of Spatial Rainfall and River Flow Distributions*, J. Geophys. Res., 95(D3), 1999-2009 (1990).

[14] Gupta, V. K., and E. Waymire, *A statistical analysis of mesoscale rainfall as a random cascade*, J. Appl. Meteorology, 12(2), 251-267 (1993).

[15] Holley, R., and E. Waymire, *Multifractal dimensions and scaling exponents for strongly bounded random cascades*, Annals Appl. Prob., 2(4), 819-845 (1992).

[16] Katz, R., *Probabilistic models* In: *Probability, Statistics, and Decision Making in Atmospheric Sciences*, A. Murphy, and R. Katz (Eds), Westview Press, Boulder CO (1985).

[17] Kahane, J. P., *Positive martingales and random measures*, Chinese Ann. Math., 8b, 1-12 (1987).

[18] Kedem, B. and L. Chiu, *Are rainrate processes self-similar?* Water Resour. Res. **23**, 1816-1818 (1987).

[19] Kolmogorov, A. N., *Local structure of turbulence in an incompressible liquid for very large Reynolds numbers*, Comptes Rendus (Doklady) de l'academie des sciences de l'URSS, 30, 301-305 (1941).

[20] Kolmogorov, A. N., *A refinement of previous hypotheses concerning the local structure of turbulence in a viscous inhomogeneous fluid at high Reynolds number*, J. Fluid Mech., 13, 82-85 (1962).

[21] LeCam, L., *A stochastic description of precipitation*, 4th Berkeley Symposium on Mathematical Statistics, and Probability, 3, Univ. of California, Berkeley, California, 165-186 (1961).

[22] Lovejoy, S., and B. B. Mandelbrot, *Fractal properties of rain and a fractal model*, Tellus, 37A, p. 209-232 (1985).

[23] Lovejoy, S., and D. Schertzer, *Generalized scale invariance in the atmosphere and fractal models of rain*. Water Resour. Res., 21(8), 1233-1250 (1985).

[24] Lovejoy, S., and D. Schertzer, *Multifractals, universality classes, and satellite and radar measurements of cloud and rain fields*, J. Geophys. Res.,95(D3), 2021-2031 (1990).

[25] Mandelbrot, B. B., *Intermittent turbulence in self-similar cascades: divergence of high moments and dimension of the carrier*, J. Fluid Mech., 62, 331-358 (1974).

[26] Mandelbrot, B. B., *The Fractal Geometry of Nature*, Freeman, San Francisco, California (1982).

[27] Over, T. and V. Gupta, *Statistical analysis of meso-scale rainfall: Dependence of a random cascade generator on large-scale forcing*, J. Appl. Meteorology, 33(12), 1526-1542 (1994).

[28] Phelan, M. J., and C. R. Goodall, *An assessment of a generalized Waymire-Gupta-Rodriquez-Iturbe model for GARP Atlantic Tropical Experimental Rainfall*, J. Geophys. Res., 95(D6), 7603-7615 (1990).

[29] Rodriguez- Iturbe, I., *Scale of fluctuation of rainfall models*, Water Resour. Res., 22, 155-375 (1986).

[30] Rodiguez-Iturbe, I., D. R. Cox, and P. S. Eagleson, *Spatial modeling of total storm rainfall*, Proc. Royal Soc. London, Ser. A. 403, 27-50 (1986).

[31] She, Z.S., and E. Leveque, *Universal scaling laws in fully developed turbulence*, Phy. Rev. Letters, 72, 336 (1994).

[32] She, Z. S., and E.C. Waymire, *Quantized energy cascade and log-Poisson statistics in fully developed turbulence*, Phy. Rev. Letters, 74(2), 262-265 (1995).

[33] Smith, J. A., and A. F. Karr, *Parameter estimation for a model of space-time rainfall*, Water Resour. Res., 21(8), p. 1251-1257 (1985).

[34] Tessier, Y., S. Lovejoy, and D. Schertzer, *Universal multifractals: Theory and observations for rain and clouds*, J. Appl. Meteorol., 32(2), 223-250 (1993).

[35] Waymire, E., *Scaling limits and self-similarity in precipitation fields*, Water Resour. Res., 21(8), 1271-1281 (1985).

[36] Waymire, E. C., V. K. Gupta, and I. Rodriguez-Iturbe, *A spectral theory of rainfall intensity at the meso-scale*, Water Resour. Res., 20(10), p. 1453-1465 (1984).

[37] Waymire, E. and S. Williams, *A general decomposition theory for random cascades*, Bull. Amer. Math. Soc., 31(2), 216-222 (1994).

[38] Waymire, E. and S. Williams, *Multiplicative cascades: Dimension spectra and dependence*, J. Fourier Analysis and Appl., Special issue in hornor of J-P Kahane, 589-609 (1995).

[39] Waymire, E. and S. Williams, *A cascade decomposition theory with applications to Markov and exchangeable cascades*, Trans. Amer. Math. Soc., 348(2), 585-632 (1995).

[40] Waymire, E. and S. Williams, *Markov cascades*, in *IMA Volume on Classical and Modern Branching Processes* ed. by K. Athreya, P. Jagers, Springer-Verlag, NY (1995c).

[41] Yaglom, A.M., *The influence of fluctuations in energy dissipation on the shape of turbulence characteristics in the inertial range*, Soviet Phys. Dolk. **2**, 26-29 (1966).

[42] Zawadzki, I. I., *Statistical properties of precipitation patterns*, J. Appl. Meteorol., 12, 459-472 (1973).

THE ROLE OF STATISTICAL MODELS IN TURBULENCE THEORY

J.R. HERRING*

1. Introduction. Stochastic models of turbulent flows replace velocity fields by random variables, whose covariances approximate those of the Navier-Stokes equations. The traditional approach is to assume the covariances are accurately given by some form of renormalized perturbation theory, (an *analytic turbulence theory*), such as the Direct Interaction Approximation (DIA) (Kraichnan (1959)). Thus the first step is to construct stochastic amplitude equations, whose covariances are exactly those of the analytic theory. Usually the model amplitude equations contain Markovian forcing terms, which simulate the nonlinear terms in the Eulerian form of Navier-Stokes. The forcing may be either white in time or red, with the lagged correlation time, prescribed by dynamical constraints. In either case, the effect of the random forcing is to destroy in the model any evidence of structures that Navier-Stokes may possess. In this sense, such models are "structureless", and their deviations from reality may give some indication of the effects of structures on covariance information. Included in such information are heat and momentum fluxes for wall-bounded flows.

From the historic perspective, stochastic models were developed *after* results of the analytic theories (such as the DIA) were known, at least in their broad outline. One reason for their development was to assure that a particular analytic theory was realizable: they were "proof by construction" that the covariance of an analytical theory is physically realizable, a point of some insecurity among theorists in the early days (1955–1964). Finally, it was realized that the stochastic models could be used to generate perturbation expansions in their own right, with the analytic theory serving as the unperturbed state (Phythian (1969), Kraichnan (1971), Herring and Kraichnan (1972)). Stochastic models may also be used as Monte Carlo simulations in complex flows, for which analytic theories may present a system too formidable for practical application. This approach has not been much used, but may be a computational advantage for certain flows, as has been found by Kaneda (1992).

In the next section, we outline the simplest of these models, and discuss one of their practical usages, that of estimating an eddy viscosity that accounts for the effects of scales of motion too small to fit into the computer. Those scales that fit into the computer are called *retained* scales, and those that do not, *sub-grid-scales* (SGS). The basic idea here is to contrive an eddy viscosity (or conductivity) which compensates for the extraction of energy by the absent modes in such a way as to maintain correct covariances

* N.C.A.R., Box 3000, Boulder CO, 80307.

for the retained modes. Such an eddy viscosity is scale-dependent. Then the Navier-Stokes with eddy-viscosity are used on the reduced wave number span in place of the full Navier Stokes. The basic ideas presented here go to the papers of Kraichnan (1976), and Chollet and Lesieur (1981).

2. Stochastic models and their relation to large eddy simulations. We write the Navier-Stokes equations in a compact form:

$$(\partial_t + \nu k^2)u_i = \sum C_{ij\ell} u_j u_\ell \tag{2.1}$$

Here, as in what follows, the indices (i, j, ℓ) are used collectively to represent the vector and wave number degrees of freedom of $u_\alpha(\boldsymbol{k})$, the αth component of the Fourier transform of $\boldsymbol{u}(\boldsymbol{x}, t)$. Thus the sum in (2.1) has the explicit form:

$$\sum_{j,\ell} = \sum_{\beta=1,\gamma=1}^{3} \int d\boldsymbol{p} d\boldsymbol{q} \delta(\boldsymbol{k} - \boldsymbol{p} - \boldsymbol{q}) \tag{2.2}$$

where the Greek letters $\alpha, \beta, \gamma (\subset (i, j, \ell))$ are reserved for vector indices. We further reserve $(\boldsymbol{k}, \boldsymbol{p}, \boldsymbol{q})$ for the corresponding wave number vectors $\subset (i, j, \ell)$. In this notation, the coefficients $C_{ij\ell}$

$$C_{ij\ell} = \sqrt{-1}\delta(\boldsymbol{k} - \boldsymbol{p} - \boldsymbol{q})(\delta_{\alpha\beta} - k_\alpha k_\beta / k^2)(q_\beta + p_\gamma) \tag{2.3}$$

correspond to Navier-Stokes. Specifically, we make the association $i = \boldsymbol{k}, \alpha, j = \boldsymbol{p}, \beta, k = \boldsymbol{q}, \gamma$. The stochastic models that generate Markovian turbulence theories have the form:

$$(\partial_t + \eta_i + \nu k^2)u_i = \sigma(t) \sum_{j,k} \tilde{C}_{ij\ell} \xi_j \xi_\ell \tag{2.4}$$

$$\tilde{C}_{ij\ell} = C_{ij\ell} \sqrt{\theta_{ij\ell}} \tag{2.5}$$

$$\eta_i = 2 \sum_{jk} \tilde{C}_{ij\ell} \tilde{C}_{jik} U_\ell \tag{2.6}$$

where $\langle \sigma(t)\sigma(t')\rangle = \delta(t - t')$, and the ξ are Gaussian random variables chosen (at every t) by $\langle \xi_i \xi_j \rangle \equiv \langle u_i u_j \rangle$. The η_i is an eddy viscosity, whose form guarantees quadratic conservation properties of $N - S$ in the ensemble mean. Finally, $\theta_{ij\ell}$ are triple-moment time-scales, whose values are determined by a comparison of the covariance equations of motion for (2.4), $(\partial_t U_i \equiv \partial_t \langle u_i^2 \rangle = \cdots)$, with that for a renormalization perturbation theory, such as the direct interaction approximation (DIA) (Kraichnan (1959), (1964)). Briefly, $\theta_{ij\ell} = 1/(\tilde{\eta}(k) + \tilde{\eta}(p) + \tilde{\eta}(q))$, where $\tilde{\eta}(k)$ is the Lagrangian decorrelation rate of the mode whose wave number is $\boldsymbol{k}$. This is estimated by $\sqrt{\int_0^k k^2 E(k)dk}$, where $E(k)$ is the energy spectrum ($= 2\pi k^2 U(k)$). From (2.4),

$$(\partial_t + 2\nu k^2 + 2\eta_i)U_i = 2\sum_{kj} C_{ij}^2 \theta_{ij\ell} U_j U_\ell \tag{2.7}$$

We will also write this equation for the evolution of the energy spectrum U_i in an alternate form:

$$(\partial_t + 2\nu k^2)U_i = 2\mathcal{T}_i \tag{2.8}$$

which defines the (modal) energy transfer function, $\mathcal{T}_i$. We also introduce the omnidirectional energy transfer function,

$$T(k,t) \equiv 2\pi k^2 \mathcal{T}(k,t) \tag{2.8$'$}$$

Both (2.1) and (2.4) inhabit a wave number span ($0 < k < \infty$). We now introduce an LES stochastic field, v_i, that inhabits

$$\mathcal{D} \equiv (0 < |\boldsymbol{k}| < k_c) \tag{2.9}$$

It satisfies an equation of identical form to (2.4), but with a different eddy viscosity, μ_i, chosen so that:

$$U_i \equiv \langle u_i^2 \rangle = V_i \equiv \langle v_i^2 \rangle, i \in \mathcal{D}. \tag{2.10}$$

$$(\partial_t + \nu k^2 + \mu_i)v_i = \sigma(t)\tilde{C}_{ij\ell}\xi_j\xi_\ell \tag{2.11}$$

This requires

$$\mu_i V_i = (\mathcal{T}_i^v - \mathcal{T}_i^u), i \in \mathcal{D} \tag{2.12}$$

where,

$$\mathcal{T}_i\{U\} = 2\sum_{j,\ell\in\mathcal{D}} C_{ij\ell}^2 \theta_{ij\ell} U_j U_\ell - 2\eta_i U_i \tag{2.13}$$

The superscripts in (2.12) identify the variances (U_i or V_i) that are used to compute $\mathcal{T}_i$. If we have a simple way of computing μ_i, we may compute on $\mathcal{D}$, using (2.11), a much less imposing system than (2.4). By so doing, we guarantee valid spectral information, provided the stochastic model (2.4) is able to represent spectra accurately. But a more important dynamical question is how long u_i and v_i will remain correlated, if their initial data are perfectly correlated. To that end we compute

$$W_i(t) \equiv \langle u_i v_i \rangle \tag{2.14}$$

From (2.4) and (2.11) the equation of motion for $W_i(t)$ is:

$$(2.15) \qquad (\partial_t + \eta_i + \mu_i)W_i = 2\sum_{j,\ell} \tilde{C}_{ij\ell} W_j W_\ell, \quad (i,j,\ell) \in \mathcal{D}$$

Where we recall $\tilde{C}_{i,j,\ell}$ from (2.5). From (2.12)., we may relate μ_i and η_i by

$$(2.16) \qquad \eta_i = \mu_i + \mathcal{R}_i,$$

where

$$(2.17) \qquad \mathcal{R}_i = 2\sum_{j',\ell'} C_{i,j'\ell'} C_{j'i\ell'} \theta_{ij'\ell'} U_{\ell'}, \quad (j',\ell') \not\subset \mathcal{D}$$

Note further that if we arbitrarily put $\mathcal{R} = 0$ in (2.16), then (2.14) would yield $W_i(t) = U_i(t)$, since U and W satisfy the same equations. Hence, $\mathcal{R}_i$ is the rate of decorrelation for each retained mode i (i.e. ($i \in \mathcal{D}$) from its true value, u_i produced by the the omission of modes $i \not\subset \mathcal{D}$ from the system. The latter are modes labeled (j',ℓ') in (2.17). The wave number domain of $(j,\ell) \notin \mathcal{D}$ is shown in Fig. 1 as the shaded region. From this figure, we may estimate the rate of decorrelation of the energy-containing modes (wave number k_0), for inertial range flows ($E(k) \sim \epsilon^{2/3}/(k_0+k)^{5/3}$), for the case in which the maximum wave number in $\mathcal{D}$ is k_c:

$$(2.18) \qquad \Omega(k_0) \sim (u_0 L)(k_0/k_c)^{4/3}$$

where $L \sim 1/k_0$. Fig. 2 shows $\mathcal{R}(k)/k^2$ for the inertial range $E(k)$ noted above. Here $\mathcal{R}$ is normalized by an energy scale eddy-viscosity, u_0^2/L, $u_0 = \sqrt{(2/3)\int_0^\infty E(k)dk}, L = \int_0^\infty E(k)dk/k/[3u_0^2] \sim k_0^{-1}$. As expected, the form of $\mathcal{R}(k)$ is that of an eddy viscosity, which remains quite small until k approaches k_c.

3. Large Reynolds number asymptotics. An examination of the energy transfer function ($T(k) \equiv 2\pi k^2 \mathcal{T}(k)$) (see (2.8)) at large Reynolds numbers reveals a useful bit of asymptotics. Fig. 3 shows the evolved, self-similar spectrum for $E(k)$, $T(k)$, and $\mu(k)$ (the Reynolds number, $R_\lambda \equiv u_0\lambda/v$, $\lambda^2 = (10/3)\int_0^\infty E(k)dk/\int_0^\infty k^2 dk E(k)$ is here ~ 1000). Fig. 3.b shows $T(k)$ for the untruncated system (solid line), together with two SGS calculations with $k_c = (5.0, \&10.0)$ (dashed and dotted lines respectively). Notice that both truncated SGS calculations agree well with the untruncated $T(k)$, over the negative lobe of $T(k)$. The positive lobe of $T(k)$ is very extensive in k; its magnitude at any k becomes progressively smaller as $R_\lambda \to \infty$. The value of k beyond which $T(k)$ is effectively negligible is the Kolmogorov wave number, $k_s \equiv (\epsilon/\nu^3)^{1/4}$, where $\epsilon \equiv 2\nu\int_0^\infty k^2 E(k)$. The area of positive and negative lobe balance, and since dissipation (the positive lobe) is well separated from inertial effects (the negative lobe), the area of either is ϵ. Thus the height of the positive lobe if $(\nu\epsilon)^{3/4}$. In

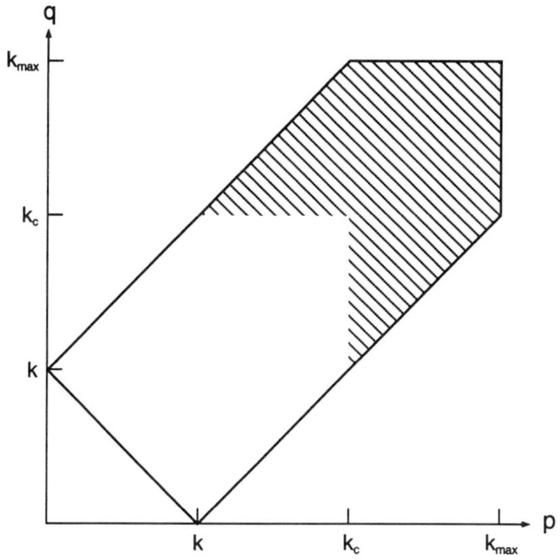

FIG. 1. *(p,q)-wave number domain corresponding to the sums in (2.17). Here, (p,q) are $|\boldsymbol{p}|$ and $|\boldsymbol{q}|$.*

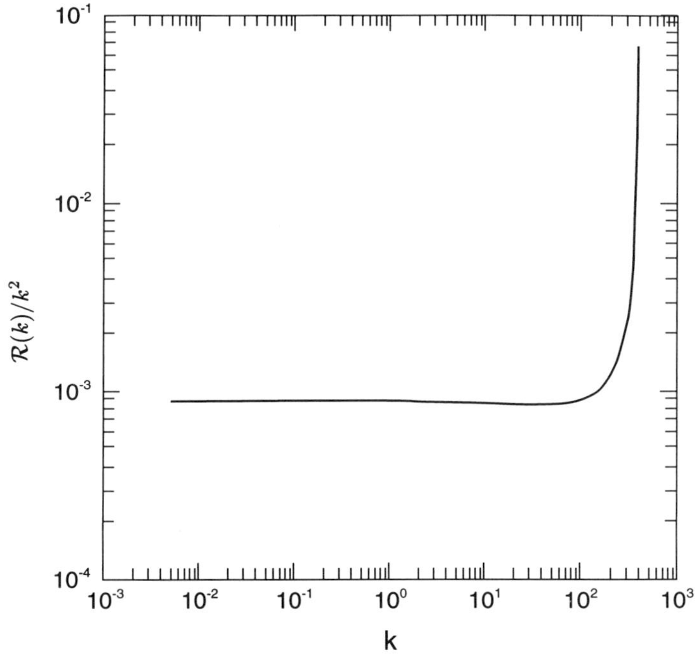

FIG. 2. *Eddy viscosity, $\mathcal{R}(k)/k^2$ according to (2.17) as a function of wave number k. Here $\mathcal{R}(k)$ is normalized as explained in the text, just after equation (2.18).*

this circumstance, if we place k_c near or just beyond the crossover wave number, k_{co}, where $T(k)$ changes sign,

$$T(k_{co}) = 0 \tag{3.1}$$

we are assured that large-scale covariance information is correct, together with its rate of change. Now consider a direct numerical simulation on $\mathcal{D} : (0 < k < k_c)$, in which we impose an eddy viscosity computed from (2.12). Our problem is that we do not know $\mathcal{T}^u$, the true energy transfer function without truncation of small scales. But if the statistical theory is a valid guide, we know $\mathcal{T}^u = \mathcal{T}^v$, in the energy-containing range and $\mathcal{T}^u \sim 0$ near the crossover wave number. This leads to the following rule for the eddy viscosity $\mu(k)$:

$$\mu(k) = 0, k \leq k_{co}$$

$$\mu(k) = T^v(k)/[2E(k)], k_{co} \leq k \leq k_c \tag{3.2}$$

Notice that, provided $k_{co}, < k_c, \mu(k)$ is computable from the DNS, and no recourse to the statistical theory now needs to be made. The results of this analysis leading to (3.2) are quite similar to ideas proposed by She and Jackson (1993) and Jiménez (1993).

4. Concluding comments. The usefulness of the stochastic models presented here depends on the validity of the underlying statistical theory which they approximate. The more elaborate of these theories (such as the DIA of Kraichnan or its Lagrangian generalization) have well-known difficulties, mainly associated with the development of structures in the real flow. As we noted earlier, the moment closure procedures such as the DIA ignore such structures. For three-dimensional, isotropic turbulence, a comparison of DNS to these theories suggests that structural effects on covariances seems to be most important in the range beyond the Kolmogorov wave number, k_s (Herring and Kerr (1993), Chen *et al.* (1993)). For two-dimensional turbulence, or flows constrained by rapid rotation, structural effects become important at much larger scales, and moment closures cannot be justified.

We have suggested in the last section that at large R_λ, a properly framed DNS (with eddy viscosity) may be employed to compute the large scales. If the computer used for DNS can contain all scales $< k_{co}$, such LES would have no need of empiricism. But k_{co} is a fraction of k_s, which $\rightarrow \infty$ as $R_\lambda \rightarrow \infty$. Nonetheless, a factor 10 in $|\boldsymbol{k}|$ translates into 10^3 in computational needs, so that the savings could be considerable.

Finally, we should mention that the idea of "bootstrapping" an LES, so that empirical coefficients are eliminated has become an active industry recently, and offers much promise for engineering computations (Germano *et al.* (1991). But for the statistical models discussed here, it is suggested

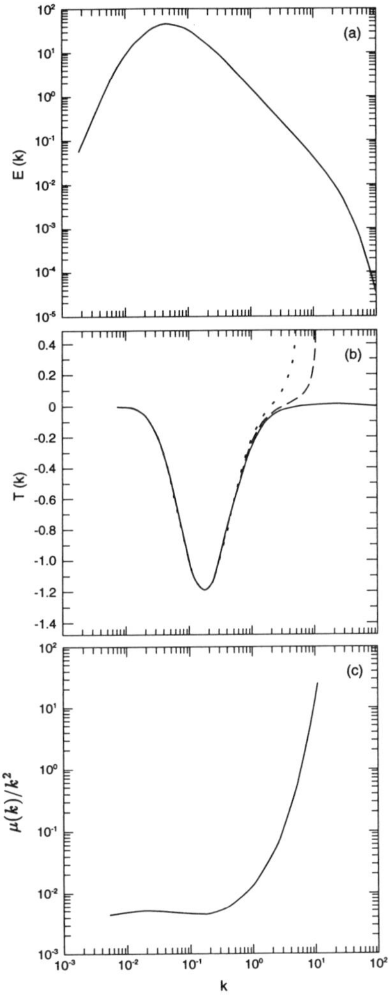

FIG. 3. *(a) Energy spectrum $E(k)$ for evolved flow with $R_\lambda \sim 1000$. (b) energy transfer spectra $T(k)$ (see (1.8′)) for full wave-number span (solid), cut-off wave number $k_c = 10$ (dashed line), and $k_c = 5$ (dotted line). (c) Eddy viscosity $\mu(k)/k^2$, from (2.12) (un-normalized).*

that a rather large wave number span (of order of a fraction of the Kolmogorov wave number) is needed to eliminate empiricism.

REFERENCES

[1] Chen, S., G. Doolan, J.R. Herring, R.H. Kraichnan, and Z.S. She, *Far-Dissipation Range of Turbulence*, Phys. Rev. Let., 70, pp. 3051–3054, (1993).

[2] Chollet, J.P. and M. Lesieur, *Parameterization of small scales of three-dimensional isotropic turbulence utilizing spectral closure*, J. Atmos. Sci, 38, pp. 2747–2757, (1981).

[3] Germano, M., M. U. Piomelli, P. Moin, and W.H. Cabot, *A dynamic subgrid-scale eddy viscosity model*, Phys. Fluids A, 3, pp. 1760–1765, (1991).

[4] Herring, J.R., and R.M. Kerr, *Small-Scale Structures in Turbulence: Their Implication for Turbulence Closures*, New Approaches and Concepts in Turbulence, Monte Verità, Th. Dracos and A. Tsinober Eds., Birkhäuser, Bale Publishers, pp. 367–376, (1993).

[5] Herring, J.R., and R.H. Kraichnan, *Comparison of some approximations for isotropic turbulence*, Lecture Notes in Physics, M. Rosenblatt and C. Van Atta, eds. Springer-Verlag, Berlin-Heidelberg-New York, 12, (1972).

[6] Jiménez, J., *Energy transfer and constrained simulations in isotropic turbulence*, Annual Research Briefs - 1993, Center for Turbulence Research, pp. 171–186 (1993).

[7] Kaneda, Y., *Application of Monte Carlo Method to the Lagrangian Renormalized Approximation*, Research Trends in Physics: Chaos and Transport in Fluids and Plasmas, W. Horton, Y. Ichikawa, I. Prigogine, and G. Zaslavsky, Eds., La Jolla International School of Physics, The Institute for Advanced Studies, (1992).

[8] Kraichnan, R.H., *The structure of isotropic turbulence at very high Reynolds numbers*, J. Fluid Mech., 5, pp. 497–543, (1959).

[9] Kraichnan, R.H., *Lagrangian-history closure approximation for turbulence*, Phys. Fluids, 8, pp. 575–598 (1964).

[10] Kraichnan, R.H., *Inertial-range transfer in two and three dimensional turbulence*, J. Fluid Mech., 47, pp. 525–535, (1971).

[11] Kraichnan, R.H., *Eddy viscosity in two- and three- dimensional turbulence*, J. Atmos. Sci., 33, pp. 1521–1536, (1976).

[12] Phythian, R., *Self-consistent perturbation series for stationary homogeneous turbulence*, J. Phys. A, 2, p. 181 (1969).

[13] She, Z.-S., and E. Jackson, *Constrained Euler System for Navier-Stokes Turbulence*, Phys. Rev. Let., 70, p. 1255, (1993).

OCEAN CIRCULATION: FLOW IN PROBABILITY UNDER STATISTICAL DYNAMICAL FORCING

GREG HOLLOWAY*

Abstract. Evolution of the ocean is considered from the view of evolving moments of probability distributions of possible oceans. The result is to anticipate statistical mechanical forcings that are qualitatively different than those considered by conventional ocean modeling. Simplified practical implementation of statistical mechanical forcings within context of conventional modeling shows significant impact upon model results with the suggestion of improvement in areas of chronic model deficiency.

1. The probable sea. Variability in the ocean spans an enormously wide range both in space and in time. From scales of 10^4 km (ocean basin size), substantial variability extends down to scales smaller than 1 cm. Time scales of oceanic variability include decades and longer while small scale turbulence, which ultimately shapes that oceanic response, involves time scales of seconds. Either in space or in time, the dynamic range spans something like 10^9. Neither can we say that any signficant part of these ranges is inconsequential or can be somehow parameterized. It is a daunting challenge, whether to observe the ocean or to construct numerical models in which we have confidence, e.g., to forecast climate change.

Progress invites a probabilistic formulation. We cannot hope to observe the ocean in all its detail nor to execute a numerical model in such detail. Perhaps we should never wish to. Conceptually, we imagine instead knowing the ocean only in probability. Let $\mathbf{Y}$ be a state vector that describes the ocean, for example a collection of velocities, temperatures, salinities, pressure, trace chemicals, biological parameters, etc., which might be known on some lattice of gridpoints throughout the world ocean. The number of elements in $\mathbf{Y}$ could be huge (zillions). In this enormously large (in dimensions) phase space, trajectory $\mathbf{Y}(t)$ might well describe some hideously complicated path, the tortuous details of which are of no practical interest. Instead we consider elemental probability $p(\mathbf{Y})dY$ that the state of the ocean may be found within neighborhood dY of any $\mathbf{Y}$. So far this hasn't helped. However, we can consider marginal $p'(\mathbf{Y}')$ on any subset of the dimensions that span $\mathbf{Y}$. Perhaps we retain only 10^3 or 10^6 or so dimensions, depending on how big may be our computer.

Still our task is too big for at least two reasons. First, the full joint *pdf* $p'(\mathbf{Y}')$ expresses more information than we can deal with, even admitting that $\mathbf{Y}'$ is enormously reduced in dimension from $\mathbf{Y}$. Second, we've no equation of evolution for $p'(\mathbf{Y}')$. Moreover, we appreciate that practically our interest is only for certain low order moments of $p(\mathbf{Y})$, i.e., $< \mathbf{Y} >$, $< \mathbf{YY} >$, ..., where $< . >$ denote expectation over $p(\mathbf{Y})$. $< \mathbf{Y} >$ is the mean state; $< \mathbf{YY} >$ includes quantities such as eddy energetics, heat

* Institute of Ocean Sciences, Sidney B.C. V8L 4B2, Canada.

transport, stress, etc. Although this may help, we still don't have equations for evolution of moments – since we don't have evolution of $p(\mathbf{Y})$. We have only the equation for evolution of $\mathbf{Y}$ in its full phase space.

What has happened, without having addressed the probabilistic formulation, is that we've attempted a "historical cheat". We've pretended that evolution of $< \mathbf{Y} >$ is just about like evolution of $\mathbf{Y}$ itself, on whatever subset of dimensions of phase space we've chosen to retain. Why should we get away with that? One way would be if the evolution equation for $\mathbf{Y}$ were linear. $< . >$ and operators in the $\mathbf{Y}$ equation would commute and we are done. But equations for evolution of $\mathbf{Y}$ are not linear, and we are obliged to make further assumptions. We may assume, e.g., $< \mathbf{YY} > = < \mathbf{Y} >< \mathbf{Y} >$. It's haphazard, of course. And it turned out in actual ocean models not to work. So we look around to fiddle other terms in the dynamic equation for $d\mathbf{Y}/dt$. Favorite fiddles have consisted of replacing coefficients of molecular viscosity and diffusivity with enormously larger coefficients (up to ten orders of magnitude larger), called 'eddy viscosities' (hoping that giving them a name may make them ok?) Practice shows that smoothing model fields with a lot of artificial viscous 'goo' does help assure that numerical integration can proceed. The unanswered question is how this result should bear upon our physically-motivated concern. Our practice is conspicuously haphazard!

Although careful solution for $< \mathbf{Y} >$, following $p(\mathbf{Y})$, may exceed our ability, we should not be surprised if terms arise in the dynamical equation for $< \mathbf{Y} >$ for which there are no analogous terms in the equation for $\mathbf{Y}$ itself. (Recall that eddy viscosity is just a rescaled analog of molecular viscosity, however the eddy viscosity coefficient may be assumed to vary in space, possibly depending upon $\mathbf{Y}$.) A result, which will appear below, is that forces occur in the $< \mathbf{Y} >$ equation that are qualitatively different from known forces in the $\mathbf{Y}$ equation. As well, direct tests in models give evidence that these hitherto-missing terms may have profound impact on model results.

2. Entropy. Progress toward approximate equations for $< \mathbf{Y} >$ is best understood in terms of the overall entropy $S = -\int p \log(p) dY$ where integration is over all the phase space. This quantity can present conceptual difficulties, especially given that $\mathbf{Y}$ is continuously variable (hence p is a probability density function) and that different components of $\mathbf{Y}$ will have different physical units (some may be temperatures, others velocities, etc.) Importantly though, what will matter are changes in S as $\mathbf{Y}$ evolves. Choices of different units only affect an added constant that does not depend upon the state of $\mathbf{Y}$.

Interest in S arises in part from the supposition that interactions among the myriad degrees of freedom present in $\mathbf{Y}$ exert a tendency toward increasing entropy: $dS/dt \geq 0$. However, when we restrict ourselves to only some subset of $\mathbf{Y}$, and when we attempt to substitute $d\mathbf{Y}/dt$ for $d <\mathbf{Y}> /dt$, we discard the tendency for $dS/dt \geq 0$. Small wonder this leads to gross model infidelities! In part we compensate the discard by means of enhanced 'eddy viscosities', as mentioned above. These reflect enhanced dissipation of resolved scales of flow, ultimately heating the fluid (increased S). Does this *ad hoc* approach reflect the way actual oceans generate entropy?

It turns out that this *ad hoc* approach (enhanced dissipation) is wrong in some absolutely fundamental ways. Moreover, it has been realized only in recent years that the resulting errors in practice are huge – like modelled ocean current systems which consistently run backwards compared with what is observed. Clearly we are motivated to rethink matters! (The good news is that some early guesses at how to improve things appear to be "working" – more or less.)

3. A simple ("thought") model. What causes the *ad hoc* approach to fail dramatically is the effect of bottom topography. To see this most easily consider first the simplified dynamics of the barotropic, quasigeostrophic vorticity equation on f-plane. (The basis for this approximate dynamics can be read in any geophysical fluid dynamics text, e.g. in Gill [8].) The equation of motion is:

$$(1) \qquad \partial \nabla^2 \Psi / \partial t + \partial(\Psi, \nabla^2 \Psi + h)/\partial(x, y) = \text{forcing} + \text{dissipation}$$

where the nearly-two-dimensional (horizontal) velocity $\mathbf{u} = (u, v)$ defines a streamfunction $\mathbf{z} \times \nabla \Psi$, where $\mathbf{z}$ is the vertical unit vector and ∇ is horizontal gradient. $\nabla^2 \Psi = \mathbf{z} \bullet \nabla \times \mathbf{u}$ is the fluid vorticity, $\partial(,)/\partial(x, y)$ is the Jacobian with respect to horizontal (Cartesian) coordinates (x, y), and $h = f(H_o - H(x, y))/H_o$ is a scaled representation of bottom topography for which $H(x, y)$ is the depth of fluid, H_o is a mean reference depth and f is the Coriolis parameter (assumed constant for present discussion). External forcing (e.g., applied torques) and internal dissipation (e.g., viscosity) are listed on the right side. This equation states the local temporal change of vorticity is due to advection of vorticity plus a topographic effect, as well as forcing and dissipation. Topography is influential due to the stretching or squashing of fluid columns as they move over varying topography h. (Validity of the equation assumes $|\nabla^2 \Psi| \ll f$ and $|h| \ll f$.) Although this equation is by no means sufficient to describe the dynamics of oceans, it is believed to capture an important part of the larger scale dynamics. (In (1) we are supposing that "larger scale" is in fact smaller than planetary radius. Thus, for example, we neglect variation of f with latitude.) Later we will come to the full equations of motion such as may be integrated in a large general circulation model (GCM).

Except under specially chosen circumstances, we can't solve (1) – due largely to nonlinearity in the Jacobian term. If we turn to brute force numerical integration, we are confounded by the continuous dependence of upon continuous coordinates x, y. For computation we are obliged to discretize (1), perhaps onto some finite difference or truncated spectral scheme. We reduce the continuously many degrees of freedom in (1) to some discrete number (which may be 10^6 or 10^7 or so, depending on one's computing resource). Again, just as in the actual planetary case, what can be the consequence of this discard? Can we, for example, "make up" by artificially increasing the dissipation on the retained modes? Unlike the actual planetary case, for equation (1) we can examine the latter question with some analytical skill.

Consider (1) with no forcing and no dissipation whatsoever. (This will be a thought experiment not meant to be realistic.) We suppose that some $\Psi(x, y)$ is given at $t = 0$ and inquire about its subsequent evolution. Suppose the flow is contained in some domain D with boundary ∂D. If ∂D is rigid and impermeable, then no flow can cross this boundary. Therefore Ψ is a constant on ∂D (supposing our domain is simply connected for this thought experiment). Expand $\Psi = \sum \Psi_n(t)\phi_n(x, y)$ on some set of orthogonal basis functions, for which convenient choices are solutions of $\nabla^2\phi_n + \kappa_n^2\phi_n = 0$. κ_n^{-1} defines a natural length scale for mode n. In principle the representation of involves infinite summation. We retain a finite number, N, of terms but may consider the limit $N \to \infty$. Now we suppose that the state vector $\mathbf{Y} = \{\Psi_n\}$ cannot be known precisely, just as in real oceans, and that our interest should be directed toward moments of probable states, thus $< \mathbf{Y} >$, $< \mathbf{YY} >$, etc. The thought question is this: If you have information about $< \mathbf{Y} >$ and $< \mathbf{YY} >$ at initial $t = 0$, what can you predict about $< \mathbf{Y} >$ and $< \mathbf{YY} >$ for subsequent t? An answer from classical statistical mechanics or, equivalently, from information theory is that for some large t one supposes the probable distribution that maximizes S subject to information available at $t = 0$ as preserved by the equation of motion (1). Integrals of the motion under (1), without forcing or dissipation, include circulation $C = \int dxdy\nabla^2\Psi$, energy $E = \int dxdy|\nabla\Psi|^2/2$ and enstrophy $Q = \int dxdy(\nabla^2\Psi + h)^2$ where $\int dxdy$ is over D. Suppose further that we have no phase-specific information at $t = 0$, hence $< \mathbf{Y} >= 0$ and our only knowledge is that there is some "random" distribution of eddies, characterized by some E and Q. At large t we expect the $p(\mathbf{Y})$ to be given by the variation

$$\delta \int d\mathbf{Y}(p\log(p) + \alpha Ep + \beta Qp + \gamma p) = 0 \tag{2}$$

where α, β and γ are Lagrange multipliers introduced to enforce the constraints to E, Q and the normalization $\int dYp = 1$. (Because we've supposed $< \mathbf{Y} >= 0$ at $t = 0$, we have $C = 0$ for present case.) Solution to (2) is the joint-normal distribution $p = \exp(-1 - \gamma - \alpha E - \beta Q)$, from which we may take moments. In particular,

$$(\alpha/\beta - \nabla^2) < \Psi >= h \tag{3}$$

where α and β depend upon E, Q, N and h in a complicated, implicit way.

The point here is to consider what a mistake we would have made by following the historical cheat, i.e. supposing we tried to substitute $d\mathbf{Y}/dt$

for $d < \mathbf{Y} > /dt$ and fudge the difference by introducing *ad hoc* dissipation. We would commence from $< \mathbf{Y} >= 0$ and solve $d < \mathbf{Y} > /dt =$ "dissip", for which the trivial result is $< \mathbf{Y} >= 0$ forever after – entirely contrary to (3) – except under such special choices of E, Q, N, h that would yield $\alpha/\beta = 0$.

4. Objections. Let us pause for some objections and comments. First, even given the idealizations in the foregoing discussion, evolution to (3) should assume some ergodic hypothesis. What if that's not so? Simply, if it is not so, then one might not evolve to (3). In absence of information to the contrary, I "guess" that evolution to (3) is overwhelmingly likely. Second, there may be other constraints that should be applied to the entropy maximization. In its continuous form, the left side of (1) conserves $\int dxdyF(\nabla^2\Psi + h)$ for any function F, thereby offering a continuous infinity of constraints. It seems though that when (1) is represented for any finite N, the continuous invariants are broken, leaving only C, E and Q. Doubtless this circumstance will be revisited as researchers devise different algorithms for representing (1); doubtless this may cause some revision to (3). However, for present purpose, the main issue is that (3) strongly indicates evolution away from $< \Psi >= 0$, just contrary to the historical cheat. Comments are that (3) is a subset of results obtained by Salmon et al. [14]. A corresponding result, replacing α/β by a parameter μ, was obtained by Bretherton and Haidvogel [2] on the argument that flows should dissipate enstrophy more effectively than energy, hence seek a minimum of enstrophy for given energy. These results are further discussed, including the limit of $N \to \infty$, by Carnevale and Frederiksen [4]. That maximum entropy / minimum enstrophy are complementary considerations (for dissipative flows) is exhibited in numerical simulations by Zou and Holloway [15].

Far a more severe objection is that the entire discussion leading to (3) has such an airy-fairy quality that it can be of no use with respect to any actual ocean or planetary fluid. Especially, the equilibrium statistical mechanical method, examining unforced, non-dissipative dynamics, considers a fluid system in closed isolation. Manifest reality is that oceans are open systems, subject to sun and wind and rain and tides and ... and subject to internal dissipation. Perhaps we are back to the historical cheat (business-as-usual)?

To carry statistical mechanics much beyond the isolated system consideration above has proven enormously tedious within context of large scale fluid dynamics. Including forcing, dissipation and time-dependent statistical evolution, one soon becomes embroiled in the unsolved turbulence problem. (That will not be reviewed here.) However, some small advances have been made which bear directly upon forcing non-zero $< \mathbf{Y} >$. Herring [9] and Holloway [10] each treated the time-dependent, dissipative version of (1) from view of turbulence theory. Holloway [11] later included also the role of large scale mean forcing and of latitudinal variation of Coriolis parameter. These results were complicated, though they clearly exhibit the systematic eddy-driving of non-zero $< \mathbf{Y} >$. Importantly, a work of Carnevale et al. [3] ties the approaches together by showing analytically that the turbulence theories used in these studies strictly support the eddy tendency for $dS/dt \geq 0$. What is limiting though is that the detailed calculations are so tedious, while depending upon their own idealizations (for example requiring perturbation quantities to satisfy periodic boundary conditions), without (to date) yielding simple formulae of practical use.

Back to historical cheat? Not yet. Because a system is not near the unforced, non-dissipative equilbrium, this does not mean that statistical mechanics no longer applies. Entropy S remains well-defined, although we may not know enough about $p(\mathbf{Y})$ to calculate S. Conceptually, at least, one recognizes that "conjugate forces" arise, driving $d < \mathbf{Y} > /dt$ with forces given by $\mathbf{L} \bullet \nabla_{\mathbf{Y}} S$ where $\nabla_{\mathbf{Y}} S$ denotes the gradient of S with respect to components of $\mathbf{Y}$, "as if" any component of $\mathbf{Y}$ might be fixed (as a constraint) and we could ask how S changes if we vary the value assigned to this component. $\mathbf{L}$ is a symmetric matrix of "Onsager coefficients". Sufficiently near unforced, non-dissipative equilbrium, say $\mathbf{Y}_{eq}$, values of $\mathbf{L}$ would depend only upon $\mathbf{Y}_{eq}$ while the conjugate forces would be proportional to $\mathbf{Y}_{eq} - \mathbf{Y}$. A discussion of this subject, combining the statistical mechanical and information theoretical viewpoints, can be read in Katz [13].

For actual oceans, $\mathbf{Y}$ is not near $\mathbf{Y}_{eq}$, so we've no basis to expect $\mathbf{L}$ to be independent of $\mathbf{Y}$ or the conjugate forces to be proportional to $\mathbf{Y}_{eq} - \mathbf{Y}$. We admit the problem of evaluating $\nabla_{\mathbf{Y}} S$ for oceans is unsolved. However, the practical issue of understanding (modeling) oceans is urgent. We don't have a choice between "right" and "wrong" but between "wrong" and "wronger", with default (business-as-usual) being the historical cheat: effectively assuming conjugate forces $-\mathbf{L} \bullet \mathbf{Y}$, where $\mathbf{L}$ expresses various *ad hoc* eddy viscosities, diffusivities and whatnot. The eddy viscous assumption drags $\mathbf{Y}$ toward a state of rest, $\mathbf{Y} = 0$. Yet we have seen above that $\mathbf{Y} = 0$ is not a higher entropy state whereas nonzero $\mathbf{Y}_{eq}$ is. We are obliged to make a plausible guess whether $\nabla_{\mathbf{Y}} S$ points more toward $\mathbf{Y} = 0$ or toward $\mathbf{Y}_{eq}$. Only recently have researchers begun exploring consequences of the latter guess, i.e. running ocean models with conjugate forces given by $\mathbf{L} \bullet (\mathbf{Y}_{eq} - \mathbf{Y})$. Here we summarize briefly some experiences under the latter guess.

Practical issues are that we must assign both $\mathbf{L}$ and $\mathbf{Y}_{eq}$. Consider first $\mathbf{Y}_{eq}$. The simple case following equation (1) assumed barotropic, quasigeostrophic flow. Model equations for actual oceans are far more realistic. What shall we assign for $\mathbf{Y}_{eq}$ which is consistent with (3) in the barotropic, quasigeostrophic limit and plausibly appropriate for realistic (three-dimensional, stratified, finite depth variation, ...) ocean model?

5. Simplification for practical implementation. In part limitations are not so severe as they may seem. Even in the early work of Salmon et al. [14], effects of stratification were considered within context of quasigeostrophic flow. A result was that for horizontal scales larger than first internal deformation radius (R_1=ratio of horizontal gravity wave speed to Coriolis parameter), absolute equilibrium flow tends to be barotropic. For realistic oceans, R_1 ranges from tens of km in open ocean, tropical regions down to only a few km at high latitudes and in marginal seas. For large scale issues such as global climate, efficient computer modeling tends toward coarser resolution (especially when integration periods may extend over centuries to millenia). Even for regional ocean models, it often is the case that R_1 is only marginally resolved, if at all. Thus, adopting a barotropic $\mathbf{Y}_{eq}$ after (3) may be not so bad. Of course the modeled $\mathbf{Y}$ will be baroclinic in response to imposed forcing.

We can simplify (3) somewhat. Define a length scale $\lambda = \sqrt{\beta/\alpha}$. We do not know a priori how large λ is (or even if it is real!) Supposing though that λ is real and moreover (to be checked post facto) that λ is smaller than the resolved length scale, we can omit ∇^2 in (3), obtaining simply $< \Psi >= \lambda^2 h$. Recall that $h = f(H_o - H(x,y))/H_o$ and we supposed $|H_o - H(x,y)| \ll H_o$ whereas in fact we have $|H_o - H(x,y)|/H_o$ of order unity. How to extend the quasigeostrophic result from (3)? We take advantage of an ambiguity of quasigeostrophy for which "streamfunction" does not distinguish velocity streamfunction $\mathbf{u} = \mathbf{z} \times \nabla\Psi$ from transport streamfunction $H\mathbf{u} = \mathbf{z} \times \nabla\Psi$, with transport streamfunction given by H_o times velocity streamfunction. For use in a three-dimensional model permitting horizontally divergent flow subject to depth-integrated volume conservation, we will want the transport streamfunction, hence $< \Psi >= \lambda^2 f(H_o - H(x,y))$. Appreciating that horizontal length scales for variation of H are much shorter than the scale of variation of f (the planetary radius), we see that the spatially varying part of $< \Psi >$ is dominated by $-\lambda^2 fH$, which we designate as Ψ^* (given that this is not in fact an "expected" flow and also recognizing the many "approximations" taken). Thus we have arrived at our assignment for $\mathbf{Y}_{eq}$. It is the velocity field $\mathbf{u}^*$ given by

$$(4) \qquad H\mathbf{u}^* = \mathbf{z} \times \nabla\Psi^* \quad \text{where} \quad \Psi^* = -\lambda^2 fH$$

$\mathbf{u}^*$ is independent of depth. To complete the assignment for $\mathbf{Y}_{eq}$ we may also need such fields as temperature, salinity, etc. Because $\mathbf{u}^*$ is independent of depth, geostrophy implies that $\mathbf{Y}_{eq}$ is characterized by horizontally uniform density for which we assume horizontally uniform temperature and salinity in $\mathbf{Y}_{eq}$.

What remains is to assign is λ^2. This is tough. Although λ^2 can be evaluated from the thought exercise leading to (3), for which depends upon E, Q, h and N (including $N \to \infty$), it is doubtful how well this applies. Here understanding is most inadequate. For the present, with regret, we consider λ a "fudge factor" which is thought "roughly" to reflect a length scale somewhat shorter than a dominant eddy vorticity length scale. This is thought to be somewhat smaller than R_1. We suppose that λ should not vary strongly spatially, as indeed we've assumed when simplifying the expression (4) for Ψ^*. In practice we seem to find "agreeable" values of ranging from a few km to a couple of tens of km, with spatial variation broadly reflecting large scale variation of R_1. It must be clear though that this point is dreadfully clumsy and desperately wants to be improved upon.

Having suffered the clumsiness of λ, we move to the next (but mercifully the last) clumsiness: **L**. Here we are clueless. We are so far from absolute equilibrium that we cannot hope to relate **L** to fluctuations near $\mathbf{Y}_{eq}$. An approach that should help to clarify **L** as well as λ is from the kind of turbulence calculations mentioned earlier. Only this has not been addressed to date. The other, more straightforward, approach is by brute force numerical simulation employing such high resolution that tendencies to drive nonzero $< \Psi >$ will be realized explicitly. This approach also remains to be explored. For the present, without further guidance, the prudent course is to proceed as simply as possible. Simplest of all is to suppose **L** is a single inverse time scale $1/\tau$. All spatial scales of motion will then relax from **Y** toward $\mathbf{Y}_{eq}$ on the same characteristic timescale τ. Arguing plausibly that shorter spatial scales should exhibit quicker response times, one might (for example) let **L** represent a Laplacian diffusion operator $A\nabla^2$. It should be possible for turbulence theory and/or high resolution numerical experiments to refine the characterization of **L**. Meanwhile, in absence of further insight, one proceeds simply with practicality in mind – all the while awaiting refinements that may come from future research.

Given the clumsy uncertainty just described, why proceed with this at all? The answer is, again, that the choice is between "wrong" and "wronger". If we are to proceed with ocean modeling, choices must be made despite uncertainty. In fact there have been a number of recent efforts to explore consequences of suggestions outlined above. We have taken business-as-usual ("real world") ocean models and simply re- directed their hypothesized eddy tendency terms. Often models employ a horizontal eddy viscosity of form $A\nabla^2\mathbf{u}$. Recognizing that this term drives toward the lower entropy state of rest ($\mathbf{u} = 0$), we "redirect" the eddy viscosity toward higher entropy as $A\nabla^2(\mathbf{u}-\mathbf{u}^*)$ with $\mathbf{u}^*$ given by (4). [A comparison between this form and results using $(\mathbf{u} - \mathbf{u}^*)/\tau$ showed little qualitative difference, as reported by Eby and Holloway [5].]

6. Practical exercises. There have been a number of practical exercises, over only the last year or so, bringing these statistical mechanical ideas into conventional ocean modeling. The earliest study (Eby and Holloway, [5]) has been expanded and elaborated by Eby and Holloway [6], using **L** in the form $A\nabla^2$ and taking $\mathbf{u}^*$ from (4) while considering global ocean circulation under climatological mean forcing. A regional study of circulation of the western Mediterranean, using $A\nabla^2$ and $\mathbf{u}^*$ from (4), is reported by Alvarez et al. [1]. At smaller scales (estuarine), Fyfe and Marinone [7] investigated residual circulation in the Strait of Georgia, British Columbia. Studies as yet unpublished include a Japan Sea study by T. Sou, a North Atlantic study by R. Gerdes, and global studies by P. Duffy and by S. Power.

These practical exercises have been performed in the manner of sensitivity studies, comparing unmodified models with models including statistical mechanical forcing. Since the models employed have already an eddy viscous operator $A\nabla^2$, the modification is only to include $\mathbf{u}^*$ from (4). Choosing $\lambda = 0$ causes $\mathbf{u}^* = 0$, whereas we believe a plausible choice for λ should be some few to few tens of km. Studies to date have focussed upon sensitivity of model results to choice of λ.

We ask two questions. First, does this stuff matter? It may be that quasi-linear dynamics of larger scales are dominant, and considerations discussed above are of little practical consequence. Second, if this stuff matters, can one see evidence of practical progress toward improved ocean models?

The answer to the first question depends, in part, on how large one chooses λ (which enters as λ^2). However, results showed that λ ranging from 3 km to 15 km have quite powerful influences. Figure 1, from Tessa Sou's study of the Japan Sea (unpubl.) shows annual mean circulation at 500 m depth in two cases: λ=0 and λ=4 km. Here the sense of the circulation is reversed. Similar results were found by Alvarez et al. [1] in their Mediterranean study. At global scale, Eby and Holloway [6] show examples of currents which are absent or reversed but which appear under statistical mechanical forcing.

The second question is more difficult because it requires us to compare model output with the poorly known ocean itself. When we sometimes get agreement between some model result and some partial observation, we cannot say if the model is "good" or if we just got lucky with a model that may be quite poor, under poorly known forcing, compared with inadequately observed ocean, yielding fortuitious "agreement". Sorting out the methodology of model validation under these circumstances is an on-going, challenging problem. Let me here, in closing, mention only qualitatively some of the ways we seem to be "improving" the models.

An equatorward tendency develops along the western margins of basins, enhancing (sometimes creating!) undercurrents opposed to surface flows. In the North Atlantic, the result is to carry water from the Labrador and Greenland-Iceland Seas down along the western margin, underthrusting a warm Gulf Stream. This is a process that seems crucial to global climate concerns yet has been chronically under-represented by ocean models. In the western Pacific, deep equatorward flow occurs along the Japan margin despite the lack of high latitude source for this deep flow. In these cases of western boundary regimes, the equatorward (statistical mechanical) tendency is opposed to the externally forced surface flow. Competition between these two tendencies (absent under business-as-usual!) helps correct another chronic problem – that of modeled western boundary currents travelling to too high latitude before turning eastward.

Statistical mechanical forcing can be seen quite clearly along eastern

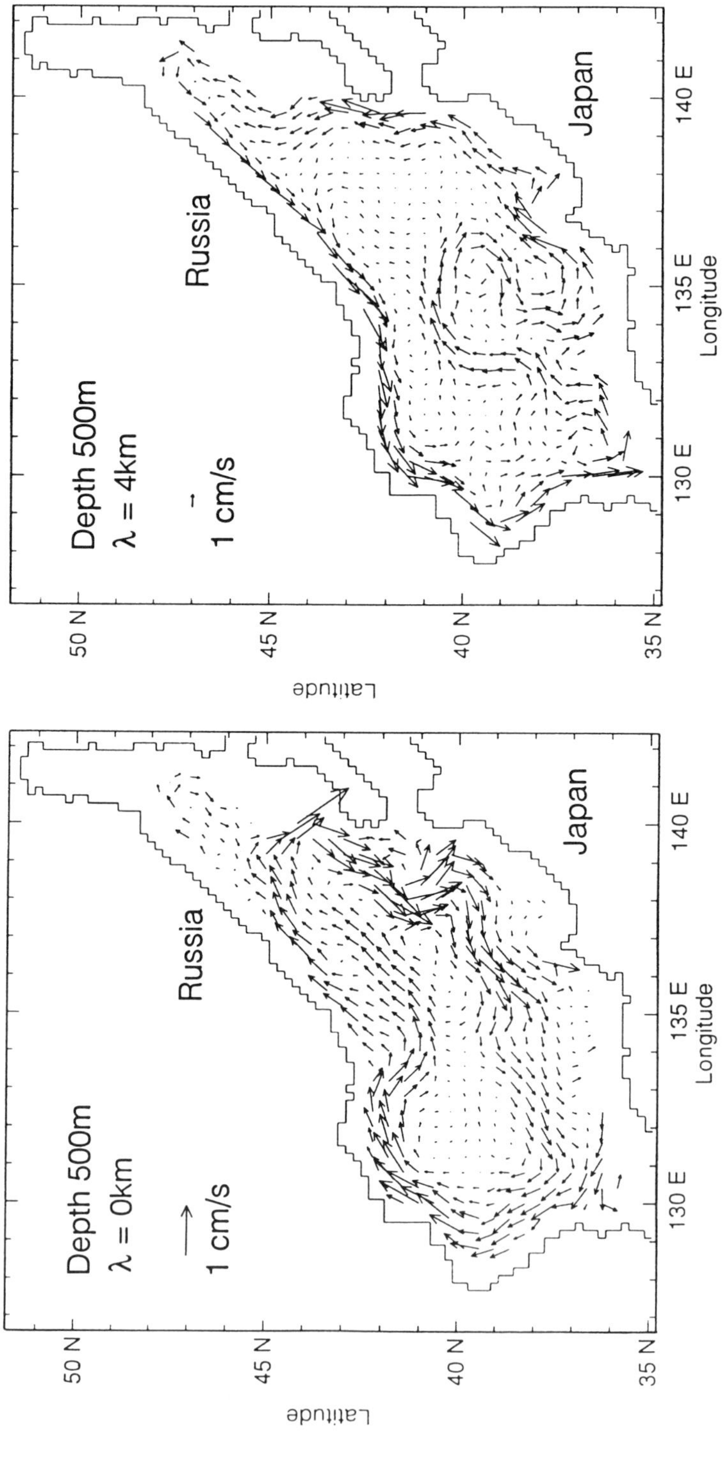
Depth 500m
λ = 4km
1 cm/s
Russia
Japan
50 N
45 N
40 N
35 N
Latitude
130 E
135 E
140 E
Longitude
Depth 500m
λ = 0km
1 cm/s
Russia
Japan
50 N
45 N
40 N
35 N
Latitude
130 E
135 E
140 E
Longitude

FIG. 1.

boundaries where it propels poleward flows that often are absent or reversed under business-as-usual. Where the poleward tendency competes with externally forced equatorward tendency, the poleward flow occurs in undercurrents such as the California or Peru-Chile undercurrents.

In the Arctic, statistical mechanical forcing drives cyclonic flow around the perimeters of basins (much as in the Mediterranean and Japan Seas), and also drives flow across the Arctic above the Lomonosov Ridge. Around Antarctica, statistical mechanical forcing drives flows westward on the Antarctic continental margin, opposed to the prevailing Antarctic Circumpolar Current which lies farther offshore.

Although effects of statistical mechanical forcing are most readily seen at depth, where externally forced flow is weaker, there are manifestations at the sea surface. In particular, by displacing the latitude of eastward flowing surface jets, there can be climatically important influences upon implied air-sea heat exchange.

7. Summary. Taking account of the underlying probabilistic character of our knowledge of the ocean and of effort to model the ocean, we recognize that forces will arise in the equations of evolution of moments of probable oceans and that these statistical mechanical forces can be qualitatively different than the forces that would appear in equations of motion of any specific ocean. Especially we identify a forcing that will drive first moments (means) away from rest. Although this forcing should appear "naturally" in well-eddy-resolved supercomputing efforts, it is not clear how great must be that supercomputing resolution before explicit eddies behave faithfully. On the other hand it seems possible to approximate this mean forcing even at far coarser resolution. Initial experiments at practical implementation have yielded interesting results. First, it is seen that inclusion of statistical mechanical forcing can have major impacts on model results. Second, but less clearly, it apears that inclusion of statistical mechanical forcing improves model performance against some hitherto chronic weaknesses.

Much remains to be done. Steps taken to yield practical implementation have been most clumsy, and desperately await careful refinement. Then the application wants to be extended. Applied at smaller scales, $\mathbf{Y}_{eq}$ is not barotropic, so $\mathbf{u}^*$ will vary with depth and new terms will appear in the density (heat and salt) equations, for example.

Acknowledgement. I am grateful to Michael Eby for effort under the practical implementation of these ideas, and to Tessa Sou for contributing Figure 1. This research has been supported in part by the Office of Naval Research (N00014-92-J- 1775)

REFERENCES

[1] Alvarez, A., J. Tintore, G. Holloway, M. Eby and J. M. Beckers, 1994, "Effect of topographic stress on the circulation in the western Mediterranean", J Geophys Res., 99, 16053-16064.

[2] Bretherton, F. P. and D. B. Haidvogel, 1976, "Two-dimensional turbulence above topography", J. Fluid Mech., 78, 129-154.

[3] Carnevale, G. F., U. Frisch and R. Salmon, 1981, "H-theorems in statistical fluid dynamics", J. Phys. A., 14, 1701-1718.

[4] Carnevale, G. F. and J. S. Frederiksen, 1987, "Nonlinear stability and statistical mechanics of flow over topography". J. Fluid Mech., 175, 157-181.

[5] Eby, M. and G. Holloway, "Experiments with a hybrid statistical mechanics/ ocean circulation model", pg 507-518 in "Statistical methods in Physical Oceanography", Proc.'Aha Huliko'a 1993, P. Muller and D. Henderson, eds., U. Hawaii

[6] Eby, M. and G. Holloway, 1994, "Sensitivity of a large scale ocean model to a parameterization of topographic stress", J Phys Oceanogr, 24, 2577-2588.

[7] Fyfe, J. and G. Marinone, 1995, "On the role of unresolved eddies in a model of the residual currents in the central Strait of Georgia, B.C.", Atmos-Ocean, 33, 613–619.

[8] Gill, A. E., 1982, Atmosphere-Ocean Dynamics, Academic Press, 662 pp.

[9] Herring, J. R., 1977, "Two-dimensional topographic turbulence", J. Atmos. Sci., 34, 1731-50.

[10] Holloway, G., 1978, "A spectral theory of nonlinear barotropic motion above irregular topography", J. Phys. Oceanogr., 8, 414-427.

[11] Holloway, G., 1987, "Systematic forcing of large-scale geophysical flows by eddy-topography interaction", J. Fluid Mech., 184 , 463-476.

[12] Holloway, G., 1992, "Representing topographic stress for large scale ocean models", J Phys Oceanogr, 22, 1033-46.

[13] Katz, A., 1967, Principles of Statistical Mechanics, W. H. Freeman & Co., San Fracisco, 188 pp.

[14] Salmon, R. G. Holloway and M. C. Hendershott, 1976, "The equilibrium statistical mechanics of simple quasi-geostrophic models", J. Fluid Mech., 75, 691-703.

[15] Zou, J. and G. Holloway, 1994, "Entropy maximization tendency in topographic turbulence", J. Fluid Mech., 263, 361-374.

RANDOM TOPOGRAPHY IN GEOPHYSICAL MODELS

V. KLYATSKIN* AND D. GURARIE†

Abstract. Two problems are discussed. One deals with Equilibrium States for Quasigeostrophic Flows with Random Topography. We consider 2D incompressible fluid flows in the quasigeostrophic approximation and show the existence of the statistical equilibria based on the joint Gaussian probability distribution of the stream function and the topography. We derive the differential equations for the 2-point correlators and find their explicit solutions. An interesting feature of our solution is the possibility of coherent structures in a fully developed inviscid turbulent flow. The second problem has to do with Topographic Rossby Waves Over Randomly Stratified bottom. The problem is reduced to the Helmholtz equation with random index of refraction. Statistics of wave field intensity are stidied in finite width layers, then the localization phenomena are shown in the entire randomly layered space. We also address the question of baroclinic influence on the equilibrium states and on the Rossby wave localization.

1. Introduction. The effects of topography on geophysical flows and waves (atmosphere and ocean) are important in the deterministic setup and were studied in numerous works. The discussions is usually based on "nice" models of the bottom topography and suitable geophysical equations. The latter include quasigeostrophic approximations (QGS) for the large scale flows and various models of linear and nonlinear geophysical waves: surface-gravity, internal, inertial, Rossby, etc. The typical bottom topography, though a deterministic function, can be viewed as a realization of the random process, so is the solution of the problem. One is then interested in the statistical description of such solutions. In the QGS-problem the statistical (ensemble) averaging results from a suitable ergodicity Ansatz, typical of "large" (infinite) dynamical systems, rather than the direct statistical sampling of different bottom realizations. Ergodicity could drive such system to a state of statistical equilibrium. In the wave-problems there is no ergodicity, but one is able to compute exact probability distribution of solutions and demonstrate the appearance of certain statistically stable features, like "wave localization" vs. "propagation".

Two problems will be reviewed and discussed in the paper

- Equilibrium states for 2-D incompressible fluid flows with Random Topography in the Quasigeostrophic approximation. We show the existence of an equilibrium ensemble based on the Gaussian properties of the joint probability distribution of the random stream function and random topography. We exhibit some interesting features of the statistical equilibria, like possibility of coherent structures in fully developed inviscid turbulent flows.

* Institute of Atmospheric Physics, 109017 Moscow, and Pacific Oceanological Institute, 690041 Vladivostok, Russia.

† Department of Mathematics, Center for Stochastic and Chaotic Processes, CWRU, Cleveland, OH 44106.

- The localization of topographic Rossby waves over randomly stratified bottom.

We study the statistics of wave field intensity first in layers of finite width, then establish the localization phenomena in the entire randomly layered space. The localization length is estimated in terms of the spectral density of depth fluctuations.

We also discuss the baroclinic effects (two-layered structure) for both models, and demonstarte its influence on QGS-equilibria and on the Rossby wave localization.

2. Equilibrium states. This section will briefly review the equilibrium QGS-states over the stationary random Gaussian topography. The state of an incompressible 2-D fluid is given by the stream function $\psi(\mathrm{r};t)$, $\mathrm{r} = (x;y)$, where the velocity field $\mathrm{u} = \left(-\frac{\partial\psi}{\partial y};\frac{\partial\psi}{\partial x}\right)$. Function ψ satisfies the quasigeostrophic equation[1]

$$\frac{\partial}{\partial t}\Delta\psi = J(\Delta\psi + h;\psi)\,;\ \psi|_{t=0} = \psi_0(\mathrm{r})\,, \tag{2.1}$$

where $J(\psi;\zeta) = \frac{\partial\psi}{\partial x}\frac{\partial\zeta}{\partial y} - \frac{\partial\psi}{\partial y}\frac{\partial\zeta}{\partial x}$ denotes the Jacobian of two functions, and the *topographic factor* h represents the combined effect of the "depth-variation over mean depth" $\delta H/H_0$, the local Coriolis parameter f_0 and the so called β-factor (the latitudinal change of Coriolis f),

$$h = f_0\frac{\delta H}{H_0} + \beta y.$$

We assume that the bottom profile $h(\mathrm{r})$ is the homogeneous and isotropic Gaussian random field with zero mean $\langle h(\mathrm{r})\rangle = 0$ and correlation

$$B_{hh}(|\mathrm{r}-\mathrm{r}'|) = \langle h(\mathrm{r})\,h(\mathrm{r}')\rangle\,.$$

Here we use the standard convention $\langle ...\rangle$ to indicate the statistical averaging over the joint h,ψ_0 -distribution.The beta-effect will be ignored in the first part.

The complicated nonlinear interaction (2.1) is expected to bring the system to a statistical equilibrium. Futhermore, the continuous process of multiple scales (wave numbers) interaction, i.e. the fluid *self-action* and the fluid-topography *inter-action*, could bring about the homogenization and isotropization of random fields. One could also expect the limiting equilibrium to become Gaussian. The problem then is to determine the statistics of the stationary ensemble in terms of the topographic input.

In the absence of topographic factor $h = 0$ system (2.1) becomes the classical 2-D Euler hydrodynamics, studied in depth by many authors (see recent survey [2] for the detailed discussion of the exact model and various approximations). Among earlier works we mention the pioneering contribution by Kraichnan [3]-[5] and paper[6]. The latter, in particular demonstrated the possibility of certain coherence in the 2-D turbulent flows.

We shall adopt the mathematical formalism of [6]-[7]. The streamfunction being determined up to a constant its statistics are described by the *structure function*

$$D_\psi(\mathrm{r}-\mathrm{r}';t)=\left\langle[\psi(\mathrm{r};t)-\psi(\mathrm{r}';t)]^2\right\rangle=2\left[B_{\psi\psi}(0;t)-B_{\psi\psi}(\mathrm{r}-\mathrm{r}';t)\right],$$

where $B_{\psi\psi}(\mathrm{r}-\mathrm{r}';t)=\langle\psi(\mathrm{r},t)\,\psi(\mathrm{r}',t)\rangle$ denotes the correlation function of field ψ. We are interested in stationary ensemble of ψ on the class of joint ψ,h-Gaussian distributions, so unknown quantities of interest are the structure-function and the cross-correlation

$$D_\psi(\mathrm{r})=\lim_{t\to\infty}D_\psi(\mathrm{r};t)\,;\ B_{\psi h}(\mathrm{r})=\lim_{t\to\infty}\langle h(\mathrm{r})\,\psi(\mathrm{r}-\mathrm{r}',t)\rangle$$

The key idea is to consider the 3-point correlators at a fixed time t and different point locations. Using the condition of stationarity

$$\frac{\partial}{\partial t}\langle\Delta\psi(\mathrm{r}_1;t)\,\Delta\psi(\mathrm{r}_2;t)\,\Delta\psi(\mathrm{r}_3;t)\rangle=\left\langle\Delta\dot\psi\Delta\psi\Delta\psi\right\rangle+\left\langle\Delta\psi\Delta\dot\psi\Delta\psi\right\rangle+\ldots=0 \tag{2.2}$$

and expressing each time derivative via quadratic (Jacobian) term of (2.1) we get in the r.h.s. of (2.2) a combination of quartic moments of ψ. Those in turn are expanded into products of 2-nd moments (2-point correlators) by the Gaussian property of ensemble ψ. The off-shot is a linear functional equation for certain function $X(p;q)$, expressed through products of derivatives of the unknown correlators $D_\psi(|\mathrm{r}|)$ and $B_{\psi h}(|\mathrm{r}|)$ (see [7] for details). Precisely,

$$\begin{aligned}X(p;q)=\tfrac{1}{pq}\tfrac{\partial^2}{\partial p\partial q}\big\{&\left[\Delta_p^2D_\psi(p)\ \Delta_qD_\psi(q)-\Delta_q^2D_\psi(q)\ \Delta_pD_\psi(p)\right]-\\&-2\left[\Delta_pB_{\psi h}(p)\,\Delta_qD_\psi(q)-\Delta_pD_\psi(p)\,\Delta_qB_{\psi h}(q)\right]\big\}.\end{aligned} \tag{2.3}$$

where Δ_p denotes the radial part of the 2-D Laplacian $\Delta_p=\partial_p^2+\frac{1}{p}\partial_p$. The functional equation takes the form

$$X(q_1;q_2)+X(q_2;q_3)+X(q_3;q_1)=0 \tag{2.4}$$

for any triplets of coordinates $q_1=|\mathrm{r}_1-\mathrm{r}_2|\,;q_2=|\mathrm{r}_2-\mathrm{r}_3|\,;q_3=|\mathrm{r}_3-\mathrm{r}_1|$. Using equation (2.4) along with the anti-symmetry of function $X(p,q)=-X(q,p)$ we could separate variables p and q and break the quadratic expression (2.3) into a system of linear differential equations for D_ψ and $B_{\psi h}$. Namely (see [7])

$$(\Delta_q+\lambda)\,\Delta_qD_\psi-2\Delta_qB_{\psi h}=0 \tag{2.5}$$

In the absence of topographic factor equation (2.5) is simplified to

$$(\Delta_q + \lambda)\Delta_q \left[D_\psi(q)\right] = 0. \tag{2.6}$$

2.1. The Euler case. We shall first discuss the Euler case ($h = 0$). There are two classes of solutions of (3.10), corresponding to positive and negative values [1] $\lambda = k_0^2 > 0$ and $\lambda = -k_0^2 < 0$. In the former case $\lambda > 0$ the 4-th order differential equation (3.10) is reduced to the second order one

$$\Delta_q \left[D_\psi(q)\right] = C\, J_0(k_0 q), \tag{2.7}$$

with the Bessel function $J_0(z)$ of the first kind. Solving the Laplace's equation we get the structure-function $D_\psi = \Delta^{-1}(J_0)$ and the appropriate spectral density

$$E(k) = E_0 \delta(k - k_0). \tag{2.8}$$

The delta-type spectral density indicates the high level of correlation of field ψ, and gives the meaning to coherence of the fully developed 2-D stochastic (turbulent) flow.

In the second case $\lambda = -k_0^2 < 0$ equation (3.10) is reduced to

$$\Delta_q \left[D_\psi(q)\right] = -C_1 K_0(k_0 q) \tag{2.9}$$

but this time the r.h.s. contains the Macdonald (3-rd kind Bessel) function K_0 with some (dimensional) parameters k_0 and C_1. The corresponding spectral density becomes ([3]-[5])

$$E(k) = \frac{E_0}{k^2 + k_0^2}. \tag{2.10}$$

A peculiar feature of (2.10) is the log-divergence of the mean kinetic energy density, due to the absent viscous dissipation in the model. The structure function $D_\psi(q)$ is obtained by integrating (2.9) subject to the boundary conditions: $D_\psi(0) = 0$; $D'_\psi(0) = 0$. Hence follows

$$D_\psi(q) = C\left[K_0(k_0 q) + \ln\left(k_0 \frac{q}{2}\right) + \gamma\right], \tag{2.11}$$

---

[1] Those could be related to the negative and positive "temperature" of the Kraichnan's energy/enstrophy theory.

with Euler constant γ. Let us remark that the 2-D and 3-D stationary spectral densities behave in a markedly different manner. The former is either localized (2.8) or the decaying type (2.10) at large k. The latter has only the white noise (δ-correlated) solution (see [8]-[10]). Let us also note that solution (2.11) grows unboundedly with q. To get a bounded solution one can interchange two differential operations in (3.10) and solve it in the form

$$(\Delta_q + \lambda)\left[D_\psi(q)\right] = 0.$$

2.2. General topographic case. Next we turn to the general topographic case (2.5). We look for solutions of (2.5) in the class of bounded functions $D_\psi(q), B_{\psi h}(q)$. Equation (2.5) could be recast in the form

$$(\Delta_q + \lambda)\left[D_\psi(q)\right] - 2B_{\psi h}(q) = 0,$$

or, using correlation function $B_{\psi\psi}$ in place of D_ψ

$$(\Delta_q + \lambda)\left[B_{\psi\psi}(q)\right] + B_{\psi h}(q) = 0. \tag{2.12}$$

Equation (2.12) is no more closed as it contains new (yet undetermined) cross-correlation function $B_{\psi h}$. To determine the latter we consider the evolution of the 3-point correlator $\langle \Delta\psi\ h\ h\rangle$, i.e.

$$\frac{\partial}{\partial t}\langle \Delta\psi(\mathrm{r}_1;t)\ h(\mathrm{r}_2)\ h(\mathrm{r}_3)\rangle = 0$$

By analogy with (2.3)-(2.4) we get a pde

$$\frac{1}{q_1 q_3}\frac{\partial^2}{\partial q_1 \partial q_3}\{\Delta_1\left[B_{\psi h}(q_1)\right] B_{\psi h}(q_3) - \Delta_3\left[B_{\psi h}(q_3)\right] B_{\psi h}(q_1) +$$
$$+ B_{hh}(q_1) B_{\psi h}(q_3) - B_{hh}(q_3) B_{\psi h}(q_1)\} = 0$$

The latter is also solved by separation of variables. As the result we get an equation similar to (2.12)

$$(\Delta_q + \lambda_1)\left[B_{\psi h}(q)\right] + B_{hh}(q) = 0 \tag{2.13}$$

with another separation constant λ_1. The combined system of two equations (2.12)-(2.13)

$$\begin{cases} (\Delta_q + \lambda)\left[B_{\psi\psi}(q)\right] + B_{\psi h}(q) = 0 \\ (\Delta_q + \lambda_1)\left[B_{\psi h}(q)\right] + B_{hh}(q) = 0 \end{cases} \tag{2.14}$$

is already closed. However, two important issues arise here.

- Existence of separation constant λ_1 (and the proper length scale) independent of constant λ of (2.12).
- Consistency of our basic hypothesis on the existence of Gaussian equilibrium ensemble $\{\psi; h\}$.

Both issues could be addressed via the characteristic functional of random fields $(\psi; h)$

$$\Phi_t[v;\kappa] = \left\langle \exp\left\{ i \int d^2\mathrm{r}\, [\psi(\mathrm{r};t) v(\mathrm{r}) + h(\mathrm{r})\,\kappa(\mathrm{r})]\right\}\right\rangle$$

$$= \langle \exp i\, \{(\psi|v) + (h|\kappa)\}\rangle\, .$$

Here $(f|g)$ denotes the standard inner product on function spaces over R^2. The characteristic functional determines all spatial statistics of fields $\{\psi(\mathrm{r};t); h(\mathrm{r})\}$ at fixed time t. The basic dynamic evolution(2.1) yields a linear variational equation for Φ_t known as the *Hopf equation* (see [8]-[10])

$$\frac{\partial}{\partial t}\Phi_t[v;\kappa] = -i\left(v \left| \Delta^{-1} J\left(\Delta\frac{\delta}{\delta v} + \frac{\delta}{\delta\kappa}; \frac{\delta}{\delta v}\right)\right.\right)\Phi_t[v;\kappa]\, .$$

The limiting (equilibrium) state $\Phi_\infty = \lim_{t\to\infty}\Phi_t$ clearly satisfies $\frac{\partial}{\partial t}\Phi_\infty = 0$. So the stationary distribution obeys the equation

$$\left(\Delta^{-1} v \left| J\left(\Delta\frac{\delta}{\delta v} + \frac{\delta}{\delta\kappa}; \frac{\delta}{\delta v}\right)\right.\right)\Phi_\infty[v;\kappa] = 0. \tag{2.15}$$

The latter was analyzed in [7]. We took the standard expression of the quadratic exponential (in terms of correlators)

$$\Phi_\infty[v;\kappa] = \exp\left\{-\tfrac{1}{2}\left[(v\,|B_{\psi\psi}|\,v) + 2\,(v \mid B_{\psi h} \mid \kappa) + (\kappa \mid B_{hh} \mid \kappa)\right]\right\}. \tag{2.16}$$

substituted it in (2.15) and utilized equations (2.14). The off-shot was

$$\begin{aligned} &\left(\Delta^{-1}v\,\middle|\,J\left(\Delta\tfrac{\delta}{\delta v} + \tfrac{\delta}{\delta\kappa}; \tfrac{\delta}{\delta v}\right)\right)[\Phi_\infty] = \\ &= (\lambda_1 - \lambda)\left(\Delta^{-1}v\,\middle|\,J\left(B_{\psi\psi}[v]; B_{\psi h}[\kappa]\right)\right)\Phi_\infty. \end{aligned} \tag{2.17}$$

Clearly the r.h.s. vanishes for arbitrary choice of $\{v;\kappa\}$ (the stationarity condition) if and only if two separation constants (2.14) are equal

$$\lambda = \lambda_1 .$$

The latter provides the sufficient condition for the existence of stationary Gaussian ensemble. Since two separation constants are equal such theory would allows a single length scale determined by λ.

We have shown that quasigeostrophic flows with random Gaussian topography do admit Gaussian stationary ensembles. Furthermore, any ensemble is obtained by augmenting the homogeneous solution ("flat bottom") of (2.12), (2.13) with a particular inhomogeneous solution related to $B_{hh}(q)$ by

$$\text{(2.18)}\qquad \begin{aligned} B_{\psi h}(q) &= (\Delta_q + \lambda)^{-1}[B_{hh}(q)] \\ B_{\psi\psi}(q) &= B^0_{\psi\psi}(q) + (\Delta_q + \lambda)^{-2}[B_{hh}(q)], \end{aligned}$$

where $B^0_{\psi\psi}$ denotes the Bessel-type correlator of the stream-field over the unperturbed (flat) bottom.

Remark 1 Formulae (2.18) and the related spectral densities were derived earlier in [12] and [15] based on the canonical "Gibbsian" approach. Namely,

$$\left\langle |\psi(k)|^2 \right\rangle = \frac{k^2}{\alpha + \beta k^2} + \frac{(\beta k^2)^2 H(k)}{(\alpha + \beta k^2)^2}$$

and

$$\langle \psi^*(k)\, h(k) \rangle = \frac{k^2}{\alpha + \beta k^2} + \frac{(\beta k^2)^2 H(k)}{(\alpha + \beta k^2)^2}$$

where $\psi(k); h(k)$ are Fourier coefficients of $\psi; h$ and $H(k)$ - the spectral density of h (Fourier transform of B_{hh}). These results are consistent with our formulae (2.10), (2.11), (2.18) but our appoach also yields the δ - correlated (coherent) case.

Remark 2 Let us notice that any deterministic stationary solution of (2.1) satisfies

$$\Delta\psi + h = F(\psi)$$

with an arbitrary function F. In the simplest linear case $F = \lambda\psi$ it becomes the Foffonoff flow [14]

$$\text{(2.19)}\qquad \Delta\psi + h = -\lambda\psi.$$

The correlation equation for (2.19) with $\psi(\mathrm{r}')$ and $h(\mathrm{r})$ is easily seen to coincide with (2.12), (2.13). This means that the Gaussian equilibrium is statistically equivalent to a statistical Fofonoff flow. Thus the quasi-geostrophic flow, a highly nonlinear process arrives (in the limit $t \to \infty$) at a statistical state where all nonlinearities effectively disappear.

2.3. The baroclinic effects and two-layer QGS flows. The quasi-geostrophic equation (2.1) describes a single layered fluid and does not take

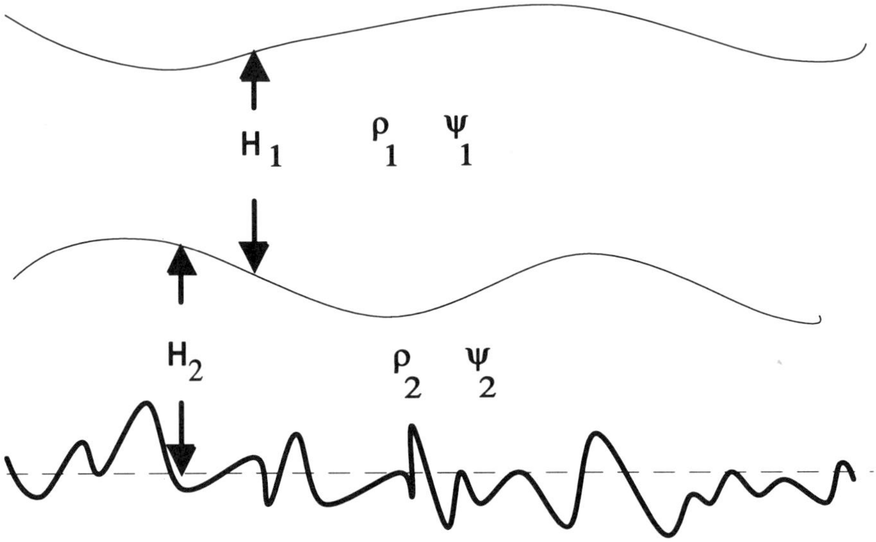

FIG. 1. *Schematic view of the two-layered flow*

into account the baroclinic effects. The latter is modeled (in the simplest approximation) by the two layered QGS-system [1] (see fig. 1)

$$\begin{cases} \frac{\partial}{\partial t}\left[\Delta\psi_1 - \alpha_1 F\left(\psi_1 - \psi_2\right)\right] = J\left(\Delta\psi_1 - \alpha_1 F\left(\psi_1 - \psi_2\right); \psi_1\right) \\ \frac{\partial}{\partial t}\left[\Delta\psi_2 - \alpha_2 F\left(\psi_2 - \psi_1\right)\right] = J\left(\Delta\psi_2 - \alpha_2 F\left(\psi_2 - \psi_1\right) + f_0\alpha_2 H; \psi_2\right) \end{cases} \tag{2.20}$$

Here parameters $\alpha_1 = \frac{1}{H_1}; \alpha_2 = \frac{1}{H_2}$ denote reciprocals of layers thicknesses; $\rho_2; \rho_1$ - their densities; f_0 -local Coriolis parameter and the topographic coefficient

$$F = f_0^2 \Big/ g\,\frac{\Delta\rho}{\rho}\ ; \text{ with } \frac{\Delta\rho}{\rho} = \frac{\rho_2 - \rho_1}{\rho_0}.$$

In the absence of topographic factor system (2.20) was considered in [12]. Here we shall consider the equilibrium Gaussian distribution whose parameters are statistically equivalent to the steady-state Fofonoff-type solution of (2.20)

$$\begin{cases} \Delta\psi_1 - \alpha_1 F\left(\psi_1 - \psi_2\right) = -\lambda_1\psi_1 \\ \Delta\psi_2 - \alpha_2 F\left(\psi_2 - \psi_1\right) + f_0\alpha_2 H = -\lambda_2\psi_2 \end{cases} \tag{2.21}$$

Such equilibria depend on two "scale" parameters $\lambda_1; \lambda_2$, so it has two different length scales. Equations (2.21) can be written in the matrix-operator form

$$\begin{bmatrix} \Delta - \alpha_1 F + \lambda_1 & \alpha_1 F \\ \alpha_2 F & \Delta - \alpha_2 F + \lambda_2 \end{bmatrix} \begin{bmatrix} \psi_1 \\ \psi_2 \end{bmatrix} = \begin{bmatrix} 0 \\ f_0 \alpha_2 H \end{bmatrix},$$

or the equivalent scalar form

$$\hat{L}\,[\psi_1] = \alpha_1 \alpha_2 F f_0 H$$

$$\hat{L}\,[\psi_2] = -\alpha_2 f_0 \,(\Delta + \lambda_1 - \alpha_1 F)\, H$$

where operator

$$\hat{L} = \Delta^2 + [\lambda_1 + \lambda_2 - F\,(\alpha_1 + \alpha_2)]\,\Delta + \{\lambda_1\lambda_2 - F\,(\lambda_2\alpha_1 + \lambda_1\alpha_2)\}\,.$$

Operator $\hat{L}$ can be factored in the product

$$(\Delta + \mu_1)\,(\Delta + \mu_2)\,, \tag{2.22}$$

wand its characteristic roots are expressed through the shifted eigenvalue parameters: $\tilde{\lambda}_{1,2} = \lambda_{1,2} - F\alpha_{1,2}$

$$\mu_{1,2} = \tfrac{1}{2}\left(\tilde{\lambda}_1 + \tilde{\lambda}_2\right) \pm \sqrt{\tfrac{1}{4}\left(\tilde{\lambda}_1 - \tilde{\lambda}_2\right)^2 + \alpha_1\alpha_2 F^2}.$$

The coherent structure (delta-type spectral density) corresponds to the positive characteristic roots (2.22). The resulting range of parameters $\tilde{\lambda}$ that give positive roots include (fig. 2).

- $\tilde{\lambda}_1\tilde{\lambda}_2 < \alpha_1\alpha_2 F^2$ - single positive root
- $\tilde{\lambda}_1\tilde{\lambda}_2 > \alpha_1\alpha_2 F^2$ and $\tilde{\lambda}_1 + \tilde{\lambda}_2 > 0$ - two positive roots.

We remark that coherent structure of the two-layer case allow two length scales to develop (region above the upper branch of the hyperbola $\tilde{\lambda}_1\tilde{\lambda}_2 = \alpha_1\alpha_2 F^2$), compared to a single scale of a simple QGS-flow (2.1).

We established the existence of Gaussian stationary ensembles for QGS-flow in the presence of random topographic factor h and computed their statistics (correlation functions) in terms of statistics of h. A peculiar feature of our solutions is a possibility of coherent states in the fully developed turbulent flow. The question of their stability is left open as well as rigorous justification of Gaussianity of the limiting distribution. We hope to return to these topics elsewhere. Let us remark that Gaussian equilibria provide a natural "background noise" in many geophysical systems, described by the QGS-type equations. Their role is similar to that of "heat noises" in the statistical Physics. In regard to non-Gaussian equilibria we mention several works ([11]-[13]), and a recent paper [16] based on the functional (Hopf) approach.

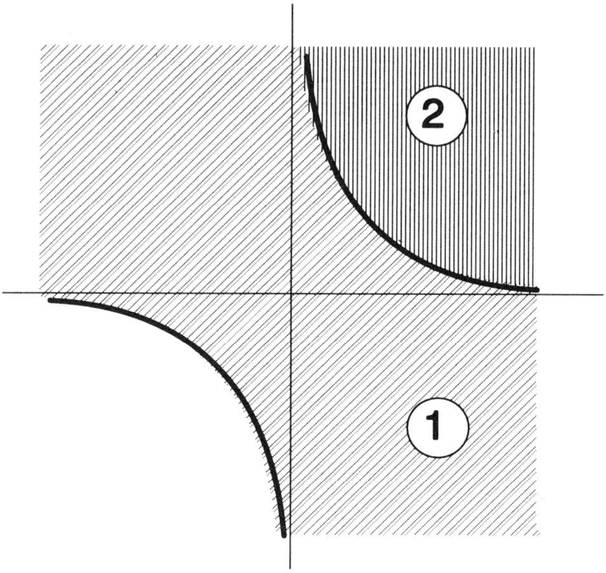

FIG. 2. *Two regions of the coherent structure in the $\lambda_1\lambda_2$-parameter plane: Region 1 corresponds to the single-scale coherent structure, region 2 the double-scale one.*

3. Topographic Rossby waves. The bottom shape as well as the beta-effect play an important role in propagation of large scale, low frequency oscillations in the Earth oceans and the atmosphere, called *Rossby waves.* The effect of topography on such waves depends to a large degree on the relation between the characteristic wave length λ and the horizontal scale of topographic inhomogeneities ℓ_h [18]. In the practically important case $\ell_h \ll \lambda$ such topography would sustain propagation of large scale waves with or without the β-effect. There are many studies where topography was considered a periodic or quasi periodic function (superposition of Fourier modes). The real topography however is often highly irregular, so there is a considerable interest in the study of low frequency waves propagating over random topography [19] -[20]. The problem of topographic Rossby waves in the case of stratified topography can be reduced to a boundary value problem for the Helmholtz equation with the random index of refraction. So the results and methods of the large body of work on wave propagation random media are applicable to such Rossby waves. In this section we shall outline an approach based on the imbedding method of [21]-[23].

3.1. Localization for one-layer Rossby waves. Rossby waves obey the standard linearized quasigeostrophic equation (2.1), which in the single-layer case takes the form

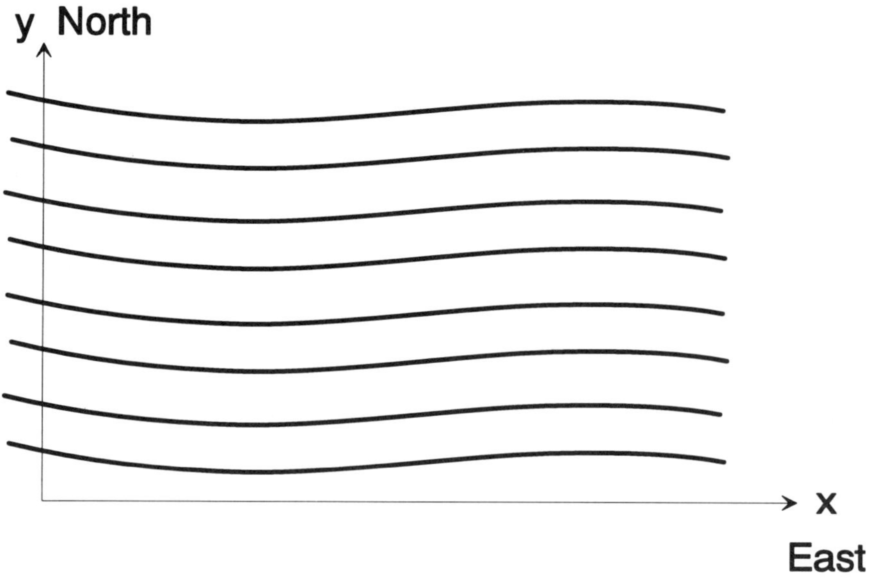

FIG. 3. *Schematic view of stratified topography*

$$\frac{\partial}{\partial t}\Delta\psi + \beta\frac{\partial\psi}{\partial x} + \frac{f_0}{H_0}J(\psi; H) = 0, \tag{3.1}$$

Here we use the notations of part 1, namely f_0 - local Coriolis parameter, β- its derivative in the vertical (latitudinal) direction, H_0 - the mean depth, and $J(\psi; H) = \frac{\partial\psi}{\partial x}\frac{\partial H}{\partial y} - \frac{\partial\psi}{\partial y}\frac{\partial H}{\partial x}$ - the standard Jacobian of two functions. Our basic assumption on the depth-function is the vertical (North-to-South) stratification (fig. 3). So H depends on a single variable y

$$H = H_0 + h(y),$$

where $h(y)$ measures random fluctuations of the bottom topography from the constant mean depth H_0.

We look for product-type solutions of (3.1) made of the Fourier mode in the x-variable (of wave number k)multiplied with a suitable y-mode

$$\tilde{\psi}(x; y; t) = \psi(y)e^{-i(\omega t+\kappa x)}.$$

Such $\tilde{\psi}$ represents a west-traveling mode for $k > 0$, $\omega > 0$. The "vertical" (north-bound) component ψ then satisfies the reduced equation

$$\frac{d^2}{dy^2}\psi(y) + \left[k^2 + \frac{\kappa f_0}{H_0\omega}\frac{dh(y)}{dy}\right]\psi(y) = 0 \tag{3.2}$$

or

$$\frac{d^2}{dy^2}\psi(y) + k^2\left[1 + \epsilon(y)\right]\psi(y) = 0, \tag{3.3}$$

where $\epsilon(y) = \frac{\kappa f_0}{H_0 \omega k^2}\frac{dh}{dy}(y)$. The parameter k^2 is obtained from the Rossby dispersion relation as a function of κ and ω

$$k^2 = \kappa\left(\frac{\beta}{\omega} - \kappa\right) > 0. \tag{3.4}$$

There is a number of papers devoted to localization phenomena for Rossby waves by random topographies. By combining analytical tools with numerical simulations they try to find the *localization length.* One such approach is based on estimates of the Liapounov exponents ([25]). Here we shall adopt a different approach and focus the primary attention on the spatial structure of point-sources (Green's functions) of equation (3.3) with suitable (radiation) boundary conditions at $\pm\infty$.

So we first consider a strip $L_0 \le y \le L$ of finite width, and write the equation for the point source at y_0 as

$$\left\{\frac{d^2}{dy^2} + k^2\left[1 + \epsilon(y)\right]\right\}\psi(y;y_0) = 2ik\delta(y - y_0). \tag{3.5}$$

Assuming constant (homogeneous) medium outside the layer we take a plane wave with the wave number k in the upper (free) half space $y > L$ and in the lower half-space $y < L_0$. The corresponding boundary conditions for (3.5) become

$$\left(\frac{d}{dy} - ik\right)\psi(y;y_0)\bigg|_{y=L} = 0; \quad \left(\frac{d}{dy} + ik\right)\psi(y;y_0)\bigg|_{y=L_0} = 0 \tag{3.6}$$

Our goal is to pass eventually to the infinite layer limit: $L_0 \to -\infty$; $L \to \infty$.

The fluctuating component of the refraction index $\epsilon(y)$ will be assumed a homogeneous random Gaussian field with zero mean $\langle\epsilon(y)\rangle = 0$ and correlation function

$$B_\epsilon(y - y') = \langle\epsilon(y)\,\epsilon(y')\rangle$$

Furthermore, we assume the variance of ϵ sufficiently small $\sigma^2 = B_\epsilon(0) \ll 1$. Function B_ϵ is characterized by its correlation radius ℓ_h, i.e. $B = B(y/\ell_h)$. The random bottom topography is described by random field $h(y)$ (relative depth fluctuations times local Coriolis parameter), whose correlation and spectral density are given by

$$B_h(z) = \langle h(y)\, h(y') \rangle\, ; \; \Phi_h(q) = \int_{-\infty}^{\infty} dz\, B_h(z)\, e^{iq\cdot z}; \; z = y - y'$$

Using (3.5) one can deduce the statistics of random process $\epsilon(y)$ from those of h. Namely,

$$(3.7) \qquad \begin{aligned} B_\epsilon(\xi) &= -\left(\frac{\kappa f_0}{H_0 \omega k^2}\right)^2 \frac{d^2}{d\xi^2} B_h(\xi) \\ \Phi_\epsilon(q) &= \left(\frac{\kappa f_0}{H_0 \omega k^2}\right)^2 q^2 \Phi_h(q) \end{aligned}$$

while the variance of ϵ

$$\sigma_\epsilon^2 = \frac{1}{2\pi} \int_{-\infty}^{\infty} dq \left(\frac{k f_0}{H_0 \omega \kappa_0^2}\right)^2 q^2 \Phi_h(q) = B_\epsilon(0).$$

If the source of the plane waves is located on the boundary we rewrite equation (3.5) with boundary conditions (3.6) as

$$\left\{\frac{d^2}{dy^2} + k^2[1+\epsilon(y)]\right\} \psi(y; y_0) = 0$$

$$(3.8) \quad \left(\frac{d}{dy} - ik\right) \psi(y; y_0)\Big|_{y=L} = -2ik; \; \left(\frac{d}{dy} + ik\right) \psi(y; y_0)\Big|_{y=L_0} = 0.$$

The boundary value problems (3.5)-(3.6) and (3.8) were studied in detail in [21]-[23]. The idea is to consider wave-field $\psi(y, L)$ as a function of the boundary parameter L, then apply the *imbedding method* formalism of [21]-[23] to transform the boundary-value problem (3.8) into the initial value problem with respect to parameter L (see Appendix). The main advantage of initial value problems is their better amenability to statistical analysis compared to boundary value problems. Here the initial value problem takes the form of a system

$$(3.9) \qquad \begin{cases} \frac{d}{dL} \psi(y; L) = 2ik\psi(y; L) + \frac{ik}{2}\epsilon(L)(1 + R_L)\, \psi(y; L) \\ \frac{d}{dL} R_L = 2ik \frac{ik}{2} \epsilon(L)(1 + R_L)^2 \end{cases}$$

where $R_L = \psi(L; L) - 1$ denotes the reflection coefficient.

Next we let $L \to \infty$, to pass to the half-space region limit and use the *diffusion approximation* of the stachastic equation (3.9). In the diffusion approximation all statistical parameters of solution are completely determined by spectral density Φ_ϵ of (3.7) evaluated at a particular value of frequency $2k$ with k defined by (3.4). Precisely

$$\Phi_\epsilon(2k) = \left(\frac{2\kappa f_0}{H_0 \omega k^2}\right)^2 \Phi_h(2k)$$

The diffusion approximation is valid under certain conditions. Namely, random fluctuations should have no affect on the dynamics of the wave motion on the scale of correlation radius ℓ_h. In other words the waves do not feel the medium inhomogeneities on scales $\leq \ell_h$ and propagate as in the free space. The diffusion approximation makes the problem statistically equivalent to a white-noise process with parameters

$$\langle \epsilon(y) \rangle = 0; B_\epsilon^{eff}(\xi) = 2\Phi_\epsilon(2k)\delta(\xi) = 2\left(\frac{2\kappa f_0}{H_0 \omega k}\right)^2 \Phi_h(2k)\delta(\xi)$$

Furthermore, if $k\ell_h \ll 1$ we get

$$B_\epsilon^{eff}(\xi) = 2\left(\frac{2\kappa f_0}{H_0 \omega k}\right)^2 \Phi_h(0)\delta(\xi); \ \Phi_h(0) \sim \sigma_h^2 \ell_h$$

which makes the process essentially independent of specific features of the random medium. Let us note a sharp contrast of this case with an opposite extreme $k\ell_h \gg 1$ (short waves relative to the medium variations). There the effective correlation strongly depends on peculiar details of random model.

Next we move the source to the boundary L of the layer, i.e. consider field $\psi(y; L)$. It was shown ([21]-[23]) that the field intensity $I(y; L) = |\psi(y; L)|^2$ of such source obeys the log-normal probability law. The log-normality has several important consequences: constant mean intensity and exponentially increasing higher moments

$$\langle I^n(y; L) \rangle = e^{n(n-1)D(L-y)}$$

Here the diffusion coefficient

$$D = \frac{k^2}{4}\Phi_\epsilon(2k) = \left(\frac{\kappa f_0}{H_0 \omega}\right)^2 \Phi_h(2k) \tag{3.10}$$

From the physical standpoint log-normality means the existence of rare but large fluctuation in almost any realization of the process. The fluctuations occur over the general trend of exponential decay for a typical realization [24]

$$I(y; L) \sim e^{-(L-y)/\ell_{loc}}$$

Such behavior is usually associated with the localization phenomena in disordered systems. Parameter ℓ_{loc} called the *localization length* is reciprocal of the diffusion rate (3.10) (see [26]-[28])

$$\ell_{loc} = \frac{1}{D}$$

Furthermore one could give an upper bound for the field intensity $I(y;L)$, valid in the entire range of variable $\xi = D(L-y)$, that implies inequality

$$I(y;L) < 4e^{-\xi/2}$$

with probability $p = \frac{1}{2}$.

Returning to the original problem of the point source in the entire space, all dynamical characteristics remain unchanged. However, their statistics, like the mean and higher moments of intensity could change significantly. Here one needs a suitable regularization of the problem, since the wave intensity $I(y;y_0) = |\psi(y;y_0)|^2 = \infty$ with probability 1. The infinitely large value of intensity means complete reflexivity of two random half-spaces: $y > y_0$ and $y < y_0$. Hence a steady-state solution could result by pumping the energy into a "finite volume" between two halves, during an infinitely long time interval. The regularization procedure would remove such singularity and produce a finite answer. We shall skip further deatails and state the conclusion.

The localization for Rossby waves was established in the presence of randomly stratified bottom topography. The localization is characterized by

- Exponentially decaying curve that represents a typical realization of Rossby random field
- Existence of an exponentially decaying bound for all realizations within any prescribed probability margin $p < 1$.

The above features are typical for single layer Rossby waves, but the situation could be very different and more comlicated when the baroclinic effects are taken into account..

3.2. Baroclinic Rossby waves: The two-layer model. Here we shall briefly discuss the two-layer model of baroclinic Rossby waves. The basic dynamic equations consist now of a coupled system[1]

$$\begin{cases} \frac{\partial}{\partial t}\left[\Delta\psi_1 - \alpha_1 F\left(\psi_1 - \psi_2\right)\right] + \beta\frac{\partial\psi_1}{\partial x} = J\left(\Delta\psi_1 - \alpha_1 F\left(\psi_1 - \psi_2\right);\psi_1\right) \\ \frac{\partial}{\partial t}\left[\Delta\psi_2 - \alpha_2 F\left(\psi_2 - \psi_1\right)\right] + \beta\frac{\partial\psi_2}{\partial x} = J\left(\Delta\psi_2 - \alpha_2 F\left(\psi_2 - \psi_1\right) + f_0\alpha_2 h;\psi_1\right) \end{cases}$$

whose parameters are defined in part 1 (see fig. 1),

$$F = f_0^2 \Big/ g\,\frac{\Delta\rho}{\rho}\ ;\ \text{with}\ \frac{\Delta\rho}{\rho} = \frac{\rho_2 - \rho_1}{\rho_0};\ \alpha_1 = \frac{1}{H_1};\ \alpha_2 = \frac{1}{H_2}$$

The corresponding Rossby waves arise from the linearized system

$$\begin{cases} \frac{\partial}{\partial t}\left[\Delta\psi_1 - \alpha_1 F\left(\psi_1 - \psi_2\right)\right] + \beta_0\frac{\partial\psi_1}{\partial x} = 0 \\ \frac{\partial}{\partial t}\left[\Delta\psi_2 - \alpha_2 F\left(\psi_2 - \psi_1\right)\right] + \left(\beta_0 + f_0\alpha_2 h\right)\frac{\partial\psi_2}{\partial x} = 0 \end{cases} \tag{3.11}$$

Representing solution in the standard Fourier mode product form $e^{-i(x\cdot k+\omega t)}$ $\psi(y)$ we get a coupled ODE system

$$\begin{cases} \frac{d^2}{dy^2}\psi_1 + k^2\psi_1 - \alpha_1 F(\psi_1 - \psi_2) = 0 \\ \frac{d^2}{dy^2}\psi_2 + k^2\psi_2 - \alpha_2 F(\psi_1 - \psi_2) + \epsilon(y)\psi_2 = 0 \end{cases} \tag{3.12}$$

with coefficients

$$k^2 = \kappa\left(\frac{\beta}{\omega} - \kappa\right) > 0 \text{ and } \epsilon(y) = \kappa f_0 h'(y)/D_2\omega$$

We consider the corresponding vector Green's functions

$$\Psi(y; y_0) = \{\psi_1(y; y_0); \psi_2(y; y_0)\}$$

for (3.12) that obey

$$\begin{cases} \frac{d^2}{dy^2}\psi_1 + k^2\psi_1 - \alpha_1 F(\psi_1 - \psi_2) = -v_1\delta(y - y_0) \\ \frac{d^2}{dy^2}\psi_2 + k^2\psi_2 - \alpha_2 F(\psi_1 - \psi_2) + \epsilon(y)\psi_2 = -v_2\delta(y - y_0) \end{cases} \tag{3.13}$$

The latter could be rewriten in the matrix form

$$\left[\frac{d^2}{dy^2} + A^2 + \epsilon\Gamma\right]\Psi(y; y_0) = -\mathrm{v}\delta(y - y_0); \ \mathrm{v} = \begin{pmatrix} v_1 \\ v_2 \end{pmatrix} \tag{3.14}$$

with matrix-coefficients

$$A^2 = \begin{bmatrix} k^2 - \alpha_1 F & \alpha_1 F \\ \alpha_2 F & k^2 - \alpha_2 F \end{bmatrix}; \ \Gamma = \begin{bmatrix} 0 & 0 \\ 0 & 1 \end{bmatrix}$$

The fundamental matrix-function Ψ solves matrix equations (3.14) with source $-E\delta(y - y_0)$ in the r.h.s. and the vector Green's pair is then given by

$$\Psi(y; y_0)\begin{pmatrix} v_1 \\ v_2 \end{pmatrix} = \begin{pmatrix} \psi_{11}v_1 + \psi_{12}v_2 \\ \psi_{21}v_1 + \psi_{22}v_2 \end{pmatrix}$$

Two columns of Ψ represent coordinate sources $\begin{pmatrix} v_1 \\ 0 \end{pmatrix}$ and $\begin{pmatrix} 0 \\ v_2 \end{pmatrix}$. We apply a linear coordinate transformation

$$\Psi \to \Phi(y; y_0) = K\Psi(y; y_0)$$

with matrix $K = \begin{bmatrix} 1 & -1 \\ a_2 & a_1 \end{bmatrix}$ depending on parameters

$$a_1 = \frac{\alpha_1}{\alpha_1 + \alpha_2}; \ a_2 = \frac{\alpha_2}{\alpha_1 + \alpha_2}$$

That would bring system (3.14) into the form

$$\left[\frac{d^2}{dy^2}+B^2+\epsilon\tilde{\Gamma}\right]\Phi(y;y_0)=-K\delta(y-y_0) \tag{3.15}$$

with another set of coefficients

$$\text{diagonal matrix } B^2=\begin{bmatrix}\lambda^2 & \\ & k^2\end{bmatrix} \text{ and } \tilde{\Gamma}=\begin{bmatrix}a_2 & -1\\ -a_1a_2 & a_1\end{bmatrix}$$

The eigenvalue parameter $\lambda^2=k^2-(\alpha_1+\alpha_2)F$ describes the baroclinic mode.

As above we first assume that the inhomogeneities occupy a finite portion of space, strip $L_0<y<L$. Then the proper boundary conditions for (3.15) are the radiation conditions

$$\left(\frac{d}{dy}-iB\right)\Phi(y;y_0)\Big|_{y=L}=0;\ \left(\frac{d}{dy}+iB\right)\Phi(y;y_0)\Big|_{y=L_0}=0$$

If the source of plane waves is moved on the boundary of the strip $y_0=L$, we use as above the jump-condition at $y_0\pm 0$ to get the boundary-value problem

$$\begin{cases}\left[\frac{d^2}{dy^2}+B^2+\epsilon\tilde{\Gamma}\right]\Phi(y;L)=0\\ \left(\frac{d}{dy}-iB\right)\Phi(y;L)\Big|_{y=L}=K;\ \left(\frac{d}{dy}+iB\right)\Phi(y;L)\Big|_{y=L_0}=0\end{cases}$$

As a consequence we get the relations for fundamental matrices

$$\Phi(y;L)=U(y;L)K \text{ and } \Psi(y;L)=K^{-1}U(y;L)K$$

Introducing yet another fundamental matrix-function U, that solves

$$\begin{cases}\left[\frac{d^2}{dy^2}+B^2+\epsilon\tilde{\Gamma}\right]U(y;L)=0\\ \left(\frac{d}{dy}-iB\right)U(y;L)\Big|_{y=L}=E;\ \left(\frac{d}{dy}+iB\right)U(y;L)\Big|_{y=L_0}=0\end{cases} \tag{3.16}$$

we can apply once again the imbedding method (Appendix) to transform (3.16) into an initial value problem with respect to the boundary parameter L,

$$\begin{cases}\frac{\partial}{\partial L}U(y;L)=2iU(y;L)+\epsilon(L)U(y;L)\tilde{\Gamma}U(L;L)\\ U(y;L)|_{L=y}=U(y;y)\end{cases}$$

Hence follows the evolution equation for U in variable L

$$\begin{cases}\frac{d}{dL}U(L;L)=E+i[U(L;L)B+BU(L;L)]+\epsilon(L)U(L;L)\tilde{\Gamma}U(L;L)\\ U(L;L)|_{L=L_0}=\frac{i}{2}B^{-1}\end{cases}$$
(3.17)

From (3.17) we can pass to the original fundamental matrix $\Psi(y;L)$ and couples its evolution equation to that of the reflection coefficient matrix $\mathrm{r}(L)$. This yields a coupled system

$$\begin{cases} \frac{\partial}{\partial L}\Psi(y;L) = i\Psi(y;L)K^{-1}BK + \frac{i}{2}\epsilon(L)\Psi(y;L)K^{-1}\tilde{\Gamma}B^{-1}(E+R(L))K \\ \frac{d}{dL}R(L) = i[R(L)B + BR(L)] + \frac{i}{2}K^{-1}B^{-1}(E+R(L))K \end{cases}$$

subject to suitable initial condition at $L = y$ and $L = L_0$.

After the imbedding scheme is implemented in the deterministic problem one can pass to the statistical solution of (3.11) via the diffusion approximation. This time however (unlike the previous one-layer case) the diffusion approximation gives 4 different coefficients namely

$$\begin{array}{ll} D_1 \sim 4k^2\Phi_h(2k) & D_2 \sim 4k^2\Phi_h(2\lambda) \\ D_3 \sim 4(\lambda+k)^2\Phi_h(\lambda+k) & D_4 \sim 4(k-\lambda)^2\Phi_h(k-\lambda) \end{array}$$

They all have the dimension of inverse length and are connected to the barotropic and baroclinic modes.

Coefficients $\{D_i\}$ play different roles in the localization process. Indeed statistical solutions could vary significantly for various types of boundary problems, like barotropic and baroclinic modes excited in each of two layers. We shall illustrate the last point in two opposite asymptotic regimes: $\alpha_1 \ll \alpha_2$ (typical for the atmosphere), and $\alpha_1 \gg \alpha_2$ (typical for the ocean).

In the former case (atmosphere) the fundamental system (3.13) is replaced by

$$\begin{cases} \frac{d^2}{dy^2}\psi_1 + k^2\psi_1 = -v_1\delta(y-y_0) \\ \frac{d^2}{dy^2}\psi_2 + \left[\lambda^2 + \epsilon(y)\right]\psi_2 + \alpha_2 F\psi_1 = -v_2\delta(y-y_0) \end{cases}$$

Clearly the wave excited in the upper layer $v_2 = 0$ would propagate as a free barotropic mode and would excite a baroclinic mode in the inner layer subjected to the random topography. On the contrary if a Rossby wave is excited in the inner layer $v_1 = 0$ it would propagate just inside the inner layer and would be localized by the random scattering.

In the second limiting case (ocean) we replace (3.13) with

$$\begin{cases} \frac{d^2}{dy^2}\psi_1 + \lambda^2\psi_1 + \alpha_1 F\psi_2 = -v_1\delta(y-y_0) \\ \frac{d^2}{dy^2}\psi_2 + \left[k^2 + \epsilon(y)\right]\psi_2 = -v_2\delta(y-y_0) \end{cases}$$

Here a wave excited in the upper layer $v_2 = 0$ would propagate therein as a free wave. But an inner layer excitation $v_1 = 0$ would propagate inside it subject to the localization effects, and also excite the baroclinic mode in the upper layer. The latter would eventually propagate as a free wave.

Our brief survey of the localization phenomena for the baroclinic Rossby waves touches on a few essential points, but leave many issues open.We hope to return to this subject elsewhere.

4. Appendix: Imbedding method for wave problems. The following 2-nd order linear differential equation or system provides appears in many physical wave propagation models and problems

$$\left\{\begin{array}{c}\frac{d^2}{dt^2}X(t)+A(t)\frac{d}{dt}X(t)+K(t)X(t)=0\\ \left(\frac{d}{dt}+B\right)X(t)\big|_{t=T}=v;\ \left(\frac{d}{dt}+C\right)X(t)\big|_{t=0}=0\end{array}\right. \tag{4.1}$$

A standard approach to (4.1) is via reduction to a first order system, but this procedure doubles the system size and makes it cumbersome. So we shall work directly with (4.1). We take the fundamental matrix $G(t;T)$ of (4.1) that obeys

$$\left\{\begin{array}{c}\left(\frac{d^2}{dt^2}+A(t)\frac{d}{dt}+K(t)\right)G(t;T)=0\\ \left(\frac{d}{dt}+B\right)G(t;T)\big|_{t=T}=E;\ \left(\frac{d}{dt}+C\right)G(t;T)\big|_{t=0}=0\end{array}\right. \tag{4.2}$$

where $E=\delta_{ij}$ - the identity matrix, and write solution $X(t)=G(t;T)v$, and differentiate it with respect to the boundary parameter T

$$\left(\frac{d^2}{dt^2}+A(t)\frac{d}{dt}+K(t)\right)\frac{\partial}{\partial T}G(t;T)=0 \tag{4.3}$$

Hence $\frac{\partial}{\partial T}G$ is expressed in terms of the fundamental matrix itself as

$$\frac{\partial}{\partial T}G(t;T)=G(t;T)\Lambda \tag{4.4}$$

where Λ corresponds to the initial value of $\frac{\partial}{\partial T}G$ at $t=T$. Alternatively (4.4) could be viewed as a differential equation in $T>t$ supplemented with the initial condition

$$G(t;T)|_{T=t}=G(t;t) \tag{4.5}$$

Applying boundary operator $\left(\frac{d}{dt}+B\right)$ to (4.4) and setting $t=T$ we get the following expression for Λ

$$\Lambda(T)=A(T)-B+\left\{K(T)-A(T)B+B^2\right\}G(T;T)$$

The initial value $G(T;T)$ in turn obeys the relation

$$\frac{d}{dT}G(T;T)=\left\{\frac{\partial}{\partial t}G(t;T)+\frac{\partial}{\partial t}G(t;T)\right\}\bigg|_{T=t}=E-BG(T;T)+G(T;T)\Lambda(T)$$

The latter can be also viewed as a differential equation with the condition derived from (4.2) at $T=0$

$$G(0;0)=(B-C)^{-1}$$

One specific example is the 1-D boundary value problem

$$\begin{cases} \left\{\frac{d^2}{dt^2} + k^2(t)\right\} X(t) = 0 \\ \left(\frac{d}{dt} - ik\right) X\Big|_{t=T} = -2ik; \ \left(\frac{d}{dt} + ik\right) X\Big|_{t=0} = 0 \end{cases}$$

Here the imbedding equation for $X(t;T)$ takes the form

$$\frac{\partial}{\partial T} X(t;T) = X(t;T)\,\Lambda; \ X(t;T)|_{T=t} = X(t;t)$$

where Λ is found to be

$$\Lambda(T) = ik + \frac{i}{2k}\left[k^2(T) - k^2\right] X(T;T)$$

Function $X(T;T)$ in turn solves the Riccati equation

$$\frac{d}{dT} X(T;T) = 2ik\left[X(T;T) - 1\right] + \frac{i}{2k}\left[k^2(T) - k^2\right] X(T;T)$$

with initial condition $X(0;0) = 0$.

Acknowledgment: The first author was supported by the International Science Foundation grant # MBPOO, and the Russian Fund for Fundamental Research, grant # 94-05-16151.

REFERENCES

[1] J. Pedlosky, *Geophysical Fluid Dynamics*, Springer, N-Y, 1982.
[2] J. Miller, P. Weichman, M. Cross, *Statistical mechanics, Euler's equation and Jupiter's red spot,* Phys. Reviews A, **45**, 2328, 1992.
[3] R. Kraichnan, *Inertial waves in two-dimensional turbulence* , Phys. Fluids, **10**, 1417-1423, 1967.
[4] R. Kraichnan, *Statistical dynamics of two-dimensional flows*, J. Fluid Mech., **67**, 155-175, 1975.
[5] R. Kraichnan, D. Montgomery, *Two-dimensional turbulence*, Rep. Prog. Phys., **43**, 547-619, 1980.
[6] V. Klyatskin, *On statistical theory of two-dimensional turbulence*, J. Appl. Math. and Mech., **33** (5), 864-866, 1969.
[7] V. Klyatskin, D. Gurarie, *Equilibrium states for quasigeostrophic flows with random topography*, Physica D, 1996.
[8] A. Monin, A. Yaglom, *Statistical fluid mechanics*, MIT Press, Cambridge MA., 1980.
[9] V. Klyatskin, *Statistical description of dynamical systems with fluctuating parameters*, Nauka, Moscow, 1975 (Russian).
[10] V. Klyatskin, *Ondes et equations stochastique dans les milieux aleatorement non homogenes,* Edition de Physique, Besançon - Cedex, 1985.
[11] R. Salmon, G. Holloway, M. Hendershott, *The equilibrium statistical mechanics of simple quasigeostrophic models*, J. Fluid Mech. **75**, 4, 691-703, 1976.
[12] G. Holloway, *Eddies, waves, circulation and mixing: statistical geofluid mechanics*, Ann. Rev. Fluid Mech., **18**, 91-147, 1986.
[13] G. Carnevale, J. Frederiksen, *Nonlinear stability and statistical mechanics of flow over topography*, J. Fluid Mech., **175**, 157-181, 1987.

[14] N. P. Fofonoff, *Steady flows in a frictionless homogeneous ocean*, J. Mar. Res. **13**, 254, 1954.

[15] J. R. Herring, *On the statistical theory of two-dimensional topographic turbulence*, J. Atm. Sci., **34** (11), 1731-1750, 1977.

[16] H. H. Shen, *Strong and weak turbulence for gravity waves and the cubic Schrödinger equation*, Nonlinear waves and Weak turbulence with applications to Oceanography and Condenced Matter Physics, Boson, Burkhauser, 97-120, 1993.

[17] E. Hopf, *Statistical hydrodynamics and functional calculus*, J. Ration. Mech. Anal., **1**, 87, 1952.

[18] P. Phines, F. Bretherton, *Topographic Rossby waves in a rough-bottom ocean*, J.Fluid Mech, 61 (3), 583-608, 1973.

[19] J.B. Keller, G. Veronis, *Rossby waves in the presence of random currents*, J. Geoph. Res., 74 (8) 1969.

[20] R.E. Thomson, *The propagation of planetary waves over a random topography*, J. Fluid Mech., 70(2), 267-285, 1975.

[21] V. Klyatskin, *Stochastic equations and waves in randomly inhomogeneous medium*, Moskwa, Nauka, 1980 (Russian).
Ondes et Équations Stochastiques dans les Milieus Aléatoirement non Homogènes, Edition de Physique, Besancon-Cedex, 1985 (French).

[22] V. Klyatskin, *The embedding method in wave propagation theory*, Moskwa, Nauka, 1986.

[23] V. Klyatskin, *Statistical theory of radiation transport in stratified random media*, Izv. of Atm. and Oceanic Physics, 27 (1), 31-44, 1991.

[24] V. Klyatskin, A. Saichev, *Statistical and dynamical localization of plane waves in randomly layered media*, Sov. Physics Usp., 35 (3), 231-247, 1992.

[25] D. Segupta, L. Piterbarg, L. Reznik, *Localization of topographic Rossby waves over random relief*, Dynamics of Atm. and Oceans, 17, 1-21, 1992.

[26] P. Sheng, B. White, Z. Zhang, G. Papanicolaou, *Wave localization and multiple scattering in randomly layered media, Scattering and localization of Classical waves in Random media*, World Sci. Publ., Singapore, 1989.

[27] S.A. Gredesscul, V.D. Freilikher, Localization and wave propagation in randomly layered media, Soviet Phys. Usp., 33 (1), 134-146, 1990.

[28] M. Guzev, V. Klyatskin, G. Popov, *Phase fluctuations and localization length in layered randomly inhomogeneous media*, Waves in Random media, v.2, (2), 117-123, 1992.
M. Guzev, V. Klyatskin, *Influence of boundary conditions on statistical characteristics in layered randomly inhomogeneous medium*, Waves in Random media, 3 (4), 307-315, 1993.

DYNAMICAL AND STATISTICAL CHARACTERISTICS OF GEOPHYSICAL FIELDS AND WAVES AND RELATED BOUNDARY-VALUE PROBLEMS

V.I. KLYATSKIN* AND W.A. WOYCZYNSKI†

Abstract. Statistical characteristics of geophysical fields and waves in random media often differ considerably from the behavior of their realizations. Practically, in each specific realization of the process one can observe features, that are completely absent in its statistical description.

In the simplest case, such features are described by the lognormal probability distribution. Our illustrations of this phenomenon include parametric stochastic resonance, dynamical and statistical energy localization for wavefields in randomly layered media, wave beam propagation in random parabolic waveguides, and diffusing tracers in random velocity fields.

Another example of this phenomenon is an appearance of certain singularities in the dynamics of individual realizations, accompanied by their absence in the statistical description. Such models are often reduced to boundary-value problems for the corresponding Fokker-Planck equations. Examples of such features of statistical solutions are provided and include a comparison of the mean exponential divergence of geometric-optical rays in a random medium with the almost sure existence of caustics on finite distances, and the phase fluctuations of plane waves in a randomly layered medium.

1. Introduction. Statistical characteristics of fields in random media often differ considerably from the behavior of their separate realizations. In addition, in each specific realization of the process one can observe certain features that are absent in the statistical description. For example, in some realizations of wavefields in layered random media one can observe the *dynamical localization* but, in several cases, the *statistical energetic localization* is absent (the averaging is made over all medium realizations) [1]. Another example is provided by the phenomenon of the relative diffusion of rays, that is an *exponential divergence of rays* on the average, in a random medium in the *geometric optics approximation*, and the simultaneous existence of *caustics* at finite distances [2-4].

In this context, it is an important question whether it is possible to find any deterministic evolution patterns which would give a fairly complete description of characteristics of particular realizations of a random field. In what follows we will demonstrate that, in some cases, it is indeed possible.

The one-dimensional probability distributions of a random process $y(t)$ are determined by their *cumulative distribution functions*

$$F(y;t) = P\Big(y(t) < y\Big) = \int_{-\infty}^{y} dy' P_t(y') = \Big\langle \theta(y - y(t)) \Big\rangle,$$

* Institute of Atmospheric Physics, Russian Academy of Sciences, Moscow 109017, and Pacific Oceanological Institute, Russian Academy of Sciences, Vladivostok 690041.

† Department of Statistics and Center for Stochastic and Chaotic Processes in Science and Technology, Case Western Reserve University, Cleveland, Ohio 44106.

where $\theta(z)$ is the Heaviside function equal to 0 for $Z < 0$, and to 1 for $Z > 0$. Function $P_t(y) = \langle \delta(y(t) - y) \rangle$ is the one-time probability density of process $y(t)$, and the brackets $\langle\ .\ \rangle$ stand for averaging over the ensemble of realizations of the stochastic process $y(t)$.

For a fixed $0 < p < 1$, a deterministic function $Z(t;p)$ will define the so-called *p-isoprobable curve* of the process $y(t)$, if its value for any given moment of time t, is a solution of the equation

$$F(Z(t;p));t) = p.$$

Mathematicians will immediately recognize $Z(t;\ .)$ as a concrete representation of random variable $y(t)$. Integrating the above equality over an arbitrary time interval (t_1, t_2), we obtain that

$$\left\langle \int_{t_1}^{t_2} dt\, \theta\Big(Z(t;p) - y(t)\Big) \right\rangle = \langle T(t_1, t_2) \rangle = p(t_2 - t_1),$$

where $T(t_1, t_2) = \sum \Delta t_k$ is the total time, inside the interval (t_1, t_2), spent by process $y(t)$ underneath the isoprobable curve $Z(t;p)$. Hence, if p is selected close to 1, the plots of the realization of the process $y(t)$ are almost always below the isoprobable curve inside an any interval (t_1, t_2). More precisely, for any small $\epsilon > 0$,

$$p = \frac{\langle T(t_1, t_2) \rangle}{t_2 - t_1}$$

$$\leq (p - \epsilon) \cdot P\left(\frac{T(t_1, t_2)}{t_2 - t_1} \leq p - \epsilon \right) + 1 \cdot P\left(\frac{T(t_1, t_2)}{t_2 - t_1} > p - \epsilon \right)$$

$$\leq p \left[\frac{1}{p} - \epsilon P\left(\frac{T(t_1, t_2)}{t_2 - t_1} \leq p - \epsilon \right) \right]$$

and, as a consequence, necessarily,

$$P\left(\frac{T(t_1, t_2)}{t_2 - t_1} \leq p - \epsilon \right) \leq \frac{1 - p}{\epsilon p}.$$

In particular, if we select $p = 0.999$ and $\epsilon = 0.1$, then the above estimate assures that, with probability better than 0.99, the realizations of $y(t)$ will stay below the isoprobable curve more than 89.9 percent of time. Also, intuitively, if $p = 1/2$, the realization of the process $y(t)$ will weave around the isoprobable curve, being on the average half of the time above and a half of the time below that curve. For these reasons, it is natural to think of the isoprobable curve $Z(t; 1/2)$ as a *typical realization* of the process $y(t)$, although the plot of $Z(t; 1/2)$ can differ significantly from the plot of any realization of the process $y(t)$.

To illustrate the above phenomena, consider an example of the log-normal random process, which arises in all fields of physics where it is necessary to describe characteristics of positive physical quantities, processes and fields. This includes a description of intensity fluctuations of optical and radio waves in random media, and an analysis of the behavior of amplitudes of signals in radio systems subject to parameter fluctuations. For all these physical problems, a lognormal process emerges as the simplest adequate model, correctly accounting for principal properties of the phenomena under investigations, which include positivity, validity of conservation laws, and parametric instability [1].

We begin by an observation that the *Wiener random process* $w(t)$ is a solution of the stochastic equation

$$\frac{d}{dt}w(t) = z(t), \qquad w(0) = 0, \tag{1.1}$$

where $z(t)$ is a Gaussian process, delta-correlated in time with parameters $\langle z(t)\rangle = 0$, $\langle z(t)z(t')\rangle = 2\delta(t-t')$ (we assume the time t is a dimensionless quantity). The Wiener process is a continuous *nonstationary* Gaussian random process, determined by parameters

$$\langle w(t)\rangle = 0, \qquad \langle w(t)w(t')\rangle = 2\min(t,t'). \tag{1.2}$$

In addition to the usual Wiener process, we will have need for a more general Wiener process with a drift. It depends on parameter α via formula

$$w(t;\alpha) = -\alpha t + w(t).$$

Process $w(t;\alpha)$ is a Markov process, and it has a Gaussian one dimensional probability density $P_t(w) = \langle\delta(w(t;\alpha) - w)\rangle$ of the form

$$P_t(w) = \frac{1}{2(\pi t)^{1/2}}\exp\left[-\frac{(w+\alpha t)^2}{4t}\right]. \tag{1.3}$$

By the usual reflection principle we can find (see e.g. [1]) that the cumulative probability distribution of the absolute maximum

$$w_{max}(\alpha) = \max_{t\in(0,\infty)} w(t;\alpha)$$

of the process $w(t;\alpha)$– is of the form

$$F(h,\alpha) = P(w_{max}(\alpha) < h) = 1 - e^{-\alpha h}. \tag{1.4}$$

The *lognormal process*

$$y(t;\alpha) = e^{-w(t;\alpha)}, \tag{1.5}$$

constructed with the help of the Wiener process, has a probability density

$$P_t(y;\alpha) = \frac{1}{2(\pi t)^{1/2} y} \exp\left[-\frac{1}{4t}\ln^2(ye^{\alpha t})\right]. \tag{1.6}$$

Its moments are easily found to be

$$\langle y^n(t;\alpha)\rangle = e^{n(n-\alpha)t}$$

and, in particular, for process $y(t) = y(t;1) = e^{w(t)-t}$ they are given by the formula

$$\langle y^n(t)\rangle = e^{n(n-1)t}.$$

Thus, the mean value of process $y(t)$ is 1 and independent of t, whereas all the other moments of $y(t)$ grow exponentially with t.

The exponential increase of higher moments of the lognormal process $y(t)$ is caused by a slow decrease of *tails* of the probability density (1.6) for $y \gg 1$. As far as the realizations of the process $y(t)$ are concerned, this means that rare but high peaks will appear while t increases. Thus, most of the time, realizations of the process $y(t)$ will remain below the level of its mean value $\langle y(t)\rangle = 1$, although its statistical moments will be mainly determined by its large jumps. That apparent contradiction between the behavior of statistical moments of the process $y(t)$ and of its realizations provides a motivation for a more detailed study of the dynamics of realizations of process $y(t)$.

In the meaning described above, curve $Z(t;1/2) = e^{-t}$ is a typical realization of the introduced earlier process $y(t)$.

A *p-majorant curve* $M_p(t)$ for process $y(t)$ is defined by the condition

$$P\Big(y(t) \le M_p(t) \text{ for all } t\Big) = p.$$

A knowledge of the distribution (1.4) of the absolute maximum of process $W(t;\alpha)$ suggests a large class of majorant curves [1]

$$M_p(t;\beta) = (1-p)^{-1/\beta} e^{-(1-\beta)t} \tag{1.7}$$

parametrized by parameter $0 < \beta < 1$. Note that, despite the facts that the statistical mean of $\langle y(t)\rangle = 1$, and that the higher moments of process $y(t)$ grow exponentially, it is always possible to find an exponentially decreasing ($\beta < 1$) majorant curve (1.7) such that realization of process $y(t)$ will stay below it at all times, with any previously set probability $p < 1$. In particular, one-half of realizations of $y(t)$ will lie below the exponentially decreasing majorant curve

$$M_{1/2}(t;1/2) = 4e^{-t/2}. \tag{1.7'}$$

An existence of exponentially decreasing majorant curves implies conclusions explaining the statistical and the dynamical behavior of realizations of process $y(t)$:

• The exponential growth of higher moments of $y(t)$ is a purely statistical effect, caused by averaging over the whole ensemble of realizations.

• The area under an exponentially decreasing majorant curve is finite. Hence, very large jumps of process $y(t)$ do not influence the areas under its realizations, which, in practice, are also finite.

The random area

$$S = \int_0^\infty dt\, y(t)$$

under realizations of process $y(t)$ has the probability density [1]

$$P(S) = S^{-2} e^{-1/S}, \tag{1.8}$$

and cumulative distribution function

$$F(s) = P(S < s) = e^{-1/s}.$$

Observe, that all the moments of the random variable S are infinite.

2. Lognormal law in physical phenomena.

2.1. Stochastic parametric resonance. In the present subsection we shall discuss the stochastic parametric excitation in an oscillatory system with small linear friction due to parameter fluctuations. The dynamical system under consideration is described by the system of equations

$$\frac{d}{dt}x(t) = y(t), \qquad \frac{d}{dt}y(t) = -2\gamma y(t) - \omega_0^2[1 + z(t)]x(t), \tag{2.1}$$

with initial values $x(0) = x_0$ and $y(0) = y_0$. Process $z(t)$ is assumed to be stationary, Gaussian, with zero mean, variance $\sigma_z^2 = \langle z^2(t)\rangle$, and correlation time τ_0. In the delta-correlated process approximation (see, for example, [5]) $z(t)$ is assumed to have parameters

$$\langle z(t)\rangle = 0, \qquad \langle z(t)z(t')\rangle = 2\sigma_z^2\tau_0\delta(t - t').$$

System (2.1) appears in many areas of physics (see e.g. [5]), and it admits a parametric excitation because random process $z(t)$ contains harmonic components of all frequencies, including values $2\omega_0/n$, $n = 1, 2, 3, \ldots$, which exactly correspond to *parametric resonance* in a system with a periodic function $z(t)$ (Mathieu equation, $\gamma = 0$).

Let us rewrite solutions of system (2.1) in the form

$$x(t) = A(t)\sin(\omega_0 t + \phi(t)), \quad y(t) = \omega_0 A(t)\cos(\omega_0 t + \phi(t)), \tag{2.2}$$

where $A(t)$ is the oscillation amplitude function and $\phi(t)$— the phase function.

If parameter ω_0/D, where $D = \sigma_z^2\omega_0^2\tau_0$ is the diffusion coefficient in the corresponding Fokker-Planck equation, is large ($\omega_0/D \gg 1$), the relatively slow change (in t) of statistical characteristics of the solution system (2.2) is accompanied by ordinary oscillations with frequency ω_0. If the latter are eliminated by averaging the appropriate statistical characteristics over the time period $T = 2\pi/\omega_0$, then the probability density of the oscillation amplitude $A(t)$ is lognormal ([5]). Its moment functions satisfy equality

$$\langle A^n(t)\rangle = A_0^n \exp\left[-n\gamma t + \frac{1}{8}n(2+n)Dt\right].$$

If condition $8\gamma < (2+n)D$ is satisfied, then the stochastic system (2.1) is subject to stochastic parametric excitation (starting with the moment function of order n).

Note that the typical realization function of the random process $A(t)$ is of the form

$$A_0 e^{-(\gamma-D/4)t}.$$

Thus, if the friction in the dynamical system (2.1) is small, but $D < 4\gamma < (1+n/2)D$, then the typical realization function decays exponentially. However, at the same time, the moment functions of the order $\geq n$ have an exponential growth.

2.2. Localization of plane waves in layered random media. In this model, a layered, randomly inhomogeneous medium occupies a layered slab $L_0 < x < L$ in the space, and an inclined plane wave

$$U(x,\boldsymbol{\rho}) = \exp[ip(L-x) + i\boldsymbol{q}\boldsymbol{\rho}],$$

where $\boldsymbol{\rho} = (y,z)$, and $p = (k^2 - \boldsymbol{q}^2)^{1/2} = k\cos\theta$, is incident on it from the region $x > L$. Within the layer, wavefield

$$U(x,\boldsymbol{\rho}) = u(x)\exp[i\boldsymbol{q}\boldsymbol{\rho}],$$

where function $u(x)$ is a solution of the boundary value problem

$$\frac{d^2}{dx^2}u(x) + \left[p^2 + k^2\varepsilon(x)\right]u(x) = 0, \tag{2.3}$$

$$\frac{i}{p}\frac{d}{dx}u(x) + u(x)\Big|_{x=L} = 2, \quad \frac{i}{p}\frac{d}{dx}u(x) - u(x)\Big|_{x=L_0} = 0,$$

where $\varepsilon(x)$ is a random process describing random layering of the medium (e.g. fluctuations of the refraction coefficient for acoustic waves, or of the dielectric permittivity for the electromagnetic waves). For the wave normal incidence to the boundary $x = L$ the problem simplifies as $\theta = 0$ and $p = k$.

The wavefield $u(x)$ also depends on parameter L, and in the framework of the *imbedding method,* the boundary problem (2.3) can be reformulated as the following initial value problem for $u(x) = u(x; L)$ with respect to parameter L [6]:

$$(2.3') \qquad \frac{\partial}{\partial L} u(x; L) = ipu(x; L) + \frac{ik^2}{2p}\varepsilon(L)(1 + R_L))u(x; L),$$

$$\frac{d}{dL} R_L = 2ipR_L + \frac{ik^2}{2p}\varepsilon(L)(1 + R_L)^2,$$

where R_L is the reflection coefficient.

In the half-space limit ($L_0 \to -\infty$), the reflection coefficient $|R_L| = 1$ with probability 1, and it has the structure $R_L = \exp(i\phi_L)$ ([6,7]). Now, the phase ϕ_L and the wavefield intensity

$$I(x; L) = |u(x; L)|^2 = 2W(x; L)(1 + \cos\phi_L)$$

satisfy the imbedding equations

$$(2.4) \qquad \frac{d}{dL}\phi_L = 2p + \frac{k^2}{p}\varepsilon(L)(1 + \cos\phi_L),$$

$$\frac{\partial}{\partial L} I(x; L) = -\frac{k^2}{p}\varepsilon(L) I(x; L)\sin\phi_L.$$

Let's discuss the case of wave normal incidence. Function $W(x; L)$ has a structure ([5],[6])

$$(2.5) \qquad W(x; L) = e^{-[q(L) - q(x)]},$$

where function $q(L)$ is described by a stochastic equation

$$\frac{d}{dL} q(L) = k\varepsilon(L)\sin\phi_L.$$

If random process $\varepsilon(L)$ is a Gaussian process with variance σ_ε^2 and correlation radius l_0 then, in the delta-correlated process approximation, random functions $q(L)$ and ϕ_L are statistically independent, and the random quantity

$$\chi(x; L) = \ln W(x; L))$$

has a Gaussian distribution with parameters

$$\langle\chi(x; L)\rangle = -D(L - x), \quad \sigma_\chi^2(x; L) = D(L - x),$$

where $D = k^2\sigma_\varepsilon^2 l_0/2$ is a diffusion coefficient in the corresponding Fokker-Planck equation. Consequently, the quantity $W(x; L)$ has a lognormal distribution with mean $\langle W(x; L))\rangle = 1$, and higher order moments

$$\langle W^n(x; L))\rangle = e^{n(n-1)D(L-x)}, \quad n = 2, \ldots$$

growing exponentially, deep into the medium *(stochastic wave parametric resonance)* [5-7] .

This case was discussed in detail in Subsection 2.1. The lognormal distribution indicates the existence in each realization of rare, but strong discontinuities of the wavefield intensity. These jumps take place against the background of exponential decay described by function

$$W(x; L)) = \exp[-(L - x)l_{loc}^{-1}], \tag{2.6}$$

where

$$l_{loc}^{-1} = -\frac{d}{dL}\langle \chi(x; L)\rangle.$$

Usually, such a phenomenon is associated with the *dynamic localization* property of physical disordered systems, and the quantity l_{loc} is then called the *localization length* [8]. In the case considered above $l_{loc} = D^{-1}$. However, all statistical moments of quantities $W(x; L)$ increase, so the *statistical energetic localization* does not occur in this case.

Each realization of function $W(x; L)$ represents an exponentially decreasing function with possibly large peaks and, in the physics of disorder systems, function $\exp(-\xi)$, $\xi = D(L - x)$, is called a typical realization of the random function $W(x; L)$. This term is justified because it is the p-isoprobable curve for random function $W(x; L)$, corresponding to $p = 1/2$. In other words, for any finite interval on the axis $\xi = D(L - x)$, on the average, over half of the interval (measurewise) function $W(x; L)$ majorizes the typical realization, and over the other half it is dominated by it.

It is also possible to find a global majorant of the wavefield intensity $W(x; L)$. In particular, one can show that, with probability $p = 1/2$,

$$W(x; L) < 4e^{-\xi/2}$$

for all values of ξ [1].

Next, consider the situation where the source of plane waves is located inside the medium layer at a point x_0, $L_0 < x_0 < L$. Then, within the layer, the wavefield is described by the boundary-value wave problem ([5], [6])

$$\frac{d^2}{dx^2}G(x; x_0) + k^2\Big[1 + \varepsilon(x)\Big]G(x; x_0) = 2ik\,\delta(x - x_0), \tag{2.7}$$

$$\frac{i}{k}\frac{d}{dx}G(x; x_0) + G(x; x_0)\Big|_{x=L} = 0, \quad \frac{i}{k}\frac{d}{dx}G(x; x_0) - G(x; x_0)\Big|_{x=L_0} = 0,$$

where random process $\varepsilon(x)$ is, in general, complex-valued. Outside the layer $\varepsilon(x) = 0$, and inside the medium slab $\varepsilon(x) = \varepsilon_1(x) + i\gamma$, where $\varepsilon_1(x) = \varepsilon_1^*(x)$ is real-valued, and the attenuation parameter $\gamma \ll 1$ describes wave absorption. Introduction of small attenuation parameter is essential here since, as we shall see later, the solution of this statistical problem is singular with respect to parameter γ.

Note, that problem (2.3) of wave incidence on a medium layer corresponds, in problem (2.7), to the location of the source on the boundary $x_0 = L$. In other words, $u(x) = G(x; L)$.

The solution of boundary-value problem (2.7) is of the form

$$G(x;x_0) = \frac{1+R_2(x_0)}{1-R_1(x_0)R_2(x_0)}u(x;x_0), \qquad x \le x_0, \tag{2.8}$$

where quantity $R_1(L) = R_L$ is a reflection factor of the plane wave incident from region $x > L$ on the medium layer, and $u(x; L)$ is a solution of problem (2.3'). Quantity $R_2(x_0)$ has a similar meaning. Moreover, in the region $x < x_0$, the field of a point source is proportional to the field of the plane wave incident from the homogeneous space $x > x_0$ on the medium layer (L_0, x_0). The influence of the sublayer (x_0, L) is reflected only by quantity $R_2(x_0)$.

Problems with reflecting boundaries, where $\partial G/\partial x = 0$, are of special significance for physical applications. If a source is placed on the reflecting boundary then $R_2(x_0) = 1$ and, consequently,

$$G_{ref}(x;x_0) = \frac{2}{1-R_1(x_0)}u(x;x_0). \tag{2.8'}$$

Here, we will focus on statistical problems for a source in the infinite space (2.8), and for a source in the infinite half-space (2.8'), when the attenuation is small ($\gamma \to 0$).

If fluctuations $\varepsilon_1(x)$ are delta-correlated, then quantities $R_1(x_0)$ and $R_2(x_0)$ are statistically independent. As a result, for the infinite space, the mean intensity of a point source (2.8) is of the form

$$\langle I(x;x_0)\rangle = \beta^{-1}\Phi_{loc}(\xi), \quad \xi = D|x-x_0|,$$

for $\beta = k\gamma/D \ll 1$ [7], where the *localization curve*

$$\Phi_{loc}(\xi) = \lim_{\beta\to 0}\beta\langle I(x;x_0)\rangle = \lim_{\beta\to 0}\frac{\langle I(x;x_0)\rangle}{\langle I(x_0;x_0)\rangle}$$

$$= 2\pi\int_0^\infty d\tau\,\tau(\tau^2+1/4)\frac{\sinh\pi\tau}{\cosh^2\pi\tau}e^{-(\tau^2+1/4)\xi}.$$

When values of ξ are small, the localization curve $\sim e^{-2\xi}$ decreases rather quickly, and when values of ξ are large ($\xi \gg \pi^2$), it decreases much more slowly according to the universal law

$$\Phi_{loc}(\xi) \approx \frac{1}{8}\pi^{5/2}\xi^{-3/2}e^{-\xi/4}.$$

The same situation occurs when a source is located on the reflective boundary, although in this case,

$$\lim_{\beta\to 0}\frac{\langle I_{ref}(x;L)\rangle}{\langle I_{ref}(L;L)\rangle} = \frac{1}{2}\Phi_{loc}(\xi), \qquad \xi = D(L-x).$$

These differences in behavior of the mean intensity for different wave problems is caused by statistical averaging. For separate realizations of random function $\varepsilon(x)$, wave realizations have a similar spatial structure regardless of the case, and differ only by multiplicative constants which are different, however, for different realizations. The difference for the mean wave intensities are caused by correlation of these constants with the main spatial structure.

2.3. Wave-beam propagation in random parabolic waveguide. In the quasioptics approximation, wave-beam propagation in the direction of the x-axis is described by a parabolic equation [5, 9]

$$2ik\frac{\partial}{\partial x}u(x,\boldsymbol{\rho}) + \Delta_\perp u(x,\boldsymbol{\rho}) + k^2\varepsilon(x,\boldsymbol{\rho})u(x,\boldsymbol{\rho}) = 0, \tag{2.9}$$

where $\boldsymbol{\rho} = \{y, z\}$, and

$$\Delta_\perp = \frac{\partial^2}{\partial y^2} + \frac{\partial^2}{\partial z^2}.$$

Random field $\varepsilon(x,\boldsymbol{\rho})$ represents the fluctuating part of the dielectric permittivity or of the refractive index, and $\boldsymbol{\rho}$ designates coordinates in the plane perpendicular to the x-axis.

In several important cases describing propagation of waves in natural waveguides (such as acoustic waves in the ocean, or radio waves in the atmosphere) $\varepsilon(x,\boldsymbol{\rho}) = -\alpha^2\boldsymbol{\rho}^2$. Note that, in the absence of random inhomogeneities, wavefield $u(x,\boldsymbol{\rho})$ assumes the form

$$u_0(x,\boldsymbol{\rho}) = u_0\exp\left\{-\frac{\boldsymbol{\rho}^2}{2a^2} - i\alpha x\right\},$$

where parameter a satisfies condition $k\alpha a^2 = 1$. This wavefield corresponds to an eigenmode of the wave equation and its amplitude does not vary during the wave propagation.

Now, suppose that the fluctuation field is given by the formula

$$\varepsilon(x,\boldsymbol{\rho}) = -\alpha^2\boldsymbol{\rho}^2 + z(x)\boldsymbol{\rho}^2,$$

where $z(x)$ is a Gaussian, delta-correlated random process with parameters

$$\langle z(x)\rangle = 0, \qquad \langle z(x)z(x')\rangle = 2\sigma^2 l\delta(x-x').$$

The solution of the wave-beam problem (2.9) can be written in the form

$$u(x,\boldsymbol{\rho}) = u_0 \exp\left\{-\frac{\boldsymbol{\rho}^2}{2a^2}A(x) + B(x)\right\},$$

so that we obtain the following expression for the wavefield intensity [5]

$$I(x,\boldsymbol{\rho}) = I(x,0)\exp\left\{-\frac{\boldsymbol{\rho}^2}{a^2}I(x,0)\right\},$$

where

$$I(x,0) = \frac{1}{2}[A(x) + A^*(x)]$$

is the variation along the x-axis of wave intensity of the unperturbed waveguide.

Thus the statistical characteristics of wave intensity are determined by the statistical characteristics of quantity A. Equation for quantity A

$$\frac{d}{dx}A(x) = -\frac{i}{\kappa a^2}[A^2(x) - \alpha^2\kappa^2 a^4] - i\kappa^2 z(x), \quad A(0) = 1,$$

is similar to equation for the wave reflection coefficient in a one-dimensional problem [5], and if parameters of the beam are matched to parameters of the waveguide, that is if $\alpha k a^2 = 1$, then we get that [5]

$$\langle I^n(x,0)\rangle = e^{Dn(n-1)x}, \tag{2.10}$$

where $D = \sigma^2 l/2\alpha^2\kappa^2$ is the diffusion coefficient in the corresponding Fokker-Planck equation. This indicates that quantity $I(x,0)$ has a lognormal distribution. Moments of the wave intensity field grow exponentially along the waveguide axis. However, as we have seen earlier, the typical realization of wavefield intensity

$$I(x,0) \sim e^{-Dx}, \quad \text{for } Dx \gg 1,$$

so that the radiation escapes from the waveguide axis for each concrete realization; this indicates an existence of dynamical localization in the x-direction.

2.4. Statistical description of diffusing tracers in random velocity fields. Governig equations for the problem of diffusing tracers in a random velocity field are of the following two types:

$$\left(\frac{\partial}{\partial t}+\boldsymbol{U}(\boldsymbol{r},t)\frac{\partial}{\partial \boldsymbol{r}}\right)q(\boldsymbol{r},t)=\kappa\frac{\partial^2}{\partial \boldsymbol{r}^2}q(\boldsymbol{r},t),\quad q(\boldsymbol{r},0)=q_0(\boldsymbol{r}),\tag{2.11}$$

$$\left(\frac{\partial}{\partial t}+\boldsymbol{U}(\boldsymbol{r},t)\frac{\partial}{\partial \boldsymbol{r}}\right)p_i(\boldsymbol{r},t)=-\frac{\partial u_k(\boldsymbol{r},t)}{\partial r_i}p_k(\boldsymbol{r},t)+\kappa\frac{\partial^2}{\partial \boldsymbol{r}^2}p_i(\boldsymbol{r},t),$$

$$\boldsymbol{p}(\boldsymbol{r},0)=\boldsymbol{p}_0(\boldsymbol{r}).\tag{2.12}$$

Equation (2.11) describes a scalar field $q(\boldsymbol{r},t)$, and equation (2.12) describes its spatial gradient $\boldsymbol{p}(\boldsymbol{r},t)=\partial q(\boldsymbol{r},t)/\partial\boldsymbol{r}$. Let us also note equation

$$\left(\frac{\partial}{\partial t}+\frac{\partial}{\partial \boldsymbol{r}}\boldsymbol{U}(\boldsymbol{r},t)\right)\rho(\boldsymbol{r},t)=\kappa\frac{\partial^2}{\partial \boldsymbol{r}^2}\rho(\boldsymbol{r},t),\quad \rho(\boldsymbol{r},0)=\rho_0(\boldsymbol{r}),\tag{2.13}$$

for the density of a passive "matter". In the above equations, κ denotes the *molecular diffusion coefficient.*

The fluid flow can be either *compressible* or *incompressible* ($\nabla\boldsymbol{U}(\boldsymbol{r},t)=0$). In the latter case equations (2.11) and (2.13) are identical. In the one-dimensional case, equations (2.12) and (2.13) are also identical, and the fluid flow is always compressible.

Equations (2.11) - (2.13) give the *Eulerian description* of the system. The probability distribution of $q(\boldsymbol{r},t)$ can not be studied directly as (2.11) contains the second order (diffusion) term in $\boldsymbol{r}$. To remedy this, one can introduce an auxiliary field $\tilde{q}(\boldsymbol{r},t)$ described by a stochastic equation

$$\left(\frac{\partial}{\partial t}+\boldsymbol{U}(\boldsymbol{r},t)\frac{\partial}{\partial \boldsymbol{r}}\right)\tilde{q}(\boldsymbol{r},t)=-\boldsymbol{\alpha}(t)\frac{\partial}{\partial \boldsymbol{r}}\tilde{q}(\boldsymbol{r},t),\quad \tilde{q}(\boldsymbol{r},0)=q_0(\boldsymbol{r}),\tag{2.14}$$

where $\boldsymbol{\alpha}(t)$ is a delta-correlated Gaussian random vector process (independent of $\boldsymbol{U}$) with parameters

$$\langle\boldsymbol{\alpha}(t)\rangle=0,\quad \langle\alpha_i(t)\alpha_j(t')\rangle=2\kappa\delta_{ij}\delta(t-t');\quad i,j=1,2,3.$$

Then, the solution of (2.11) corresponds to ensemble averaging of (2.14) relative to the $\boldsymbol{\alpha}$ - process, that is [10]

$$q(\boldsymbol{r},t)=\langle\tilde{q}(\boldsymbol{r},t)\rangle_{\boldsymbol{\alpha}}.\tag{2.15}$$

Formula (2.15) gives a *path integral representation* of solution (2.11).

The first order stochastic partial differential equation (2.14) can be solved by the *method of characteristics* which replaces (2.14) by a system of ordinary differential equations

$$\frac{d}{dt}\boldsymbol{r}(t)=\boldsymbol{U}(\boldsymbol{r}(t),t)+\boldsymbol{\alpha}(t),\qquad \boldsymbol{r}(0)=\boldsymbol{\xi};$$

$$\frac{d}{dt}\tilde{q}(t) = 0, \qquad \tilde{q}(0) = q_0(\boldsymbol{\xi}). \tag{2.16}$$

Solution of the equations (2.16) depends on the initial parameter ξ

$$\boldsymbol{r}(t) = \boldsymbol{r}(t|\boldsymbol{\xi}); \qquad \tilde{q}(t) = \tilde{q}(t|\boldsymbol{\xi}).$$

That is the *Lagrangian description.* Eliminating parameter $\boldsymbol{\xi}$ in system (2.16) we obtain the *Eulerian description* of the concentration field

$$\boldsymbol{\xi} = \boldsymbol{\xi}(t, \boldsymbol{r}); \qquad \tilde{q}(\boldsymbol{r}, t) = \tilde{q}\Big(t|\boldsymbol{\xi}(t, \boldsymbol{r})\Big).$$

In order to trace the main features of statistical solutions of the diffusing tracer problem, we restrict ourselves by the simplest case: a one-dimensional problem with zero-mean velocity and without molecular diffusion. In this case we have equations

$$\left(\frac{\partial}{\partial t} + u(x,t)\frac{\partial}{\partial x}\right) q(x,t) = 0, \qquad q(x,0) = q_0(x); \tag{2.17}$$

$$\left(\frac{\partial}{\partial t} + \frac{\partial}{\partial x}u(x,t)\right) \rho(x,t) = 0, \qquad \rho(X,0) = \rho_0(x) \tag{2.18}$$

instead of equations (2.11) and (2.13). The above one-dimensional problem always describes a compressible fluid flow, and spatial concentration gradient $p(x,t) = \partial q(x,t)/\partial x$ is also described by equation (2.18).

The method of characteristics applied to equation (2.17) gives the corresponding Lagrangian description

$$\frac{d}{dt}x(t|\xi) = u(x(t|\xi), t), \qquad x(0|\xi) = \xi;$$

$$\frac{d}{dt}q(t|\xi) = 0, \qquad q(0|\xi) = q_0(\xi). \tag{2.19}$$

Hence, $q(t|\xi) = q_0(\xi)$. The divergence described by quantity $j(t|\xi) = |\partial u(t|\xi)/\partial \xi|$ plays an important role if transferred to Eulerian description. In view of (2.19), it satisfies equation

$$\frac{d}{dt}j(t|\xi) = \frac{\partial u(x,t)}{\partial x}j(t|\xi), \quad j(0|\xi) = 1. \tag{2.20}$$

To describe statistical properties of $x(t|\xi)$ and $j(t|\xi)$, we will have need of function

$$\Phi_t(x, j|\xi) = \delta\Big(x(t|\xi) - x\Big)\delta\Big(j(t|\xi) - j\Big),$$

satisfying Liouville's equation

$$\frac{\partial}{\partial t}\Phi_t(x,j|\xi) = -\left[\frac{\partial}{\partial x}u(x,t) + \frac{\partial}{\partial j}j\frac{\partial u(x,t)}{\partial x}\right]\Phi_t(x,j|\xi),$$

$$\Phi_0(x,j|\xi) = \delta(x-\xi)\delta(j-1). \tag{2.21}$$

For the sake of simplicity we shall assume random velocity field $u(x,t)$ to be Gaussian, homogeneous and isotropic in space, and stationary in time, with parameters

$$\langle u(x,t)\rangle = 0, \qquad B(x-x',t-t') = \langle u(x,t)u(x',t')\rangle.$$

We shall also keep the assumption that the field is delta-correlated in time, in which case correlation function $B(x,t)$ may be approximated by the expression

$$B(x,t) = 2B^{eff}(x)\delta(t), \qquad 2B^{eff}(x) = \int_{-\infty}^{\infty} d\tau\, B(x,\tau). \tag{2.22}$$

By averaging equation (2.22) over the realization ensemble of field $u(x,t)$, we obtain the Fokker-Planck equation for joint probability density $P_t(x,j|\xi) = \langle\Phi_t(x,j|\xi)\rangle$ of the "particle" position and of its divergence ([11], [12])

$$\frac{\partial}{\partial t}P_t(x,j|\xi) = D_1\frac{\partial^2}{\partial x^2}P_t(x,j|\xi) + D_2\frac{\partial^2}{\partial j^2}j^2P_t(x,j|\xi),$$

$$P_0(x,j|\xi) = \delta(x-\xi)\delta(j-1), \tag{2.23}$$

where diffusion coefficients D_i are determined by equalities

$$D_1 = B^{eff}(0), \qquad D_2 = -\frac{\partial^2}{\partial x^2}B^{eff}(x)\Big|_{x=0}.$$

It is clear from equation (2.23) that the diffusion of a "particle" does not depend upon divergence statistics, and that it is described by a Gaussian probability distribution with parameters

$$\langle x(t|\xi)\rangle = \xi, \qquad \sigma_x^2(t) = \left\langle[x(t|\xi) - \langle x(t|\xi)\rangle]^2\right\rangle = 2D_1t,$$

i.e. it corresponds to the usual Brownian motion. Probability distribution of the divergence j is lognormal, and

$$\langle j(t)\rangle = 1, \qquad \langle j^n(t)\rangle = e^{D_2n(n-1)t}, \tag{2.24}$$

i.e. the mean value of divergence is constant and moments of order ≥ 2 grow exponentially in time.

Note, that quantity $\tilde{\rho}(t) = 1/j(t)$, which has the meaning of particle density, satisfies in the Lagrangian description equation

$$\frac{d}{dt}\tilde{\rho}(t) = -\frac{\partial u(x,t)}{\partial x}\tilde{\rho}(t), \quad \tilde{\rho}(0) = 1,$$

and one can prove that it has a lognormal probability distribution with the moment functions

$$\langle \tilde{\rho}^n(t) \rangle = e^{D_2 n(n+1)t}.$$

Thus, the mean density of passive tracer grows exponentially in time together with its higher moments.

This "paradoxical" behavior of statistical characteristics of divergence and of particle density, which both have moment functions simultaneously growing in time, is due to properties of the lognormal probability distribution. Thus a typical realization of the random divergence field may be represented by an exponentially decaying curve

$$j(t) = e^{-D_2 t},$$

and there exists exponential majorants for realizations of random process $j(t)$. In particular, with probability $p = 1/2$,

$$j(t) < 4e^{-D_2 t/2}$$

for any time t. Similarly, a typical realization of the density is of the form

$$\tilde{\rho}(t) = e^{D_2 t},$$

and there exist exponential minorants so that, with probability $p = 1/2$,

$$\tilde{\rho}(t) > \frac{1}{4}e^{D_2 t/2}.$$

The presented above estimates for the statistics of $j(t)$ and $\tilde{\rho}(t)$ indicate the presence in their realizations of jumps over the typical realization as the particles are being compressed and form clusters located in mostly low density zones.

Let us consider now the Eulerian description of our problem and introduce probability density functions

$$P_{t,x}(q) = \langle \delta(q(x,t) - q) \rangle, \qquad P_{t,x}(\rho) = \langle \delta(\rho(x,t) - \rho) \rangle,$$

which are described by the Fokker-Planck equations [12]

$$(2.25) \qquad \frac{\partial}{\partial t}P_{t,x}(q) = D_1 \frac{\partial^2}{\partial x^2}P_{t,x}(q), \qquad P_{0,x}(q) = \delta(q_0(x) - q);$$

$$\frac{\partial}{\partial t}P_{t,x}(\rho) = D_1\frac{\partial^2}{\partial x^2}P_{t,x}(\rho) + D_2\frac{\partial^2}{\partial \rho^2}\rho^2 P_{t,x}(\rho),$$

$$P_{0,x}(\rho) = \delta(\rho_0(x) - \rho). \tag{2.26}$$

The solution of (2.25) corresponds to the spatial diffusion of the initial distribution. In the simplest case of homogeneous initial condition, the probability distribution does not depend on x and $P_t(q) = \delta(q - q_0)$.

If the initial density $\rho_0(x) = \rho_0$ is constant then the probability distribution does not depend on x either, and equation (2.26) takes a simplified form

$$\frac{\partial}{\partial t}P_t(\rho) = D_2\frac{\partial^2}{\partial \rho^2}\rho^2 P_t(\rho), \qquad P_0(\rho) = \delta(\rho_0 - \rho). \tag{2.27}$$

The solution of (2.27) corresponds to the lognormal distribution and

$$\langle \rho(x,t)\rangle = \rho_0, \qquad \langle \rho^n(x,t)\rangle = \rho_0^n e^{D_2 n(n-1)t}. \tag{2.28}$$

From (2.27) and (2.28) one can obtain a typical realization

$$\rho(x,t) = \rho_0 e^{-D_2 t}$$

of the random field at any fixed point in space. The Eulerian statistics are related to density fluctuations relative to this curve, which confirms their cluster character.

The spatial gradient of concentration $p(x,t) = \partial q(x,t)/\partial x$ is described by an equation which coincides with the equation for media density. In this case, the joint probability density $P_{t,x}(q,p) = \langle \delta(q(x,t) - q)\delta(p(x,t) - p)\rangle$ for values of $q(x,t)$ and of $p(x,t)$ is also described by equation

$$\frac{\partial}{\partial t}P_{t,x}(q,p) = D_1\frac{\partial^2}{\partial x^2}P_{t,x}(q,p) + D_2\frac{\partial^2}{\partial p^2}p^2 P_{t,x}(q,p),$$

$$P_{0,x}(q,p) = \delta\Big(q_0(x) - q\Big)\delta\Big(\frac{\partial}{\partial x}q_0(x) - p\Big),$$

from which one obtains the joint moment functions

$$\Big\langle q^n(x,t)p^m(x,t)\Big\rangle \sim e^{D_2 m(m-1)t}.$$

Hence, at a fixed point in space, the statistics of concentration gradient are formed by jumps with respect to a typical, exponentially decaying realization.

The peculiarities of the statistical solutions described above are entirely caused by the compressibility of the one dimensional flow (the case of a multidimensional compressible flow is considered in [13]). The situation changes completely in the case of a multidimensional incompressible fluid flow. As in the case of the one dimensional flow, we will limit ourselves to sufficiently large scale motions, when the impact of molecular diffusion can be neglected, and will assume that the flow has zero mean. In this case, the concentration q of the passive tracer and its gradient $\boldsymbol{p}$ satisfy equations

$$\left(\frac{\partial}{\partial t}+\boldsymbol{u}(\boldsymbol{r},t)\frac{\partial}{\partial \boldsymbol{r}}\right)q(\boldsymbol{r},t)=0,\quad q(\boldsymbol{r},0)=q_0(\boldsymbol{r}), \tag{2.29}$$

$$\left(\frac{\partial}{\partial t}+\boldsymbol{u}(\boldsymbol{r},t)\frac{\partial}{\partial \boldsymbol{r}}\right)p_i(\boldsymbol{r},t)=-\frac{\partial u_k(\boldsymbol{r},t)}{\partial r_i}p_k(\boldsymbol{r},t),$$

$$\boldsymbol{p}(\boldsymbol{r},0)=\boldsymbol{p}_0(\boldsymbol{r})=\frac{\partial q_0(\boldsymbol{r})}{\partial \boldsymbol{r}}. \tag{2.30}$$

Consider the Eulerian description and introduce, as before, function

$$\Phi_{t,r}(q,\boldsymbol{p})=\delta\Big(q(\boldsymbol{r},t)-q\Big)\delta\Big(\boldsymbol{p}(\boldsymbol{r},t)-\boldsymbol{p}\Big), \tag{2.31}$$

which determines the one-point joint distribution of fields q and $\boldsymbol{p}$ at a given point in space, and which satisfies Liouville's equation

$$\left(\frac{\partial}{\partial t}+\boldsymbol{u}(\boldsymbol{r},t)\frac{\partial}{\partial \boldsymbol{r}}\right)\Phi_{t,r}(q,\boldsymbol{p})=\frac{\partial u_k(\boldsymbol{r},t)}{\partial r_i}\frac{\partial}{\partial p_i}p_k\Phi_{t,r}(q,\boldsymbol{p}), \tag{2.32}$$

with the initial condition

$$\Phi_{0,r}(q,\boldsymbol{p})=\delta(q_0(\boldsymbol{r})-q)\delta(\nabla_r q_0(\boldsymbol{r})-\boldsymbol{p}). \tag{2.32$'$}$$

Furthermore, we assume that the velocity field $\boldsymbol{u}(\boldsymbol{r},t)$ is Gaussian, homogeneous and isotropic in space, and stationary in time. Its statistical properties are completely determined by its covariance tensor

$$B_{ij}(\boldsymbol{r}-\boldsymbol{r}',t-t')=\Big\langle u_i(\boldsymbol{r},t)u_j(\boldsymbol{r}',t')\Big\rangle,\quad \langle\boldsymbol{u}(\boldsymbol{r},t)\rangle=0.$$

For the sake of simplicity we will limit ourselves, as we did earlier, to the delta-correlated in time approximation of the field $\boldsymbol{u}(\boldsymbol{r},t)$, when the covariance function B_{ij} can be approximated by the expression

$$B_{ij}(\boldsymbol{r},t)=2B_{ij}^{eff}(\boldsymbol{r})\delta(t),\quad 2B_{ij}^{eff}(\boldsymbol{r})=\int_{-\infty}^{\infty}dt\,B_{ij}(\boldsymbol{r},t). \tag{2.33}$$

Let us introduce the spectral density of the energy of the (incompressible) fluid flow by the formula

$$B_{ij}^{eff}(\boldsymbol{r})=\int d\boldsymbol{\kappa}\,E(\boldsymbol{\kappa})\left(\delta_{ij}-\frac{\kappa_i\kappa_j}{\boldsymbol{\kappa}^2}\right)\exp(i\boldsymbol{\kappa}\boldsymbol{r}), \tag{2.34}$$

and average equation (2.32) with respect to the ensemble of realizations of the random field $\boldsymbol{u}$. As a result, we obtain the Fokker-Planck equation

$$\frac{\partial}{\partial t} P_{t,r}(q,\boldsymbol{p}) = D_1 \frac{\partial^2}{\partial \boldsymbol{r}^2} P_{t,r}(q,\boldsymbol{p}) \tag{2.35}$$

$$+ D_2 \left\{ -2\frac{\partial}{\partial \boldsymbol{p}}\boldsymbol{p} + (N+1)\frac{\partial^2}{\partial \boldsymbol{p}^2}\boldsymbol{p}^2 - 2\left(\frac{\partial}{\partial \boldsymbol{p}}\boldsymbol{p}\right)^2 \right\} P_{t,r}(q,\boldsymbol{p}),$$

for the joint one-point probability density $P_{t,r}(q,\boldsymbol{p}) = \langle \Phi_{t,r}(q,\boldsymbol{p}) \rangle$, where N is the dimension of space (that is $N = 3$ or 2), and constants D_i are diffusion coefficients for the Fokker-Planck equation defined by formulas

$$D_1 = \frac{N-1}{N} \int d\boldsymbol{\kappa}\, E(\boldsymbol{\kappa}), \quad D_2 = \frac{1}{N(N+2)} \int d\boldsymbol{\kappa}\, \boldsymbol{\kappa}^2 E(\boldsymbol{\kappa}). \tag{2.36}$$

Note, that in this case, the solution of equation (2.35) can be written in the form

$$P_{t,r}(q,\boldsymbol{p}) = \int d\boldsymbol{\xi}\, P(\boldsymbol{r}-\boldsymbol{\xi},t) P_{t,\xi}(q,\boldsymbol{p}), \tag{2.37}$$

where

$$P(\boldsymbol{r},t) = \exp\left(tD_1 \frac{\partial^2}{\partial \boldsymbol{r}^2}\right) \delta(\boldsymbol{r}-\boldsymbol{\xi})$$

is the corresponding probability density of the particle trajectory (Gaussian in our case) in the Lagrangian representation, and where the quantity $P_{t,r}(q,\boldsymbol{p})$ is the joint probability density of the tracer concentration and its gradient. In the Lagrangian description it satisfies equation

$$\frac{\partial}{\partial t} P_{t,r}(q,\boldsymbol{p}) = D_2 \left\{ -2\frac{\partial}{\partial \boldsymbol{p}}\boldsymbol{p} + (N+1)\frac{\partial^2}{\partial \boldsymbol{p}^2}\boldsymbol{p}^2 - 2\left(\frac{\partial}{\partial \boldsymbol{p}}\boldsymbol{p}\right)^2 \right\} P_{t,r}(q,\boldsymbol{p}). \tag{2.35'}$$

Equality (2.37) is a consequence of the Liouville Theorem which states that, for an incompressible flow, the Jacobian of transformation from Lagrangian to Eulerian coordinates

$$j(\boldsymbol{r},t|\boldsymbol{\xi}) = \operatorname{Det} \left\| \frac{\partial r_i(t|\boldsymbol{\xi})}{\partial \xi_j} \right\| \equiv 1.$$

Equation (2.35') also gives formulas for moments of the absolute value of the tracer concentration gradient (in Lagrangian representation)

$$\left\langle |\boldsymbol{p}(t|\boldsymbol{\xi})|^n \right\rangle = |\boldsymbol{p}_0(\boldsymbol{\xi})|^n \exp\{D_2 n(N+n)(N-1)t\}. \tag{2.38}$$

As a consequence, random quantity $|\boldsymbol{p}(t|\boldsymbol{\xi})|$ has a lognormal probability distribution with parameters

$$\langle\chi(t)\rangle = D_2 N(N-1)t, \quad \sigma_\chi^2(t) = 2D_2(N-1)t, \tag{2.39}$$

where $|\boldsymbol{p}| = |p_0(\boldsymbol{\xi})| \exp\chi(t)$.

In view of general properties of the lognormal distribution, in this case (as different from the one dimensional case) the typical realization of process $|\boldsymbol{p}(t)|$ is exponentially increasing in time, or more exactly

$$|\boldsymbol{p}(t|\boldsymbol{\xi})| \sim |p_0(\boldsymbol{\xi})| \exp\{D_2 N(N-1)t\}. \tag{2.40}$$

The realizations are dominated by large excursions above this typical curve. In addition, as we indicated above, there exist various probabilistic minorants for these realizations.

In the Eulerian description, the exponential growth in time displayed in (2.38) is caused by the ordinary diffusion in space of the Brownian type.

Paper [13] provides a vivid geometric interpretation of equality (2.38). Function $\delta(q(\boldsymbol{r},t)-q)$, which enters in definition (2.31), determines a surface (in the 3D case) or a contour (in the 2D case) of constant concentration, and as a result, quantities

$$A_n(t) = \int d\boldsymbol{r}\, |\boldsymbol{p}(\boldsymbol{r},t)|^{n+1} \delta(q(\boldsymbol{r},t)-q) = \oint |p(\boldsymbol{r},t)|^n dS \tag{2.41}$$

determine the averaged (in the geometric sense) value of the magnitude of the tracer concentration gradient on this surface (or contour). In this case, quantity (2.41) for $n=0$ defines the area of the surface (length of the contour) [14-17]

$$A_0(t) = \int d\boldsymbol{r}\, |\boldsymbol{p}(\boldsymbol{r},t)| \delta(q(\boldsymbol{r},t)-q) = \oint dS, \tag{2.41'}$$

and values of the tracer concentration gradient, averaged of the volume V bound by this surface, are given by formula

(2.41″)

$$\boldsymbol{A}(t) = \int_V \nabla q(\boldsymbol{r},t) dV = q \int d\boldsymbol{r}\, \boldsymbol{p}(\boldsymbol{r},t) \delta(q(\boldsymbol{r},t)-q) = q \oint \frac{\boldsymbol{p}(\boldsymbol{r},t)}{|\boldsymbol{p}(\boldsymbol{r},t)|} dS$$

Equation (2.41-41") can be rewritten in terms of the distribution function $\Phi_{t,r}(q,\boldsymbol{p})$ in the form

$$A_n(t) = \int d\boldsymbol{r} \int d\boldsymbol{p}\, |\boldsymbol{p}(\boldsymbol{r},t)|^{n+1} \Phi_{t,r}(q,\boldsymbol{p}), \tag{2.42}$$

$$\boldsymbol{A}(t) = q \int d\boldsymbol{r} \int d\boldsymbol{p}\, \boldsymbol{p}(\boldsymbol{r},t) \Phi_{t,r}(q,\boldsymbol{p}),$$

and the corresponding mean values are determined by the probability density $P_{t,r}(q, \boldsymbol{p})$ via formulas

$$\langle A_n(t)\rangle = \int d\boldsymbol{r} \int d\boldsymbol{p}\, |\boldsymbol{p}(\boldsymbol{r},t)|^{n+1} P_{t,r}(q,\boldsymbol{p}), \tag{2.42'}$$

$$\langle \boldsymbol{A}(t)\rangle = q \int d\boldsymbol{r} \int d\boldsymbol{p}\, \boldsymbol{p}(\boldsymbol{r},t) P_{t,r}(q,\boldsymbol{p}).$$

Differentiating (2.42') with respect to time, and utilizing equation (2.35) and (2.38), we obtain explicit expressions

$$\langle A_n(t)\rangle = A_n(0)\exp\{D_2(n+1)(N+n+1)(N-1)t\}, \tag{2.43}$$

$$\langle \boldsymbol{A}(t)\rangle = \boldsymbol{A}(0).$$

In particular, for $n = 0$, we get that

$$\langle S(t) = \begin{cases} S_0 \exp(8D_2 t), & \text{for } N = 3; \\ l_0 \exp(3D_2 t), & \text{for } N = 2. \end{cases} \tag{2.44}$$

This means that the average surface area (or contour length) of the surface (contour) of constant concentration grow exponentially in time. At the same time, in view of Liouville Theorem, the volume bounded by this surface is preserved. This fact, together with (2.43) indicates that we are witnessing here emergence of the fractal structure of of the surface of constant concentration (see e.g. [18]).

From the physical viewpoint, solutions (2.38) and (2.40) essentially differ from one dimensional solutions. The emergence for large times, at an arbitrary point in space, of structures with finer and finer spatial scales, makes it necessary to take into account the molecular diffusion at some instant of time.

The above discussion indicates that the knowledge of the behavior of moment functions for concentration and its gradient, or for the spatial density, is insufficient for a detailed description of the passive tracer diffusion. What is needed here is a more complete study of the full probability distributions. In the general, three dimensional case, the molecular diffusion term is the cause of principal difficulties and one has to resort to approximate methods. Some of these approaches have been tried in recent papers [19-21].

3. Boundary-value problems for Fokker-Planck equation.

3.1. Probability distribution of the maximum of the Wiener process. As a simple but instructive example of the boundary-value problem for the Fokker-Planck equation we will consider derivation of the probability distribution (1.4) following a method employed in [1].

For the purposes of this subsection, the Wiener process $w(t;\alpha)$ is described by a stochastic differential equation

$$\frac{dw(t;\alpha)}{dt} = -\alpha + z(t), \quad w(0,\alpha) = 0, \tag{3.1}$$

where $z(t)$ is a Gaussian process, delta-correlated in dimensionless (as in the introduction) time, with parameters

$$\langle z(t) \rangle = 0, \quad \langle z(t) z(t') \rangle = 2\delta(t-t').$$

Process $w(t,\alpha)$ is a Markov process, and its probability density $P_t(w) = \langle \delta(w(t,\alpha) - w \rangle$ satisfies the Fokker-Planck equation

$$\frac{\partial}{\partial t} P_t(w) = \alpha \frac{\partial}{\partial w} P_t(w) + \frac{\partial^2}{\partial w^2} P_t(w), \quad P_0(w) = \delta(w). \tag{3.2}$$

Its solution has a form of Gaussian distribution (1.3).

Now, in addition to the initial condition, let us complement equation (3.2) with a boundary condition

$$P_t(w = h) = 0, \quad t > 0, \tag{3.3}$$

breaking the realizations of the process $w(t;\alpha)$ at the moment when they reach the boundary h. The solution of the boundary problem (3.2-3) will be denoted by $P_t(w;h)$. For $w < h$, it describes the probability distribution of values of those realizations of the process $w(t;\alpha)$ that have survived up to time t; that is, during the entire time they have never reached the boundary h. Consequently, the probability density $P_t(w;h)$ integrates not to 1 but to the probability of $t > t^*$, where t^* is the time when process $w(t;\alpha)$ reaches the boundary h for the first time:

$$\int_{-\infty}^{h} dw\, P_t(w;h) = P(t < t^*). \tag{3.4}$$

Let us introduce the cumulative distribution function and the probability density of the random time moment when the boundary is attained for the first time:

$$F(t;\alpha,h) = 1 - P(t < t^*) = 1 - \int_{-\infty}^{h} dw\, P_t(w;h), \tag{3.5}$$

$$P(t;\alpha,h) = \frac{\partial}{\partial t} F = -\frac{\partial}{\partial w} P_t(w;h) \Big|_{w=h}. \tag{3.5$'$}$$

If $\alpha > 0$, then on the average process $w(t;\alpha)$ moves away from the boundary h as t increases. When $t \to \infty$, the probability $P(t < t^*)$ in (3.4) converges to the probability of an event that process $w(t;\alpha)$ never reaches the boundary h. In other words,

$$\lim_{t\to\infty}\int_{-\infty}^{h} dw\, P_t(w;h) = P(w_m < h), \tag{3.6}$$

where

$$w_m(\alpha) = \max_{t\in(0,\infty)} w(t;\alpha) \tag{3.7}$$

is the global maximum of the process $w(t;\alpha)$. Thus, it follows from (3.6) and (3.4) that the cumulative distribution function of the global maximum w_m has a form

$$F(h;h) = P(w_m < h) = \lim_{t\to\infty}\int_{-\infty}^{h} dw\, P_t(w;h). \tag{3.8}$$

For example, having solved the boundary problem (3.2-3) by the reflection method, we obtain that

$$P_t(w;h) = \frac{1}{2\sqrt{\pi t}}\left(\exp\left[-\frac{(w+\alpha t)^2}{4t}\right]\right. \\ \left. - \exp\left[-h\alpha - \frac{(w-2h+\alpha t)^2}{4t}\right]\right). \tag{3.9}$$

Substituting this expression into (3.5), we find that the probability density of the first hitting time t^* of the boundary h by process $w(t;\alpha)$ is given by the formula

$$P_t(w;\alpha,h) = \frac{1}{2t\sqrt{\pi t}}\exp\left[-\frac{(h+\alpha t)^2}{4t}\right] \tag{3.10}$$

Integrating (3.9) with respect to w and allowing $t \to \infty$ we obtain the cumulative distribution function

$$F(h;\alpha) = 1 - e^{-\alpha h}$$

of the global maximum w_m, as claimed in (1.4).

3.2. Phase fluctuations of plane wave in layered random medium. Localization of plane waves' normal incidence to the layer boundary was already discussed in Subsection 2.2. The situation is different for the inclined incidence of plane wave on a slab of the medium. To study this case, instead of angle ϕ_L, we shall introduce function $z_L = \tan\phi_L/2$, which has singular points. The dynamical equation

$$\frac{d}{dL}z_L = p(1+z_L^2) + \frac{k^2}{p}\varepsilon(L). \tag{3.11}$$

for z_L follows from (2.4). As before, function $\varepsilon(L)$ is assumed to be a Gaussian, delta-correlated random process. Then, the steady-state probability density

$$P(z) = \lim_{L_0 \to -\infty} P_L(z),$$

(the case of the half-space) is described by equation

$$-\kappa \frac{d}{dz}(1+z^2)P(z) + \frac{d^2}{dz^2}P(z) = 0,$$

where parameter

$$\kappa = p^3/2k^2D = \frac{\alpha}{2}\cos^3\theta, \quad \alpha = k/D.$$

A further analysis of this equation largely depends on the specific boundary conditions with respect to z, which determine the type of problems under study. *Thus, if we consider function z_L to be discontinuous and defined for all values of L, then the divergence of its values to $-\infty$ on one side of some points is accompanied by their divergence to $+\infty$ on the other side of the same points.* In this situation the boundary condition is

$$J(z)|_{z=-\infty} = J(z)|_{z=+\infty},$$

where

$$J(z) = \lim_{L_0 \to -\infty} J_L(z) = -\kappa(1+z^2)P(z) + \frac{d}{dz}P(z)$$

is the steady state of the flux of probability density, where $P_L(z)$ and $J_L(z)$ satisfy a Fokker -Planck equation

$$\frac{\partial}{\partial t}P_L(z) = \frac{\partial}{\partial z}J_L(z).$$

Hence [22],

$$P(z) = J(\kappa)\int_z^\infty d\xi \, \exp\left\{-\kappa\xi[1+\xi^2/3+z(z+\xi)]\right\}, \tag{3.12}$$

$$J^{-1}(\kappa) = \left(\frac{\pi}{k}\right)^{1/2}\int_0^\infty d\xi \, \xi^{-1/2}e^{-\kappa(\xi+\xi^3/12)}.$$

Asymptotically, for $\kappa \gg 1$, the sought probability density of z is $P(z) = 1/\pi(1+z^2)$, $-\infty < z < \infty$, and it corresponds to the uniform probability distribution of $\phi : P(\phi) = 1/2\pi$, $0 \le \phi \le 2\pi$.

Probability distribution (3.12) allows us to calculate various quantities connected with fluctuations of the reflection coefficient phase. For example, on the boundary $x = L$ the mean value of the wavefield intensity

$$\langle I(L;L)\rangle = 2\langle 1+\cos\phi_L\rangle = 2(3)^{1/6}\Gamma(2/3)\kappa^{1/3}, \quad \kappa \ll 1.$$

For the sliding incidence, when $\theta \to \pi/2$, the quantity $R_L \to -1$, and on the boundary $x = L$ wavefield $u(L) = 1 + R_L$ tends to zero. This result indicates that the medium behaves in a mirror-like fashion—an effect that is linked to a discontinuity of function $\varepsilon(x)$ at the boundary $x = L$. If the jump is small then it contributes little to the statistics of the phase in incidence angles θ ($\kappa \gg 1$). On the other hand, for the sliding incidence, this jump appears to be like an infinite barrier and contributes significantly to the statistics. Hence, the probability distribution of the reflection coefficient phase contains information about how the wave scatters on a jump of $\varepsilon(L)$ at the boundary, and on random inhomogeneities inside the medium, without separating these effects.

Now, consider a medium with an adjusted right-hand boundary so that $\varepsilon(x) = \varepsilon(L)$ for $x > L$. In this case $k_L = p$, and the effect of discontinuity at $x = L$ can be excluded. The boundary problem is then reduced to the following initial-value problem with respect to L:

$$\frac{\partial}{\partial L}u(x;L) = ipu(x;L) + \frac{k^2\xi(L)}{2p^2}(1-R_L)u(x;L),$$

$$\frac{d}{dL}R_L = 2ipR_L + \frac{k^2\xi(L)}{2p^2}(1-R_L^2),$$

where $\xi(L) = \partial\varepsilon(L)/\partial L$ [23]. It is easy to see what is the impact of the nonlinear term on the equation of the reflection coefficient. In the case of a random medium occupying half-space $x < L$, for the sliding incidence, we have that values of the phase of reflection coefficient $R_L = \exp[i\varphi_L] \to \pm 1$. Hence, these two points contribute most to the statistical characteristics linked with the phase.

For the finite slab with $\varepsilon(x)$ with no discontinuity at L, substituting $R_L = \exp[i\varphi_L]$, we obtain equations

$$\frac{d}{dL}\varphi_L = 2p - \frac{k^2\xi(L)}{2p^2}\sin\varphi_L, \quad \frac{d}{dL}z_L = p(1+z_L^2) - \frac{k^2\xi(L)}{2p^2}z_L$$

for φ_L and $z_L = \tan\varphi_L/2$. And in the case of half-space, there is the steady-state probability density, which for $\kappa \gg 1$ becomes the uniform probability distribution of the phase. In this case, for the mean value of the wave intensity on the boundary $x = L$ we have $\langle I(L;L)\rangle = 2$ as $\kappa \ll 1$, which means that statistical weights of phase values for which $R_L = \pm 1$

are equal, although the probability distribution $P(\varphi)$ differs considerably from the uniform one [23].

In conclusion, we consider statistical moment functions of wavefield intensity $\langle I^n(x, L)\rangle = \langle |u(x; L)|^{2n}\rangle$. From (2.4), we obtain that

$$\langle I^n(\xi)\rangle = 2^n \pi^{1/2} \frac{(2n-3)!!}{(n-1)!} \xi^{-1/2}, \quad \xi \gg 1, \kappa \ll 1, \tag{3.13}$$

where $\xi = D(L-x)$ [22]. Hence, for the sliding incidence of the plane wave on a layered medium, quantities $\langle I^n(\xi)\rangle$ decrease according to the universal $\xi^{-1/2}$ law at a large distance inside the medium.

For the adjusted boundary, one can obtain expression

$$\langle I^n(\xi)\rangle \cong \frac{2^{n+1}(2n-3)!!}{(n-1)!} \frac{e^{-n\xi}}{\pi[1 + 4\pi^{-2}(\ln \kappa + \gamma)^2]\kappa}, \tag{3.14}$$

where γ is the Euler constant, and which is valid for $\kappa \ll 1$ and $\xi \gg 1$ (see [23]). This primary difference in the behavior of statistical characteristics for continuous and discontinuous $\varepsilon(x)$ is explained by the discontinuity at $x = L$. Thus, the asymptotic behavior of (3.14) is a "purely" statistical" effect. We excluded the influence of boundary $x = L$ linked with a discontinuity of $\varepsilon(x)$, and have taken into account scattering of the wave off random inhomogeneities.

4. Appearance of caustics in random media. In the parabolic equation approximation the wavefield is described by equation (2.9) and, for the purpose of this section, we will write it in the form

$$u(x, \rho) = A(x, \rho)e^{iS(x,\rho)}.$$

In the geometric optics approximation, when $k \to \infty$, the analysis of amplitude and phase fluctuations is greatly simplified, and equations for phase and wave intensity assume the following form:

$$k\frac{\partial}{\partial x}S(x, \boldsymbol{\rho}) + (\nabla_\perp S)^2 = k^2\varepsilon(x, \boldsymbol{\rho}), \quad k\frac{\partial}{\partial x}I(x, \boldsymbol{\rho}) + \nabla_\perp(I\nabla_\perp S) = 0, \tag{4.1}$$

where ($\nabla_\perp = \partial/\partial\boldsymbol{\rho}$). Solving these partial differential equations by the method of characteristics (rays), we can introduce a function $\boldsymbol{R}(x)$ such that, substituting $\boldsymbol{p}(x, \boldsymbol{\rho}) = \frac{1}{k}\nabla_\perp S$, we obtain a system

$$\frac{d}{dx}\boldsymbol{R}(x) = \boldsymbol{p}(X), \quad \frac{d}{dx}\boldsymbol{p}(x) = \frac{1}{2}\frac{\partial}{\partial \boldsymbol{R}}\varepsilon(x, \boldsymbol{R}) \tag{4.2}$$

of closed equations. It must be solved with the initial condition at $x = 0$, determining a parameterization of rays.

The equation for intensity can be rewritten in the form of an equation along the characteristics:

$$\frac{d}{dx}I(x) = -\frac{1}{k}I(x)\Delta_{\boldsymbol{R}}S(x,\boldsymbol{R}). \tag{4.3}$$

Let us introduce functions

$$u_{ij}(x,\boldsymbol{\rho}) = \frac{1}{k}\frac{\partial^2}{\partial\rho_i\partial\rho_j}S(x,\boldsymbol{\rho}),$$

which describes the *curvature of the phase front* $S(x,\boldsymbol{\rho}) = \text{const}$. These functions satisfy equations

$$\frac{d}{dx}u_{ij}(x) + u_{ik}(x)u_{kj}(x) = \frac{1}{2}\frac{\partial^2}{\partial R_i\partial R_j}\varepsilon(x,\boldsymbol{R}), \tag{4.4}$$

$$\frac{d}{dx}I(x) = -I(x)u_{ii}(x).$$

In the two-dimensional case ($R = y$), equations (3.16)–(3.18) become much simpler and take the form

$$\frac{d}{dx}y(x) = p(x), \quad \frac{d}{dx}p(x) = \frac{1}{2}\frac{\partial}{\partial y}\varepsilon(x,y), \tag{4.5}$$

$$\frac{d}{dx}I(x) = -I(x)u(x), \quad \frac{d}{dx}u(x) = -u^2(x) + \frac{1}{2}\frac{\partial^2}{\partial y^2}\varepsilon(x,y).$$

Note that in absence of medium inhomogeneities ($\varepsilon = 0$), rays appear to be straight lines, and integration of equations (3.19) gives

$$u(x) = \frac{u_0}{1+u_0x}, \qquad I(x) = \frac{I_0}{u_0}u(x). \tag{4.6}$$

If the initial condition u_0 is < 0, then $u(x_0) = -\infty$, $I(x_0) = \infty$ at the point $x_0 = -1/u_0$; this means that the solution has an *explosive character.* In presence of inhomogeneities $\varepsilon(x,y)$ such singular points exist regardless of the sign of u_0, which means that in the statistical problem singular points appear at finite distances. Their appearance is a result of random focusing of the wavefield which is caused by the fact that the phase front curvature and the wavefield intensity become infinite. This *random focusing* of the wavefield in the randomly inhomogeneous medium causes an appearance of *caustics.*

In the two-dimensional case, the phase curve curvature is described by equation (3.19). For a Gaussian, homogeneous, isotropic and delta-correlated fluctuation $\varepsilon(x,y)$ with

$$\langle\varepsilon(x,y)\varepsilon(x',y'))\rangle = \delta(x-x')A(y-y'),$$

it's probability density is described by the Fokker-Planck equation

$$(4.7) \quad \frac{\partial}{\partial x} P_x(u) = \frac{\partial}{\partial u} u^2 P_x(u) + \frac{D}{2} \frac{\partial^2}{\partial u^2} P_x(u), \quad P_0(u) = \delta(u - u_0),$$

where diffusion coefficient

$$D = \frac{1}{4} \frac{\partial^4}{\partial y^4} A(y) \Big|_{y=0} .$$

Equation (3.21) has been studied in [2,3], where it was shown that random process $u(x)$ becomes infinite at finite distance $x(u_0)$, which is determined by the initial condition u_0. In this case, the mean value

$$(4.8) \qquad \langle x(u_0) \rangle = \frac{2}{D} \int_{-\infty}^{u_0} d\xi \, e^{2\xi^3/3D} \int_{\xi}^{\infty} d\eta \, e^{-2\eta^3/3D}$$

and, consequently,

$$D^{1/3} \langle x(\infty) \rangle = 6.27, \quad D^{1/3} \langle x(0) \rangle = \frac{2}{3} D^{1/3} \langle x(\infty) \rangle = 4.18.$$

Quantity $\langle x(0) \rangle$ describes the mean distance of the focus appearance for the initial plane wave, and quantity $\langle x(\infty) \rangle$ describes the mean distance between two subsequent foci.

Further analysis of equation (3.21) depends in an essential way on the specific boundary condition with respect to u. Thus, as in Subsection 3.1, if we consider function $u(x)$ to be discontinuous and defined for all values of x, and its blowup to $-\infty$ for $x \Rightarrow x_0 - 0$ accompanied by a blowup to $+\infty$ for $x \to x_0 + 0$, then the boundary condition is

$$J(x, u) \, |_{u \to \infty} = J(x, u)|_{u \to -\infty} ,$$

where

$$(3.21') \qquad J(x, u) = u^2 P_x(u)) + \frac{D}{2} \frac{\partial}{\partial u} P_x(u)$$

is the flux of probability density.

In this case, there exists a steady-state probability distribution

$$P_\infty(u) = J e^{-2u^3/3D} \int_0^\infty d\xi \, e^{2u^3/3D},$$

defined by equation

$$J = u^2 P_\infty(u) + \frac{D}{2} \frac{d}{du} P_\infty(u),$$

where steady-state probability flux density is

$$J = 1/\langle x(\infty) \rangle.$$

Asymptotic behavior of $P_\infty(u)$ for large values of u can be shown to be

$$P_\infty(u) \sim \frac{1}{\langle x(\infty)\rangle}\frac{1}{u^2},$$

and it depends on the jumps of function $u(x)$, and on its behavior near the discontinuity (caustic) points x_k which is of the form (3.20)

$$u(x) \sim \frac{1}{x - x_k}.$$

Note, that according to (3.20) the wave intensity normalized by the intensity of incident wave I_0 has the following structure in the neighborhood of the discontinuities:

$$I(x) \sim \frac{x_k}{|x - x_k|}.$$

In this case, the asymptotics of probability distribution $z(x) = I^2(x)$, for sufficiently large x and z, is described by expression

$$P_x(z) \sim \sum_{k=0}^{\infty}\left\langle \delta\left(\frac{x_k^2}{(x-x_k)^2} - z\right)\right\rangle = \frac{x}{z\sqrt{z}}\sum_{k=0}^{\infty}\langle\delta(x - x_k)\rangle$$

$$= \frac{x}{z\sqrt{z}}\frac{1}{2\pi}\int_{-\infty}^{\infty} d\kappa\, e^{-i\kappa x}\frac{\langle e^{i\kappa x_0}\rangle}{1 - \langle e^{i\kappa x}\rangle},$$

where $\langle e^{i\kappa x_0}\rangle$ is the characteristic function of distance to the first caustic, and $\langle e^{i\kappa x}\rangle$ is the characteristic function of distance between two subsequent caustics. Thus, if $x \gg \langle x(\infty)\rangle$, then the probability distribution of z has asymptotic form

$$P_x(z) \sim \frac{x}{z\sqrt{z}}\frac{1}{\langle x(\infty)\rangle},$$

and, as a result, probability distribution for the intensity itself is of the form

$$P_x(I) = \frac{2x}{I^2}\frac{1}{\langle x(\infty)\rangle},$$

so that, in particular, it has a power decay as I increases.

Another type of boundary conditions arises when curve $u(x)$ blows up to $-\infty$ as x approaches point x_0. This corresponds to the condition that the probability flux density $J(x,u)$ must converge to zero when $u \Rightarrow \infty$, which can be rephrased as conditions

$$J(x,u) \to 0 \text{ as } u \to \infty; \qquad P_x(u) \to 0 \text{ as } u \to -\infty.$$

Probability density of the caustic location is then ([2])

$$p(x) = \lim_{u \to -\infty} J(x, u). \tag{4.9}$$

To derive the asymptotic dependence of probability density $p(x)$ on parameter $D \to 0$, we shall use for equation (3.21) a standard procedure (see e.g. [24]) for analysis of parabolic equations with a small parameter in the highest derivative. Let us write the solution of equation (3.21) in the form

$$P_x(u) = C(D) \exp\left\{-\frac{1}{D}A(x, u) - B(x, u)\right\}. \tag{4.10}$$

Constant $C(D)$ is determined by the condition that, for $x \to 0$, the probability distribution of the plane incident wave must have the form

$$P_x(u) \sim \frac{1}{(2\pi D x)^{1/2}} \exp\left\{-\frac{u^2}{2Dx}\right\}. \tag{3.23'}$$

This gives that $C(D) \simeq D^{1/2}$.

Note, that expressing $P_x(u)$ in the form (3.23) immediately permits determination of structural dependence of $p(x)$ on x by dimensional analysis ([4]). Indeed, quantities u, D and $P_x(u)$ have, respectively, dimensions $[u] = x^{-1}$, $[D] = x^{-3}$, $[P_x(u)] = x$. Consequently, in view of (3.21'), function $p(x)$ has a structure

$$p(x) \sim C_1 D^{-1/2} x^{-5/2} e^{-C_2/Dx^3},$$

and the only remaining problem is to calculate positive constants C_1 and C_2. This has been done in [2], and the resulting formula is

$$p(x) \sim 3\alpha^2 (2\pi D)^{-1/2} x^{-5/2} \exp\{-\alpha^4/6Dx^3\}, \tag{4.11}$$

where $\alpha = 1.85$. The condition of applicability of (3.25) is $Dx^3 \ll 1$. However, as was shown in [2] by numerical modelling, expression (3.25) also accurately describes the probability distribution of appearance of random foci when $Dx^3 \simeq 1$.

Note, that for a three-dimensional problem, dimensional considerations give the following form of the probability density of appearance of caustics [4]: $p(x) = \alpha D^{-1} x^{-4} \exp(-\beta/Dx^3)$, where α and β are numerical constants. Their values, $\alpha = 1.74$ and $\beta = 0.66$, have been computed in [3].

Appendix: Fokker-Planck equation for dynamical systems. Assume that vector-valued function $\boldsymbol{\xi}(t)$ satisfies a system of dynamical equations

$$\frac{d}{dt}\boldsymbol{\xi}(t) = \boldsymbol{v}(\boldsymbol{\xi}, t) + \boldsymbol{f}(\boldsymbol{\xi}, t), \qquad \boldsymbol{\xi}(0) = \boldsymbol{\xi}_0, \tag{A.1}$$

where function $\boldsymbol{v}(\boldsymbol{\xi}, t)$ is deterministic functions and $f_i(\boldsymbol{\xi}, t)$ is a random field of $(n+1)$ variables, which satisfies two conditions:

(a) $f_i(\boldsymbol{x}, t)$ is a Gaussian random field in the $(n+1)$-dimensional $(\boldsymbol{x}, t)$-space;

(b) $\langle f_i(\boldsymbol{x}, t)\rangle = 0$.

Statistical characteristics of field $f_i(\boldsymbol{x}, t)$ are fully determined by the correlation tensor

$$B_{ij}(\boldsymbol{x}, t; \boldsymbol{x}', t') = \langle f_i(\boldsymbol{x}, t) f_j(\boldsymbol{x}', t')\rangle.$$

The quantities $\xi_i(t)$ represent nonanticipating functionals that depend only on the values of $f_j(\boldsymbol{x}, t')$ for $t' \leq t$, that is

$$(A.1') \qquad \frac{\delta \xi_i(t)}{\delta f_j(\boldsymbol{x}, t')} = 0 \text{ if } t' > t.$$

However, for $t'' > t$, statistical dependence between $\xi_i(t)$ and the values of $f_j(\boldsymbol{x}, t'')$ is possible. Indeed, the latter are correlated with values $f_j(\boldsymbol{x}, t')$ for $t' < t$. Obviously, a correlation between $\xi_i(t)$ and the subsequent values $f_i(\boldsymbol{x}, t'')$ exists only if $t'' - t \sim \tau_0$, where τ_0 denotes the correlation radius of the field $f_i(\boldsymbol{x}, t)$ with respect to variable t.

For a large number of actual physical processes the characteristic correlation radius of $\xi_i(t)$ has order of magnitude T, where T reflects the natural scales of the dynamical problem without the fluctuation term. In this case, parameter τ_0/T is small and can be used in the construction of an asymptotic solution. In the first order approximation with respect to the small parameter τ_0/T we take $\tau_0 \to 0$ (so called delta-correlation approximation). Then, there is neither a functional dependence nor a statistical dependence between the values of $\xi_i(t')$ for $t' < t$, and the values of $f_j(\boldsymbol{x}, t'')$ for $t'' > t$. The first-order approximation leads to an introduction of the effective correlation tensor

$$(A.1'') \qquad B_{ij}(\boldsymbol{x}, t; \boldsymbol{x}', t') = 2\delta(t - t') F_{ij}(\boldsymbol{x}, \boldsymbol{x}', t),$$

where

$$F_{ij}(\boldsymbol{x}, \boldsymbol{x}', t) = \frac{1}{2}\int_{-\infty}^{\infty} dt'\, B_{ij}(\boldsymbol{x}, t; \boldsymbol{x};, t').$$

Thus we pass from the original field $\boldsymbol{f}$ to a Gaussian random field which is delta-correlated with respect to t.

Consider the probability density

$$(A.2) \qquad P_t(\boldsymbol{x}) = \langle \delta(\boldsymbol{x} - \boldsymbol{\xi}(t))\rangle$$

of the solution $\boldsymbol{\xi}(t)$ of the system of equations (A.1). If we differentiate (A.2) with respect to t, taking into account (A.1), we get equation

$$(A.3) \qquad \frac{\partial}{\partial t} P_t(\boldsymbol{x}) = -\frac{\partial}{\partial x_k}\Big(v_k(\boldsymbol{x}, t) P_t(\boldsymbol{x})\Big) - \frac{\partial}{\partial x_k}\Big\langle f_k(\boldsymbol{x}, t)\delta(\boldsymbol{x} - \boldsymbol{\xi}(t)\Big\rangle.$$

Equation (A.3) can be rewritten in the form

$$(A.4) \qquad \frac{\partial}{\partial t}P_t(\boldsymbol{x}) = -\frac{\partial}{\partial x_k}\Big(v_k(\boldsymbol{x},t)P_t(\boldsymbol{x})\Big)$$

$$+\frac{\partial}{\partial x_k}\int d\boldsymbol{x}'\int_0^t d\tau\Big\langle f_k(\boldsymbol{x},t)f_1(\boldsymbol{x}',\tau)\Big\rangle\frac{\partial}{\partial x_m}\Big\langle\delta(\boldsymbol{x}-\boldsymbol{\xi}(t))\frac{\delta\xi_m(t)}{\delta f_1(\boldsymbol{x},\tau)}\Big\rangle.$$

Here, we have used the *Furutsu-Novikov formula* [25, 26]

$$\Big\langle f_k(\boldsymbol{x},t)R[\boldsymbol{f}(\boldsymbol{x}',\tau)]\Big\rangle$$

$$= \int d\boldsymbol{x}'\int_0^t d\tau\Big\langle f_k(\boldsymbol{x},t)f_1(\boldsymbol{x}',\tau)\Big\rangle\Big\langle\frac{\delta}{\delta f_1(\boldsymbol{x},\tau)}R[\boldsymbol{f}(\boldsymbol{x}',\tau)]\Big\rangle,$$

which gives the covariance of a Gaussian random field $\boldsymbol{f}(\boldsymbol{x},t)$ and its arbitrary functional $R[\boldsymbol{f}]$. Notice, that condition (A.1') permitted to restrict the integration over τ from $\tau = 0$ to $\tau = t$. The above equation implies that the probability density for the solution $\boldsymbol{\xi}$ at instant t is determined by the interdependence between $\boldsymbol{\xi}(t)$ and the field $\boldsymbol{f}(\boldsymbol{x}',\tau)$ for all values of τ in the interval $(0,t)$. In general, $P_t(\boldsymbol{x})$ does not satisfy a closed differential equation.

If the correlation function of field $\boldsymbol{f}(\boldsymbol{x},t)$ is asymptotically approximated by (A.1"), then the integration over τ reduces to the substitution $\tau = t$, and only values $\delta\xi_m(t)/\delta f_1(\boldsymbol{x}',t)$ occur. In this case, these values can be expressed in terms of $\boldsymbol{\xi}(t)$:

$$\frac{\delta\xi_m(t)}{\delta f_l(\boldsymbol{x}',t-0)} = \delta_{ml}\delta(\boldsymbol{\xi}-\boldsymbol{x}'),$$

and we get the Fokker-Planck equation

$$\frac{\partial}{\partial t}P_t(\boldsymbol{x}) + \frac{\partial}{\partial x_k}\Big\{\Big[v_k(\boldsymbol{x},t) + A_k(\boldsymbol{x},t)\Big]P_t(\boldsymbol{x})\Big\}$$

$$(A.5) \qquad -\frac{\partial^2}{\partial x_k\partial x_1}\Big[F_{kl}(\boldsymbol{x},\boldsymbol{x};t)P_t(\boldsymbol{x})\Big] = 0,$$

where

$$A_k(\boldsymbol{x},t) = \frac{\partial}{\partial x_1'}F_{kl}(\boldsymbol{x},\boldsymbol{x}')\Big|_{\boldsymbol{x}'=\boldsymbol{x}}.$$

Equation (A.5) has to be solved with the initial condition $P_0(\boldsymbol{x}) = \delta(\boldsymbol{x}-\boldsymbol{\xi}_0)$, or with a more general initial condition $P_0(\boldsymbol{x}) = W_0(\boldsymbol{x})$.

Let us return to the dynamical system (A.1), and derive an equation for the joint probability density

$$(A.6)\qquad \mathcal{P}_m(\boldsymbol{x}_1, t_1; \ldots; \boldsymbol{x}_m, t_m) = \Big\langle \delta(\boldsymbol{\xi}(t_1) - \boldsymbol{x}_0) \cdot \ldots \cdot \delta(\boldsymbol{\xi}(t_m) - \boldsymbol{x}_m) \Big\rangle$$

which depends on m different moments of time $t_1 < t_2 < \ldots < t_m$. After differentiation of (A.6) with respect to t_m and a subsequent application of the dynamical equation (A.1), the definition of the F_{ij} yields an equation

$$\frac{\partial}{\partial t_m}\mathcal{P}_m(\boldsymbol{x}_1, t_1; \ldots; \boldsymbol{x}_m, t_m) + \sum_{i=1}^{n} \frac{\partial}{\partial x_{mi}} \Big\{ \Big[V_i(\boldsymbol{x}_m, t_m) + A_i(\boldsymbol{x}_m, t_m) \Big] \mathcal{P}_m \Big\}$$

$$(A.7)\qquad = \sum_{i=1}^{n} \frac{\partial}{\partial x_{mi}} \sum_{j=1}^{n} \frac{\partial}{\partial x_{mj}} \Big[F_{ij}(\boldsymbol{x}_m, \boldsymbol{x}_m; t_m) \mathcal{P}_m \Big],$$

similar to the Fokker-Planck equation (A.5). There is no summation over index m here. The initial condition for (A.7) can be derived from (A.6). If we suppose that $t_m = t_{m-1}$ in (A.6), then we get that

$$\mathcal{P}_m(\boldsymbol{x}_1, t_1; \ldots; \boldsymbol{x}_{m-1}, t_{m-1}; \boldsymbol{x}_m, t_{m-1})$$

$$(A.8)\qquad = \delta(\boldsymbol{x}_m - \boldsymbol{x}_{m-1}) \mathcal{P}_{m-1}(\boldsymbol{x}_1, t_1; \ldots; \boldsymbol{x}_{m-1}, t_{m-1}),$$

and we can look for a solution of (A.7) of the form

$$\mathcal{P}_m(\boldsymbol{x}_1, t_1; \ldots; \boldsymbol{x}_m, t_m)$$

$$(A.9)\qquad = p(\boldsymbol{x}_m, t_m | \boldsymbol{x}_{m-1}, t_{m-1}) \mathcal{P}(\boldsymbol{x}_1, t_1; \ldots; \boldsymbol{x}_{m-1}, t_{m-1}).$$

Since all the derivatives in (A.7) are with respect to $t_m, \boldsymbol{x}_m$, a substitution of expression (A.9) into (A.7) and (A.8) yields the following equation for the transition probability density:

$$\frac{\partial}{\partial t} p(\boldsymbol{x}, t | \boldsymbol{x}_0, t_0) + \frac{\partial}{\partial x_k} \Big\{ \Big[v_k(\boldsymbol{x}, t) + A_k(\boldsymbol{x}, t) \Big] p(\boldsymbol{x}, t | \boldsymbol{x}_0, t_0) \Big\}$$

$$(A.10)\qquad = \frac{\partial}{\partial x_k \partial x_1} \Big[F_{kl}(\boldsymbol{x}, \boldsymbol{x}; t) p(\boldsymbol{x}, t | \boldsymbol{x}_0, t_0) \Big],$$

with initial condition

$$(A.11)\qquad p(\boldsymbol{x}, t | \boldsymbol{x}_0, t_0) \Big|_{t \to t_0} = \delta(\boldsymbol{x} - \boldsymbol{x}_0).$$

Here, variables $\boldsymbol{x}_m, t_m$ are replaced by $\boldsymbol{x}$, t and variables $\boldsymbol{x}_{m-1}$, t_{m-1} by $\boldsymbol{x}_0, t_0$ (that is $p(\boldsymbol{x}, t | \boldsymbol{x}_0, t_0) = \langle \delta(\boldsymbol{x} - \boldsymbol{\xi}(t) | \boldsymbol{\xi}(t_0) = \boldsymbol{x}_0) \rangle$).

Applying (A.9) $(m-1)$ times, we arrive at the relation

$$\mathcal{P}_m(\boldsymbol{x}_1, t_1; \ldots; \boldsymbol{x}_m, t_m)$$

$$= p(\boldsymbol{x}_m, t_m | \boldsymbol{x}_{m-1}, t_{m-1}) \ldots p(\boldsymbol{x}_2, t_2 | \boldsymbol{x}_1, t_1) P_{t_1}(\boldsymbol{x}_1), \tag{A.12}$$

where $P_{t_1}(\boldsymbol{x}_1)$ is the probability density determined by equation (A.10) and depending on the single time instant t_1. Formula (A.12) expresses the multipoint probability density as a product of transition probability densities. Therefore, $\boldsymbol{\xi}(t)$ is a *Markov process.*

In order to determine conditions of applicability of the Fokker-Planck equation we assume that the time correlation radius τ_0 of field $f_j(\boldsymbol{x}, t)$ is finite. Then, equation (A.5) for the probability density can be replaced by equation

$$\hat{E} P_t(\boldsymbol{x}) = -\frac{\partial}{\partial x_k} S'_k(\boldsymbol{x}, t), \tag{A.13}$$

where $\hat{E}$ denotes the operator appearing in the left-hand side of equation (A.13). Instead of $F_{kl}(\boldsymbol{x}, \boldsymbol{x}', t)$ we set

$$\tilde{F}_{kl}(\boldsymbol{x}, \boldsymbol{x}', t) = \int_0^t dt'\, B_{kl}(\boldsymbol{x}, t; \boldsymbol{x}', t'). \tag{A.14}$$

Term $S'_k(\boldsymbol{x}, t)$ contains corrections of the probability density due to the finite positive value of τ_0 and it imposes, in general, limitations on the intensity of fluctuations of field $\boldsymbol{f}$. If $\tau_0 \to 0$, then the right-hand side of (A.13) tends to zero and we obtain equation (A.5) again.

So, in order to describe the statistical characteristics of solution (A.1) in the the delta-correlation random field approximation (Fokker-Planck equation), it is necessary, but in general not sufficient, that parameter τ_0/T be small. However, in each concrete problem, a more detailed study can to be carried out to determine the validity of the delta-correlated approximation.

We would like to reemphasize that the delta-correlated random process approximation is by no means a simple replacement of $f_j(\boldsymbol{\xi}, t)$ in (A.1) by a random field with the correlation function (A.1"). It means a limiting procedure where the radius of correlation τ_0 of field $\boldsymbol{f}(\boldsymbol{\xi}, t)$ tends to zero, and mean quantities like $\langle f_j(\boldsymbol{\xi}, t)\phi(\boldsymbol{\xi})\rangle$ tend asymptotically to the values of corresponding functionals of the delta-correlated in time field . The demand $t \gg \tau_0$ points to the fact that the delta-correlated process approximation is suitable for a time behavior description of such mean values only on time intervals when system (A.1) is influenced by a great number of independent random sources. Average description of the system behavior during times at most equal to τ_0 can be considered as satisfactory. At the same time, an analysis of stochastic equations (such as (A.1)) containing random forces like $\boldsymbol{f}(\boldsymbol{\xi}, t)$, requires their continuity to guarantee smooth time dependence of solution $\boldsymbol{\xi}(t)$.

Example. Following [27], consider the *Langevin equation*

$$\frac{d}{dt}X(t) = -\lambda X(t) + f(t), \quad X(t_0) = 0, \tag{A.15}$$

where $f(t)$ is assumed to be a zero-mean Gaussian random process with covariance function $\langle f(t)f(t')\rangle = B_f(t - t')$. For a given realization of external forces $f(t)$, the solution of the equation (A.15) has the form

$$X(t) = \int_{t_0}^{t} d\tau\, f(\tau)e^{-\lambda(t-\tau)}.$$

Consequently, $X(t)$ is a Gaussian random process. Its one-point probability density $P_t(x) = \langle\delta(X(t) - x)\rangle$ satisfies an exact equation

$$\frac{\partial}{\partial t}P_t(x) = \lambda\frac{\partial}{\partial x}xP_t(x) + \int_{t_0}^{t} d\tau\, B_f(t-\tau)e^{-\lambda(t-\tau)}\frac{\partial^2}{\partial x^2}P_t(x), \tag{A.16}$$

with the initial condition $P_{t_0} = \delta(x)$. When $t_0 \to -\infty$, then process $x(t)$ converges to a Gaussian random function with parameters

$$\langle X(t)\rangle = 0, \quad \sigma_x^2 = \langle X^2(t)\rangle = \frac{1}{\lambda}\int_0^{\infty} d\tau\, B_f(\tau)e^{-\lambda\tau}. \tag{A.16$'$}$$

In particular, for covariance function of the form

$$B_f(\tau) = \sigma_f^2 e^{-|\tau|/\tau_0},$$

we obtain

$$\langle X(t)\rangle = 0, \quad \langle X^2(t)\rangle = \sigma_f^2\tau_0/\lambda(1 + \lambda\tau_0) \tag{A.17}$$

and, in the limit $\tau_0 \to 0$,

$$\langle X^2(t)\rangle = \sigma_f^2\tau_0/\lambda. \tag{A.18}$$

Multiply equation (A.15) by $X(t)$. Assuming $X(t)$ to be a nice enough function of time, we obtain equality

$$X(t)\frac{d}{dt}X(t) = \frac{1}{2}\frac{d}{dt}X^2(t) = -\lambda X^2(t) + f(t)X(t).$$

Averaging it over the ensemble of realizations of functions $f(t)$ we obtain equation

$$\frac{1}{2}\frac{d}{dt}\langle X^2(t)\rangle = -\lambda\langle X^2(t)\rangle + \langle f(t)X(t)\rangle. \tag{A.19}$$

A direct calculation shows that the stationary solution of (A.19),

$$\langle X^2(t)\rangle = \frac{1}{\lambda}\langle f(t)X(t)\rangle,$$

corresponding to $t_0 \to -\infty$, coincides with (A.18). This gives that

$$\langle f(t)X(t)\rangle = \frac{\sigma_f^2\tau_0}{1+\lambda\tau_0}\tau_0 \to \sigma_f^2\tau_0,$$

as $t_0 \to -\infty$ and $\tau_0 \to 0$. Since $\delta X(t)/\delta f(t-0) = 1$, the same result is obtained for $\langle f(t)X(t)\rangle$ if one utilizes formula

$$\langle f(t)X(t)\rangle = \int_{-\infty}^{t} d\tau\, B_f^{eff}(t-\tau)\Big\langle \frac{\delta}{\delta f(\tau)}X(t)\Big\rangle$$

including the "effective" correlation function

$$B_f^{eff}(\tau) = 2\sigma_f^2\,\tau_0\,\delta(\tau).$$

As we mentioned before, some of probabilistic properties of the above equations in the delta-correlated process approximation coincide with probabilistic properties of a Markov process. However, it should be understood that is true only for statistical means and equations for them. In particular, for the Langevin equation (A.15), realizations of the process $X(t)$ and the corresponding Markov process are qualitatively different. The latter satisfies the equation (A.15) with ideal "white noise" $f(t)$ on the right-hand side which has the correlation function $B_f(\tau) = 2\sigma_f^2\tau_0\delta(\tau)$; the equation itself should be understood in the generalized sense because the ideal Markov process is not differentiable in the ordinary sense. At the same time, process $X(t)$, whose statistical properties are sufficiently described by a "white noise" approximation, remains differentiable in the ordinary sense and the ordinary rules of calculus apply. For example,

$$X(t)\frac{d}{dt}X(t) = \frac{1}{2}\frac{d}{dt}X^2(t), \tag{A.20}$$

and, in particular,

$$\langle X(t)\frac{d}{dt}X(t)\rangle = 0. \tag{A.21}$$

On the other hand, if $X(t)$ is an ideal Markov process satisfying, in the generalized sense, the Langevin equation (A.15) with the "white noise" on the right-hand side, equality (A.20) loses its validity, and the relation

$$\langle X(t)\frac{d}{dt}X(t)\rangle = -\lambda\langle X^2(t)\rangle + \langle f(t)X(t)\rangle \tag{A.22}$$

depends on the definition of the mean values. Indeed, if we treat (A.22) as the limit of equality

$$(A.23)\qquad \langle X(t+\delta)\frac{d}{dt}X(t)\rangle = -\lambda\langle X(t)X(t+\delta)\rangle + \langle f(t)X(t+\delta)\rangle$$

when $\delta \to 0$, the result will be qualitatively different depending on whether $\delta \to 0+$, or $\delta \to 0-$. If $\delta \to 0+$, then we have

$$\lim_{\delta\to 0+} \langle f(t)X(t+\delta)\rangle = 2\sigma_f^2\,\tau_0$$

and, taking into account (A.18), equality (A.23) yields equation

$$(A.24)\qquad \langle X(t+0)\frac{d}{dt}X(t)\rangle = \sigma_f^2\,\tau_0.$$

If $\delta \to 0-$, then $\langle f(t)X(t+0)\rangle = 0$, because of the dynamic causality condition, and equality (A.23) takes the form

$$(A.25)\qquad \langle X(t-0)\frac{d}{dt}X(t)\rangle = -\sigma_f^2\,\tau_0.$$

Comparing (A.21) with (A.24) and (A.25) we see that for the ideal Markov process described by the Langevin equation with the "white noise", usually called the *Ornstein-Uhlenbeck process,* we have

$$\langle X(t+0)\frac{d}{dt}X(t)\rangle \neq \langle X(t-0)\frac{d}{dt}X(t)\rangle \neq \langle \frac{1}{2}\frac{d}{dt}X^2(t)\rangle.$$

Note, that equalities (A.24) and (A.25) can be also obtained from the correlation function

$$\langle X(t)X(t+\tau)\rangle = \frac{\sigma_f^2\tau_0}{\lambda}e^{-\lambda|\tau|},$$

of process $X(t)$.

This concludes our discussion of the question of the delta-correlated approximation of a random process. Throughout this paper the phrase "dynamical system (equation) with delta-correlated parameter fluctuations" was to mean the asymptotic case in which the correlation radius of these parameters is small in comparison with all characteristic scales of the problem.

Acknowledgments. This work has been supported, in part, by grants from the U.S. Office of Naval Research, Grant No. MBP00 from the International Science Foundation, and Project No. 94-05-16151 of the Russian Fund of Fundamental Researches.

REFERENCES

[1] V.I. KLYATSKIN AND A.I. SAICHEV, Statistical and dynamical localization of plane waves in randomly layered media, *Soviet Physics Usp.*, Vol. **35** (3), pp. 231-247, 1992.

[2] V.A. KULKARNYY AND B.S. WHITE, Focusing of rays in a turbulent inhomogeneous medium, *Phys. Fluids*, Vol. **25**(10), pp. 1770-1784, 1982.

[3] B.S. WHITE, The stochastic caustic, *SIAM J. Appl. Math.*, Vol. **44**(1), pp. 127-149, 1983.

[4] V.I. KLYATSKIN, Caustics in random media, *Waves in Random Media* Vol. **3**(2), pp. 93-100, 1993.

[5] V.I. KLYATSKIN, *Stochastic Equations and Waves in Randomly Inhomogeneous Medium*, Nauka, Moscow, 1980 (in Russian); *Ondes et Équations Stochastiques dans les milieus Aléatoirement non Homogènes*, Eddition de Physique, Besançon-Cedex, 1985 (in French).

[6] V.I. KLYATSKIN, *The Imbedding Method in Wave Propogation Theory*, Nauka, Moscow, 1986 (in Russian); *The Imbedding Method in Statistical Boundary-Value Wave Problems*, in *Progress in Optics*, Vol. XXXIII, Ed. E. Wolf, North-Holland, Amsterdam 1994.

[7] V.I. KLYATSKIN, The statistical theory of radiative transfer in layered random media, *Izvestiya, Atmospheric and Oceanic Physics*, Vol. **27** (1), pp. 31-44, 1991.

[8] I.M. LIFSHITS, S.A. GREDESKUL AND L.A. PASTUR, *Introduction to the Theory of Disordered Solids*, John Wiley, New York, 1988.

[9] S.M. RYTOV, YU. A. KRAVTSOV AND V.I. TATARSKII, *Principles of Statistical Radiophysics*, Vol. 1-4, Springer-Verlag, Berlin, 1987-1989.

[10] S. CHANDRASEKHAR, Stochastic problems in physics and astronomy, *Rev. Modern Physics*, Vol. 15, pp. 1-89, 1943.

[11] A.I. SAICHEV, Chaotic motion of particles flows, *Dynamics of Systems*, Vol. **1** (1), pp. 1-31, 1993.

[12] V.I. KLYATSKIN, Statistical description of diffusion of tracers in random velocity fields, *Physics Uspekhi*, Vol. **37** (5), 1994.

[13] A.I. SAICHEV AND W.A. WOYCZYNSKI, Probability distributions of passive tracers in randomly moving media, in *Stochastic Models in Geosystems*, Springer-Verlag, pp. 1-43, 1995.

[14] S.D. RICE, Mathematical analysis of random noise, *Bell Syst. Tech. J.*, vol. 23, p. 282, 1944; vol. 24, p. 46, 1945.

[15] M.S. LONGUET-HIGGINS, The statistical analysis of a random moving surface, *Philos. Trans. R. Soc. London*, Ser. A249, p. 321, 1957.

[16] M.S. LONGUET-HIGGINS, Statistical properties of an isotropic random surface, ibid. ser. A250, p. 157, 1957.

[17] P. SWERLING, Statistical properties of the contours of random surfaces, *IRE Trans. Inf. Theory*, Vol IT-8, p. 315, 1962.

[18] M.B. ISICHENKO, Percolation, statistical topography, and transport in random media, *Reviews of Modern Physics*, vol. 64 (4), pp. 961-1043, 1992.

[19] YA. G. SINAI AND V. YAKHOT, Limiting probability distributions of a passive scalar in a random velocity field, *Phys. Rev. Lett.*, vol. 63 (18), pp. 1962-1964, 1989.

[20] H. CHEN, S. CHEN AND R.H. KRAICHNAN, Probability distribution of a

stochastically advected scalar field, *Phys. Rev. Letters*, Vol. 63 (24), pp. 2657-2660, 1989.

[21] Y. Kimura and R.H. Kraichnan, Statistics of an advected passive scalar, *Phys. Fluids*, Vol. A5 (9), pp. 2264-2277, 1993.

[22] M.A. Guzev, V.I. Klyatskin and G.V. Popov, Phase fluctuations and localization length in layered randomly inhomogeneous media, *Waves in Random Media*, Vol. **2** (2), pp. 117-123, 1992.

[23] M.A. Guzev and V.I. Klyatskin, Influence of boundary conditions on statistical characteristics of wavefield in layered randomly inhomogeneous medium, *Waves in Random Media*, Vol. **3** (4), pp. 307-315, 1993.

[24] C.W. Gardiner, *Handbook of Stochastic Methods for Physics, Chemistry and the Natural Sciences*, Springer-Verlag, Berlin, 1985.

[25] K. Furutsu, On the statistical theory of electromagnetic waves in a fluctuating medium, *J. Res.* NBS, Vol. D-67, p. 303, 1963.

[26] E.A. Novikov, Functionals and the random-force method in turbulence theory, *Sov. Phys. JETP*, Vol. 20 (5), pp. 1290-1294, 1964.

[27] V.I. Klyatskin, Approximations by delta-correlated random processes and diffusive approximation in stochastic problems, in *Mathematics of Random Media*, eds. W. Kohler, B.S. White, Lectures in Appl. Math., Vol. 27, pp. 447-476, 1991.

LOCALIZATION OF LOW FREQUENCY ELASTIC WAVES

W. KOHLER*, G. PAPANICOLAOU† AND B. WHITE‡

Abstract. We consider the propagation of compressional (P) and vertical shear (SV) waves in a plane-stratified medium with a stochastic microstructure. For such a medium, localization theory applies: waves are exponentially attenuated with propagation distance solely by the mechanism of random multiple scattering. We compute here the localization length and another deterministic length, called the equilibration length, which is associated with the equilibration of shear and compressional energy on transmission through a large slab. These characteristic lengths are seen to be reciprocals of the Lyapunov exponents for this system. We also provide expressions for the probability density of P to SV wave energy on transmission through a large slab, and for the fraction of energy which is mode converted on backscatter from a random half-space.

1. Introduction. In this paper we consider the statistics of waves generated when a plane, time-harmonic elastic wave is incident on a plane-stratified medium with a stochastic microstructure. In such a medium, three distinct wave modes are possible: compressional (P), vertical shear (SV) and horizontal shear (SH). It can be shown [1] that the calculation of SH waves can be decoupled from the calculation of P and SV, giving a problem that is mathematically equivalent to the calculation of the statistics of an acoustic wave in a plane-stratified medium. This acoustic problem, for media with stochastic microstructure, has been studied extensively [2], and much is known about reflection from, and transmission through a slab of material. In particular, such waves are localized. That is, because of random multiple scattering, the waves are attenuated exponentially with propagation distance. The reciprocal of the attenuation constant, called the localization length, is deterministic, and is implicated in other related phenomena. For instance, when an acoustic pulse is backscattered from the random medium, the power spectrum of the random backscatter can be predicted from knowledge of the form of the localization length as a function of frequency [2].

We will outline below the calculations and results for P and SV waves, which are coupled for oblique angles of incidence. A more complete treatment is in [3], which includes numerical calculations and comparisons to Monte-Carlo simulations. We compute the localization length and another deterministic length, called the equilibration length, which is associated with the equilibration of shear and compressional energy on transmission through a large slab. These characteristic lengths are seen to be reciprocals of the two positive Lyapunov exponents for this system. We also provide expressions for the probability density of P to SV wave energy on trans-

* Department of Mathematics, Virginia Polytechnic Institute and State University, Blacksburg, VA 24061.

† Department of Mathematics, StanfordUniversity, Stanford, CA 94305.

‡ Exxon Research and Engineering Company, Route 22 East, Annandale, NJ 08801.

mission through a large slab, and for the fraction of energy which is mode converted on backscatter from a random half-space.

2. P/SV waves in stratified media. Let ρ be the density and let λ, μ be the Lame parameters of an elastic medium occupying the half space $x_3 = z > 0$. We consider that the medium is stratified, so that ρ, λ, μ are functions of the depth coordinate, z, only. We express the assumption that the medium has a random microstructure by introducing the small parameter $\epsilon > 0$, and assuming that $\rho = \rho(z/\epsilon^2)$, $\lambda = \lambda(z/\epsilon^2)$, $\mu = \mu(z/\epsilon^2)$ are stationary random functions of z, varying on the microscale which is of order ϵ^2.

We consider time-harmonic plane waves, so that the displacement $\boldsymbol{u}$ and the stress tensor $\overline{\tau}$ are given by

$$\begin{aligned} \boldsymbol{u} &= e^{-i\overline{\omega}(t-px_1)}\hat{\boldsymbol{u}}(z) \\ \overline{\tau} &= e^{-i\overline{\omega}(t-px_1)}\hat{\tau}(z) \end{aligned} \tag{2.1}$$

where $\overline{\omega}$ is angular frequency and p is the horizontal slowness. We choose a state vector

$$\boldsymbol{X} = (-i\overline{\omega}\hat{u}_3, \hat{\tau}_{13}, \hat{\tau}_{33}, -i\overline{\omega}\hat{u}_1)^T. \tag{2.2}$$

then it can be shown [1,6] that P/SV waves satisfy the equation

$$\frac{d\boldsymbol{X}}{dz} = -i\overline{\omega}M\boldsymbol{X} \tag{2.3}$$

where the 4×4 matrix M is of the form

$$M = \begin{bmatrix} 0 & M_1 \\ M_2 & 0 \end{bmatrix} \tag{2.4}$$

with real symmetric 2×2 blocks M_1, M_2

$$M_1 = \begin{bmatrix} \frac{1}{\lambda+2\mu} & \frac{\lambda p}{\lambda+2\mu} \\ \frac{\lambda p}{\lambda+2\mu} & \rho - \frac{4p^2\mu(\lambda+\mu)}{\lambda+2\mu} \end{bmatrix} \tag{2.5}$$

$$M_2 = \begin{bmatrix} \rho & p \\ p & \frac{1}{\mu} \end{bmatrix} \tag{2.6}$$

Now for ϵ small, the microstructure can be averaged to give "effective medium theory", i.e. equations of the form (2.3), but with the averaged matrices

$$\overline{M} = E[M] = \begin{bmatrix} 0 & \overline{M}_1 \\ \overline{M}_2 & 0 \end{bmatrix}$$

$$\overline{M}_1 = E[M_1] = \begin{bmatrix} \gamma_1 & \gamma_2 p \\ \gamma_2 p & \overline{\rho} - \gamma_3 p^2 \end{bmatrix} \tag{2.7}$$

$$\overline{M}_2 = E[M_2] = \begin{bmatrix} \overline{\rho} & p \\ p & \frac{1}{\overline{\mu}} \end{bmatrix}$$

where $E[\ ]$ denotes expected value and

$$\text{(2.8)} \qquad \begin{gathered} \overline{\rho} = E[\rho], \quad \overline{\mu} = \left(E\left[\tfrac{1}{\mu}\right]\right)^{-1} \\ \gamma_1 = E\left[\tfrac{1}{\lambda+2\mu}\right], \gamma_2 = E\left[\tfrac{\lambda}{\lambda+2\mu}\right], \gamma_3 = E\left[\tfrac{4\mu(\lambda+\mu)}{\lambda+2\mu}\right]. \end{gathered}$$

It can be shown that, provided the propagation distance is not too long, the effective medium equations give a good approximation to the solution of (2.3) for small ϵ [4,5]. Thus localization phenomena do not occur on this scale, where wave phenomena are well approximated by a homogeneous (i.e. constant, non-random coefficient) medium.

To see localization we must propagate through many wavelengths, by introducing the scaling

$$\text{(2.9)} \qquad \overline{\omega} = \frac{\omega}{\epsilon}.$$

In this scaling, the wavelength is of order ϵ, which is both much larger than the order ϵ^2 microscale and much smaller than the order one propagation distance. Thus, the wave is low frequency (i.e. the randomness is sub-wavelength), and the propagation distance is many wavelengths. It will be seen that with the scaling (2.9), the fluctuation matrices

$$\text{(2.10)} \qquad \begin{gathered} \hat{M} = M - \overline{M} \\ \hat{M}_j = M_j - \overline{M}_j, j = 1, 2 \end{gathered}$$

play an important role as $\epsilon \downarrow 0$. The fully scaled equations then become

$$\text{(2.11)} \qquad \frac{d\boldsymbol{X}}{dz} = \frac{-i\omega}{\epsilon}[\overline{M} + \hat{M}(z/\epsilon^2)].$$

3. Ups and downs. Equations (2.11) can be decomposed into up and down-going P and S waves, with respect to the effective medium, by introducing a constant matrix L that diagonalizes $\overline{M}$. Details are in [3], which follows the method of Ursin [6]. Let

$$\text{(3.1)} \qquad \boldsymbol{Y} = L^{-1}\boldsymbol{X} = \begin{bmatrix} \boldsymbol{U} \\ \boldsymbol{D} \end{bmatrix}$$

where $\boldsymbol{U}$ is a 2-vector of up-going P and S waves (first and second components, respectively), and $\boldsymbol{D}$ is a 2-vector of downgoing P and S waves. Then it can be shown that

$$\text{(3.2)} \qquad \frac{d\boldsymbol{Y}}{dz} = \frac{-i\omega}{\epsilon}\left\{\Lambda_1 + \nu(z/\epsilon^2)\right\}\boldsymbol{Y}$$

where

$$\text{(3.3)} \qquad \begin{gathered} \Lambda_1^* = \begin{bmatrix} \Lambda & 0 \\ 0 & -\Lambda \end{bmatrix} \\ \Lambda = \begin{bmatrix} \alpha_p & 0 \\ 0 & \alpha_s \end{bmatrix} \end{gathered}$$

and α_p, α_s are the vertical P and S wave slownesses, respectively, in the effective medium. The fluctuation matrix ν is of the form

$$\nu = \begin{bmatrix} \nu_1 & -\nu_2 \\ \nu_2 & -\nu_1 \end{bmatrix} \tag{3.4}$$

where ν_1, ν_2 are real symmetric matrices of the form

$$\begin{aligned} \nu_1 &= \begin{bmatrix} \delta_1 & \delta_2 \\ \delta_2 & \delta_3 \end{bmatrix} \\ \nu_2 &= \begin{bmatrix} \delta_4 & \delta_5 \\ \delta_5 & \delta_6 \end{bmatrix} \end{aligned} \tag{3.5}$$

Here $\delta_j = \delta_j(z/\epsilon^2), j = 1, 2, ...6$ are mean zero stationary stochastic processes which vary on the microscale, ϵ^2. An interpretation of these processes may be obtained by considering their role in equation (3.2), coupling the various components of $\boldsymbol{Y}$: $\delta_1, \delta_2, \delta_3$ act as local transmission coefficients, coupling P-P, P-S, S-S waves, respectively, for waves propagating in the same (i.e. either in the up or down) direction. $\delta_4, \delta_5, \delta_6$ act as local reflection coefficients, coupling P-P, P-S, S-S waves, respectively, for waves propagating in opposite directions.

Important statistical parameters are the scattering strengths of δ_j

$$\sigma_{jj} = \int_0^\infty E[\delta_j(0)\delta_j(z)]dz, \quad j = 1, ..., 6. \tag{3.6}$$

In particular, we will use the two parameters

$$\begin{aligned} \Sigma &= \tfrac{(\sigma_{22} - \sigma_{55})}{(\sigma_{22} + \sigma_{55})} + \tfrac{1}{2}\tfrac{(\sigma_{44} + \sigma_{66})}{(\sigma_{22} + \sigma_{55})} \\ \gamma &= \tfrac{(\sigma_{44} - \sigma_{66})}{(\sigma_{22} + \sigma_{55})}. \end{aligned} \tag{3.7}$$

In what follows, we will look at fundamental solution matrices of (3.2), often called propagator matrices, or transfer matrices. We consider a finite slab of random material of width $\bar{z}$, occupying $0 \le z \le \bar{z}$, with a wave incident on $z = 0$ and transmitted through $z = \bar{z}$. Details for setting up this boundary value problem are in [3]. For this purpose it is convenient to introduce propagators which start at the transmission end of the slab. Thus we let $\zeta = -z$ and define the propagator matrix $\overline{Q}$ as the solution of

$$\begin{aligned} \tfrac{d\overline{Q}}{d\zeta} &= \tfrac{i\omega}{\epsilon}\left\{\Lambda_1 + \nu(-\zeta/\epsilon^2)\right\}\overline{Q} \\ \overline{Q}|_{\zeta=-\bar{z}} &= I \end{aligned} \tag{3.8}$$

where I is the 4×4 identity matrix.

Another useful form of this propagator matrix, as explained in the next section, is

$$Q = e^{-i\frac{\omega}{\epsilon}\Lambda_1\zeta}\overline{Q}e^{-i\frac{\omega}{\epsilon}\Lambda_1\bar{z}} \tag{3.9}$$

which satisfies

$$\frac{dQ}{d\zeta} = \frac{i\omega}{\epsilon}\eta(\zeta/\epsilon^2, \zeta/\epsilon)Q \qquad (3.10)$$
$$Q|_{\zeta=-\overline{z}} = I$$

with

$$\eta = \begin{bmatrix} \eta_1 & -\eta_2^* \\ \eta_2 & -\eta_1^* \end{bmatrix}. \qquad (3.11)$$

Here $*$ denotes complex conjugate, and

$$\begin{aligned} \eta_1 &= e^{-i\frac{\omega}{\epsilon}\Lambda\zeta}\nu_1 e^{i\frac{\omega}{\epsilon}\Lambda\zeta} \\ \eta_2 &= e^{i\frac{\omega}{\epsilon}\Lambda\zeta}\nu_2 e^{i\frac{\omega}{\epsilon}\Lambda\zeta} \end{aligned} \quad . \qquad (3.12)$$

Note that η_1 is Hermitian, while η_2 is complex symmetric. It can be shown that Q has the following symmetry properties (as does $\overline{Q}$):

$$Q = \begin{bmatrix} A & B^* \\ B & A^* \end{bmatrix} \qquad (3.13)$$

with A, B 2×2 matrices. Also

$$Q^{-1} = \begin{bmatrix} A^{*T} & -B^{*T} \\ -B^T & A^T \end{bmatrix}. \qquad (3.14)$$

Reflection and transmission through the random slab are governed by the matrices A, B and the properties of the homogeneous media within which the slab is embedded, i.e. the medium occupying the region $z < 0$, above the random slab, and the medium occupying $z > \overline{z}$, below the random slab. We report here only the results for the "matched case", i.e. when the properties of the media enclosing the random slab are the same as those of the effective medium. Thus, in this case, there are no effective impedance mismatches at the interfaces where waves enter and leave the random slab. More general results are in [3].

For the matched case, reflection and transmission are governed by the 2×2 matrices Γ and τ respectively, where

$$\begin{aligned} \Gamma &= B^* A^{*-1} \\ \tau &= A^{*-1}. \end{aligned} \qquad (3.15)$$

Then reflected and transmitted fields are produced by multiplying the 2-vector representing the incident field by Γ and τ, respectively.

4. Limits and Lyapunov exponents. The two forms of the propagator matrix, Q and $\overline{Q}$ are each useful for different purposes. Note that the coefficients in (3.8) are stationary random functions of position. Thus $\overline{Q}$ satisfies the Oseledec theorem [7,8], as described below, which provides us with the existence of Lyapunov exponents, and other structural facts pertinent to large slab lengths. However the other form, Q, is more useful for detailed calculations. This is because the coefficients on the right hand side of equation (3.10) have mean zero. The large, mean zero, rapidly fluctuating coefficients in this equation allow use of an approximate "white noise" diffusion equation as $\epsilon \downarrow 0$. Of course Q and $\overline{Q}$ are easily related through the deterministic transformation (3.9).

Let $\overline{\zeta} = \zeta + z$. Then the Oseledec theorem provides the a.s. convergence

$$\lim_{\overline{\zeta}\to\infty} [\overline{Q}(\overline{\zeta})^{*T}\overline{Q}(\overline{\zeta})]^{\frac{1}{2\overline{\zeta}}} = F. \tag{4.1}$$

The eigenspaces of F are random, but the (real, non-negative) eigenvalues

$$e^{\Omega_1} \geq e^{\Omega_2} \geq e^{\Omega_3} \geq e^{\Omega_4} \tag{4.2}$$

are not random. Ω_j are the Lyapunov exponents of $\overline{Q}$.

Let $l_j, \tilde{W}_j, j = 1, 2, 3, 4$ be the (real) eigenvalues and the eigenvectors, respectively, of the Hermitian, positive semi-definite matrix $[\overline{Q}(\overline{\zeta})^{*T}\overline{Q}(\overline{\zeta})]$, with $l_1 > l_2 > l_3 > l_4$. Using the symmetry relations (3.13), (3.14), one can show that

$$\tilde{W}_j = \frac{1}{\sqrt{2}} \begin{bmatrix} \tilde{V}_j \\ \tilde{V}_j^* \end{bmatrix} \tag{4.3}$$

with

$$\begin{aligned} \tilde{V}_3 &= i\tilde{V}_2, \quad \tilde{V}_4 = i\tilde{V}_1 \\ l_3 &= \tfrac{1}{l_2}, \quad l_4 = \tfrac{1}{l_1}. \end{aligned} \tag{4.4}$$

Furthermore, $\tilde{V}_1, \tilde{V}_2$ are an orthonormal basis of C^2.

Now the Lyapunov exponents are

$$\Omega_j = \lim_{\overline{\zeta}\to\infty} \frac{1}{(2\overline{\zeta})} ln(l_j) \tag{4.5}$$

so that (4.4) implies that there are only two positive Lyapunov exponents

$$\Omega_1 > \Omega_2 > 0 \tag{4.6}$$

and that the other two are their negatives

$$\Omega_3 = -\Omega_2, \quad \Omega_4 = -\Omega_1. \tag{4.7}$$

The Lyapunov exponents could now in principle be calculated using the $\epsilon \downarrow 0$ limit of (3.10) which could be obtained from a "white noise" limit theorem of the type first proposed by Khasminskii [9] and generalized by others [10]. We could then obtain a weak convergence to a Markov diffusion process. However, as it is written, equation (3.10) is for a 4×4 matrix, giving 16 complex stochastic processes, and hence a diffusion in 32-dimensional real space. We therefore utilize various reductions of this equation before taking the $\epsilon \downarrow 0$ limit.

First, the largest exponent, Ω_1, can be obtained from the growth rate of an arbitrarily chosen vector solution of equation (3.10). Furthermore, using the symmetry relations, we need only consider vectors of the form $[V, V^*]^T$, where V is a 2-vector. This gives a problem of diffusion in 4-dimensional space which can be solved in quadrature. The result is that

$$\begin{aligned}\Omega_1 = \tfrac{\omega^2}{2}\Big\{&-2\left(\sigma_{55} - \tfrac{1}{4}[\sigma_{44}+\sigma_{66}]\right)\textstyle\int_{-\infty}^{\infty}\operatorname{sech}^2(R)\overline{P}_R(R)dR + \\ &(\sigma_{44}-\sigma_{66})\textstyle\int_{-\infty}^{\infty}\tanh(R)\overline{P}_R(R)dR + 4\left(\sigma_{55}+\tfrac{1}{4}[\sigma_{44}+\sigma_{66}]\right)\Big\}\end{aligned} \tag{4.8}$$

Here the probability density, $\overline{P}_R(R)$ is given, for $\Sigma > 1, \gamma \neq 0$, by

$$\overline{P}_R(R) = \frac{\left(\gamma\left[\Sigma+\sqrt{\Sigma^2-1}\right]^{\frac{\gamma}{(\sqrt{\Sigma^2-1})}}\right)}{\left(\left[\Sigma+\sqrt{\Sigma^2-1}\right]^{\frac{\gamma}{(\sqrt{\Sigma^2-1})}}-1\right)} \cdot \left[\frac{e^{2R}+\Sigma-\sqrt{\Sigma^2-1}}{e^{2R}+\Sigma+\sqrt{\Sigma^2-1}}\right]^{\left(\frac{\gamma}{2\sqrt{\Sigma^2-1}}\right)} \cdot \frac{1}{(\cosh(2R)+\Sigma)}. \tag{4.9}$$

Corresponding expressions for $\Sigma \leq 1$ *or* $\gamma = 0$ are given in [3].

Ω_2 is somewhat harder to get, since it produces an exponentially smaller growth rate than Ω_1. However, again using symmetries, coupled with a singular value decomposition of Q, we find that

$$|\det A|^2 = \frac{1}{16}\left(l_1 + \frac{1}{l_1} + 2\right)\left(l_2 + \frac{1}{l_2} + 2\right) \tag{4.10}$$

so that

$$\Omega_1 + \Omega_2 = \lim_{\zeta = \overline{z} \to \infty} \frac{1}{\overline{\zeta}} \operatorname{Re}\{ln(\det(A))\}. \tag{4.11}$$

From equation (4.11), and an appropriate $\epsilon \downarrow 0$ diffusion equation, we obtain that

$$\Omega_1 + \Omega_2 = \omega^2[2\sigma_{55} + \sigma_{44} + \sigma_{66}]. \tag{4.12}$$

5. Results and conclusions. In the same way that we obtained explicit results for the Lyapunov exponents, we can calculate the statistics of reflection and transmission through a large slab. We sketch here the case of the matched medium, as defined above, when these phenomena are governed by the matrices Γ and τ of equations (3.15). Results for more general boundary conditions are in [3].

It can be shown that as the slab length is increased, the ratio of transmitted S wave energy to transmitted P wave energy equilibrates to a random quantity

$$\frac{\mathcal{E}_s}{\mathcal{E}_R} = e^{2R} \tag{5.1}$$

where R has distribution $\overline{P}_R(R)$ as given, for $\Sigma > 1, \gamma \neq 0$, in equation (4.9). This equilibration takes place over a length scale

$$z_{\text{equil}} = \frac{1}{\Omega_1} \tag{5.2}$$

which we call the equilibration length. The transmitted energy then decays to zero at an exponential rate Ω_2. Thus the localization length is

$$z_{\text{loc}} = \frac{1}{\Omega_2}. \tag{5.3}$$

We next consider reflection from a random half space. We assume that the incident field is a monochromatic wave consisting of a single mode (i.e. either P or S). The reflected field will then, in general, be a mixture of both P and S. That is, some of the energy will be mode converted into the type other than that of the incident field. The fraction of energy which is reflected in the same mode as that of the incident field is represented as

$$|\Gamma_{11}|^2 = \tanh^2(\theta/2). \tag{5.4}$$

That is, $1 - |\Gamma_{11}|^2$ is the fraction of energy which is mode converted on reflection. It can be shown that θ is a random variable with probability density (for $\Sigma \neq 1$)

$$\overline{P}(\theta) = \sqrt{\frac{1+\Sigma}{2}} \frac{(e^{2\theta} - 1)}{(e^{2\theta} + 2\Sigma e^{\theta} + 1)^{\frac{3}{2}}}. \tag{5.5}$$

From (5.4) and (5.5) algebraic expressions can be obtained for the moments of $|\Gamma_{11}|^2$. For these moments, and the corresponding formulas for the mismatched medium case, the reader is referred to [3].

We therefore obtain quite explicit results for the full complement of reflection, transmission, mode conversion and localization phenomena.

REFERENCES

[1] R. Burridge, *Some Mathematical Topics in Seismology*, Courant Inst. Math. Sciences, New York (1976).

[2] M. Asch, W. Kohler, G. Papanicolaou, M. Postel, and B. White, *Frequency content of randomly scattered signals*, SIAM Rev. 33, 519–625 (1991).

[3] W. Kohler, G. Papanicolaou, and B. White, *Localization and mode conversion for elastic waves in randomly layered media*, Wave Motion, Submitted (1994).

[4] R.Z. Khasminskii, *On stochastic processes defined by differential equations with a small parameter*, Theory Prob. Applic. 11, 211-228 (1966).

[5] B. White and J. Franklin, *A limit theorem for stochastic two-point boundary value problems of ordinary differential equations*, Comm. Pure Appl. Math. 32, 253–275 (1979).

[6] B. Ursin, *Review of elastic and electromagnetic wave propagation in horizontally layered media*, Geophysics 48, 1063–1081 (1983).

[7] V. Oseledec, *A multiplicative ergodic theorem. Lyapunov characteristic numbers for dynamical systems*, Trans. Moscow Math. Soc. 19, 197–231 (1968).

[8] J. Cohen, H. Kesten and C. Newman, *Oseledec's multiplicative ergodic theorem: a proof*, in *Random Matrices and their Applications*, Contemporary Math. 50, Amer. Math. Soc., Providence, 23-30 (1986).

[9] R.Z. Khasminskii, *A limit theorem for the solutions of differential equations with random right-hand sides*, Theory Prob. Applic. 11, 390–406 (1966).

[10] G. Papanicolaou and W. Kohler, *Asymptotic theory of mixing stochastic ordinary differential equations*, Comm. Pure Appl. Math. 27, 641–668 (1974).

STOCHASTIC FORCING OF OCEANIC MOTIONS

PETER MÜLLER*

Abstract. Most oceanic motions are forced by stresses and heat and fresh water fluxes at the air-sea interface. When these forcing fields are assumed to be prescribed and stochastic and when the oceanic response is assumed to be linear then the forcing problem reduces to a set of stochastically forced linear oscillators. This set can in principle be decoupled. The asymptotic response of a linear oscillator to stationary random forcing is well understood and depends on whether the oscillator is stable, unstable or neutral. The explicit decoupling requires additional simplifying assumptions as exemplified by the stochastic forcing of surface gravity waves, internal gravity waves, Rossby waves and sea-surface temperature anomalies. A powerful diagnostic tool is a coherence map which desribes the coherence between the oceanic response at one location and the atmospheric forcing at another location as a function of separation for different frequancies. A simple model of the stochastic forcing of barotropic Rossby waves by fluctuations in the atmospheric windstress reproduces basic features of observed coherence maps. The model expecially accounts for the qualitative changes that occur when different oceanic variables are considered or when the frequency is changed.

1. Introduction. Except for the tides, oceanic motions are forced by the stresses and heat and fresh water fluxes at the air-sea interface. When the forcing fields are assumed to be

- solely governed by the atmosphere,
- prescribed and
- stochastic due to the inherent instability and turbulence mechanisms within the atmosphere, and

when the response of the ocean is assumed to be

- governed by linear dynamics

then the forcing problem reduces to a set of stochastically forced linear oscillators.

In the first part of the paper we review the asymptotic response of a set of linear oscillators to stationary random forcing. The response depends on whether the oscillators are stable, unstable or neutral. Discrete sets of eigenfrequencies must be distinguished from continuous sets.

The set of linear oscillators can in principal be decoupled. Explicit decoupling requires additional simplifying assumptions. In the second part we give the standard assumptions and derivations that decouple the forcing problem for surface gravity waves, internal gravity waves and Rossby waves. The forcing of sea-surface temperature anomalies is also disussed. These oceanic wave motions cover time scales from seconds to months.

The third part discusses in detail the stochastic forcing of barotropic Rossby waves by fluctuations in the atmospheric windstress. The governing equations and the atmospheric forcing field are discussed. The concept of a coherence map is introduced. It is shown that a simple stochastic forcing model can reproduce basic features of observed coherence maps.

* Department of Oceanography, University of Hawaii, Honolulu, Hawaii 96822 U.S.A.

We conclude with a perspective on simple linear stochastic forcing models.

2. Stochastically forced oscillators. Oceanic motions are described by the Navier-Stokes equations or some approximation to them. In the linear limit these equations take the form

$$\dot{y}_j(\mathbf{x},t) + \sum_{k=1}^{N} D_{jk} y_k(\mathbf{x},t) = f_j(\mathbf{x},t) \qquad j = 1,\dots,N \tag{2.1}$$

where the vector $\mathbf{y} = (y_1,\dots,y_N)$ describes the state of the ocean. Its components include the velocity components, the pressure, the density and possibly other variables. Explicit examples are given in section 3. The independent variables are position $\mathbf{x}$ and time t. The dot denotes the time derivative, D_{jk} linear spatial differential operators and the vector $\mathbf{f}(\mathbf{x},t)$ the prescribed stochastic forcing. We have to distinguish between bounded and unbounded spatial domains.

2.1. Bounded domain. In a spatially bounded domain the homogeneous form of (2.1) usually has a discrete and complete set of eigensolutions

$$\mathbf{y}_\nu(\mathbf{x},t) = \mathbf{h}_\nu(\mathbf{x}) e^{-i\omega_\nu t} \qquad \nu = 1,2,\dots \tag{2.2}$$

with spatial eigenmode $\mathbf{h}_\nu(\mathbf{x})$ and eigenfrequency ω_ν. The state vector $\mathbf{y}(\mathbf{x},t)$ and the forcing vector $\mathbf{f}(\mathbf{x},t)$ can then be expanded into eigenmodes

$$\begin{pmatrix} \mathbf{y}(\mathbf{x},t) \\ \mathbf{f}(\mathbf{x},t) \end{pmatrix} = \sum_{\nu=1}^{\infty} \begin{pmatrix} \eta_\nu(t) \\ \rho_\nu(t) \end{pmatrix} \mathbf{h}_\nu(\mathbf{x}) \tag{2.3}$$

where $\eta_\nu(t)$ and $\rho_\nu(t)$ are the time dependent expansion coefficients. Projection of (2.1) onto the eigenmodes results in

$$\dot{\eta}_\nu(t) + i\omega_\nu \eta_\nu(t) = \rho_\nu(t) \qquad \nu = 1,2,\dots \tag{2.4}$$

The problem is therefore reduced to a set of decoupled forced linear oscillators. The solution is given by

$$\eta_\nu(t) = \eta_\nu^0 e^{-i\omega_\nu t} + \int_0^t dt' \rho_\nu(t') e^{-i\omega_\nu (t-t')} \tag{2.5}$$

where η_ν^0 is the initial condition.

It is now assumed that the initial conditions η_ν^0 are random variables and that the forcing coefficients $\rho_\nu(t)$ are stationary random functions. The solution $\eta_\nu(t)$ then becomes a random function as well. The analysis is generally performed in terms of moments. Only first and second moments are considered here. Without loss of generality the first moments of the initial conditions and forcing coefficients are assumed to be zero

$$<\eta_\nu^0> = <\rho_\nu(t)> = 0 \qquad \nu = 1, 2, \dots \tag{2.6}$$

which implies

$$<\eta_\nu(t)> = 0 \tag{2.7}$$

Here cornered brackets denote the ensemble average. With regards to the second moments, one assumes that the covariance matrix

$$R_{\eta_\nu^0 \eta_\mu^0} = <\eta_\nu^{0*} \, \eta_\mu^0> \tag{2.8}$$

exists, that the covariance function

$$R_{\rho_\nu \rho_\mu}(\tau) = <\rho_\nu^*(\tau) \, \rho_\mu(t+\tau)> \tag{2.9}$$

exists and is integrable, and that initial conditions and forcing are uncorrelated

$$<\rho_\nu^*(t) \, \eta_\mu^0> = 0 \tag{2.10}$$

Under these assumptions one can calculate the covariance function

$$R_{\eta_\nu \eta_\mu}(t, \tau) = <\eta_\nu^*(t) \, \eta_\mu(t+\tau)> \tag{2.11}$$

of the response. The random function $\eta_\nu(t)$ is asymptotically stationary if this covariance function exists and is independent of t as t approaches infinity. This is the case only for stable oscillators when the imaginary part of the eigenfrequency is smaller than zero. However, as was shown by Hasselmann (1962) in an oceanographic context, the normalized random functions

$$\begin{aligned}
&\tilde{\eta}_\nu = \eta_\nu && \text{for } \mathrm{Im}\{\omega_\nu\} < 0 && \nu \in M_1 \\
&\tilde{\eta}_\nu = \eta_\nu e^{i\omega_\nu t} && \text{for } \mathrm{Im}\{\omega_\nu\} > 0 && \nu \in M_2 \\
&\tilde{\eta}_\nu = \eta_\nu / \sqrt{t} && \text{for } \mathrm{Im}\{\omega_\nu\} = 0 && \nu \in M_3
\end{aligned} \tag{2.12}$$

are asymptotically stationary with covariance functions

$$\lim_{t \to \infty} R_{\tilde{\eta}_\nu \tilde{\eta}_\mu}(t, \tau) = \tag{2.13}$$

$$\begin{cases} -\dfrac{1}{i(\omega_\nu^* - \omega_\mu)} \displaystyle\int_0^\infty ds \left[R_{\rho_\nu\rho_\mu}(\tau + s)e^{i\omega_\nu^* s} + R_{\rho_\nu\rho_\mu}(\tau - s)e^{-i\omega_\mu s}\right] & \\ & \text{for } \nu, \mu \in M_1 \\ R_{\eta_\nu^0\eta_\mu^0} + \dfrac{1}{i(\omega_\nu^* - \omega_\mu)} \displaystyle\int_0^\infty ds \left[R_{\rho_\nu\rho_\mu}(s)e^{i\omega_\mu s} + R_{\rho_\nu\rho_\mu}(-s)e^{-i\omega_\nu^* s}\right] & \\ & \text{for } \nu, \mu \in M_2 \\ e^{-i\omega_\nu \tau} \displaystyle\int_{-\infty}^{+\infty} ds R_{\rho_\nu\rho_\mu}(s)e^{i\omega_\nu s} & \text{for } \omega_\nu = \omega_\mu \text{ and } \nu, \mu \in M_3 \\ 0 & \text{otherwise.} \end{cases}$$

The stable ($\nu \in M_1$), unstable ($\nu \in M_2$) and neutral modes ($\nu \in M_3$) are mutually uncorrelated. The neutral modes are only correlated if $\omega_\nu = \omega_\mu$. The covariance function for the unstable case is constant, i. e. independent of the lag τ, and forcing and initial conditions play equivalent roles. The covariance functions of the original variables grow exponentially in time for unstable modes and linear in time for neutral modes.

Frequency cross-spectra $S(\omega)$ are defined via the Fourier transformations

$$(2.14) \qquad \begin{aligned} S(\omega) &= \frac{1}{2\pi} \int_{-\infty}^{+\infty} d\tau e^{-i\omega\tau} R(\tau) \\ R(\tau) &= \int_{-\infty}^{+\infty} d\omega e^{i\omega\tau} S(\omega) \end{aligned}$$

and are explicitly given by

$$(2.15) \quad S_{\tilde{\eta}_\nu\tilde{\eta}_\mu}(\omega) = \begin{cases} \dfrac{S_{\rho_\nu\rho_\mu}(\omega)}{(\omega + \omega_\nu^*)(\omega + \omega_\mu)} & \text{for } \nu, \mu \in M_1 \\ \left[R_{\eta_\nu^0\eta_\mu^0} + \displaystyle\int_{-\infty}^{+\infty} d\omega' \frac{S_{\rho_\nu\rho_\mu}(\omega')}{(\omega' + \omega_\nu^*)(\omega + \omega_\mu)}\right] \delta(\omega) & \\ & \text{for } \nu, \mu \in M_2 \\ 2\pi S_{\rho_\nu\rho_\mu}(\omega)\delta(\omega + \omega_\nu) \text{ for } \omega_\nu = \omega_\mu \text{ and } \nu, \mu \in M_3 & \end{cases}$$

The cross-spectrum of the response in the stable case equals the cross-spectrum of the forcing divided by a resonance denominator. In the unstable case, all the covariance is concentrated at zero frequency. In the neutral case all the covariance is concentrated at the eigenfrequency and proportional to the cross-spectrum of the forcing at that frequency. Only the forcing at the eigenfrequencies matters.

2.2. Unbounded domain. When the spatial domain is unbounded the eigenfrequencies usually form a continuous set. In this case it is hard to obtain results without further specifying the problem. Here we consider the equation of motion

$$\ddot{y}(\mathbf{x},t) + D_1[\dot{y}(\mathbf{x},t)] + D_2[y(\mathbf{x},t)] = f(\mathbf{x},t) \tag{2.16}$$

where $y(\mathbf{x},t)$ is a variable describing the oceanic response, $f(\mathbf{x},t)$ the stochastic forcing and D_1 and D_2 linear spatial differential operators. We will only consider the stable and neutral case. The unstable case does not require stochastic forcing and is generally formulated in terms of random initial conditions.

The basic assumption in the unbounded case is that the problem is spatially homogeneous. Specifically, one assumes that

(i) the differential operators D_1 and D_2 are constant
(ii) the forcing $f(\mathbf{x},t)$ is a homogeneous and stationary random function with zero mean, autocovariance function

$$R_f(\mathbf{r},\tau) = <f^*(\mathbf{x},t)\, f(\mathbf{x}+\mathbf{r},t+\tau)> \tag{2.17}$$

and wavenumber-frequency autospectrum

$$S_f(\mathbf{k},\omega) = \frac{v_1}{(2\pi)^3} \iiint_{-\infty}^{+\infty} d\mathbf{x} d\tau e^{-i(\mathbf{k}\cdot\mathbf{x}+\omega\tau)} R_f(\mathbf{r},\tau) \tag{2.18}$$

For such homogeneous and stationary processes the wavenumber-frequency Fourieramplitudes $f(\mathbf{k},\omega)$ are uncorrelated and are related to the spectrum by

$$<f^*(\mathbf{k},\omega)\, f(\mathbf{k}',\omega')> = S_f(\mathbf{k},\omega)\delta(\mathbf{k}-\mathbf{k}')\delta(\omega-\omega') \tag{2.19}$$

Spatial Fourier transformation of (2.16) yields

$$\ddot{y}(\mathbf{k},t) + d_1(\mathbf{k})\dot{y}(\mathbf{k},t) + d_2(\mathbf{k})y(\mathbf{k},t) = f(\mathbf{k},t) \tag{2.20}$$

where $d_s(\mathbf{k})$ $(s = 1, 2)$ are functions. Again the problem is reduced to a set of decoupled stochastically forced linear oscillators, a continuous set in this case.

Assume that the homogeneous equation has the two eigensolutions

$$y_s(\mathbf{k},t) = e^{-i\omega_s(\mathbf{k})t} \qquad s = 1,2 \tag{2.21}$$

with $\omega_1(\mathbf{k}) \neq \omega_2(\mathbf{k})$ for all $\mathbf{k}$. The solution of the inhomogeneous equation is then given by

$$\begin{aligned} y(\mathbf{k},t) = {} & \frac{i\omega_2 y^0 + \dot{y}^0}{i(\omega_2 - \omega_1)} e^{-i\omega_1 t} + \frac{i\omega_1 y^0 + \dot{y}^0}{i(\omega_1 - \omega_2)} e^{-i\omega_2 t} \\ & + \int_{-\infty}^{+\infty} d\omega \frac{1}{\omega_1 - \omega_2} \left(\frac{e^{i\omega t} - e^{-i\omega_1 t}}{\omega_1 + \omega} - \frac{e^{i\omega t} - e^{-i\omega_2 t}}{\omega_2 + \omega} \right) f(\mathbf{k},\omega) \end{aligned} \tag{2.22}$$

where $y^0 = y^0(\mathbf{k})$ and $\dot{y}^0 = \dot{y}^0(\mathbf{k})$ are the initial conditions. For stable systems ($\text{Im}\{\omega_s(\mathbf{k})\} < 0$ for $s = 1,2$ and all $\mathbf{k}$) all terms proportional to $\exp\{-i\omega_s t\}$ can be dropped asymptotically and one obtains the asymptotic solution

$$y(\mathbf{k},t) = \int_{-\infty}^{+\infty} d\omega \frac{f(k,\omega)}{(\omega_1 + \omega)(\omega_2 + \omega)} e^{i\omega t} \tag{2.23}$$

or

$$y(\mathbf{k},\omega) = \frac{f(k,\omega)}{(\omega_1 + \omega)(\omega_2 + \omega)} \tag{2.24}$$

The response is a spatially homogeneous and asymptotically stationary process. Its wavenumber-frequency spectrum is given by

$$S_y(\mathbf{k},\omega) = \frac{S_f(\mathbf{k},\omega)}{|\omega_1 + \omega|^2 |\omega_2 + \omega|^2} \tag{2.25}$$

For the neutral case ($\text{Im}\{\omega_s(\mathbf{k})\} = 0$ for $s = 1,2$ and all $\mathbf{k}$) the wavenumber spectrum grows linear in time and one defines a "growth rate" by

$$\lim_{t\to\infty} \frac{1}{t} S_y(\mathbf{k},t,\tau) = 2\pi \sum_{s=1,2} e^{-i\alpha_s(\mathbf{k})\tau} \frac{S_f(\mathbf{k},\omega = -\alpha_s(\mathbf{k}))}{(\alpha_2 - \alpha_1)^2} \tag{2.26}$$

where $\alpha_s(\mathbf{k}) = \text{Re}\{\omega_s(\mathbf{k})\}$.

Calculation of the wavenumber-frequency spectrum (2.25) for the stable and of the growth rate (2.26) for the neutral case are the most common calculations done in oceanography. The calculations involve specification of the eigenfrequencies $\omega_s(\mathbf{k})$ and of the forcing spectrum $S_f(\mathbf{k},\omega)$. Often integrations over part of wavenumber-frequency space are carried out to obtain marginal spectra. Problems are that the wavenumber-frequency structure of the forcing and the damping mechanisms within the ocean are not well known. Most often one assumes a horizontally unbounded ocean of finite depth and employs a Fourier decomposition in the horizontal and a discrete eigenmode decomposition in the vertical.

3. Examples of reductions. Here we give the standard assumptions and derivations that reduce the equations of motion to a set of decoupled forced oscillators for some well-known wave systems in the ocean. The time scales of these wave systems range from seconds to months.

3.1. Surface gravity waves. The first example is the stochastic forcing of surface gravity waves by random fluctuations of the atmospheric pressure field. Consider a horizontally unbounded infinitely deep ocean of constant density. Assume the motion to be incompressible, irrotational and inviscid. Linear motions are then described by a velocity potential $\varphi(\mathbf{x},t)$ that satisfies Laplace equation

$$\Delta\varphi = 0 \qquad \text{for } -\infty < z < 0 \tag{3.1}$$

and the boundary conditions

$$\ddot{\varphi} + g\partial_z\varphi = -\frac{1}{\rho_0}\dot{p}_a \qquad \text{at } z = 0 \tag{3.2}$$

$$\partial_z\varphi = 0 \qquad \text{for } z \to -\infty \tag{3.3}$$

Here z is the vertical coordinate, ρ_0 the constant density, g the gravitational acceleration and p_a the atmospheric pressure. The surface boundary condition (3.2) is the linear version of the inviscid dynamic boundary condition which states that the pressure is continuous across the air-sea interface. The homogeneous equation has eigensolutions of the form

$$\varphi_s(\mathbf{x},t) = e^{kz}e^{i(\mathbf{k}\cdot\mathbf{x}-\omega_s(\mathbf{k})t)} \qquad s = +,- \tag{3.4}$$

where

$$\omega_s(\mathbf{k}) = s\sqrt{gk} \tag{3.5}$$

and $\mathbf{k} = (k_x, k_y)$ and $k = |\mathbf{k}|$. These eigensolutions represent surface gravity waves. Equation (3.5) is the dispersion relation for short gravity waves. Typical periods are of the order of seconds.

When a Fourier transformation is applied to the horizontal spatial dependence the dynamic boundary condition (3.2) reduces to

$$\ddot{\varphi}(\mathbf{k}, z = 0, t) + gk\varphi(\mathbf{k}, z = 0, t) = -\frac{1}{\rho_0}\dot{p}_a(\mathbf{k}, t) \tag{3.6}$$

and represents a forced oscillator equation of the type discussed in the previous section. If the atmospheric pressure is assumed to be a stationary random function the growth rate of the wavenumber spectrum is, according to (2.26), given by

$$\lim_{t\to\infty} \frac{1}{t} S_\varphi(\mathbf{k}, t) = 2\pi \sum_{s=+,-} \frac{S_{p_a}(\mathbf{k}, \omega = -\omega_s)}{4\rho_0^2} \tag{3.7}$$

Here $S_\varphi(\mathbf{k}, t)$ is the variance of the Fourier component $\varphi(\mathbf{k}, z = 0, t)$ and contains contributions from waves traveling both in positive and negative **k**-direction. The variance of waves traveling in positive **k**-direction grows according to

$$\lim_{t\to\infty} \frac{1}{t} S_\varphi^+(\mathbf{k}, t) = \frac{\pi}{2\rho_0^2} S_{p_a}(\mathbf{k}, \omega = -\omega_+) \tag{3.8}$$

This growth rate is proportional to the growth rate of the surface wave energy spectrum and was originally considered by Phillips (1957). He also assumed that spatial atmospheric pressure fluctuations are frozen into and carried by a mean wind **U**. The forcing spectrum then takes the form

$$S_{p_a}(\mathbf{k}, \omega) = S_{p_a}(\mathbf{k})\delta(\omega + \mathbf{k}\cdot\mathbf{U}) \tag{3.9}$$

and leads to

$$\lim_{t\to\infty} \frac{1}{t} S_\varphi^+(\mathbf{k}, t) = \frac{\pi}{2\rho_0^2} S_{p_a}(\mathbf{k})\delta(\omega_+ - \mathbf{k}\cdot\mathbf{U}) \tag{3.10}$$

Only waves whose phase speed in the direction of the wind equals the wind speed are excited or, equivalently, only waves in directions

$$\cos\varphi = \pm\frac{\sqrt{gk}}{kU} \tag{3.11}$$

are excited. This bimodel directional distribution is indeed observed at initial stages of wave growth. Other mechanisms become relevant at later stages.

3.2. Internal gravity waves. Consider an incompressible stratified Boussinesq fluid in a horizontally unbounded ocean of constant depth H rotating with a constant angular frequency $f_0/2$. In the hydrostatic or long wave approximation the linear, inviscid, non-diffusive equations of motion are given by

$$\partial_t u - f_0 v = -\frac{1}{\rho_0}\partial_x p + F_x \tag{3.12}$$

$$\partial_t v + f_0 u = -\frac{1}{\rho_0}\partial_y p + F_y \tag{3.13}$$

$$0 = -\partial_z p - \rho g \tag{3.14}$$

$$\partial_t \rho - \frac{\rho_0}{g} N^2(z) w = 0 \tag{3.15}$$

$$\partial_x u + \partial_y v + \partial_z w = 0 \tag{3.16}$$

with boundary conditions

$$w = 0 \qquad \text{at } z = 0, -H \tag{3.17}$$

Here u and v are the horizontal velocity components, w the vertical component, x and y the horizontal spatial coordinates, z the vertical coordinate, p the pressure, f_0 the Coriolis frequency and ρ_0 a constant reference pressure. The density ρ is the deviation from a background density gradient $\tilde{\rho}(z)$ which defines the Brunt-Väisälä frequency

$$N^2(z) = -\frac{g}{\rho_0}\frac{d}{dz}\tilde{\rho}(z) \tag{3.18}$$

The first two equations represent the horizontal momentum equation, the third equation the vertical momentum balance in the hydrostatic approximation. Equation (3.15) represents the density evolution and equation (3.16) is the incompressibility constraint. The boundary conditions assume the rigid lid approximation which eliminates the surface gravity waves considered in the previous section. The body forces F_x and F_y in the horizontal momentum equations represent the windstress distributed over a shallow surface layer. Often $F_x, F_y \sim \delta(z)$ is assumed.

Equations (3.12) to (3.17) describe two kinds of wave motions: gravity and Rossby waves. The Rossby waves carry the potential vorticity of the flow. Gravity waves do not carry any potential vorticity. Projection onto the gravity mode results in

$$\begin{aligned}(\partial_t\partial_t + f_0^2)\partial_z\partial_z w(\mathbf{x},t) + N^2(z)(\partial_x\partial_x + \partial_y\partial_y)w(\mathbf{x},t)&\\ = N^2(z)Q(\mathbf{x},t)&\end{aligned} \tag{3.19}$$

where

$$Q(\mathbf{x},t) = -\frac{1}{N^2(z)}\partial_z\left[\partial_t(\partial_x F_x + \partial_y F_y) + f_0(\partial_x F_y - \partial_y F_x)\right] \tag{3.20}$$

The homogeneous equation has separable solutions where the vertical structure is given by the eigenfunctions $\phi_n(z)$ $(n = 1, 2, \ldots)$ of the eigenvalue problem

$$\begin{aligned} \frac{f_0^2}{N^2(z)}\frac{d}{dz}\frac{d}{dz}\phi_n(z) + \frac{1}{R_n^2}\phi_n(z) &= 0 \\ \phi_n(z) &= 0 \qquad \text{at } z = 0, -H \end{aligned} \tag{3.21}$$

The eigenvalue R_n is called the Rossby radius of deformation. The eigenfunctions form a complete system. Projection of (3.19) onto these eigenfunctions results in

$$\begin{aligned} (\partial_t\partial_t + f_0^2)w_n(x,y,t) - f_0^2 R_n^2(\partial_x\partial_x + \partial_y\partial_y)w_n(x,y,t) \\ = -f_0^2 R_n^2 Q_n(x,y,t) \end{aligned} \tag{3.22}$$

where $w_n(x,y,t)$ and $Q_n(x,y,t)$ are the expansion coefficients. A final projection onto horizontal wavenumber components yields

$$\ddot{w}_n(\mathbf{k},t) + f_0^2(1 + k^2R_n^2)w_n(\mathbf{k},t) = -f_0^2 R_n^2 Q_n(\mathbf{k},t) \tag{3.23}$$

which is again of the form considered in section 2. The solution of the homogeneous equation are long internal gravity waves with dispersion relation

$$w_n^s(\mathbf{k}) = s f_0\sqrt{1 + k^2R_n^2} \qquad s = +,- \tag{3.24}$$

Typical periods are of the order of hours. Rubenstein (1994) has recently used equation (3.23) to calculate response spectra. He assumes that the spatial mesoscale windstress spectrum suggested by Overland and Wilson (1984) is frozen into and advected by a mean wind. He also assumes a special damping mechanism such that the solution remains separable.

3.3. Rossby waves. Projection of equations (3.12) to (3.17) onto the Rossby wave mode yields the quasi-geostrophic potential vorticity equation

$$\partial_t(\partial_x\partial_x + \partial_y\partial_y + \partial_z \frac{f_0^2}{N^2(z)}\partial_z)\psi(\mathbf{x},t) + \beta_0\partial_x\psi(\mathbf{x},t) = \hat{Q}(\mathbf{x},t) \tag{3.25}$$

where

$$\hat{Q}(\mathbf{x},t) = \partial_x F_y - \partial_y F_x \tag{3.26}$$

Here $\psi(\mathbf{x},t)$ is the quasi-geostrophic streamfunction which satisfies the boundary condition

$$\partial_t\partial_z\psi(\mathbf{x},t) = 0 \qquad \text{at } z = 0, -H \tag{3.27}$$

The potential vorticity equation includes effects of meridional changes of the Coriolis parameter. These are characterized by the beta parameter β_0 and provide the restoring forces for planetary Rossby waves. This beta effect is absent in the f-plane approximation that was used for internal gravity waves.

The vertical eigenmodes are now defined by the Sturm-Liouville problem

$$\begin{aligned} \frac{d}{dz}\frac{f_0^2}{N^2(z)}\frac{d}{dz}A_n(z) + \frac{1}{R_N^2}A_n(z) &= 0 \qquad n = 0,1,2,\ldots \\ \frac{d}{dz}A_n(z) &= 0 \qquad \text{at } z = 0, -H \end{aligned} \tag{3.28}$$

and projection onto vertical normal modes and horizontal wavenumber components yields

$$\dot{\psi}_n(\mathbf{k},t) + i\omega_n(\mathbf{k})\psi_n(\mathbf{k},t) = -\frac{\hat{Q}_n(\mathbf{k},t)}{k^2 + R_n^{-2}} \tag{3.29}$$

where

$$\omega_n(\mathbf{k}) = -\frac{\beta_0 k_x}{k^2 + R_n^{-2}} \tag{3.30}$$

is the dispersion relation of planetary Rossby waves. Typical frequencies are weeks for the barotropic ($n = 0$) mode and months and years for the baroclinic ($n = 1, 2, 3, \ldots$) modes.

Equation (3.29) was used by Frankignoul and Müller (1979) to calculate growth rates for a specific model spectrum of the atmospheric windstress and by Müller and Frankignoul (1981) to calculate response spectra using the same forcing spectrum and various damping mechanisms. The forcing of barotropic Rossby waves will be considered in more detail in section 4.

3.4. Mid-latitude sea-surface temperature anomalies. A stochastic forcing model for climate variability has been suggested by Hasselmann (1976). Climate variability has often been ascribed to changes in the external parameters of the climate system (such as the orbital parameters of the earth) or to internal instability mechanisms (such as baroclinic instabilities). The stochastic forcing model ascribes changes in the climate system to continual random forcing by short time scale weather fluctuations. This random forcing causes an ever increasing variability of the climate system and the task becomes to find negative feedback mechanisms that limit the growth of climate variations rather than positive instability mechanisms that cause climate variations.

Frankignoul and Hasselmann (1977) considered mid-latitude sea-surface temperature (SST) anomalies as a particular example. They assumed SST anomalies ΔT to be governed by

$$\dot{\Delta T} = \frac{Q}{\rho_0 c_p h} - \lambda \Delta T \tag{3.31}$$

where c_p is the specific heat. This equation is a linearized version of the heat equation in an isothermal surface layer of depth h. Q is the heat flux through the surface. All advection and all dissipation processes are parametrized by a single positive relaxation rate λ.

Equation (3.31) represents a single stable oscillator. The spectrum of the SST anomalies is therefore

$$S_{\Delta T}(\omega) \sim \frac{S_Q(\omega)}{\omega^2 + \lambda^2} \tag{3.32}$$

The heat flux variations are dominated by weather fluctuations with a correlation scale τ of a few days. The spectrum of the heat flux is therefore white at periods longer than τ. The predicted SST anomaly spectrum (3.32) is therefore flat at frequencies $\omega \ll \lambda$ and decays as ω^{-2} for frequencies $\omega \gg \lambda$ but still smaller than τ^{-1}. This behaviour is indeed observed for mid-latitude SST anomalies. Fits to observed spectra give $\lambda = O((6 \text{ months})^{-1})$. The simple stochastic forcing (3.32) also predicts correctly the correlation between SST and atmospheric variables. Modifications and generalizations of the above simple model are reviewed by Frankignoul (1985).

4. Forcing of barotropic Rossby waves. Here we look at the stochastic forcing of barotropic Rossby waves in more detail. These Rossby waves have horizontal length scales of the order of $500km$ and time scales of the order of weeks.

4.1. Barotropic potential vorticity equation. For the barotropic ($n = 0$) mode the potential vorticity equation (3.29) reduces to

$$\dot{\psi}_0(\mathbf{k}, t) + i(\omega_0(\mathbf{k}) - id(\mathbf{k}))\psi_0(\mathbf{k}, t) = -\frac{\hat{Q}_0(\mathbf{k}, t)}{k^2} \tag{4.1}$$

where

$$\omega_0(\mathbf{k}) = -\frac{\beta k_x}{k^2} \tag{4.2}$$

is the dispersion relation and $d(\mathbf{k})$ a parametrization of the unknown dissipation processes. Typical examples include Rayleigh friction where $d(\mathbf{k}) = r$ and lateral friction where $d(\mathbf{k}) = Ak^2$. Here r is a characteristic damping rate and A the lateral viscosity coefficient. Note that $R_0^{-1} = 0$ in the rigid approximation. The forcing is explicitly given by

$$\hat{Q}_0 = \frac{1}{\rho_0 H}\mathbf{z} \cdot (\boldsymbol{\nabla} \times \boldsymbol{\tau}) \tag{4.3}$$

where $\boldsymbol{\tau}$ is the atmospheric windstress, H the depth of the ocean and $\mathbf{z}$ the vertical unit vector.

If the forcing is statistically homogeneous and stationary the asymptotic response has Fourier coefficients

$$\psi_0(\mathbf{k}, \omega) = H_0(\mathbf{k}, \omega)Q_0(\mathbf{k}, \omega) \tag{4.4}$$

where

$$H_0(\mathbf{k}, \omega) = -\frac{1}{i(\omega + \tilde{\omega}_0)k^2} \tag{4.5}$$

is the response function and $\tilde{\omega}_0 = \omega_0 - id$. Response spectra are then given by

$$S_{\psi_0}(\mathbf{k}, \omega) = H^*(\mathbf{k}, \omega)H(\mathbf{k}, \omega)S_{Q_0}(\mathbf{k}, \omega) \tag{4.6}$$

where $S_{Q_0}(\mathbf{k}, \omega)$ is proportional to the wavenumber-frequency spectrum of the windstress curl.

4.2. The windstress curl spectrum. On the scales of barotropic Rossby waves the windstress curl can be calculated from FNOC (Fleet Numerical Oceanographic Center) winds. Spectra of the windstress curl have been calculated by Chave et al. (1991) and are fairly well represented by the model spectrum of Frankignoul and Müller (1979)

$$S_{\mathrm{curl}\tau}(\mathbf{k},\omega) = \frac{1}{2} S_{\mathrm{curl}\tau}(0) \frac{S_{\mathrm{curl}\tau}(k)}{2\pi k} \tag{4.7}$$

where

$$S_{\mathrm{curl}\tau}(k) = \frac{1}{k_c - k_b} \begin{cases} 1 & \text{for } k_b < k < k_c \\ 0 & \text{otherwise} \end{cases} \tag{4.8}$$

Here k_b and k_c are low and high cut-off wavenumbers. The model spectrum is white in frequency space and white and isotropic in wavenumber space. $S_{\mathrm{curl}\tau}(0)$ is the white noise level. These features are also basic to the observed spectra of Chave et al. (1991).

The model windstress spectrum has been used to calculate response spectra (e.g. Müller and Frankignoul, 1981). Results are equivocal. They depend to a certain extent on parameters such as $d(k)$, k_b and k_c which are not well established.

4.3. Coherence Maps. Correlations or coherence between atmospheric and oceanic variables establish a more direct relation between forcing and response. In wavenumber-frequency space the coherence squared between two variables x and y is defined by

$$C_{xy}^2(\mathbf{k},\omega) = \frac{|S_{xy}(\mathbf{k},\omega)|^2}{S_{yy}(\mathbf{k},\omega) S_{xx}(\mathbf{k},\omega)} \tag{4.9}$$

where the (cross) spectra are given by

$$S_{xy}(\mathbf{k},\omega)\delta(\mathbf{k}-\mathbf{k}')\delta(\omega-\omega') = \langle x^*(\mathbf{k},\omega) y(\mathbf{k}',\omega')\rangle \tag{4.10}$$

For the stochastic forcing model (4.4) the coherence $C_{Q_0\psi_0}(\mathbf{k},\omega)$ is identical to one. In separation-frequency space the coherence is defined by

$$C_{xy}^2(\mathbf{r},\omega) = \frac{|S_{xy}(\mathbf{r},\omega)|^2}{S_{yy}(\mathbf{r}=0,\omega) S_{xx}(\mathbf{r}=0,\omega)} \tag{4.11}$$

where the (cross) spectra are given by

$$S_{xy}(\mathbf{r},\omega)\delta(\omega-\omega') = \langle x^*(\mathbf{x},\omega) y(\mathbf{x}+\mathbf{r},\omega')\rangle \tag{4.12}$$

The cross spectra in wavenumber-frequency and separation-frequency space are related by the Fourier transformation

$$(4.13) \qquad S_{xy}(\mathbf{r},\omega) = \int_{-\infty}^{+\infty} d\mathbf{k} e^{i\mathbf{k}\cdot\mathbf{r}} S_{xy}(\mathbf{k},\omega)$$

and hence contain the same information. However, the definition (4.11) must be employed in experimental settings where either the forcing or the response is not given at all points in space and the spatial Fourier components necessary in the definition (4.9) cannot be determined. A plot of $C_{xy}(\mathbf{r},\omega)$ as a function of $\mathbf{r}$ for given frequency ω is called a coherence map. For the stochastic forcing model (4.4) the coherence squared in seperation-frequency space is given by

$$(4.14) \qquad C_{Q_0\psi_0}(\mathbf{r},\omega) = \frac{\left| \int d\mathbf{k} e^{i\mathbf{k}\cdot\mathbf{r}} H(\mathbf{k},\omega) S_{Q_0}(\mathbf{k},\omega) \right|^2}{\int d\mathbf{k} |H(\mathbf{k},\omega)|^2 S_{Q_0}(\mathbf{k},\omega) \int d\mathbf{k} S_{Q_0}(\mathbf{k},\omega)}$$

and has been calculated by Lippert and Müller (1995) using the windstress model spectrum (4.7). The coherence maps between the oceanic pressure p, the zonal velocity component u and the meridional velocity component v and the atmospheric windstress curl are reproduced in Figure 1 for two different frequencies. The oceanic variables p, u and v are related to the stream function ψ_0 by

$$(4.15) \qquad p = \rho_0 f_0 \psi_0 \ , \quad u = -\partial_y \psi_0 \ , \quad v = \partial_x \psi_0$$

The calculated coherence maps show decaying periodic structures with local and non-local primary maxima. The patterns change with frequency and depend on the variables for which the coherence is calculated. The periodicity and decay scales are proportional to the distance at which the autocorrelation function of the windstress curl first crosses zero. The location of the maxima depends on the symmetry of the response function in wavenumber space. This symmetry depends on the oceanic variable and changes with frequency.

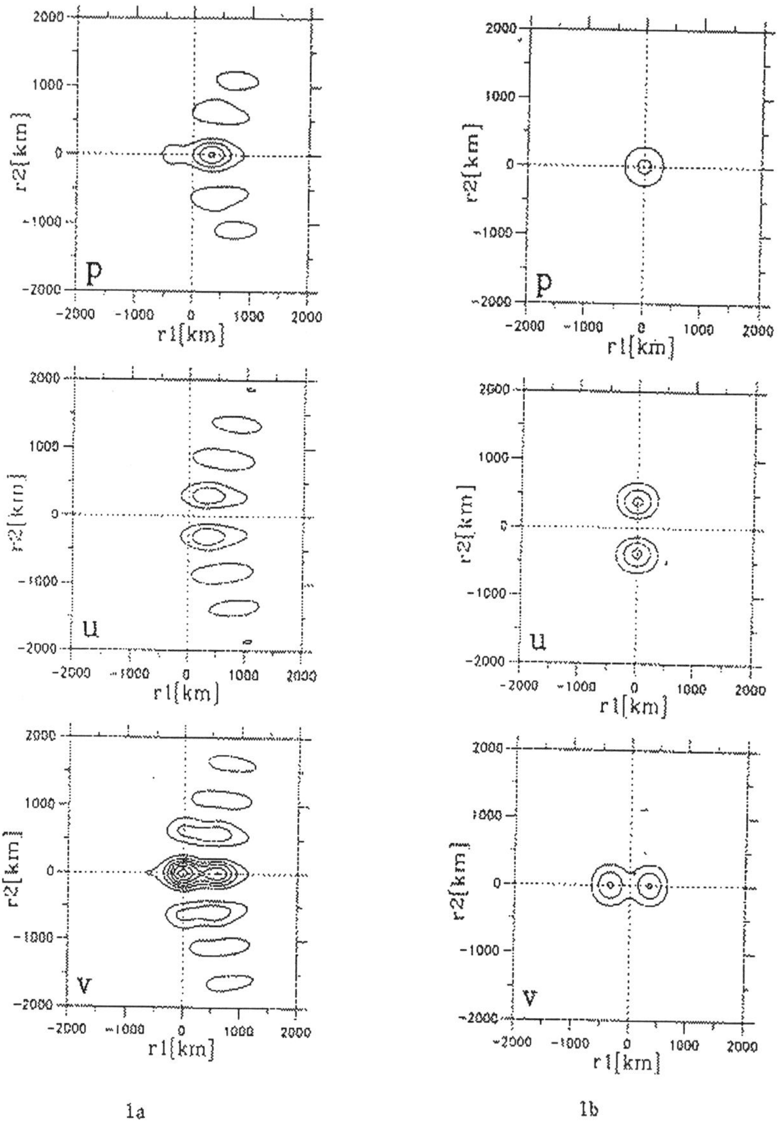

Figure 1. Squared coherence maps between the windstress curl and p (top), u (middle) and v (bottom). (a) is for $\omega = 1.4 \cdot 10^{-6} s^{-1} = 2\pi/50d$. The contour interval is 0.02. (b) is for $\omega = 10^{-5} s^{-1} = 2\pi/7d$. The contour interval is 0.1. (From Lippert and Müller (1995)).

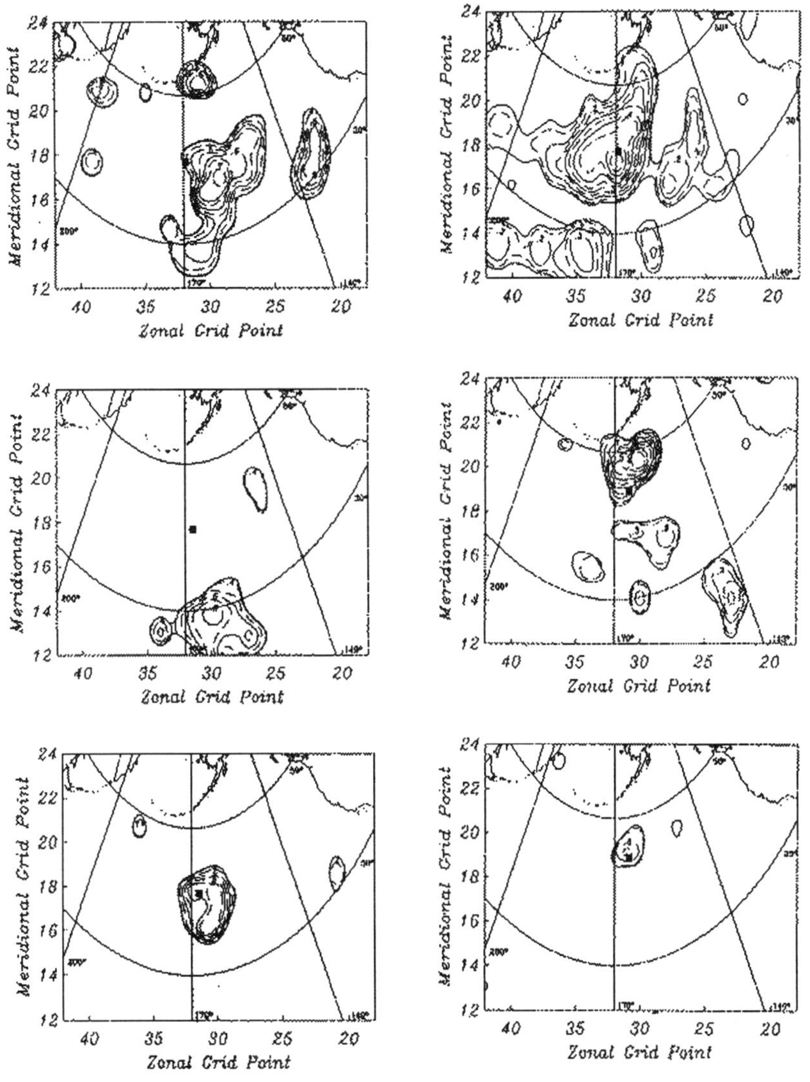

2a

2b

Figure 2. Squared coherence maps between the bottom pressure (top), the zonal barotopic velocity (middle) and the meridional barotopic velocity (bottom) at the BEMPEX site (marked by the solid sqare) and FNOC windstress curl over the North Pacific. (a) is for a period of 28 days (pressure) and 38 days (velocities). (b) is for a period of 2 days (pressure) and 6 days (velocities). The contour interval is 0.1 in all cases. Only values exceeding 0.15 are contoured. These values are statistically greater than zero at the 95% confidence level. (From Luther et al. (1990) and Chave et al. (1992)).

4.4. Model-data comparison. Figure 2 shows analogous maps that have been calculated by Chave et al. (1992) from BEMPEX (Barotropic Electromagnetic and Pressure Experiment) data and FNOC winds. The calculated maps reproduce some of the basic features seen in the observed coherence maps. First of all, the model predicts that the spatial scales seen in observed coherence maps are comparable to the scales of the windstress curl autocorrelation function. This in indeed the case. The observed coherence maps show spatial scales similar to those of the FNOC windstress curl autocorrelation function as e.g. given in Chave at al. (1991). The model also reproduces more specific features. Compare the maps in Figure 2a which are taken at a period of 28 and 38 days with the map in Figure 1a which is calculated for a period of 50 days. Both sets of maps show a coherence maximum to the east in the p-curlτ map, two maxima, one to the north-east and one to the south-east, in the u-curlτ map, and a local maximum in the v-curlτ map. The geographical directions are here interpreted with respect to the geostrophic contours which slope east-southeast in the vicinity of the BEMPEX array. Figure 2b which is taken at a period of 2 and 6 days shows that the maximum in the p-curlτ coherence becomes local as does the coherence in Figure 1b calculated for a period of 7 days.

Of course, the calculated maps cannot be expected to reproduce the observed coherence maps in any detail. The calculated maps represent the response of a simple base state ocean to a highly idealized wind field. Too many relevant features of the real ocean and the real wind field have been disregarded in the idealization process leading to the model.

5. Summary and conclusions. Oceanic motions are forced by stresses and heat and fresh water fluxes at the air sea interface. When the forcing is assumed to be prescribed and stochastic and when the oceanic response is assumed to be governed by linear dynamics then the forcing problem reduces to a set of stochastically forced linear oscillators. Additional simplifying assumptions about the geometry are usually employed to make the problem algebraically tractable and to allow an explicit decoupling of the oscillators.

The stochastically forced linear oscillator is mathematically well understood. The asymptotic response to statistically stationary forcing depends on the stability of the oscillator. Forcing problems usually consider stable or neutral systems. In the stable case response spectra are calculated, in the neutral case growth rates. These calculations require the spectrum of the forcing field and the response function which depends on the parametrization of advection and dissipation effects in the stable case. Often the advection/dissipation parameters are obtained by fitting the stochastic forcing model to observed data.

The stochastic forcing models are extremely simple. Their main purpose is to elucidate basic mechanisms rather than to model realistic flow fields. In climate research the stochastic forcing models provided an alternative perspective to the traditional models that explained climate variability by changes in external parameters or by instability processes. Though the stochastic forcing models are extremely simple they nevertheless capture basic features of observations. This was demonstrated for coherence maps between oceanic and atmospheric variables in the case of barotropic Rossby waves. Though the model cannot reproduce the rich and complicated structure seen in observed maps it can account for and rationalize some basic trends that occur when different variables and frequencies are considered.

When stochastic forcing is applied to oceanic general circulation models the response is in general complex but certain aspects can again be "explained" by a simple linear stochastic forcing model. Overall, the stochastically forced linear oscillator has proven to be a powerful conceptual model to understand aspects of motions in real and simulated oceans.

Acknowledgements. This work was supported by the Office of Naval Research.

REFERENCES

[1] CHAVE, A. D., D. S. LUTHER AND J. H. FILLOUX, *Variability of the windstress curl over the North Pacific: Implications for the oceanic response*, J. Geophys. Res., **96**, 18361–18379, 1991.

[2] CHAVE, A. D., D. S. LUTHER AND J. H. FILLOUX, *The barotropic electromagnetic and pressure experiment. 1: Barotropic current response to atmospheric forcing*, J. Geophys. Res., **97**, 9565–9593, 1992.

[3] FRANKIGNOUL, C., *Sea surface temperature anomalies, planetary waves, and air-sea feedback in the middle latitudes*, Rev. Geophys., **23**, 357–390, 1985.

[4] FRANKIGNOUL, C. AND K. HASSELMANN, *Stochastic climate models, part 2. Applicaton to sea-surface temperature anomalies and climate variability*, Tellus, **29**, 289–305, 1977.

[5] FRANKIGNOUL, C. AND P. MÜLLER, *Quasi-geostrophic response of an infinite β-plane ocean to stochastic forcing by the atmosphere*, J. Phys. Oceanogr., **9**, 104–127, 1979.

[6] HASSELMANN, K., *Über zufallserregte Schwingungssysteme*, ZAMM, **42**, 465–476, 1962.

[7] HASSELMANN, K., *Stochastic climate models, part 1. Theory*, Tellus, **28**, 473–485, 1976.

[8] LIPPERT, A. AND P. MÜLLER, *Direct atmospheric forcing of geostrophic eddies, part II: Coherence maps*, J. Phys. Oceanogr., 25, 106–121, 1995.

[9] LUTHER, D. S., A. D. CHAVE, J. H. FILLOUX AND P. F. SPAIN, *Evidence for local and nonlocal barotropic responses to atmospheric forcing during BEMPEX*, Geophys. Res. Lett., **17**, 949–952, 1990.

[10] MÜLLER, P. AND C. FRANKIGNOUL, *Direct atmospheric forcing of geostrophic eddies*, J. Phys. Oceanogr., **11**, 287–308, 1981.

[11] OVERLAND, J. E. AND J. G. WILSON, *Mesoscale varability in marine winds at mid-latitude*, J. Geophys. Res., **89**, 10599–10614, 1984.

[12] PHILLIPS, O. M., *On the generation of waves by turbulent winds*, J. Fluid Mech., **2**, 417–445, 1957.

[13] RUBENSTEIN, D., *A spectral model of wind-forced internal waves*, J. Phys. Oceanogr., **24**, 819–831, 1994.

RADIATIVE TRANSFER IN MULTIFRACTAL ATMOSPHERES: FRACTIONAL INTEGRATION, MULTIFRACTAL PHASE TRANSITIONS AND INVERSION PROBLEMS

CATHERINE NAUD*, DANIEL SCHERTZER* , AND SHAUN LOVEJOY†

Abstract. This paper is devoted to studying the inhomogeneity of the radiation field resulting from propagation through a multifractal cloud field by relating the orders of singularities and codimensions of both fields. This direct relationship is of fundamental importance for climate studies, whereas the inverse problem is fundamental for remote sensing. We point out similarities between smoothing by scattering and fractional integration, showing they are exactly analogous for certain cases: 1-D medium and plane parallel atmospheres (with a few extra hypotheses). We therefore deduce that there is a limited range of singularities susceptible to exhibiting identical multifractal characteristics before any inversion. The lower bound ($\gamma_{D'}$) is defined by a first order multifractal phase transition which occurs when the dimension D′ of the fractional integration is insufficient to smooth out the singularities of the cloud field, whereas the upper bound (γ_s) is defined by a second-order phase transition and corresponds to the limitations induced by the finite size of the samples. These two critical singularities drastically reduce the range of relevant singularities and justify some essentially ad hoc procedures used in multifractal estimation.

1. Introduction. In meteorology, climatology and remote sensing, fundamental uncertainties are related to cloud modelling and analysis. It is especially difficult to describe the complex interactions between radiative transfer, cloud microphysics and cloud dynamics: none of the three major aspects of clouds—their optical properties, their spatial and temporal variability—nor the associated precipitation are well understood. A particular aspect of interest here is that estimates of the total albedo depends largely on the variation of cloudiness in atmosphere. One of these problems, "the cloud absorptivity paradox" is still not resolved. The basic problem can be stated as follows: comparison of radiation budget estimates at the top and bottom of the atmosphere leads to the indirect inference that absorption considerably exceeds the largest values obtainable from theory, when the latter assumes homogeneity and no absorption. It is obvious that remote sensing needs a model of atmospheric variability to handle the observations, in order to deal with the inverse problem : how to deduce liquid water content variability from radiances variability.

The same holds for climate studies (see for review Somerville and Gauthier, 1994) and indeed recently, Cess et al. and Ramanathan et al. (1995) cite observations of an "anomalous absorption" of radiation in cloudy skies in comparison with the values predicted by usual models (homogeneous

* Universite Pierre et Marie Curie, Boite 99, 4 place Jussieu, 75252 Paris Cedex 05, France.

† Physics Department, McGill University, 3600 University St, Montreal, P.Q. H3A 2T8, Canada.

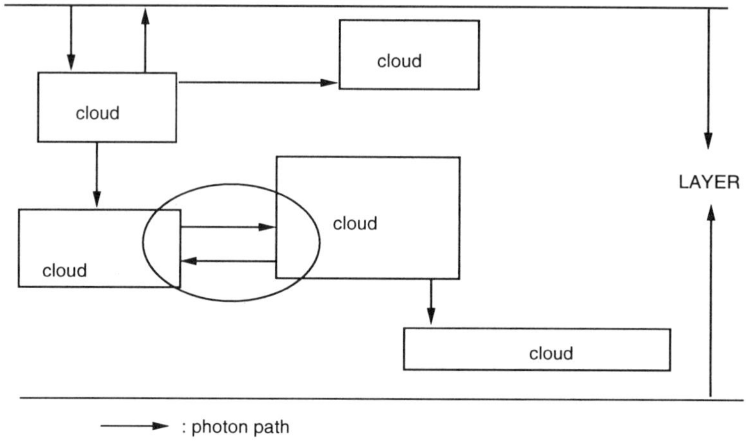

FIG. 1. *Illustration of photon trapping between high inhomogeneities of the cloud field : optical path as well as effective absorption increase considerably.*

atmosphere). If true, this would lead to large uncertainties in assessing climate change. Byrne et al. (1995), in order to explain this anomalous absorption, propose a simple model of Broken Clouds and measure an increase in the photon mean free path in comparison with the value calculated for a homogeneous atmosphere. The modelled media is a layer filled with a mixture of clouds and portions of clear sky. They argue that due to photons travelling horizontally (see Figure 1), columns of clear sky absorb more light than in the corresponding completely clear sky. These "holes" significantly increase the mean free path of photons and thus increase the absorption of the layer, i.e. photons are trapped due to cloud inhomogeneities. They show that there should be a new description of clouds and suggest the multifractal model. Our purpose is not to explicitly solve this problem of Anomalous Absorption but instead to study the effect of liquid water concentration variability on the light scattering, in the ideal case of perfect scattering. The object of this paper is to argue that homogeneous models are simply not relevant in relating the highly variable properties of clouds and radiation fields: however smoothed, the intensity of clouds' multiply scattered radiation fields reflects this extreme variability. Unfortunately, classical methods—of both radiative transfer and dynamical modelling—are limited to studying such relationships on an arbitrary scale (often considered as "characteristic"), although the interactions between clouds and radiation occur on a wide range of scale. We have argued for many years that the extreme variability of the radiation fields can be best understood in a multifractal framework (Gabriel et al. 1988, Lovejoy and Schertzer 1990, 1991, Schertzer and Lovejoy 1988, 1991). Indeed, the (scalar) multifractal model of cloud fields, as discussed below, is capable of respecting the clouds' texture, clustering, bands and intermittency, and the non-linear nature of the true dynamical processes at all scales. The simplest relevant dynamical model corresponds to a stochastic model of passive clouds, passively advected by a turbulent velocity field, using coupled cascade processes, non-linearly conserving the fluxes of energy and concentration variance (Schertzer, Lovejoy 1987, Wilson et al. 1991, Pecknold et al. 1993, 1995).

In order to propagate light in such an atmosphere, different analytical and numerical methods have been used. There have been some attempts for (mostly) 2D- cloud fields to capture some of the radiative effects of clouds (Gabriel et al. 1986 (for 3D), Lovejoy et al. 1989, 1990, Gabriel et al. 1990, Davis et al. 1990, 1991, 1992) : by evaluating the global radiative responses such as transmittance (T) or reflectance (R) for conservative scattering, using discrete angle phase functions in the (continuous angle) Radiative Transfer Equation (Chandrasekhar 1950), i.e. D.A. (Discrete Angle) Radiative Transfer, on respectively homogeneous, monofractal and some multifractal fields. More recently, for 1-D multifractal clouds, transport has been described by the diffusion equation (Silas et al. 1993, Lovejoy et al. 1995) and by radiative transfer studying the scattering statistics of a photon, dependent on the heterogeneous optical depth of the medium (Lovejoy et al. 1995). Work on related models includes Evans 1993, Barker 1992 and Cahalan 1989.

Our goal is to ascertain the consequences of the radiative transfer equation on multifractal statistics in a simplified 2D multifractal field, first by simulation and second by analytical calculation. We give some general considerations on multifractal fields and simulations in section 2, and on the radiative transfer equation in section 3. In section 4, we present the main argument that relates the radiative transfer equation to a fractional integration and then test it on a plane parallel multifractal atmosphere in section 5. Section 6 is mainly concerned with 2D-cuts of a multifractal atmosphere (although most of the results also apply for 3D) and in order to generalize the relation between the radiative transfer equation and fractional integration, we present the consequences of a D′-integration on a D-cut of a multifractal field.

2. General description of the multifractal model.

2.1. Physical basis. First we present the physical basis of the multifractal model. As advocated by Schertzer and Lovejoy (1987), it is important to first consider the fundamental and rather well defined case of passive scalar clouds. Although real clouds are not truly passive, their statistics may in fact be quite similar (Bromsalen 1994), and in any case, it is the simplest relevant nonlinear model. Such clouds result from the passive scalar advection of water concentration (ρ) by a velocity field ($\underline{v}$) in the limit of vanishing viscosity and diffusivity. The dynamical equations are the incompressible Navier Stokes equations and the equation of passive advection. They both conserve the fluxes of energy (density ε) and scalar variance (density χ) while effecting a transfer to smaller scales by a cascade process (introduced by Richardson 1922) down to the inner viscous scale:

$$\varepsilon = -\frac{\partial \underline{v}^2}{\partial t} = const \tag{2.1}$$

$$\chi = -\frac{\partial \rho^2}{\partial t} = const \tag{2.2}$$

The energy flux is mainly transferred from one scale to a neighboring scale. Considering the real space fluctuations (increments) at scale ℓ in the inertial range (viscosity scale $= \eta < \ell < L =$ outer scale) of the fields $\underline{v}$ and ρ, and considering the fluxes as rather homogeneous, we have scaling laws according to on the one hand *Kolmogorov* 1941, and on the other *Obukhov* 1949 and *Corrsin* 1951:

$$\Delta v(\ell) = \varepsilon^{1/3}\ell^{1/3} \tag{2.3}$$

$$\Delta \rho(\ell) = \varphi^{1/3}\ell^{1/3} \tag{2.4}$$

where

$$\varphi = \chi^{3/2}\varepsilon^{-1/2} \tag{2.5}$$

is the flux resulting from the non linear interaction between the velocity and water concentration.

A crucial point of criticism concerning this first approach to turbulence has been that it was assumed that the energy transfer itself is not a fluctuating quantity. In a more refined scaling theory, Kolmogorov (1962) and Obukhov (1962) also considered high inhomogeneity in the energy transfer rate ε. In order to study this question of inhomogeneity of ε and χ, we will use cascade processes which, by iterating a scale invariant step, systematically reduce the scale of homogeneity to zero. For convenience we introduce a new variable, the scale ratio $\lambda = L/\ell$ $(1 < \lambda < \Lambda = L/\eta)$. At a given resolution λ, the corresponding intermediate quantities χ_λ and ε_λ are highly variable (intermittent) but scale invariant. The flux can be rewritten:

$$\varphi_\lambda \approx \chi_\lambda^{3/2}\varepsilon_\lambda^{-1/2} \tag{2.6}$$

and equation (2.4):

$$\Delta \rho_\lambda \approx \varphi_\lambda^{1/3}\lambda^{-1/3} \tag{2.7}$$

Statistical moments of χ_λ and ε_λ exhibit multiple scaling:

$$\begin{aligned} \langle \varepsilon_\lambda^q \rangle &\approx \lambda^{K(q)} \\ \langle \chi_\lambda^q \rangle &\approx \lambda^{K(q)} \end{aligned} \tag{2.8}$$

$K(q)$ is a convex function and these relations do not affect the validity of (eq. 2.3) and (eq. 2.4), i.e. intermittency does not change the relationships between $\underline{v}$ and ε, and ρ and χ.

The nonlinear dependence on q through $K(q)$ corresponds to the multiple scaling and expresses the fact that generally, as discussed below, the most intense and weakest regions will scale differently.

2.2. General considerations on multifractal fields. A multifractal field is associated with an infinite hierarchy of singularities γ ($\varepsilon_\lambda \approx \lambda^\gamma$) (Schertzer and Lovejoy, 1987). When it is stochastic, their frequency of occurrence can be described by codimensions $c(\gamma)$, i.e. at a scale ratio λ, the probability (Pr) of the fluctuations of the field diverging faster than λ^γ scales as $\lambda^{-c(\gamma)}$ (Schertzer and Lovejoy, 1987):

$$Pr(\varepsilon_\lambda \geq \lambda^\gamma) \approx \lambda^{-c(\gamma)} \tag{2.9}$$

In order to model a multiple scaling (multifractal) field, we seek a field ε_λ with resolution scale λ, satisfying equation (2.8), i.e

$$\langle \varepsilon_\lambda^q \rangle = \lambda^{K(q)} \tag{2.10}$$

Therefore $K(q) \ln \lambda$ is the second Laplacian characteristic function (or cumulant generating function) of $\Gamma_\lambda = \ln \varepsilon_\lambda$, which is the generator of the process.

The functions $c(\gamma)$ and $K(q)$ are related to each other, using a Legendre transform (Parisi and Frish 1985):

$$K(q) = \max_\gamma (q\gamma - c(\gamma)) \tag{2.11}$$

$$c(\gamma) = \max_q (\gamma q - K(q)) \tag{2.12}$$

By mixing different processes of the same type, if we seek the limit when $\lambda \to \infty$, we converge by iterations to a universality class. In order to obtain universality we require generators that are both stable and attractive under addition. Extremal and stable Levy variables (Levy 1925), characterized by the Lévy index α ($0 \leq \alpha \leq 2$), respect these properties. Then $K(q)$ and $c(\gamma)$ have the following formulae (Schertzer, Lovejoy 1987):

$$K(q) + Hq = \frac{C_1}{\alpha - 1}(q^\alpha - q) \tag{2.13}$$

$$c(\gamma - H) = C_1\left(\frac{\gamma}{C_1 \alpha'} + \frac{1}{\alpha}\right)^{\alpha'} \tag{2.14}$$

$$\frac{1}{\alpha} + \frac{1}{\alpha'} = 1 \text{ for } \alpha \neq 1 \tag{2.15}$$

In the particular case where $\alpha = 1$ we have:

$$K(q) + Hq = C_1 q \log(q) \tag{2.16}$$

$$c(\gamma - H) = C_1 \exp(\frac{\gamma}{C_1} - 1) \tag{2.17}$$

Parameters designated H, C_1 and α are of fundamental significance (Schertzer and Lovejoy 1987) and define the local multifractal hierarchy around the mean field ($q = 1$). They have the following signification:

- H describes the *deviation from conservation* of the flux:

$$\langle \Delta \rho_\lambda \rangle \approx \lambda^{-H} \tag{2.18}$$

  where $H = 0$ for conservative fields.
- C_1 describes the *mean inhomogeneity* as it is the codimension of the mean singularity: $C_1 = c(C_1 - H)$. In the case of conservative fluxes it is also the order of the mean singularity (and simultaneously the fixed point of $c(\gamma)$).
- α represents the degree of multifractality, given by the convexity of $c(C_1)$ around the mean singularity $(C_1 - H)$, measured by the radius of curvature: $R_c(\gamma = C_1 - H) = 2^{3/2}\alpha C_1$ which increases with the range of singularities (starting from zero with the monofractal β-model).

2.3. Simulation of multifractal clouds. In order to create a multifractal cloud respecting the symmetries of passive scalar advection, both discrete or continuous cascades can be used. The more realistic continuous cascade processes will be used here.

The quantity of interest here is the passive scalar $\Delta\rho_\lambda$ which is related directly to φ_λ through equation (2.7). The field that we produce is the flux φ_λ, and then $\Delta\rho_\lambda$ can be simulated by introducing the extra scaling $\lambda^{1/3}$ to the field $\varphi_\lambda^{1/3}$ by fractional integration (power law filtering).

In order to obtain multiscaling, we require the generator to be a noise with a possible weighting function, having the following properties (Schertzer and Lovejoy, 1987):

1) The spectrum of the field must scale as k^{-1}, in order to obtain the scaling behavior: this is a $\log\lambda$ divergence of $K(q)$.
2) The generator must be band-limited to wave-number within $[1, \lambda]$; this is to ensure that for scales smaller than λ^{-1}, the field will be smooth; λ^{-1} will therefore be the resolution of the field.
3) The probability distribution of the generator must fall off faster than exponentially for positive fluctuations. This is to ensure convergence of $K(q)$ for $q > 0$.
4) It must be normalized so that $K(1) = 0$. This is the condition of conservation of the mean of the field at varying scales; $< \varepsilon_\lambda >= 1$.

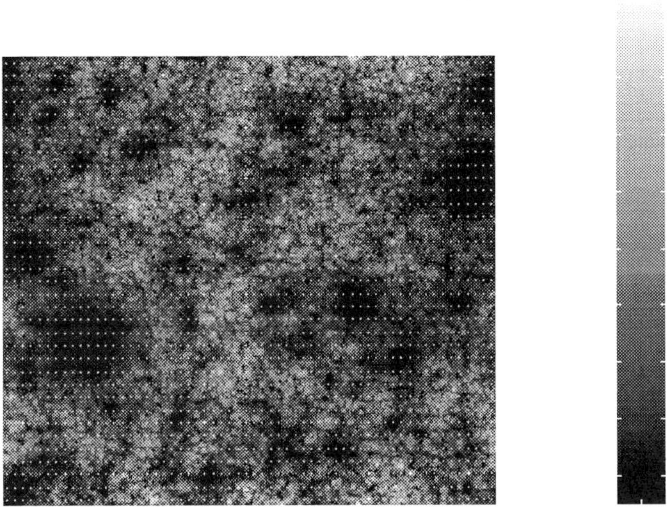

FIG. 2. *Simulation of a 2D multifractal cloud density field with* $\alpha = 1.35, C_1 = 0.75$ *and* $H = 1/3$ *(*C_1 *is chosen arbitrarily but* $\alpha = 1.35$ *and* $H = 1/3$ *have been estimated by Tessier et al. 1993). To increase the contrast we show the logarithm of the original field.*

In order to perform simulations, we start with a gaussian white noise or an extremal asymmetric Lévy distribution with large negative fluctuations in a finite bandwidth $[1, \lambda]$: the subgenerator $\gamma_\lambda(x)$. In Fourier space, we filter it by $\hat{f}(\underline{k})$ $(= |\underline{k}|^{-d/\alpha})$ in order to make a "1/f noise" and to obtain the generator $\hat{\Gamma}_\lambda(\underline{k}) = \hat{f}(\underline{k})\hat{\gamma}_\lambda(\underline{k})$ (by convention, the Fourier transform of a given quantity h is denoted $\hat{h}$). The conserved quantity $\varphi_\lambda(\underline{x})$ is then the exponentiation of $\Gamma_\lambda(\underline{x})$ and the multifractal density $\Delta\rho_\lambda$ is then obtained after having taken the 1/3 power of φ_λ and having filtered it (in Fourier space) by $|\underline{k}|^{-1/3}$.

Wilson et al. (1991) and Pecknold et al. (1993) developed efficient algorithms yielding such multifractal clouds. We used them to simulate a 2D-cut of a multifractal cloud density field, and it is represented on Figure 2 with $\alpha = 1.35, C_1 = 0.75$ and $H = 1/3$ with respect to turbulence. It simulates a cloud with axes along vertical and horizontal.

3. General considerations on radiative transfer.

3.1. Theory. The radiative field is characterized by a monochromatic intensity (radiance) $I(\underline{x}, \underline{s})$ at a point $\underline{x}$ and in a direction $\underline{s}$, in a medium with an absorption coefficient κ and a density field $\rho(\underline{x})$. The Radiative Transfer Equation is then (Chandrasekhar 1950):

$$\underline{s} \cdot \underline{\nabla} I(\underline{x}, \underline{s}) = -\kappa\rho(\underline{x})[I(\underline{x}, \underline{s}) - j(\underline{x}, \underline{s})] \tag{3.1}$$

$j(\underline{x}, \underline{s})$ is the source function and for a scattering[1] atmosphere, it can be written in the form:

$$j(\underline{x}, \underline{s}) = \frac{1}{4\pi} \iint \sigma(\underline{s}, \underline{s}') I(\underline{x}, \underline{s}') d\omega_{\underline{s}'} \tag{3.2}$$

where $\underline{s}$, $|\underline{s}| = 1$, is a unit vector specifying some direction through a point $\underline{x}$, $d\omega_{\underline{s}'}$ the solid angle, and $\sigma(\underline{s}, \underline{s}')$ the scattering coefficient corresponding to the fraction of intensity scattered from one direction $\underline{s}$ to another $\underline{s}'$ with the following condition of normalization:

$$\int \sigma(\cos\theta) \frac{d\omega}{4\pi} \leq 1 \; with \; \cos\theta = \underline{s} \cdot \underline{s}' \tag{3.3}$$

In the case of perfect scattering, the scattering coefficient is normalized to unity and the absorbed radiation reappears totally as scattered radiation. We also have to introduce some relevant quantities depending on the radiance. First the total intensity:

$$J(\underline{x}) = \frac{1}{4\pi} \int I(\underline{x}, \underline{s}) d\omega_{\underline{s}} \tag{3.4}$$

also the net flux:

$$\underline{F}(\underline{x}, \underline{s}) = \frac{1}{\pi} \int I(\underline{x}, \underline{s}) \underline{s} d\omega_{\underline{s}} \tag{3.5}$$

and the K-tensor[2]:

$$\underline{\underline{K}}(\underline{x}) = \frac{1}{4\pi} \int I(\underline{x}, \underline{s}) \underline{s} \otimes \underline{s} d\omega_{\underline{s}} \tag{3.6}$$

Using these quantities we can also express the Radiative Transfer Equation in its integrated form, considering an isotropic (all directions are identically distributed) and perfect scattering (absorbed light is totally reemitted) case. Integrating eq. (3.1) over s, and using definition (3.5) yields equation (3.7), and taking the tensor product of eq. (3.1) and $\underline{s}$, then integrating over s, with the aid of eq. (3.6), yields equation (3.8):

$$div \underline{F} = 0 \tag{3.7}$$

$$div \underline{\underline{K}} = -\frac{\kappa\rho}{4} \underline{F} \tag{3.8}$$

[1] The contributions to the source function are only due to scattering.

[2] which generalizes the K-integral of Chandrasekhar (1950) ($\otimes$ is the tensorial product).

3.2. The radiative transport calculations. The two major problems in order to simulate the R.T.E. are the differentiation due to the gradient of intensity and the inherent continuity of the scattering function. The gradient differentiation can be discretized using finite difference approximations, however these approximations are only valid in a medium without large fluctuations, i.e. $\delta\tau \ll 1$, where:

$$\delta\tau = \kappa\rho\delta z \tag{3.9}$$

is the elementary optical depth (or thickness) of the cloud between positions z and $z + \delta z$. Since multifractal fields have enormous dynamic ranges (diverging as $\lambda \to \infty$) it is therefore important to use numerical schemes which are stable for occasionally large values of $\delta\tau$.

A classical way to discretize the scattering function was developed by Chandrasekhar (1950), inspired by Schuster (1905) and Schwartzschild (1906), who divided the radiative field in 2n streams along 2n directions and obtained 2n linear equations. Whereas Chandrasekhar (1950) developed a discretization of directions in order to get a quadrature, i.e. the best approximation in a given sense for any order n, cruder approximations were considered in toy models. For instance, D.A. radiative transfer (Lovejoy et al. 1990) corresponds to the case n=1 or n=2, the phase function describing scattering only through 90° for n=2.

More elaborated discretizations of directions in the radiative transfer equation are obtained by Legendre polynomials expansion (Chandrasekhar 1950, in the case of plane parallel atmospheres) or more generally, with spherical harmonics (Appendix C). The solution is a sum of Legendre polynomials of increasing degree which give angular dependency, weighted by coefficients dependent on the position in the cloud. The net advantage of this method is a drastic simplification of the scattering term in the equation. Another approach is to expand the Green's function of the radiative transfer equation with respect to the (rather trivial) Green's function of its linearization, i.e. without the scattering term. Despite appealing features, this approach, discussed in Appendix B, still faces some theoretical difficulties.

We now consider the problem of finite differences. The spatially explicit discretization for the intensity at a position $\underline{x}+\delta\underline{x}$ is (the discrete scattering coefficient is written $\sigma_{s,s'}$, s and s' being different directions):

$$I(\underline{x} + \delta\underline{x}, s) = I(\underline{x}, s) - \delta\tau I(\underline{x}, s) + \sum_{s'} \delta\tau\sigma_{s,s'} I(\underline{x}, s') \tag{3.10}$$

This scheme will lead generally to inconsistencies. Indeed, it is rather easy to understand when considering a D.A radiative transfer on a 2D cut. We have only 4 possible directions: $s, -s, \pm s_\perp$. The elementary (for a given $\delta\tau$) transmission T, reflexion R and diffusion S then have the corresponding form:

$$\mathrm{T} = 1 - \delta\tau(1 - \sigma_{s,s}) \tag{3.11}$$

FIG. 3. *Transmission field for a cloud density field with* $\alpha = 1.35, C_1 = 0.75$ *and* $H = 1/3$ *(highest values are black and smallest white).*

$$\mathrm{R} = \delta\tau\sigma_{s,-s} \tag{3.12}$$

$$\mathrm{S} = \delta\tau\sigma_{s,s_\perp} \tag{3.13}$$

It is easy to remark that in case of large fluctuations, $\delta\tau \to \infty$, the transmittance will become negative, reflection and diffusion will diverge. In order to avoid divergences[3] induced by finite difference approximations, we use the semi-implicit scheme (Borde 1991, Borde et al. 1993):

$$\begin{aligned} I(\underline{x}, s) &= I(\underline{x} - s\delta\underline{x}, s) - \delta\tau(1 - \sigma_{s,s})I(\underline{x}, s) \\ &\quad + \sum_{s' \neq s} \delta\tau\sigma_{s,s'} I(\underline{x} - s'\delta\underline{x}, s') \end{aligned} \tag{3.14}$$

We represent in Figures 3 and 4, the radiative field transmitted and reflected for the multifractal cloud density field represented in Figure 2 with $\alpha = 1.35$, $C_1 = 0.75$ and $H = 1/3$. Initial conditions are a transmission at unity from above ($I_- = 1$) and the three other directions are zero ($I_+ = I_{\pm\perp} = 0$). The propagation is vertical.

[3] In order to cancel the negative transmittances, A.Davis (private communication) noted that while using this scheme for a multifractal cloud, he put these negative transmittances to zero.

FIG. 4. *Reflection field for simulated cloud density field with* $\alpha = 1.35, C_1 = 0.75$ *and* $H = 1/3$.

4. General argument. For largely homogeneous media, there are the classical optically thin and thick limits. In the case of extreme inhomogeneity, as in multifractal atmospheres, thin and thick "parts" of the medium are so entangled that the two limits might not exist independently and the two could exist at different scales. However, the distinction could be relevant by performing a similar analysis in a scale invariant way, i.e. directly on singularities. For instance, most of the transfer will occur in areas where the singularities will be rather low, whereas most of the scattering will occur when the singularities are rather high. Therefore one may expect the existence of a critical singularity separating 'transparent' (low order) singularities where the radiance field mostly flows through, keeping the trivial scaling of the source flux, and for 'opaque' (high order) singularities where scattering becomes more and more effective, yielding a more and more non trivial scaling for the radiance field. The field becomes more and more singular with respect to volume integration. By directly considering single scattering statistics on multifractal clouds, this has been quantified in Lovejoy et al. 1995.

It turns out that these rather general arguments are rather similar to what happens for a fractional integration, as detailed in section 6—yielding a first order multifractal phase transition which occurs when the dimension D′ of the fractional integration is no longer smoothing out the singularities of the integrated field, whereas the finite size of the sample induces a second order phase transition.

This is not surprising since in at least two cases, the radiance inhomogeneities are obtained by integration over cloud inhomogeneities (section 5).

5. Multifractal plane parallel and 1D atmospheres.

5.1. Multifractal 1D atmosphere. Going back to section 3, we consider the radiative transfer equation (eq. 3.1) with emphasis on its integrated form (eq. 3.7 and eq. 3.8). If we take a one dimensional field with radiation entering at the top of it, I_+ being the upgoing intensity and I_- being the downgoing one, these equations become:

$$\frac{dF(z)}{dz} = 0 \tag{5.1}$$

$$\frac{dK(z)}{dz} = -\kappa\rho(z)F(z) \tag{5.2}$$

with

$$F(z) = I_- - I_+ \tag{5.3}$$

and

$$K(z) = I_- + I_+ \tag{5.4}$$

Thus if we combine (5.3) and (5.4) with (5.1), we find that the K-function is proportional to the intensity, irrespective of direction. Integrating (5.2), we obtain the result that intensities are a 1D-integration of the density field.

5.2. Multifractal plane-parallel atmosphere. To illustrate the general argument, we derive the relationship between the intensity field and the cloud density field, and the corresponding conditions necessary for a plane parallel field (which is two dimensional and homogeneous along the horizontal and heterogeneous along the vertical).

Thus every quantity in the Radiative Transfer Equation only depends on the vertical coordinate z, and its integrated form becomes (v stands for vertical):

$$F_v = const \tag{5.5}$$

$$\frac{\partial K_v}{\partial z} = -\kappa\rho F_v \tag{5.6}$$

The flux being constant we can easily integrate eq. (5.6). The right hand side depends on:

$$\Delta\tau = \int_{z}^{z+\Delta z} \kappa\rho(z')dz' \tag{5.7}$$

and we obtain a simple relationship:

$$\Delta K_v \propto \Delta\tau \tag{5.8}$$

Using the Discrete Angle radiative transfer scheme we can express the relationships between the different quantities in a very simple manner:

$$\Delta F_v = 0 = \Delta I_- - \Delta I_+ \tag{5.9}$$

$$\Delta K_v = \Delta I_+ + \Delta I_- = 2\Delta I_\pm \tag{5.10}$$

The following relationship is then straightforward:

$$\Delta I_v \propto \Delta K_v \tag{5.11}$$

Another way to have a third equation completing the first two integrated R.T.E's is to consider an expansion of the scattering coefficient and intensities as a series of Legendre polynomials (see section 3.2 and Appendix C):

$$\sigma(\underline{s}, \underline{s}') = \sum_{\ell} \omega_\ell P_\ell(\cos\alpha) \tag{5.12}$$

$$I(z, \underline{s}) = \sum_{n} P_n(\cos\theta) I_n(z) \tag{5.13}$$

with θ the angle between $\underline{s}$ and the vertical and α the angle between $\underline{s}$ and $\underline{s}'$. The three first orders of functions $I_n(z)$ depend respectively on J, F and K (see equations 3.4, 3.5, 3.6). With this method, Chandrasekhar (1950) pointed out a solution of $I(z, s)$ as a function of $F(z, s)$ and of τ in conservative cases. It corresponds to the solution of least anisotropy:

$$I(\tau, \mu) = \frac{3}{4}\, F[(1 - \frac{1}{3}\omega_1)\tau + \mu] \tag{5.14}$$

with $\mu = \cos\theta = \underline{s} \cdot \underline{s}'$. These two methods, the first combining (5.8) and (5.11) and the second directly with (5.14) give us the same relationship between I_v and τ:

$$\Delta I_v \propto \Delta\tau \tag{5.15}$$

The vertical variation of the light intensity transferred through a plane-parallel cloud is proportional to the corresponding variation of the optical depth $\Delta\tau$ which is a 1D-integration of the density field ρ (eq. 5.8).

6. Full multifractal atmosphere.

6.1. General presentation. We study now a 2D-cut[4] for convenience in computation and presentation, with no loss of generality in 3D (this time we conserve the heterogeneity along the two directions) and we consider first the analytical aspect.

We might be interested in horizontal averages of the different quantities in order to restrict the dependence on the vertical coordinate only. Integrating the R.T.E along the horizontal (indicated by overbars) we obtain:

$$\frac{\partial \overline{F_v}}{\partial z} = 0 \tag{6.1}$$

$$\frac{\partial \overline{K_v}}{\partial z} = -\kappa \overline{\rho F_v} \tag{6.2}$$

Although equation (6.1) is similar to equation (5.5), equation (6.2) is much more complicated than equation (5.6) due to the non-linearity introduced by the correlation term $\overline{\rho F_v}$, and there is no simple analytical technique to determine any general solution.

Going back to the general argument, we hypothesize that for any density field these equations should have the same consequences as a fractional integration, i.e. a radiative field should have the same statistics as those of an integrated field. We have also to be sure that it is related to the full original multifractal statistics. Thus we study the influence on statistics of a (fractional) D′-integration on a D-cut of a multifractal field.

6.2. Non trivial consequences of a D′-integration on a D-cut of a multifractal field. Whereas the exponents $c(\gamma)$ and $K(q)$ are preserved for any D′-cut of a multifractal process observed in D-space D > D′, they are not preserved for a D′-integration. This can be understood with the notions of "bare" and "dressed" quantities, which are different due to the divergence of high order statistical moments for the latter (Schertzer, Lovejoy 1987). For the same resolution λ, the "bare" field is the result of a multiplicative cascade partially developed from large scales down to λ, whereas the "dressed" field is obtained by integration over λ of a completed cascade. In the limit $\lambda \to \infty$, the bare quantity ε_λ becomes singular and is implicitly given by the measure $\Pi_\lambda(A)$ (which converges):

$$\Pi_\lambda(A) = \int_A \varepsilon_\lambda d^D \underline{x} \quad (\lambda = \frac{L}{\ell}) \tag{6.3}$$

where A is a set of dimension D. Corresponding dressed quantities $\varepsilon_{\lambda(d)}$ (d for dressed) are expressed from the definition of integrated fluxes of the

[4] Physically, a vertical cross section of the atmosphere.

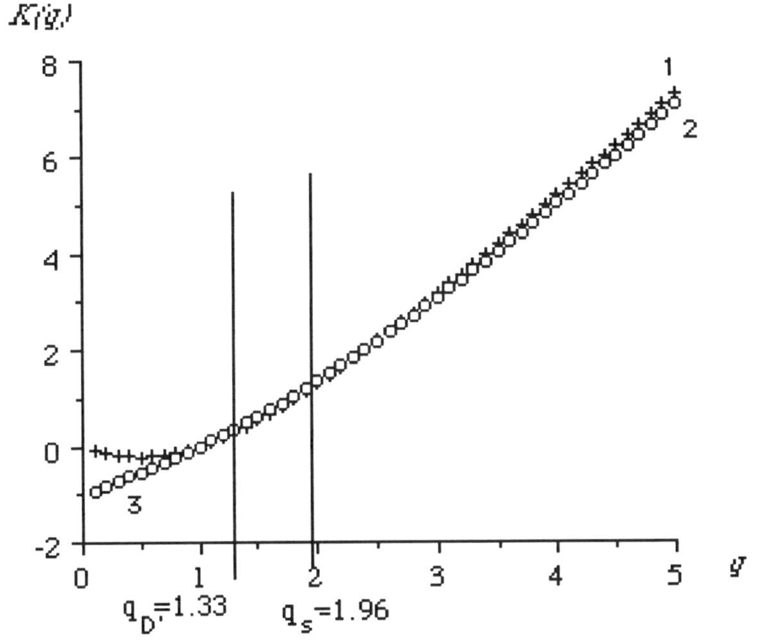

FIG. 5. *$K(q)$ for the D'-integrated field and the D-field with $\alpha = 2$ and $C_1 = 0.75$ $(H = 0)$; with D=2 and D'=1.*
Slopes:
1 $K(q) = -3,0112 + 2,0581q$
2 $K_{D'}(q) + q - 1 = -3,0220 + 2,0156q$
3 $K_{D'}(q) + q - 1 = .1.0263 + 1.0087q$ *for* $q < q_{D'}$
Captions: $^+$ *D-integrated field;* $^\circ$ *D'-field*

bare quantity ε_λ:

$$\varepsilon_{\lambda(d)} = \frac{\Pi_\infty(B_\lambda)}{vol(B_\lambda)} \tag{6.4}$$

where $\mathrm{vol}(B_\lambda) = \lambda^{-D}$ is the D-dimensional volume of a ball of size λ^{-1} and by definition, for a set $A, \Pi_\infty(A) = \lim_{\lambda\to\infty} \Pi_\lambda(A)$. These dressed quantities will display a divergence of moments above a critical order q_D, defined by:

$$K(q_D) = D(q_D - 1) \tag{6.5}$$

This situation corresponds to a hyperbolic behaviour of the probability distribution tails (extreme events) and the dressed characteristic function

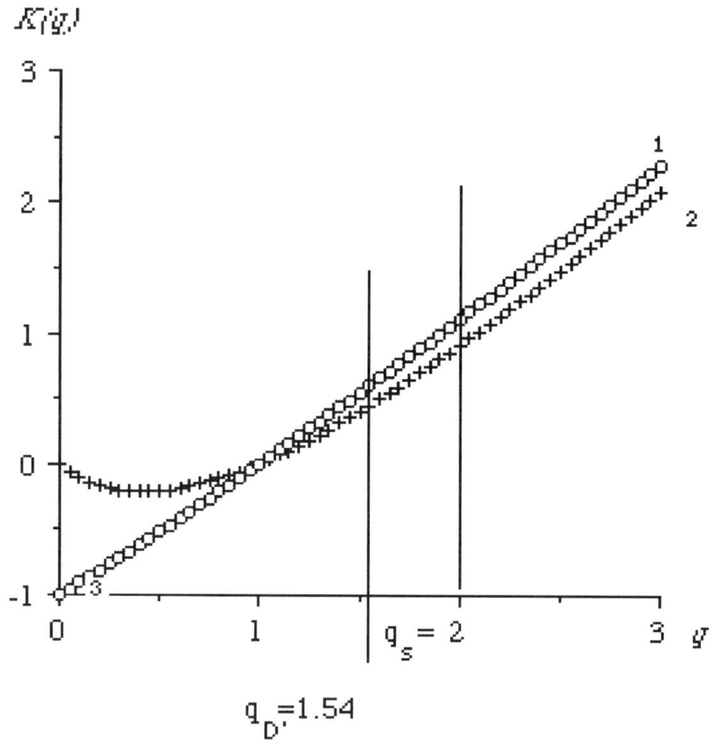

FIG. 6. *K(q) for the D'-integrated and D-field with* $\alpha = 1.35$ *and* $C_1 = 0.75$ $(H = 0)$. *D=2 and D'=1. Slopes:*
1 $K_{D'}(q) + q - 1 = -1,2464 + 1,1758q$
2 $K(q) = -1,4711 + 1,1827q$
3 $K_{D'}(q) + q - 1 = -1,0103 + 0,99967q$ *for* $q < q_{D'}$
Captions: $^+$ *D-integrated field;* $^\circ$ *D'-field*

of moments $K_{(d)}(q)$ is of constant slope above q_D (Schmitt et al. 1994), whereas below q_D, it is identical to the bare characteristic function of moments. This critical order leads a discontinuity in the first derivative of $K_{(d)}(q)$ and corresponds to a first order multifractal phase transition[5].

In our case, instead of integrating the bare field ε_λ (on a set A of dimension D) over a given scale $\ell = \frac{L}{\lambda}$, we choose an arbitrary order of integration D$'$ which can be whether integer or fractional with D$'$ < D.

[5] This terminology is borrowed from statistical thermodynamics and corresponds to an analogy between the moment order q and the inverse temperature β, as well as between $K(q)$ and the Massieu potential $\sum(\beta)$. See Schertzer and Lovejoy 1995 and references therein, for discussions on first and second order of multifractal phase transitions, as well as their consequences.

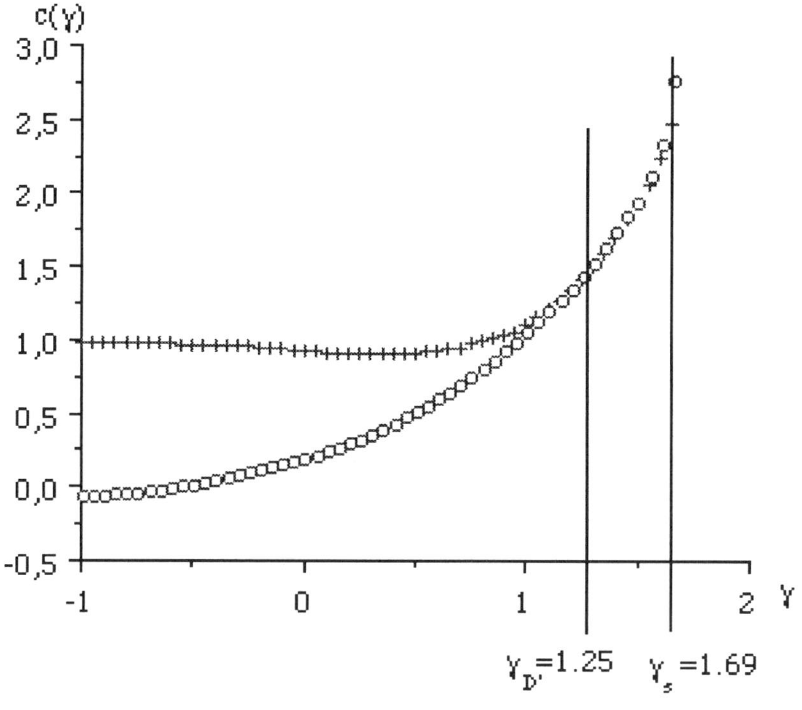

FIG. 7. *Codimensions of the D-field and the D'-integrated field for $\alpha = 2$ and $C_1 = 0.75$, ($H = 0$), for D=2 and D'=1. Captions:*
+ D'-integrated field; ° D-field

Characterizing the original field ε_λ by $K(q)$ and $c(\gamma)$, we correspondingly characterize the field obtained after D′-integration by $K_{D'}(q)$ and $c_{D'}(\gamma)$. We consider the case where D′ is an integer and then we decompose A as follows: $A = A' \times B$, where A' is a set of dimension D′ and B is the set of dimension D-D′. The field of interest here can be obtained as:

$$\Pi_\lambda(A') = \sum_{A'} \varepsilon_\lambda \lambda^{-D'} \tag{6.6}$$

Recalling the general expression for trace-moments (Schertzer, Lovejoy, 1987):

$$Tr_A \varepsilon_\lambda^q = \int_A \langle \varepsilon_\lambda^q \rangle d^{qD} x \approx \lambda^{K(q)-(q-1)D} \tag{6.7}$$

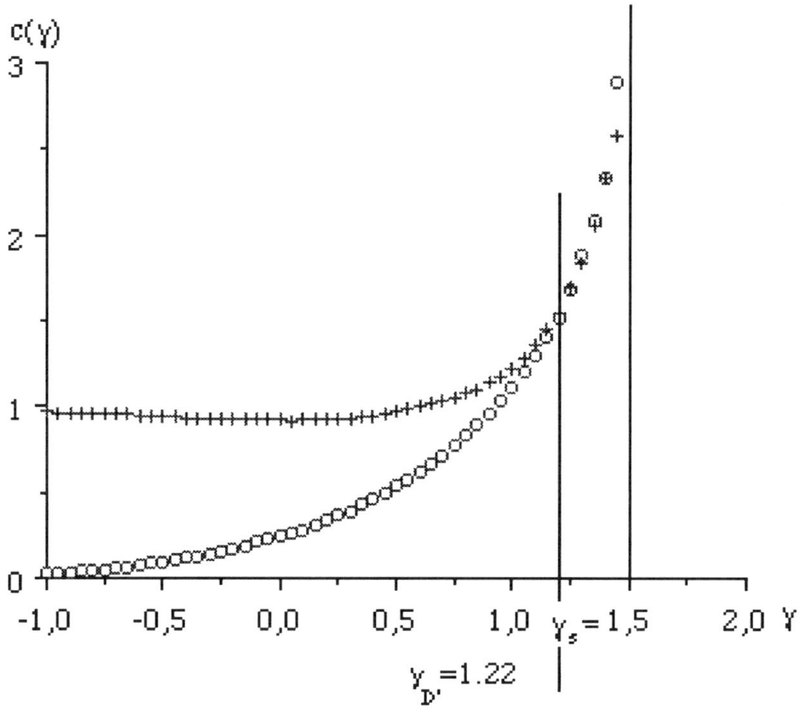

FIG. 8. *Codimensions of the D-field and the D'-integrated field with* $\alpha = 1.35$ *and* $C_1 = 0.75$ $(H = 0)$. *D=2 and D'=1. Captions:*
$^+$ *D'-integrated field;* $^\circ$ *D-field*

Here, the relevant trace-moment should be :

$$Tr_{A'}(\varepsilon_\lambda^q) \approx \lambda^{K(q)-D'(q-1)} \tag{6.8}$$

The behaviour of the moment of interest $\langle \Pi_\lambda^q(A') \rangle$ is given by this trace-moment on the one hand, and on the other can define a trace-moment as in equation (6.7):

$$\begin{aligned} Tr_B(\Pi_\lambda(A'))^q &\approx \sum_B \lambda^{-q(D-D')} \langle (\Pi_\lambda(A'))^q \rangle \\ &\approx \lambda^{-(D-D'(q-1)} \langle (\Pi_\lambda(A'))^q \rangle \end{aligned} \tag{6.9}$$

By analogy with the general theory (eq. 6.5), we should have a critical moment $q_{D'}$ given by the divergence of the trace-moment of equation (6.8) in the limit $\lambda \to \infty$:

$$K(q_{D'}) = D'(q_{D'} - 1) \tag{6.10}$$

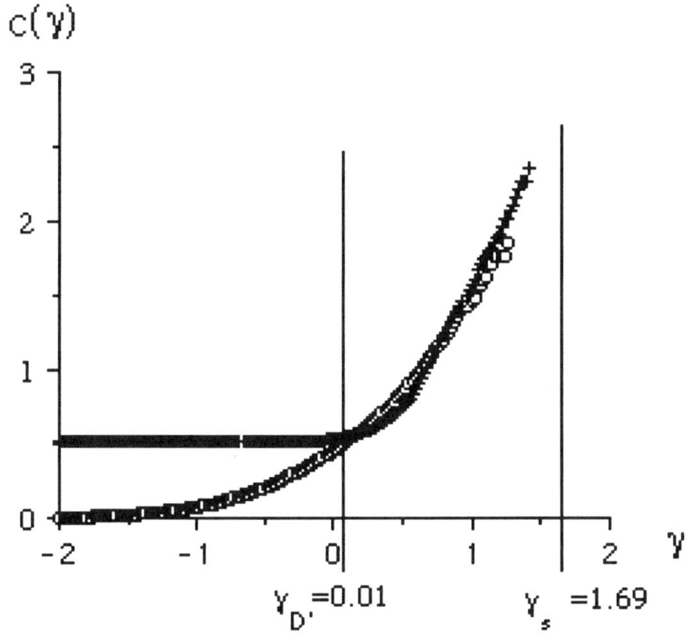

FIG. 9. *Codimensions of the D-field and the fractional D'-integrated field with* $\alpha = 2$ *and* $C_1 = 0.75$ $(H = 0)$. *for D=2 and D'=0.5. Captions:* + *D'-integrated field;* ° *D-field*

Above this exponent, the trace-moment diverges.

By definition, we should have :

$$\langle(\Pi_\lambda(A'))^q\rangle \approx \lambda^{K_{D'}(q)} \tag{6.11}$$

Now if we suppose that above the moment of order $q_{D'}$, the moment $\langle\Pi_\lambda^q(A')\rangle$ is almost equivalent to the trace-moment defined in equation (6.8), and also that the trace on A is equivalent to taking first a trace on A' and then on B, we should obtain:

$$Tr_B(\Pi_\lambda(A')^q) \approx Tr_A(\varepsilon_\lambda^q) \tag{6.12}$$

and finally that:

$$K_{D'}(q) - (D - D')(q - 1) = K(q) - D(q - 1) \tag{6.13}$$

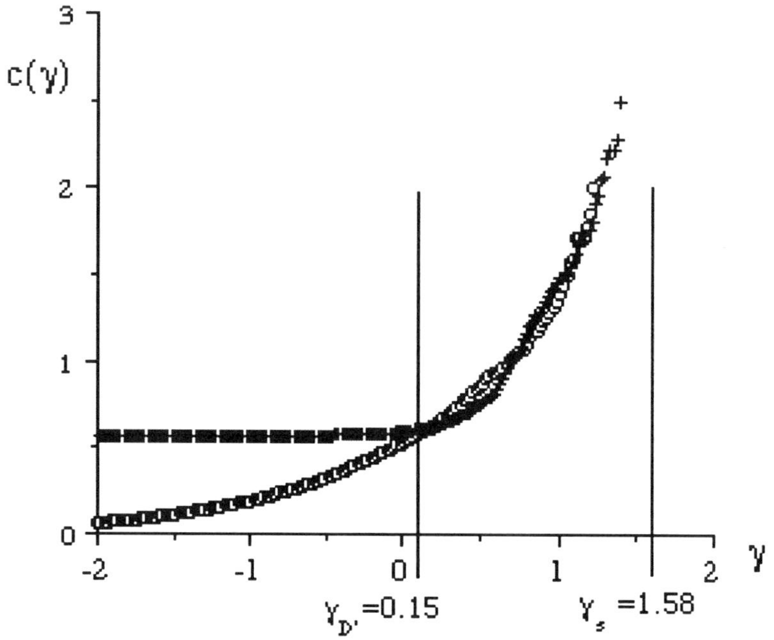

FIG. 10. *Codimensions of the D-field and the D'-integrated field with* $\alpha = 1.35$ *and* $C_1 = 0.75$ $(H = 0)$. *D=2 and D'=0.5. Captions:*
+ *D'-integrated field;* ° *D-field*

Then

$$K_{D'}(q) = K(q) - D'(q-1) \quad \text{for all}\, q > q_{D'} \tag{6.14}$$

For $q < q_{D'}$, the traces converge which implies that the moments converge as well and should lose their dependency on λ. Thus equation (6.11) implies $K_{D'}(q) = 0$. The main consequence is that the moments calculated for the D'-integrated field are smoothed out below the $q_{D'}$-moment.

The D-cut ε_λ also exhibits a critical singularity γ_s (Schertzer et al. 1991), due to finite sample size limitations. This singularity is the maximum one attainable and is given by :

$$c(\gamma_s) = D + \frac{\log(N_s)}{\log(\lambda)} \tag{6.15}$$

where Ns is the total number of samples. This expression gives by Legendre transform (eq. 2.11 and 2.12) a critical order of moments q_s. The order q_s induces a discontinuity in the second derivative of the characteristic function of moments and represents a second order multifractal phase transition.

As a conclusion we can say that: below a critical order of moment $q_{D'}$, dependent only on the order of fractional integration and representing a first order multifractal phase transition, $K_{D'}(q)$ is null; above, $K_{D'}(q)$ is simply related to $K(q)$, however, above another critical order of moment q_s, dependent on the sample size and representing a second order multifractal phase transition, $K(q)$ and $K_{D'}(q)$ become linear.

Finally, these two critical orders form the range of order of moments that give information about the statistics of the D-cut from the statistics of the D′-integrated field.

In order to illustrate this, for D=2 and D′=1, we create a D-cut of a multifractal field (100 realizations), in a conservative case ($H = 0$) for convenience and no loss of generality, and $C_1 = 0.75$. We integrate it, then calculate the moments of both fields and their respective $K(q)$. We represent $K(q)$ and $K_{D'}(q) + D'(q-1)$ (as defined previously) versus q in Figures 5 for $\alpha = 2$, and 6 for $\alpha = 1.35$. For both α, we have a perfect correspondence with theory and we can notice that the range of q's (Δq) between q_D and q_s is very small (for $\alpha = 2, \Delta q = 0.3$ and for $\alpha = 1.35, \Delta q = 0.46$). The usual technique to estimate α and C_1 (the DTM or Double Trace Moment) (Lavallée et al. 1991) calculates and analyses moments of the field after having raised it to a power η at each scale. The range of η chosen should be large enough to preserve accuracy but is also limited by divergence of moments. Equivalently, the limited range of relevant q's reduces the range of relevant η's. In this case, the DTM is not accurate enough as $\Delta\eta$ is of order of Δq. With the Legendre transform, we get the correspondingly critical singularities $\gamma_{D'}$ and γ_s. We can also represent this relationship with the codimensions $c(\gamma)$ and $c_{D'}(\gamma)$:

$$\gamma < \gamma_{D'} \Longrightarrow c_{D'}(\gamma) = 0 \tag{6.16}$$

$$\gamma > \gamma_{D'} \Longrightarrow c_{D'}(\gamma) = c(\gamma + D') - D'$$

as shown on Figures 7 (for $\alpha = 2$) and 8 ($\alpha = 1.35$). Once again we calculate the codimensions $c(\gamma)$ and $c_{D'}(\gamma)$ and compare them. One can notice that the behavior for small γ's makes the codimension representation more striking than that of $K(q)$.

The calculations for fractional integration (D′ is no longer an integer) are more tricky because we are compelled to perform a summation on all the embedded space, i.e. some renormalisation procedures have to be set properly. However it appears from performing simulations that the behaviour of $c_{D'}(\gamma)$ and $K_{D'}(q)$ are rather similar. As codimensions exhibit in a clearer manner the expected properties, we represent in Figures 9 and 10 the comparison of codimensions of a field with D=2 and the codimensions of this field D′-integrated with D′ = 0.5. Figure 9 shows a field with

$\alpha = 2$ and Figure 10 with $\alpha = 1.35$.

We have yet to compare the codimensions of simulated radiative fields and fractionally integrated fields, which would require extensive calculations on a supercomputer, we would also have to determine the order of the closest fractional integration.

The bad news is that direct application of the DTM on an integrated field gives wrong values of α and C_1. However, the good news is that the advantage of fractional integration is that it allows a reverse transformation (differentiation, see Appendix A and Lavallée et al. 1993 for application on data) which restores the original statistics of the field. Using this transformation on any integrated field returns its validity to the DTM technique.

7. Conclusion. In order to preserve its heterogeneity, its scale invariance and its physical properties, we describe the atmosphere within a theoretical framework: the universal multifractal model. Due to its universality property, it depends only on three fundamental parameters C_1, α and H, already experimentally measured for many fields in the atmosphere (Tessier et al. 1993, Schmitt et al. 1992, 1993, 1994). It seems to be a pertinent way to deal with all the problems involving clouds that persist in climatology and for remote sensing, in the sense of dealing directly with singularities of the heterogeneous water concentration field.

We focused on the case of perfect scattering, as this phenomenon should be strongly related to the problem of anomalous absorption and inversion methods for radiances. In order to propagate light in this medium, we analytically set up different expressions of the solution of the radiative transfer equation; first a discrete angle scheme, which has also been used for simulation, and also a formulation of the radiance with Legendre polynomials, spherical harmonics and Green's functions. The general argument is that the radiance field is a fractional integration of the cloud density field. It is shown as being verified for 1D and plane parallel atmospheres and we would like to generalize it for 2D and 3D cases.

We showed that the idea that (fractional) D$'$-integration on a D-cut of a multifractal field : conserves the codimension of the mean C_1, the Lévy index α, and increments the conservation parameter H, is too crude.

D$'$-integration has important consequences when applied on a multifractal field. It smoothes the low singularities and reduces the range of singularities necessary to preserve the validity of our usual estimation techniques for C_1, α and H. This phenomenon occurs until a critical singularity corresponding to a first order phase transition. Another singularity due to sample size limitations, representing a second order multifractal phase transition, is the upper limit of the range of singularities following the desired statistics.

Physically it means that if the effect of the radiative transfer equation is similar to a fractional integration, one can only observe directly the large concentrations of water in clouds, whatever the scale of observation.

However, considering fractional integration, this effect can be eliminated by fractional differentiation which returns the initial statistics, a method that could explain the fact that one can analyze the statistics of the gradients of radiances instead of the radiances themselves. More generally, the determination of the critical singularities should help to overcome many dead-locks encountered until now in inversion methods for radiances.

In order to refine our study, especially if we want to compare fractional integration and the radiative transfer equation, we would have to make large simulations on supercomputers. Analytically, a more rigorous foundation could be found by using Green's functions. Finally an important part of further study will be dedicated to anomalous absorption.

8. Acknowledgments. We acknowledge stimulating discussions with Y. Chigirinskaya, C. Gautier, D. Lavallee, N. Byrne, R. Sanchez, F. Schmitt and R. Somerville. Support from (ARM) DOE #DE–FG0390 contract is gratefully acknowledged.

A. Appendix: Fractional integration and differentiation. Fractional integration and differentiation correspond to extensions to non- integer orders (H) of integrations (I^H) or differentiations (D^H). We denote D^H by $\partial^H/\partial t^H$, where t is the argument of the function. These extensions are rather straightforward in Fourier space or one-dimensional functions, since integrations—up to a constant of integration discussed below—or differentiations of integer order n, correspond respectively to division or multiplication by $(ik)^n$ where k is the wave number (Fourier transforms of physical space quantities will be denoted by a tilde ($\sim$):

$$\tilde{\mathrm{I}}^{-H}(f) = \tilde{\mathrm{D}}^{H}(f) = (ik)^H \tilde{f}(k) \tag{A.1}$$

In the usual physical space, for a non-integer H we obtain an ordinary (i.e. for integer order n) differentiation (D^n, positive n) or integration (I^{-n}, negative n) of a convolution:

$$\mathrm{I}^{-H} f = \mathrm{D}^H f = \frac{1}{\Gamma(n-H)} \mathrm{D}^n (f * x^{n-H-1}) \tag{A.2}$$

Here Γ is the Euler Gamma function:

$$\Gamma(t) = \int_0^\infty e^{-u} u^{t-1} du \quad t > 0 \tag{A.3}$$

(A.2) is more general than (A.1), since it is written directly in physical space, but introduces ambiguities in the definition on non-integer integration or differentiation because they will clearly depend on the domain of definition of convolutions.

The same techniques can be extended to functions defined on R^d, however the analysis becomes more complex because various combinations of partial derivatives are now possible. Nevertheless one can consider the following strongly isotropic extension:

$$\text{(A.4)} \qquad \mathrm{I}_d^{-H} f = \mathrm{D}_d^H f = \frac{1}{\Gamma(n-H)} \mathrm{D}_d^n (f * |x|^{n-H-d})$$

$$\text{(A.5)} \qquad \tilde{\mathrm{I}}_d^{-H}(f) = \tilde{\mathrm{D}}_d^H f = |k|^H \tilde{f}(k)$$

which in fact corresponds to fractional powers of the Laplacian (or of the Poisson solver):

$$\text{(A.6)} \qquad \mathrm{I}_d^{-H} f = \mathrm{D}_d^H f = (-\Delta)^{H/2}$$

B. Appendix: Green's function approach. The strongly nonlinear dependence on directions introduced by multiple scattering can be perceived by first introducing the Green's function G_0 of the partial differential part of the transfer equation (on a spatial domain D and of boundary ∂D):

$$\text{(B.1)} \qquad \underline{u}.\nabla G_0(\underline{x}, \underline{x}'; \underline{u}, \underline{u}') + \kappa\rho G_0(\underline{x}, \underline{x}'; \underline{u}, \underline{u}') = \delta_{\underline{x}-\underline{x}'}\delta_{\underline{u}-\underline{u}'}$$

which allows us to write the transfer equation in a purely integral form:

$$\text{(B.2)} \qquad I(\underline{x}, \underline{u}) = \int_{\partial D \times S_d} G_0(\underline{x}, \underline{x}'; \underline{u}, \underline{u}')\kappa\rho(\underline{x}')j_0(\underline{x}', \underline{u}')d^{d-1}\underline{x}'d^{d-1}\underline{u}'$$

$$+ \int_{D \times S_d} G_0(\underline{x}, \underline{x}'; \underline{u}, \underline{u}')\kappa\rho(\underline{x}')j(\underline{x}', \underline{u}')d^{d-1}\underline{x}'d^{d-1}\underline{u}'$$

$$\text{(B.3)} \qquad I(\underline{x}, \underline{u}) = G_0 * (\kappa\rho j_0) + G_0 * (\kappa\rho j)$$

where $j_0(\underline{x}, \underline{u})$ denotes the boundary sources, $j(\underline{x}, \underline{u})$ the scattering sources (eq. 3.2) and $*$ denotes (generalized) convolutions over positions and/or directions. By iteration of equations (3.1) and (B.3) we are lead to the following Von Neumann series:

$$\text{(B.4)} \qquad I(\underline{x}, \underline{u}) = G_0 * (\kappa\rho j_0) + G_0 * (\kappa\rho\sigma) * G_0 * (\kappa\rho j_0) + \cdots$$

$$.. + G_0 * [(\kappa\rho\sigma) * G_0]^{*n} * (\kappa\rho j_0) + \cdots$$

which displays the complexity of the multiple scattering. This approach is rather formal, since on the one hand the convergence of the series may be questioned and on the other a part of the solution depends on boundary conditions. We are studying radiative properties of a cloud field occupying a domain D of large horizontal extension, with boundary conditions corresponding to a null horizontal flux:

$$\text{(B.5)} \qquad \underline{F}_h(\underline{x}) = 0 \; x \in \partial D$$

or at least (e.g. by imposing cyclic conditions):

(B.6) $$\int_{\partial D} \underline{F}(\underline{x}')d^{d-1}\underline{x}' = 0$$

On the contrary on a certain sub-domain $\partial D'(\partial D \supset \partial D')$:

(B.7) $$\underline{F}_v(\underline{x}) \neq 0 \; x \in \partial D'$$

whose unknown values are part of the solution.

C. Appendix: Legendre polynomials and spherical harmonics. Chandrasekhar (1950), in case of plane-parallel atmosphere, proposed to express the scattering function and the radiance in terms of Legendre polynomials. We consider:

(C.1) $$\sigma(\underline{s}, \underline{s}') = \sum_{\ell} \omega_\ell P_\ell(\cos\alpha) \; with \; \cos\alpha = \underline{s}.\underline{s}'$$

$P_\ell(\cos\alpha)$ is a Legendre polynomial of order ℓ for cos α. Using the addition theorem for spherical harmonics we get:

(C.2) $$\sigma(\cos\alpha) = \sum_{\ell} \omega_\ell \sum_{m=-\ell}^{\ell} \frac{4\pi}{2\ell+1} \mathrm{Y}_\ell^m(\theta,\varphi)\mathrm{Y}_\ell^{m^*}(\theta',\varphi')$$

with unit vectors $\underline{s}$ and $\underline{s}'$ expressed in spherical coordinates, respectively (θ,φ) and (θ',φ') (their modulus is 1), where $\mathrm{Y}_\ell^m(\theta,\varphi)$ represents a spherical harmonic depending on two angles (θ,φ) and of order ℓ, and $\mathrm{Y}_\ell^{m^*}(\theta',\varphi')$ represents the complex conjugate of a spherical harmonic, of order ℓ and angles (θ',φ'). We choose to express the intensity in terms of spherical harmonics:

(C.3) $$I(\underline{x},\theta,\varphi) = \sum_{\ell}\sum_{m} I_{\ell,m}(\underline{x})Y_\ell^m(\theta,\varphi)$$

where the position dependence is represented by $\underline{x}$ and the direction by (θ,φ). Thus we have a coefficient $I_{\ell,m}(\underline{x})$ that depends only on position and which we can express as:

(C.4) $$I_{\ell,m}(\underline{x}) = \iint d(\cos\theta)d\varphi I(\underline{x},\theta,\varphi)\mathrm{Y}_e^{m^*}(\theta,\varphi)$$

Then replacing these new expressions of I and σ we get the following expression for the scattering term A of equation (3.1):

(C.5) $$\frac{1}{4\pi}\iint d(\cos\theta')d\varphi' I(\underline{x},\theta',\varphi')\sigma(\theta,\varphi;\theta'\varphi') =$$

$$\sum_{\ell} \frac{\omega_\ell}{2\ell+1} \sum_{m=-\ell}^{\ell} \mathrm{Y}_\ell^m(\theta,\varphi) \iint d(\cos\theta')d\varphi' I(\underline{x},\theta',\varphi')\mathrm{Y}_\ell^{m^*}(\theta',\varphi')$$

Thus from (C.3) we find:

$$\text{(C.6)} \qquad \mathrm{A} = \sum_{\ell} \frac{\omega_\ell}{2\ell+1} \sum_{m=-\ell}^{\ell} \mathrm{Y}_\ell^m(\theta,\varphi) I_{\ell,m}(\underline{x})$$

Now we consider the radiative transfer equation (3.1), multiplying each term by $\mathrm{Y}_{\ell'}^{m'^*}(\theta,\varphi)$ and integrating over $\cos\theta$ and φ :

$$\text{(C.7)} \qquad \iint d\cos\theta d\varphi(\underline{s}\cdot \underline{gradI}(\underline{x},\theta,\varphi))\mathrm{Y}_{\ell'}^{m'^*}(\theta,\varphi) = -\kappa\rho(\underline{x})(1-\frac{\omega_{\ell'}}{2\ell'+1})I_{\ell',m}(\underline{x})$$

We can see that this new expression noticeably simplifies the right hand side term, however the left hand side part becomes rather complicated, due to the angular dependence in the vector $\underline{s}$. It is possible to express this term as a function of the coefficients $I_{\ell,m}(\underline{x})$ by virtue of the spherical harmonics recurrence formulae. Defining the following coefficients:

$$\text{(C.8)} \qquad a_{\ell,m} = \sqrt{\frac{(\ell+m)(\ell-m)}{(2\ell+1)(2\ell-1)}}$$

$$\text{(C.9)} \qquad b_{\ell,m} = \sqrt{\frac{(\ell+m)(\ell+m-1)}{(2\ell+1)(2\ell-1)}}$$

We can express:

$$\text{(C.10)} \qquad \cos\theta \mathrm{Y}_\ell^{m^*}(\theta,\varphi) = a_{\ell+1,m}\mathrm{Y}_{\ell+1}^{m^*}(\theta,\varphi) + a_{\ell,m}\mathrm{Y}_{\ell-1}^{m^*}(\theta,\varphi)$$

$$\text{(C.11)} \qquad \sin\theta\cos\theta \mathrm{Y}_\ell^{m^*}(\theta,\varphi) = \frac{1}{2}\Big(-b_{\ell,m}\mathrm{Y}_{\ell-1}^{m-1'}(\theta,\varphi) + b_{\ell,-m}\mathrm{Y}_{\ell-1}^{m+1^*}(\theta,\varphi) - b_{\ell+1,m+1}\mathrm{Y}_{\ell+1}^{m+1^*}(\theta,\varphi) + b_{\ell+1,-m+1}\mathrm{Y}_{\ell+1}^{m-1^*}(\theta,\varphi)\Big)$$

$$\text{(C.12)} \qquad \sin\theta\cos\varphi \mathrm{Y}_\ell^{m^*}(\theta,\varphi) = \frac{1}{2}\Big(-b_{\ell,m}\mathrm{Y}_{\ell-1}^{m-1^*}(\theta,\varphi) + b_{\ell,-m}\mathrm{Y}_{\ell-1}^{m+1^*}(\theta,\varphi) - b_{\ell+1,m+1}\mathrm{Y}_{\ell+1}^{m+1^*}(\theta,\varphi) + b_{\ell+1,m+1}\mathrm{Y}_{\ell+1}^{m-1^*}(\theta,\varphi)\Big)$$

Thus the final expression of the radiative transfer equation is:

$$
\begin{aligned}
\text{(C.13)} \qquad & \frac{1}{2}\frac{\partial}{\partial x}\Big[- b_{\ell,m} I_{\ell-1,m-1}(\underline{x}) + b_{\ell,-m} I_{\ell-1,m+1}(\underline{x}) \\
& - b_{\ell+1,m+1} I_{\ell+1,m+1}(\underline{x}) + b_{\ell+1,-m+1} I_{\ell+1,m-1}(\underline{x})\Big] \\
& + \frac{1}{2i}\frac{\partial}{\partial y}\Big[- b_{\ell,m} I_{\ell-1,m-1}(\underline{x}) - b_{\ell,-m} I_{\ell-1,m+1}(\underline{x}) \\
& + b_{\ell+1,m+1} I_{\ell+1,m+1}(\underline{x}) + b_{\ell+1,-m+1} I_{\ell+1,m-1}(\underline{x})\Big] \\
& + \frac{\partial}{\partial z}\Big[a_{\ell+1,m} I_{\ell+1,m}(\underline{x}) + a_{\ell,m} I_{\ell-1,m-1}(\underline{x})\Big] \\
& = -\kappa\rho(\underline{x})(1 - \frac{\omega_\ell}{2\ell+1}) I_{\ell,m}(\underline{x})
\end{aligned}
$$

This expression is still rather complicated and in order to perform simulations with this formula, we need to find other recurrence formulae.

REFERENCES

Barker H.W. and J.A. Davies, 1992, *Solar radiative fluxes for stochastic, scale invariant Broken Cloud Fields*, J. atmos. Sciences, 49, 1115–1126.

Borde R., 1991, *Rayonnement dans les nuages multifractals*, rapport de stage de D.E.A., Universite de Clermont-Ferrand II, France.

Borde R., E. Gougec, C. Moraillon, D. Schertzer, 1993, *Multifractal relationship between cloud and radiation singularities, exact and asymptotic results*, Annales Geophysicae, preprint volume.

Bromsalen G., 1994, *Radiative Transfer in lognormal multifractal clouds and analysis of cloud liquid water data*, MSc, Mcgill University, Montreal (Quebec), Canada.

Byrne R.N., R.C.J. Somerville, B. Subasilar, 1995, *Broken-cloud enhancement of solar radiation absorption*, J. Atmos. Sci., in press.

Cahalan R.F., J. H. Joseph, 1989, *Fractal statistics of cloud fields*, Mon. Wea. Rev., 117, 261–272.

Chandrasekhar S., 1950, *Radiative Transfer*, Oxford University Press, New York. (Reprinted by Dover, New York, 1960).

Cess R.D., M.H. Zhang, P. Minnis, L. Corsetti, E.G. Dutton, B.W. Forgan, D.P. Garber, W.L. Gates, J.J. Hack, E.F. Harrison, X. Jing, J.T. Kiehl, C.N. Long, J.-J. Morcrette, G.L. Potter, V. Ramanathan, B. Subasilar, C.H. Whitlock, D.F. Young, Y. Zhou, 1995, *Absorption of solar radiation by clouds: Observations versus model*, Science, 267, 496–499.

Corrsin S., 1951, *On the spectrum of isotropic temperature fluctuations in an isotropic turbulence*, J.Appl. Phys, 22, 469–473.

Davis A., S. Lovejoy, P.Gabriel, D. Schertzer, G.L. Austin, 1990, *Discrete Angle Radiative Transfer Part III: Numerical results on homogeneous and fractal clouds*, J. Geophys. Res., 95, 11729–11742.

Davis A., S. Lovejoy, D. Schertzer, 1991, *Discrete Angle Radiative Transfer in a multifractal medium*, Ed. V.V. Varadan, SPIE, 1558, 37–59.

Davis A., S. Lovejoy, D. Schertzer, 1992, *Supercomputer simulation of radiative transfer inside multifractal cloud models*, I.R.S. 92, A. Arkin et al. Eds., 112–115.

Evans K. F., 1993, *A general solution for stochastic radiative transfer*, G.R.L., 20, 19, 2075–2078.

Gabriel P., S. Lovejoy, G.L. Austin, D. Schertzer, 1986, *Radiative Transfer in extremely variable fractal clouds*, 6th conference on atmospheric radiation, AMS, Boston, 230–234.

Gabriel P., S. Lovejoy, D. Schertzer, G.L. Austin, 1988, *Multifractal analysis of resolution dependence in satellite imagery*, J. Geophys. Res. 15, 1373–1376.

Gabriel P., S. Lovejoy, A.Davis, D. Schertzer, G.L. Austin, 1990, *Discrete Angle Radiative Transfer Part II: Renormalization approach to scaling clouds*, J. Geophys. Res., 95, 11717–11728.

Kolmogorov A.N., 1941, *Local structure of turbulence in an incompressible liquid for very large Reynolds numbers*, Proc. Acad. Sci. URSS, Geochem. Sect, 30, 299–303.

Kolmogorov A.N., 1962, *A refinement of previous hypothesis concerning the local structure of turbulence in viscous incompressible fluid at high Reynolds numbers.* J. Fluid. Mech. 13, 83, 349.

Lavallée D., 1991,*Multifractal techniques: Analysis and simulation of turbulent fields*, PhD thesis, Mcgill University, Montreal (Quebec), Canada.

Lavallée D., S. Lovejoy, D. Schertzer, P. Ladoy, 1993, *Nonlinear variability of landscape topography, multifractal analysis and simulation*, in Fractals in geography, eds L. de Cola and N. Lam.

Lévy P., 1925, *Calcul des Probabilités*, Gauthier Villars, Paris.

Lovejoy S., D. Schertzer, 1989, *Fractal clouds with discrete Angle Radiative transfer.* I.R.S. 88 Eds. C. Lenoble and J.F. Geylyn, Deepak publishing, 99–102.

Lovejoy S., P. Gabriel, A.Davis, D. Schertzer, G.L. Austin, 1990, *Discrete Angle Radiative Transfer Part I: Scaling and similarity, universality and diffusion*, J. Geophys. Res., 95, 11699–11715.

Lovejoy S., D. Schertzer, 1990, *Multifractals, Universality classes and satellite and radar measurements of cloud and rain fields.* J. Geophys. Res. 95, 2021.

Lovejoy S., D. Schertzer, 1991, *Multifractal Analysis techniques and the rain and cloud fields from* 10^{-1} *to 106m.* In *Nonlinear Variability in Geophysics: Scaling and Fractals*, Kluwer, Schertzer D. and Lovejoy S., 111–144.

Lovejoy S., B. Watson, D. Schertzer, G. Bromsalen, 1995, *Scattering in multifractal media.* Proc of particle transport in Stochastic Media. L. Briggs Ed., American Nuclear Society, Portland, Or., April 30-May 4 1995, 750–760.

Lovejoy S. D. Schertzer, P. Silas, 1995, *Diffusion on one dimension multifractals.* Submitted to Phys. Rev.Lett.

Obukhov A., 1949, *Structure of the temperature field in a turbulent flow*, IZV. Akad. Nauk. SSSR. SER. Geogr. IGeofiz. 13, 55–69.

Obukhov A., 1962, *Some specific features of atmospheric turbulence.* J. Geophys. Res. 67, 3011.

Parisi G. and Frish U., 1985, *A multifractal model of intermittency in turbulence and predictability in geophysical fluid dynamics and climate dynamics*, North holland, Ghil M., Benzi R. and Parisi G., pp 84, 88, pp 111, 144.

Pecknold S., S. Lovejoy, D. Schertzer, C. Hooge and J. F. Malouin, 1993, *The simulation of universal multifractals.* In *Cellular Automata: Prospects in astronomy and astrophysics.* World Scientific, Perdang J. M. and A. Lejeune.

Ramanathan V., B. Subasilar, G.J. Zhang, W. Conant, R.D. Cess, J.T. Kiehl, H. Grassl, L. Shi, 1995, *Warm pool heat budget and shortwave cloud forcing: a missing physics?*, Science, 267, 499–503.

Richardson L.F., 1922, *Weather prediction by numerical processes*, republished by Dover 1965.

Schertzer D., S. Lovejoy, 1987, *Physical modelling and analysis of rain and clouds by anisotropic scaling multiplicative processes*, J. Geophys. Res. D, 92, 8, 9693–9714.

Schertzer D., S. Lovejoy, 1988, *Multifractal simulations and analysis of clouds by multiplicative processes*, Atmos. Res. 21, 337–361.

Schertzer D., S. Lovejoy, 1991, *Nonlinear geodynamical variability: multiple singularities, universality and observables.* In *Nonlinear Variability In Geophysics: scaling and fractals.* Kluwer, Schertzer D. and S. Lovejoy, pp 41, 82.

Schertzer D., S. Lovejoy, 1995, *The multifractal phase transition route to self-organised criticality.* In Physics reports, to appear.

Schertzer D., S. Lovejoy, D. Lavallée, F. Schmitt, 1991, *Universal Hard multifractal Turbulence: Theory and Observations.* In *Nonlinear Dynamics of structures.* World Scientific, Sagdeev R.Z. et al., 213.

Schmitt F., D. Lavallée, D. Schertzer, S. Lovejoy, 1992, *Empirical determination of universal multifractal exponents in turbulent velocity field*, Phys. Rev. Lett., 68, 305.

Schmitt F., D. Schertzer, S. Lovejoy, Y. Brunet, 1993, *Estimation of universal multifractal indices for atmospheric turbulent velocity fields*, Fractals, vol 1, 3, 568–575.

Schmitt F., D. Schertzer, S. Lovejoy, Y. Brunet, 1994, *Empirical Study of Multifractal Phase Transitions in Atmospheric turbulence*, N.P.G., 1, 95–104.

Schuster A., 1905, Astrophys. J., 21, 1.

Schwartzschild, 1906, Göttinger. Nachrichten ,41.

Silas P. et al., 1993, *Single phase diffusion in multifractal porous rock*, proceedings *Hydrofractals* 1993, 1–6.

Silas P., 1994, *Diffusion on one dimensional multifractal*, MSc, Mcgill University, Montreal (Quebec), Canada.

Somerville R.C.J., C. Gauthier, 1994, *Climate-Radiation Feedbacks: The Current State of the Science, in Elements of Change*, Eds. S.J. Hassol and P. Norris, Aspen Global Chang Institute.

Tessier Y., S. Lovejoy, Schertzer D., 1993, *Universal multifractals: theory and observations for rain and clouds.* J. Appl. Meteor. 32, 2, 223–250.

Wilson S., D. Schertzer, S. Lovejoy, 1991, *Physically based modelling by multiplicative cascade processes.* In *Nonlinear Variability in Geophysics: scaling and fractals*, Kluwer, Schertzer D. and S. Lovejoy, 185–208.

THE MORPHOLOGY AND TEXTURE OF ANISOTROPIC MULTIFRACTALS USING GENERALIZED SCALE INVARIANCE

S. PECKNOLD*, S. LOVEJOY* , AND D. SCHERTZER†

Abstract. Although scale invariance is a basic geodynamic symmetry, there is no reason to expect it to be isotropic; that the corresponding fractals or multifractal fields be self-similar. The analysis of various geophysical fields has indeed shown that although they are typically scaling multifractals, they are indeed anisotropic involving differential rotation, stratification and more complex scale changes. The framework for the analysis and simulation of such fields is generalized scale invariance (GSI), which comprises two elements — the generator and the unit ball. Although the generator is more fundamental, the shape of the possibly highly anisotropic unit ball will also affect the texture and morphology of the corresponding fields. While full nonlinear GSI is difficult to examine, linear approximations are always valid over finite ranges of scale; we will therefore focus on the latter. Building on earlier work using linear GSI and second order (convex/elliptical) balls, we establish a method of constructing higher order families necessary for modeling qualitatively different morphologies, especially non-convex ones. We illustrate the method using fourth order polynomials and multifractal simulations.

1. Introduction. Scale invariant dynamics are generally associated with multifractal fields, and indeed over twenty geophysical fields have been explicitly shown to be multifractal (for a summary and intercomparison, see [Lovejoy and Schertzer, 1995] and [Pecknold et al., 1993]). Since scaling is a statistical symmetry and geophysics deals with fields involving large numbers of degrees of freedom, stochastic geodynamical models provide a natural framework for their study. Indeed, stochastic scaling models have many advantages over the usual deterministic ones, not least because they can be valid over arbitrarily large ranges of scale and thus avoid the ad hoc homogeneity and parametrization assumptions that plague standard approaches based on integrations of (nonlinear) partial differential equations representing the dynamics.

Active fields of research into multifractality in geophysics include turbulence in the temperature [Schmitt et al., 1993] and the wind fields [Meneveau et al., 1990; Schmitt et al., 1992; Schmitt et al., 1993;Chigirinskaya et al., 1994; Lazarev et al., 1994], rainfall and cloud fields [Schertzer and Lovejoy, 1985b; Lovejoy et al., 1987; Gabriel et al., 1988; Gupta and Waymire, 1993; Tessier et al., 1993a], topography [Lovejoy and Schertzer, 1990; Lavallée et al., 1993; Lovejoy et al., 1995], ocean surfaces [Tessier et al., 1993b], sea ice [Falco et al., 1995] and the seismic moment field [Hooge, 1993; Hooge et al., 1994]. These multifractal fields are the generic re-

* Department of Physics, McGill University 3600 University Street, Montreal, Quebec, H3A 2T8, Canada.

† Université Pierre et Marie Curie, Bôite 99, 4 place Jussieu, 75252 Paris Cedex 05, France.

sult of multiplicative cascades and exhibit fractal structures, extreme variability/intermittency (including catastrophic events associated with self-organized criticality; [Schertzer and Lovejoy, 1995b]), as well as other realistic features.

In its simplest form, scale-invariance is self-similar; the resulting multifractals are isotropic exhibiting no preferred direction in space. However, no geophysical field is ever exactly isotropic: for example, due to gravity they are almost invariably stratified in the vertical direction. In addition to the vertical, other preferred directions can arise due to the Coriolis or other forces; sometimes they can even be imposed by external boundary conditions (see [Davis et al., 1992] for an example). The result is stratification, differential rotation and/or other scale and orientation dependent features, which due to the scaling are associated with texture and morphology.

In spite of its obvious importance, the study — both theoretical and empirical — of anisotropic scale invariance is still in its infancy. For example, Generalized Scale Invariance (GSI) which is the general theoretical framework for handling scaling anisotropy — and in fact for defining the very notion of scale in such a system — is quite recent [Lovejoy and Schertzer, 1985; Schertzer and Lovejoy, 1985b; Schertzer and Lovejoy, 1987; Schertzer and Lovejoy, 1989; Schertzer and Lovejoy, 1991b]. The primary exception is the simplest scale dependent anisotropy, "self-affinity", which involves only compressions of structures along coordinate axes. Whereas there exists a rather general framework of anisotropy, the basic ingredient being the Lie algebra of the possibly stochastic generator of the scaling anisotropy, little has been done to explore it. So far, work done on the already rather general problem — involving both differential rotation as well as differential stratification — has been the development of two analysis methods for estimating GSI parameters: the "Monte Carlo rotating ellipse technique, [Lovejoy et al., 1992; Pflug et al., 1992; Pflug et al., 1993] and the scale invariant generator technique," [Lewis et al., 1995], as well as the development of both fractal [Lovejoy and Schertzer, 1985], and multifractal [Pecknold et al., 1993] models incorporating GSI.

Outside of the GSI framework, (with the exception of the special case of self- affinity) there have been only a few attempts to deal with anisotropy. A recent one [Veneziano et al., 1995] uses a discrete scale ratio version of linear GSI. The latter involves scaling only over integer powers of this ratio; not over arbitrary scale ratios, furthermore, it cannot be obviously extended to the nonlinear case. Another proposal is the ad hoc introduction of differing scaling exponents varying as functions of direction. This method of handling scaling anisotropy has been independently proposed by several authors [Fox and Hayes, 1985; Van Zandt et al., 1990; Pilkington and Todoeschuk, 1993]. Although at first sight this is appealing, it turns out to be quite incompatible with scaling - it necessarily involves absolute, rather than relative, notions of scale. Thus any underlying dynamics will be fundamentally dependent, rather than independent, of size. The physical motivation for GSI is that it must be the nonlinear dynamics of a system that determines the appropriate notion of scale; the latter should not be imposed from without in an ad hoc manner as is usually the case (for example when Euclidean metrics and self-similarity is assumed). Not too surprisingly GSI has some similarities with (general) relativity in which the distribution of mass and energy determine the metric. However, in GSI the (anisotropic) notion of scale is based on equivalence between different fluxes (e.g; buoyancy fluxes and kinetic fluxes [Schertzer and Lovejoy, 1984; Lazarev et al., 1994]) and therefore is based on measure notions rather than metric ones. Indeed, in GSI, the notion of scale is defined primarily by a scale changing group (with generator $\mathbf{G}$), admitting at least one invariant family of balls B_λ generating a measure of size — and therefore of scale ratio λ — usually taken to be the measure/volume of the balls B_λ and respecting an order relation between these balls (i.e. in an ensemble, metric or measure sense large balls contain smaller balls). It may happen that the unit ball B_1 is or can be chosen as isotropic, and the corresponding scale is therefore called a "sphero-scale," but this is not a general rule.

Although in general, the full (non-linear) GSI will be required to completely describe geophysical fields, over a sufficiently small range of scales one may always approximate this by a linearization; "linear GSI" (this is analogous to the flat-space/special relativity approximation). Therefore, in the following, we limit our discussion to linear GSI in which the generator $\mathbf{G}$ is a matrix. Furthermore, we restrict our attention to two dimensions (mostly horizontal cross-sections are discussed here, although the results also apply to vertical and to space-time cross-sections).

Although the generator is the most important element of GSI since it determines the way that basic structures change with scale, nevertheless, the shape of the unit ball (which in general will itself be anisotropic) also has a large effect on the appearance of the field at a given resolution. In previous empirical work on cloud radiances [Pflug et al., 1991], and on simulations [Pecknold et al., 1993], the simplest linear GSI system was studied in which the frontier of the unit ball (and hence all the balls) were circles or ellipses (i.e. defined by a quadratic form). Empirically, this seemed adequate for handling most clouds (see Figures 1.1a and 1.1b for an example); although certain structures such as cyclones were found to poorly represented. Still other systems such as synthetic aperture radar ice fields (see Figures 1.2a and b) have thin, elongated fault/fissure type structures which cannot be obtained using linear GSI with quadratic balls. It is plausible that these features in the ice field correspond to squarish and nonconvex balls (below we use fourth order balls to model this). In any case, it seems clear that it would be advantageous to go beyond elliptical balls. Other possible generalizations (not explored here) include the maintenance of the scalar framework but use of nonlinear (possibly stochastic) generators. Alternatively — recognizing the vector and tensor nature of the dynamics — a vector/tensor "Lie cascade," [Schertzer and Lovejoy, 1995a] framework could be used. For technical reasons outlined below, the next simplest family of balls of interest here are those defined by a quartic form. Our exploration of these fourth order polynomial balls is significantly more difficult than for the corresponding second order case partially because the parameter space is much larger, but primarily because of the difficulty of choosing balls which define the scale uniquely: imposing the "no crossing" condition. Fourth and higher order polynomials are sufficiently complex that we must develop a fairly general technique in order to handle them. The approach can be summarized as follows: first for simplicity, we limit our study to a subset of fourth order polynomials expressible in terms of combinations of quadratic forms. Using quadratic basis matrices, we then obtain simple expressions for the way the balls change with scale as well as a convenient parametrization which enables their shapes to be readily analyzed. Finally, the crossing condition can be simply expressed.

Once the basic families of balls have been described, we illustrate the typical morphologies and textures by making the corresponding anisotropic multifractal simulations. Although this is necessarily insufficient to describe all possible types of anisotropies that may occur, given that most complex geophysical systems are vector and tensor in nature (rather than being limited to cascades of scalar quantities), we will nevertheless see that many interesting phenomena may be described and modeled in this way.

2. Linear GSI. We will now detail the elements of linear GSI, a special case of nonlinear GSI, the outline and basic results of which were developed in [Schertzer and Lovejoy, 1985b]. The basic ingredient for GSI is the scale changing operator T_λ, which for any given scale ratio $\lambda \geq 1$ reduces balls B_λ of scale ratio λ' (the ratio of the system size to the size of the ball), to balls $B_{\lambda\lambda'}$ of scale ratio $\lambda\lambda'$,

$$T_\lambda B_{\lambda'} = B_{\lambda\lambda'} \tag{2.1}$$

T_λ relates the statistical properties at one scale to another and involves only the scale ratio (there is no characteristic "size"), so that $\underline{x}_{\lambda\lambda'} = T_\lambda \underline{x}_{\lambda'}$, where $\underline{x}_{\lambda\lambda'}$, and $\underline{x}_{\lambda'}$ are vectors on $\partial B_{\lambda\lambda'}$ and $\partial B_{\lambda'}$ (their frontiers), respectively. In general, the family of balls are defined as open subsets of our space that form a basis for the topology of that space [Schertzer and Lovejoy, 1985b]. Here, consider the frontier of the balls (see eq. 2.3), rather than the balls themselves, so that each vector belongs uniquely to one frontier.

To be completely defined, GSI needs more than a scale changing operator; it also requires a definition of the unit scale, i.e. B_1, as well as a definition of how to measure the scale. More precisely, we require:

i) The scale changing operator T_λ which is a one parameter multiplicative group, which we will note: $T_\lambda = \lambda^{\mathbf{G}}$, where $\mathbf{G}$ is the generator of the group[1] In the approximation of linear GSI, $\mathbf{G}$ is an $n \times n$ matrix and

[1] The standard convention of $T_\lambda = \lambda^{-\mathbf{G}}$, due to the physical interpretation of a

$\underline{x}$ is an n-dimensional vector [Lovejoy and Schertzer, 1985; Schertzer and Lovejoy, 1985a]. When $\mathbf{G}$ is a more general (nonlinear) generator, we must define by differential equations using the fact that $\mathbf{G}$ defines infinitesimal transformations.

ii) The unit ball B_1 corresponding to the unit vectors. The balls are other scales are obtained from it by operating with T_λ. With the group properties of T_λ, for any $\lambda_1\lambda_2 = \lambda$, then

$$\begin{aligned} B_\lambda &= T_{\lambda_2} B_{\lambda_1} = T_{\lambda_2} T_{\lambda_1} B_1 \\ &= T_{\lambda_1} B_{\lambda_2} = T_{\lambda_1} T_{\lambda_2} B_1 \end{aligned} \tag{2.2}$$

If an isotropic ball (i.e., circle or sphere) exists, we call the corresponding scale the "spheroscale." In general, B_1 will be defined by an implicit equation:

$$\begin{aligned} B_1 &= \{\underline{x} : f_1(\underline{x}) < 1\}; \\ \partial B_1 &= \{\underline{x} : f_1(\underline{x}) = 1\} \end{aligned} \tag{2.3}$$

Thus,

$$\begin{aligned} &B_\lambda = \{\underline{x} : f_\lambda(\underline{x}) < 1\}, \\ &\forall\ \underline{x} :\ f_\lambda(\underline{x}) = f_1(T_\lambda^{-1}\underline{x}) \end{aligned} \tag{2.4}$$

iii) An increasing function of λ, here given by f_λ; this assures that the frontiers of the balls do not intersect (section 3.4) and that they uniquely define the scale. In [Schertzer and Lovejoy, 1985b], a somewhat more general framework using open balls was proposed (using a measure of the size of the ball rather than the function f).[2] This implies that

$$\frac{\partial f_\lambda}{\partial \lambda} > 0 \tag{2.5}$$

Note that this positivity requirement is necessary to ensure that T_λ corresponds to a scale reduction. Given 2D linear GSI, we may decompose

---

cascade from large to small scales, assumes a reduction in the size of balls with increasing scale ratio. Given $\underline{x}_\lambda = T_\lambda \underline{x}_1$, it is easy [Schertzer and Lovejoy, 1991b; Pflug et al., 1993] to determine that the corresponding scale changing operator $\widetilde{T}_\lambda$ in Fourier space is the one parameter group admitting the transpose of $\mathbf{G}$ as generator:

$$\begin{aligned} &\underline{x}_\lambda = T_\lambda \underline{x}_1; \underline{k}_\lambda = \widetilde{T}_\lambda \underline{k}_1 \\ &T_\lambda = \lambda^{\mathbf{G}} \Rightarrow \{\widetilde{T}_\lambda = \lambda^{-\widetilde{\mathbf{G}}}; \widetilde{\mathbf{G}} = \mathbf{G}^T\} \end{aligned}$$

A reduction in real space thus corresponds to a dilation in Fourier space with the transpose of the generator.

[2] The use of open balls has the advantage that they generate a topology of our space; in addition, the definition of convex balls, which will be useful to us, is based upon the convexity of their non-empty interior.

G into pseudo-quaternion elements [Lovejoy and Schertzer, 1985; Schertzer and Lovejoy, 1985b; Pecknold et al., 1993; Pflug et al., 1993]:

$$\mathbf{G} = d\mathbf{1} + e\mathbf{I} + f\mathbf{J} + c\mathbf{K} \ , \tag{2.6}$$

where

$$\mathbf{1} = \begin{pmatrix} 1 & 0 \\ 0 & 1 \end{pmatrix} \qquad \mathbf{I} = \begin{pmatrix} 0 & -1 \\ 1 & 0 \end{pmatrix}$$
$$\mathbf{J} = \begin{pmatrix} 0 & 1 \\ 1 & 0 \end{pmatrix} \qquad \mathbf{K} = \begin{pmatrix} 1 & 0 \\ 0 & -1 \end{pmatrix}$$

A useful parameter for the description of the overall type of anisotropy present in the system will be given by

$$a^2 = c^2 + f^2 - e^2 \tag{2.7}$$

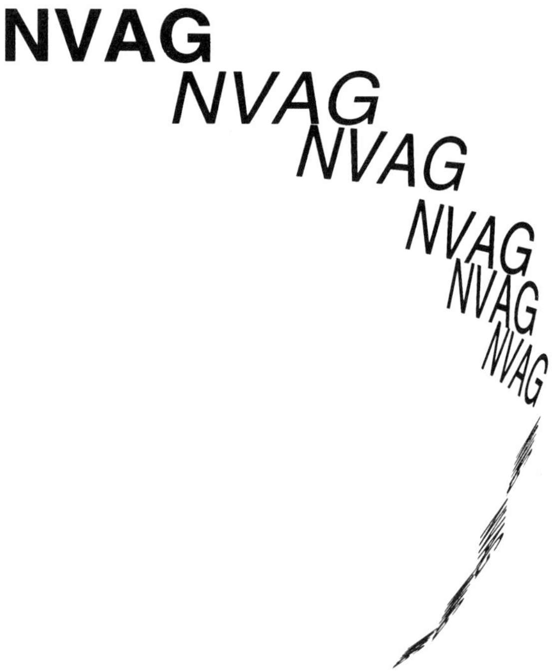

The symbol NVAG is operated upon with an anisotropic scale changing operator, and the result is shown for various scales.

In the case that $a^2 < 0$, we say that the system is rotation dominant; as the scale changes, the balls pass through an infinite degree of rotation (although for a finite total scale ratio, only a finite amount of rotation is possible). If $a^2 > 0$, we call the system stratification dominant; in a like manner, an indefinitely large "stretching" of the unit ball is permitted, with only a limited amount of rotation. The effect of operating on the vectors of a field with the T_λ operator is demonstrated in the figure above. The effect of the parameters chosen ($c = 0.3, f = -0.5, e = 0.8$) is to rotate and compress the word "NVAG" as the scale to which the vector corresponds is reduced. An isotropic scale changing operator would uniformly shrink the word, with the copies converging toward the center of the reduction.

Thus, we are able to describe the type of linear scaling anisotropy present in a 2D system by determining the values of the parameters describing **G**. We are able to reduce these four parameters to three, since the value of the parameter d may be renormalized by a different choice of measurement of scale (i.e. a different function f). Since, for example, we do not expect any overall stratification in horizontal cross-sections of the atmosphere: we may take $d = 1$, implying that scale is proportional to the square root of the area of the corresponding balls. In other cases, such as for example the vertical direction of the atmosphere compared to the horizontal direction, there is an overall stratification. In these cases it is more convenient to use the horizontal dimensions to define distance, and we obtain a nontrivial *elliptical* dimension, $d_{el} : d_{el} = Tr(\mathbf{G})$. Thus the parameter d is half of the elliptical dimension. In the case of three-dimensional atmospheric dynamics, this quantity has been estimated to be $d_{el} = 23/9(= 2.555)$ [Schertzer and Lovejoy, 1985a; Chigirinskaya et al., 1994; Lazarev et al., 1994]. Similarly, in (x, y, z, t) space, rain has been estimated to have $d_{el} = 2.89$ [Lovejoy et al., 1987; Tessier et al., 1993a]. In our case, where we expect no overall stratification, $d_{el} = 2$.

3. Polynomial balls.

3.1. Discussion. In linear GSI, T_λ is simply a linear transformation; we seek a simply connected family of balls which is invariant under linear coordinate transformations, i.e. polynomials. In order to simplify both the analysis and simulation of fields in view of the extensive use of Fourier techniques, we consider even order polynomials with the additional invariance under inversion, $\underline{x} \rightarrow -\underline{x}$. The simplest examples are the quadratic forms, characterized by the equation

$$f_1(\underline{x}) = \underline{x}^T \mathbf{A}_1 \underline{x} = 1 \qquad (3.1)$$

$$\mathbf{A}_1 = \begin{pmatrix} A_{00} & A_{01} \\ A_{01} & A_{11} \end{pmatrix}$$

where $\underline{x} = (x, y)$ is a positive vector on the frontier of the unit ball, and $\mathbf{A}_1$ is a symmetric 2×2 matrix describing the unit ball. The lack of a subscript on the position vectors will henceforth be taken to mean vectors on the frontier of the unit ball unless otherwise specified.

Only ellipses have been discussed since hyperbolae and parabolae do not form closed curves.[3] Hence, we have the constraints $A_{00}, A_{11} > 0$ and $A_{00}A_{11} - A_{01}^2 > 0$. This equation thus describes an ellipse or circle. According to eq. 2.1 a generalized magnification by factor λ of any vector $\underline{x}_1$ of the frontier of the unit ball B_1 satisfies:

$$\underline{x}_\lambda = T_\lambda \underline{x}_1 = \lambda^{-\mathbf{G}} \underline{x}_1 \tag{3.2}$$

Defining $f_\lambda(\underline{x}) = \underline{x}^T \mathbf{A}_\lambda \underline{x}$, we have:

$$\mathbf{A}_\lambda = (T_\lambda^{-1})^T \mathbf{A}_1 T_\lambda^{-1} = \lambda^{-\mathbf{G}^T} \mathbf{A}_1 \lambda^{-\mathbf{G}} \tag{3.3}$$

The case in both linear and nonlinear GSI was discussed extensively in [Schertzer and Lovejoy, 1985b]; see section 3.4.

3.2. A basis of scaling symmetric matrices. To understand the transformation of these matrices, it is convenient (although not indispensable) to decompose $\mathbf{A}_1$ into symmetric (quadratic) basis matrices $\mathbf{m}_1, \mathbf{m}_2$, $\mathbf{m}_3$

$$\mathbf{A}_1 = \sum_{i=1}^{3} \alpha_{i,1} \mathbf{m}_i \tag{3.4}$$

These 3 basis symmetric matrices are defined by being scaling for the T_λ transformation:

$$\lambda^{\mathbf{G}^T} \mathbf{m}_i \lambda^{\mathbf{G}} = \lambda^{\wedge_i} \mathbf{m}_i \tag{3.5a}$$

Hence the differential action of T_λ on $\mathbf{A}$ corresponding to the integral one given by eq. 3.2 is $\lambda \frac{dA_\lambda}{d\lambda} = sym(A_\lambda \mathbf{G})$, this scaling property corresponds to an eigenvalue problem:

$$\wedge_i \, \mathbf{m}_i = \mathbf{G}^T \mathbf{m}_i + \mathbf{m}_i \mathbf{G} = sym(\mathbf{m}_i \mathbf{G}) = \mathbf{m}_i' \tag{3.5b}$$

with

$$\wedge_1 = 2d, \quad \wedge_2 = 2d - 2a, \quad \wedge_3 = 2d + 2a$$

On this basis, we have corresponding to eq. 3.4, $\mathbf{A}' = \text{sym}(\mathbf{AG}) = \sum_i \alpha_i' \mathbf{m}_i$, with $\alpha_i' = \wedge_i \alpha_i$.

Expressing the $\mathbf{m}_i$ using the pseudoquaternion representation for $\mathbf{G}$ (see eq. 2.3 above) yields the following:

$$\mathbf{m}_1 = \begin{pmatrix} e+f & -c \\ -c & e-f \end{pmatrix} \quad \mathbf{m}_2 = \begin{pmatrix} (e+f)^2 & -(e+f)(c+a) \\ -(e+f)(c+a) & (c+a)^2 \end{pmatrix}$$

$$\mathbf{m}_3 = \begin{pmatrix} (e+f)^2 & -(e+f)(c-a) \\ -(e+f)(c-a) & (c-a)^2 \end{pmatrix} \tag{3.6}$$

[3] The latter can still have a well-defined interior and may ultimately find some application.

With these matrices[4], $\mathbf{A}_\lambda$ are then just given by:

$$\mathbf{A}_\lambda = \lambda^{\mathbf{G}^T}\mathbf{A}_1\lambda^{\mathbf{G}} = \sum_i \alpha_{i,\lambda}\mathbf{m}_i \tag{3.7}$$
$$\alpha_{i,\lambda} = \alpha_i\lambda^{\Lambda_i}$$

For simplicity of notation, in what follows the second (scale ratio) subscript on the parameters $\alpha_1, \alpha_2, \alpha_3$ will be suppressed for the case of the unit matrix $\mathbf{A}_1$.

Of course when $a^2 < 0$, a is imaginary and the above representation for $\mathbf{m}_2, \mathbf{m}_3$ will no longer be useful since it will be complex. In order to deal with this rotation dominance case, note that $\mathbf{m}_2, \mathbf{m}_3$ are complex conjugates (as are Λ_2, Λ_3), and introduce new basis matrices $(\mathbf{m}_4, \mathbf{m}_5)$ which are respectively the real and imaginary parts of $\mathbf{m}_2$:

$$\mathbf{m}_4 = \begin{pmatrix} (e+f)^2 & -(e+f)c \\ -(e+f)c & c^2 - |a|^2 \end{pmatrix} \quad \mathbf{m}_5 = \begin{pmatrix} 0 & -(e+f)|a| \\ -(e+f)|a| & 2c|a| \end{pmatrix} \tag{3.8}$$

and correspondingly

$$\begin{pmatrix} \alpha_{4,\lambda} \\ \alpha_{5,\lambda} \end{pmatrix} = \lambda^{-2d}\begin{pmatrix} \cos 2|a|u & \sin 2|a|u \\ -\sin 2|a|u & \cos 2|a|u \end{pmatrix}\begin{pmatrix} \alpha_4 \\ \alpha_5 \end{pmatrix}$$

where $u = \log\lambda$, and the scale ratio subscript is again suppressed here and henceforth on the unit scale parameters. Note that in this basis,

$$\alpha_4' = 2d\alpha_4 + 2|a|\alpha_5$$
$$\alpha_5' = -2|a|\alpha_4 + 2d\alpha_5$$

Since we will require both the scale changing properties of $\mathbf{A}$ and $\mathbf{A}'$, this quadratic basis is very convenient.

3.3. Elliptical coordinates. In order to understand the action of T_λ on the polynomial balls, it is now useful to perform a change of variables $(x, y) \to (r', \phi)$. Details are to be found in Appendix A. The change into the resulting elliptical coordinates gives the following simplifications for the second order polynomials. In the case of stratification dominance, $a^2 > 0$, we obtain:

$$r'^2 \sin 2\phi = \underline{x}^T\mathbf{m}_1\underline{x}$$
$$r'^2(1 - \cos 2\phi) = \underline{x}^T\mathbf{m}_2\underline{x} \tag{3.9}$$
$$r'^2(1 - \cos 2\phi) = \underline{x}^T\mathbf{m}_3\underline{x}$$

[4] The eigenvalue type problem (eq. 3.5b) has unique eigenvalues but not unique eigenvectors, so that the $\mathbf{m}_i$ are not unique. An equally good set is $\mathbf{s}_u \propto (\mathbf{G} - (d-\mu)\mathbf{1})^T(\mathbf{G} - (d-\mu)\mathbf{1}), \mu = \pm a; \mathbf{s}_\mu \propto (\boldsymbol{G} - d)^T\boldsymbol{I}, \mu = 0$.

Where r' is the "elliptical radius" (the curve $r' =$ constant is an ellipse), ϕ the polar elliptical angle. The corresponding quadratic forms are thus given by

$$\underline{x}^T \mathbf{A}_\lambda \underline{x} = r'^2(\alpha_{2,\lambda} + \alpha_{3,\lambda} + (\alpha_{2,\lambda} - \alpha_{3,\lambda})\cos 2\phi + \alpha_{1,\lambda}\sin 2\phi) \tag{3.10}$$

while for the case of rotation dominance a different elliptical coordinate system is used, yielding:

$$\begin{aligned} r'^2 &= \underline{x}^T \mathbf{m}_1 \underline{x} \\ r'^2 \cos 2\phi &= \underline{x}^T \mathbf{m}_4 \underline{x} \\ r'^2 \sin 2\phi &= \underline{x}^T \mathbf{m}_5 \underline{x} \end{aligned} \tag{3.11}$$

and the corresponding quadratic form:

$$\underline{x}^T \mathbf{A}_\lambda \underline{x} = r'^2(\alpha_{1,\lambda} + \alpha_{4,\lambda}\cos 2\phi + \alpha_{5,\lambda}\sin 2\phi) \tag{3.12}$$

Since the coefficients $\alpha_{i,\lambda}$ transform trivially with scale, this representation is convenient: the above clearly describes a sinusoidal variation of the contour around the basic ellipse given by $r' =$ constant. Higher order polynomials obtained by using these quadratic forms as building blocks will also be straightforward to express, involving higher order (Fourier) terms in ϕ.

3.4. The non-intersection condition. Every vector over which the field is defined must be associated with only one scale, that is, it must have only one size. A further set of constraints is thus placed on the GSI and unit ball parameters, by the requirement that the balls do not intersect at different scales (i.e. that equation 2.5 holds and the notion of "scale" is uniquely defined. In determining the constraints imposed by this non-crossing condition, it is only necessary to show non-intersection at a given scale. Thus, if we consider the balls defined by

$$\partial B_\lambda = \{\underline{x} : \ f_\lambda(\underline{x}) = 1\} \tag{3.13}$$

with, similarly to eq. 3.1,

$$f_\lambda(\underline{x}) = \underline{x}^T \mathbf{A}_\lambda \underline{x} \tag{3.14}$$

Then the condition that there be no crossing is simply that

$$\frac{\partial f_\lambda(x)}{\partial \lambda} \propto \underline{x}^T \mathbf{A}'_\lambda \underline{x} > 0 \tag{3.15}$$

In this case the non-crossing condition is straightforward; a spectral (eigenvalue) approach shows that it suffices that the real part of the eigenvalues of $\mathbf{A}' = \mathrm{sym}(\mathbf{AG})$ be positive [Schertzer and Lovejoy, 1985b]. Furthermore, since $\underline{x}^T \mathbf{A}'_\lambda \underline{x} = (\lambda^{\mathbf{G}} \underline{x})^T \mathbf{A}_1 (\lambda^{\mathbf{G}} \underline{x})$, if the condition holds at one scale it will hold any other. It is instructive however to derive this result using the quadratic basis matrices in elliptical coordinates; from (3.9) and (3.11), we find by inspection that the above condition for non-crossing corresponds to:

$$\begin{aligned} &\frac{\alpha_1^2}{\alpha_2\alpha_3} < 4\left(1-\frac{a^2}{d^2}\right); \qquad \alpha_2(d-a)+\alpha_3(d+a)>0; \quad a^2>0 \\ &\frac{\alpha_1^2}{\alpha_4^2+\alpha_5^2} > \left(1+\frac{|a|^2}{d^2}\right); \qquad \alpha_1>0 \qquad a^2<0 \end{aligned} \tag{3.16}$$

which involves both the GSI parameters and the values of α_i. Note that the critical non-crossing conditions here are indeed scale-invariant; hence, if there is no crossing at a given scale, there will be no crossing at any other scale.

We may simplify in the case where a spheroscale exists. A positive spectrum of sym $(\mathbf{AG})$ in the pseudo-quaternion representation with:

$$\mathbf{A} = \delta\mathbf{1} + \varphi\mathbf{J} + \gamma\mathbf{K} \tag{3.17}$$

$$sym(\mathbf{AG}) = D\mathbf{1} + F\mathbf{J} + C\mathbf{K}$$

is obtained if $D > 0,\ D^2 > C^2 + F^2$ where:

$$\begin{aligned} D &= \varphi f + \gamma c + \delta d \\ F &= \delta f - \gamma e + \varphi d \\ C &= \varphi e + \delta c + \gamma d \end{aligned} \tag{3.18}$$

If a scale exists at which $\mathbf{A} = \mathbf{1}$ we obtain a circle; this scale is the spheroscale, and when it exists the restriction is only on the GSI parameters:

$$d > 0;\ d^2 > c^2 + f^2 \tag{3.19}$$

4. Simulations.

4.1. Universal multifractals. We will be interested in simulations of anisotropic multifractals defined by fourth order polynomial balls. Although here we are applying the preceding discussion to the simulation of multifractals, it is equally applicable to monofractals (e.g. generalizations of Brownian motion). We will be particularly interested in the "universal' multifractals. These are the stable, attractive result of nonlinear mixing/interacting multifractal processes. These multifractal fields ε_λ have generators $\log\varepsilon_\lambda$ which are Lévy noises with parameter α. The scaling of the statistical moments of the field, i.e. their dependence upon the scale ratio λ, is determined by the $K(q)$ function, defined as:

$$\lambda^{K(q)} = \langle \varepsilon_\lambda^q \rangle \tag{4.1}$$

where the brackets indicate statistical averaging. In the case of universal multifractals, the (scale-independent) $K(q)$ function is given by:

$$K(q) = \frac{C_1}{\alpha - 1}(q^\alpha - q) \tag{4.2}$$

given that $\alpha \neq 1$, where $0 < \alpha \leq 2$. C_1 is the codimension of the mean of the field, characterizing the sparseness of the mean, and the Lévy index α characterizes the degree of multifractality [Schertzer and Lovejoy, 1987]. As $\alpha \to 0$, $K(q)$, becomes linear and we obtain the monofractal β-model [Novikov and Stewart, 1964; Yaglom, 1966; Mandelbrot, 1974; Frisch et al., 1978]; $\alpha = 2$ corresponds to the "lognormal model (with a Gaussian generator).

For nonconservative fields [Schertzer and Lovejoy, 1987; Schertzer and Lovejoy, 1991a], such as density fields ρ related to a conservative multifractal field ε_λ in a scaling manner ($|\Delta\rho_\lambda| = \varepsilon_\lambda \lambda^{-H}$), there is a third fundamental parameter, H, which is a measure of the degree of (scale by scale) non-conservation of the field:

$$\langle|\Delta\rho_\lambda|\rangle = \lambda^{-H} \tag{4.3}$$

The value of H can be determined from the scaling of the power spectrum and the moment scaling function. $H = 0$ for conservative multifractals which are the direct outcome of multiplicative cascade processes. In general H is the degree of (fractional) integration required to obtain the observed field from the conservative field; for example if ε is the cube root of the energy flux in turbulence and ρ the velocity field, then $H = 1/3$.

5. Fourth-order polynomials.

5.1. Discussion. We now wish to extend our discussion of quadratic balls to higher order polynomial balls, with the particular intent of generating non-convex and non-elliptical shapes which are impossible with quadratic balls. Taking the fourth order case as our example, the general form of our fourth order polynomials involves eight parameters (the coefficients of $x^4, y^4, x^2y^2, x^3y, y^3x, x^2, y^2, xy$). Our goal is not to study the most general case; this is unnecessarily complex. It is involves solutions to general quartic equations which are difficult to deal with, and in any case will not always yield closed curves. Rather, we seek a subset of the fourth-order polynomials which is broad enough to display the qualitative features of interest while simultaneously remaining easy to handle empirically and theoretically. It is convenient to consider the family defined by

$$f_1(\underline{x}) = (\underline{x}^T \mathbf{A}_1 \underline{x})^2 + (\underline{x}^T \mathbf{B}_1 \underline{x})^2 = 1 \tag{5.1}$$

where $\mathbf{A}_1, \mathbf{B}_1$ are symmetric 2×2 matrices, and the subscript on the vector $\underline{x}$ is again suppressed as it is on the unit ball. This is a closed family under the GSI transformation which has the additional advantage that the basic characteristic shapes can be understood from the relation of $\mathbf{A}_1, \mathbf{B}_1$ to the usual quadratic forms (e.g. if det $\mathbf{A}_1 > 0$, we have an ellipse, det $\mathbf{A}_1 < 0$, a hyperbola etc.). This gives a six-parameter family of curves, with parameters $\alpha_1, \alpha_2, \alpha_3, \beta_1, \beta_2, \beta_3$ (or $\alpha_1, \alpha_4, \alpha_5, \beta_1, \beta_4, \beta_5$ for rotation dominant cases), satisfying the previous relations (3.2)–(3.10). The α parameters are defined by eq. (3.4) and depend on the $\mathbf{A}_1$ matrix, while the β parameters are analogous to these for the $\mathbf{B}_1$ matrix. We note that the previously considered family of quadratic balls is included in this family as a subset, with either α_i or β_i being all zero (the only new effect of the additional solutions for the square in this case would be to render the quadratic hyperbolic cases closed; these cases are not very interesting, however, since — except in the trivial case of isotropic reductions — they necessarily cross. The remaining elliptical shapes are identical to those in the previously considered quadratic case.) It remains to be seen which of these have $\partial f_\lambda / \partial \lambda > 0$, i.e. have uniquely defined scales and correspond to reductions when operated on by T_λ. Applying condition 2.5 to the function f defined in eq. 5.1, we obtain the following condition:

$$\frac{\partial f_\lambda(\underline{x})}{\partial \lambda} \propto (\underline{x}^T \mathbf{A}_\lambda \underline{x})(\underline{x}^T \mathbf{A}'_\lambda \underline{x}) + (\underline{x}^T \mathbf{B}_\lambda \underline{x})(\underline{x}^T \mathbf{B}'_\lambda \underline{x}) > 0 \tag{5.2}$$

One way to proceed would be to directly extend the results of the quadratic case; i.e. to require that each factor in the sum above is always positive. We note that for $d > 0$, $d^2 > a^2$, (or equivalently, if $\mathrm{Re}[\sigma(\mathbf{G})] > 0$, where $\sigma(\mathbf{G})$ is the spectrum of $\mathbf{G}$),and if $\underline{x}^T \mathbf{A}' \underline{x} > 0 \ \forall \ \underline{x}$, then $\underline{x}^T \mathbf{A} \underline{x} > 0 \ \forall \ \underline{x}$. We can thus extend the quadratic case using a direct spectral approach (i.e. considering the sign of the eigenvalues of sym($\mathbf{AG}$) and sym($\mathbf{BG}$)). The benefit of this spectral approach is that it is valid in any dimension of space, not requiring special basis matrices. In this case, though, since positivity of each term is somewhat unnecessarily restrictive (it is a sufficient but not necessary condition), this will lead only to a subset of noncrossing balls, in fact generally different from the ones found below. Since it involves general quartic equations, it does not yield a simple criterion for describing the shapes of the balls, as discussed in section 5.4. Alternatively, the spectral approach can in principle be extended for non-positive spectra of $\mathbf{A}$ and $\mathbf{B}$, or more generally it could be based on quartics defined by fourth rank tensors (rather than the second rank tensors used here); this will be developed in a subsequent paper. In order to obtain a manageable family of non-convex balls, we therefore use a different approach by considering sub-classes (corresponding to subgroups of the transformation of f) easily obtained from the quadratic basis matrices. These sub-classes can be readily obtained by setting the various α's and β's $= 0$. Noting that the cases where $\mathbf{A} = 0$, or $\mathbf{B} = 0$, or all α's but one, all β's but one $= 0$ reduce to convex quadratic balls, we consider four parameter cases with $\alpha_1 = \beta_1 = 0$[5]. All other four-

[5] In terms of determinants the case $a^2 < 0$, $\alpha_1 = \beta_1 = 0$ corresponds to det $\mathbf{A} <$

parameter cases either cross or lead to no appreciable simplification with respect to the full six parameter problem. We note that the four-parameter space for defining unit balls involves only three-parameter families of balls, since any unit ball enlarged by T_λ can be taken as a unit ball. A specific family of balls can now conveniently be represented by defining vectors $\underline{r}_2 = (\alpha_2, \beta_2), \underline{r}_3 = (\alpha_3, \beta_3), (\underline{r}_4 = (\alpha_4, \beta_4), \underline{r}_5 = (\alpha_5, \beta_5)$ in the rotation dominant case). We now define θ to be the angle between the vectors, r the ratio of their magnitudes, and $R^2 = \sqrt{(\alpha_{2,1}^2 + \beta_{2,1}^2)(\alpha_{3,1}^2 + \beta_{3,1}^2)}$ for the stratification dominant case and $R^2 = r_{4,1}^2 + r_{5,1}^2$ for the rotation dominant case as scale invariant measures of magnitude of the vectors (see appendix A).

For stratification dominance, as we change scales, r is proportional to λ^{-4a}, and θ is constant, hence the vectors $\underline{r}_2, \underline{r}_3$ are simply isotropically reduced. In polar (r, θ) space the system evolves along a ray. Furthermore, using the elliptical coordinates, as we change scales, the necessary and sufficient no-crossing conditions are simply that[6]

$$\begin{aligned} &d \geq a \\ &\cos\theta \geq 0 \end{aligned} \tag{5.3}$$

For rotation dominance, we find that the expression:

$$\frac{\sin\theta}{r + r^{-1}}$$

is scale invariant. If we again take (r, θ) as polar coordinates, then the constancy of the above expression defines a circle which is traced out as the system is reduced. Furthermore, necessary and sufficient conditions such that $\partial f_\lambda / \partial \lambda > 0$ (no crossing occurs) are that the vectors $\underline{r}_4$ and $\underline{r}_5$ must be chosen such that:

$$\begin{aligned} &d \geq |a| \\ &\frac{\sin\theta}{r + r^{-1}} \geq \frac{|a|}{2\sqrt{d^2 + |a|^2}} \end{aligned} \tag{5.4}$$

If we take an initial vector ($\underline{r}_4$ say) as a unit vector lying in the x direction, then the allowed (noncrossing) vectors $\underline{r}_5$ will lie within a circle radius $d/|a|$ centred on the y axis at a distance $\sqrt{1 + (d/|a|)^2}$ from the origin. These conditions are scale invariant: it is necessary and sufficient that it be respected at an arbitrary scale. Examples of various possible balls are given in Figures 2 and 3. Note that the requirement that $d \geq |a|$ in both cases effectively restricts us to generators not very far from the

0, $\det \mathbf{B} < 0$ i.e. this family is disjoint to the corresponding one obtained by the above (spectral) sufficiency condition.

[6] Note that for $\alpha_1 = \beta_1$, this condition is equivalent to $\det \mathbf{A} + \det \mathbf{B} > 0$.

identity; however empirical estimates of $|a|$ in various geophysical fields (clouds, topography, ice) show that this does not seem to be a serious practical restriction.

Now that we have found noncrossing quartics, it is useful to write them in more familiar form. To start with, note that in the pseudo-quaternion representation (eq. 3.17) the condition $\alpha_1 = 0$ implies (for any value of a^2) that:

$$e\delta = f\gamma - c\varphi \tag{5.5}$$

It is not too hard to show that the condition $\alpha_1 = 0$ remains valid under any linear coordinate transformation. Hence for $a^2 > 0$, any matrices $\mathbf{A}, \mathbf{B}$ which satisfy $\alpha_1 = 0, \beta_1 = 0$ can be put in the form:

$$f = Lu^4 + Mu^2\nu^2 + N\nu^4 \tag{5.6}$$

with

$$\begin{aligned} u &= r' \sin 2\phi \\ \nu &= r' \cos 2\phi \end{aligned} \tag{5.7}$$

Using the above conditions on the α's and β's we require $L > 0$, $N > 0$, and $\sqrt{LN} \geq M > 0$, then no crossing occurs and $\partial f/\partial\lambda > 0$.

A similar expression of the rotation dominant case ($a^2 < 0$) may be obtained; in this case we have

$$f = P(u^4 + \nu^4) + Qu\nu(\nu^2 - u^2) + Ru^2\nu^2 \tag{5.8}$$

with the no crossing conditions becoming $P > 0$, $R > 0$, $4P(R+2P) > Q^2$ and:

$$\frac{\sqrt{4P(R+2P) - Q^2}}{R + 2P} \geq \frac{|a|}{2\sqrt{d^2 + |a|^2}} \tag{5.9}$$

5.2. Stratification dominant examples. Figure 2 shows a series of stratification dominant balls for varying GSI parameters and unit-ball parameters. Figure 4 shows the corresponding real fields. In all these cases, as well as in the examples described below, the fields are simulated using the techniques described in Pecknold et al. (1993), with $\alpha = 1.7, C_1 = 0.1$, and $H = 0.5$. These parameters correspond to those determined for landscape topography [Lavallée et al., 1993], and were chosen primarily for ease of visualization — the codimension of the mean, C_1, is low, thus the mean field is not too sparse, and a moderate degree of smoothing is given by the value of H chosen. The GSI parameters chosen for the stratification dominant examples all gave similar values for a^2, (about 0.4), and display fields where the c parameter dominates, where the f parameter dominates, and where c, f, and e are similar. We note that "typical" values found in cloud data are of about this size [Pflug et al., 1993], as the parameters c and f are generally limited as in equation (2.7), and the e parameter is typically also less than 1.

The equation for our balls in the stratification dominant case where $\alpha_1 = \beta_1 = 0$ is given by:

$$r'^{-4}(\phi) = \lambda^{-4d} R^2 \left(\frac{3}{2} \left(r + \frac{1}{r} \right) + \cos\theta + 2 \left(\frac{1}{r} - r \right) \cos 2\phi + \left(\frac{1}{2} \left(r + \frac{1}{r} \right) - \cos\theta \right) \cos 4\phi \right) \tag{5.10}$$

Here, the parameters R and θ are scale-invariant. The scale dependence arises from the λ^{-4d} dependence of r, as well as the λ^{-4d} overall dependence. We may also note that the $\cos 2\phi$ term is antisymmetric under the exchange $r \rightarrow r^{-1}$. The other terms are symmetric under this exchange, which corresponds to magnifying rather than reducing scales. We can observe that, with the choice of $\cos\theta = 1$, the higher order terms vanish as $r \rightarrow 1$, yielding ellipses at the appropriate scale. The choice of $\cos\theta = 0$ yields a system with the greatest degree of variation from the elliptical shape, as we see in section 5.4. The systems which are non-convex, i.e. those whose $\cos 4\phi$ term is large in comparison to the angle-independent term, are disallowed, as they result when $\cos\theta$ is negative.

Figure 2.1 shows the effect of having a large R^2 parameter; we see that due to the dominance of the α_2 term for $a > 0$ (Eq. 3.7), for virtually all scales the contours are nearly straight. This effect is also to be noticed on fields with the same type of generator: except at the largest scales (corresponding to the small balls), the fields generated with this are extremely stratified (see Figure 4.1). Even when this is not the case, very 'streaky' fields are characteristic of the narrow rectangular small contours on 2.5 and 2.6 (corresponding to simulated fields 4.5 and 4.6).

Numbers 2.2–2.3 and 2.4–2.5 and the corresponding fields in Figure 5 show the effect of changing the initial value of r (i.e. $\lambda = 1$) on the shape of the balls, which is primarily one of increasing the apparent stratification of the field. The shape of the balls remains similar even with the changes in r and in the GSI generator. The structures in the corresponding fields seem to be primarily affected in the mid-range of scales: the small scale microstructures retain a similar appearance, as they correspond to the higher-frequency components of the balls in Fourier space, which are similar.

Additionally, Figures 2.7–2.8 and 4.7–4.8 together with 2.2 and 4.2 show the effect of changing θ. We note that, as we see in section 5.4, the range of scales over which the system is "flattened" or "squirish" decreases as θ decreases.

5.3. Rotation dominant examples. Using the change of variables as in the second order case, we find that in r', ϕ coordinates the lines $r' =$ constant are again ellipses. Taking the special case with $\alpha_1 = \beta_1 = 0$, we eventually obtain the following for our balls:

$$r'(\phi)^{-4} = \lambda^{-4d}\frac{R^2}{2}(1 + R'\cos(4\phi - \zeta)) \tag{5.11}$$

where:

$$R'^2 = 1 - \left(\frac{2\sin\theta}{r + r^{-1}}\right)^2$$

$$R^2 = \alpha_{4,1}^2 + \alpha_{5,1}^2 + \beta_{4,1}^2 + \beta_{5,1}^2$$

and:

$$\tan\zeta = \frac{2\cos\theta}{r^{-1} - r}$$

where $r = |\underline{r}_5|/|\underline{r}_4|$ as before, and θ is the angle between $\underline{r}_5$ and $\underline{r}_4$. Thus this is an oscillation about an ellipse whose amplitude and phase depends on the ratio r, and angle θ. We note that as the scale ratio changes, R' is invariant, as is R; hence the basic anisotropy is fixed. Only the phase ζ changes, $\zeta \to \zeta - 4|a|u$, and the isotropic change of scale of r', given by the λ^{4d} term.

The anisotropy of this fourth order effect is determined by the ratio of the difference in the maximum and minimum r'^{-4} to the average R'; i.e the maximum anisotropy (the smaller the angle θ, the nearer $r(\lambda = 1)$ to 1), about the basic ellipse is obtained with parameters α, β which are directly limited by the crossing condition; the latter is a direct limitation of the amount of anisotropy possible about the basic ellipse. We find, however, in section 5.4, that the basic shape for rotation dominance is in fact (depending on R'), quite nonconvex, rather than elliptical.

Figure 3 shows the balls for varying unit ball parameters as well as for varying GSI generators, while Figure 5 shows the corresponding fields (simulated as above). The GSI parameters here are chosen to show slight rotation (Figs. 3.1–3.7) and a large degree of rotation dominance (Figs. 3.8, 3.9). Characterizing the curves as in Appendix A, we see that the maximum anisotropy about the ellipse is given for small θ and for $r(\lambda = 1)$ close to 1. This, as well as the observation that the basic anisotropy does not change for changing scale ratios, is confirmed by examining Figures 3.1 to 3.6. We note that all of these are nonconvex. In the rotation dominant case, however, the basic shape of the unit ball appears to depend more on the GSI generator than in the stratification dominant case, as may be noted by comparing Figure 3.8 to Figure 3.1. The differences in the balls are not as noticeable between 3.7 and 3.9; in this case, with $\theta = 1.5$ (that is, nearly $\pi/2$), and R' is near 0 (See Eq. 5.11), and the fields are convex (almost elliptical in appearance). Nonetheless, this difference is clearly noticeable in the simulated fields. The less strongly rotating field, Figure 5.7, is more similar in texture to the other fields having the same generator, although the cross-hatching effect is less extreme than in the cases with the largest anisotropy. Figure 5.9 on the other hand is similar to Figure 5.8, and is more reminiscent of the ice field of Figure 1.2, with large bodies (floes) separated by the thinner fissure-like areas.

5.4. Characterization of the shapes. We now consider the morphology/shapes of the balls defined in the previous sections. Let us first discuss the interesting possibility of obtaining nonconvex closed fourth order curves. Consider the curve given by

$$I(\phi) = r'^{-4}(\phi) = (a + b\cos 2\phi + c\cos 4\phi) \tag{5.12}$$

where $a > 0$, $c > 0$, and a, b, c constrained so that $I > 0$. A convex ball will have a positive radius of curvature for all ϕ; a nonconvex ball will have a negative radius for at least some ϕ. The condition for convexity is:

$$I(4I + I'') - \frac{3}{16}I'^2 > 0 \tag{5.13}$$

Hence the direction of curvature changes only if this inequality is violated. Noting that $I > 0, I'^2 > 0$, it is obvious that a sufficient condition for nonconvexity is:

$$4I + I'' = a - 3c\cos 4\phi < 0 \tag{5.14}$$

i.e. $c > a/3$. Furthermore, when $b = 0$ (as in the rotation dominance case, eq. (5.11), the above condition is both necessary and sufficient and implies that if the scale-independent parameter $R' > \frac{1}{3}$, the balls are nonconvex; otherwise, they are convex. In the stratification dominant case the condition is that $\cos\theta < 0$: these crossing balls will be nonconvex. Further calculation shows conversely that all of the noncrossing cases ($\cos\theta > 0$) are convex. The case with $\cos\theta = 0$ corresponds to the existence of points of zero curvature at all scales (the curves also have points of osculation): nevertheless there is no change of sign of the curvature, and the balls are convex.

In these convex stratification dominant family of balls, there is nonetheless an interesting change in shape which occurs over a finite range of scales. Consider the extrema of the $r'(\phi)$ curves (as long as the Jacobian of the transformation of $DI/d\phi \rightarrow dr/d\theta$ - where r, θ are the usual polar coordinates - is non-zero it is sufficient to consider the number of zeroes of $dI/d\phi$). Here, the zeroes of I' are given by:

$$\sin 2\phi = 0, \ \cos 2\phi = -b/4c \tag{5.15}$$

Since for $0 \leq \phi < 2\pi$ the former equation always has four solutions. Elliptical balls will have two maxima and two minima. However, if $|b| < 4c$

then the system will have a total of 8 solutions, four maxima and four minima. This occurs when:

$$\cos\theta < r < \frac{1}{\cos\theta} \tag{5.16}$$

Thus, since r changes with scale change as λ^{-4a}, the system will pass through a set of scales at which it is flattened from the basic elliptical shape, becoming a "square ellipse" (see e.g. Fig. 2.5). The extreme cases of $\cos\theta = 0$ and $\cos\theta = 1$ are respectively flat and elliptical over the entire range of scales.

6. Conclusions. We have examined several subclasses of the particular family of fourth-order ball defined by Equations 5.1, 5.4 and 5.5, which when used together with the formalism of generalized scale invariance describe a number of interesting textures which may be relevant in various geophysical fields. One of the motivations for investigating this type of unit ball was to explain and examine the non-convex and other non-elliptical Fourier space contours noted in the case of ice fields, corresponding to the fissures and floes that characterize these physical fields. We have found that certain similar types of structure have been noticed in cases of rotation dominance, and that given the extension of these unit-ball families described in Section 5.1, parametrizations may be found for these structures in stratification dominant cases as well. These may not correspond exactly to the parameters in analyzing the ice data. Indeed, the work described here was in part motivated by the need to obtain empirical estimates of fourth order unit balls which would respect the non-crossing condition. There are some difficulties in analyzing the ice data for the shape of its unit ball (due in part to the large spread in the Fourier space contours of a real field; this is noise due to the fact that we analyze a single realization). Analysis procedures for these types of balls have been developed, but results as yet are inconclusive.

It has been demonstrated as well that the nature of the unit ball has great bearing on the morphology of the resulting field, in some cases apparently more so than the GSI generator; for example, it has been shown that given different parameters describing the unit ball, rotation dominant and stratification dominant fields can look similar over the finite range of scales we observe. Of course, it is impossible to choose the unit ball entirely independently of the GSI parameters: the generator of the field constrains the types of unit balls that are possible and still respect the non-crossing condition.

Within the restriction to fourth order balls, the question of course arises as to whether the examples that have been considered here are sufficiently general and inclusive, i.e. whether other useful qualitatively different fourth order polynomial balls exist. The full parameter space we are dealing with here is of course very large. In fact, an obvious step would be to increase the number of parameters by extending this to higher order polynomials;

the method outlined here readily lends itself to generalizations. However, this would seem to be unnecessary. The type of shapes expected from our fourth order polynomial parametrization of the unit ball are limited to fourth order oscillations about the second order (elliptical) shape. The sum of the pair of fourth order polynomials describing our unit ball thus allows (in the rotation dominant case) for nonconvex balls, square-like balls, and the somewhat elliptical balls (with or without some of the other types of effect). All these qualitative types of behaviour have been noted. Further increasing the order of the polynomial used to describe the unit ball would also seem to be unwarranted — the addition of higher order oscillations about the basic shape (limited by non-crossing conditions) would not seem to justify the increase in complexity. We expect that the families developed here will be useful complements to the elliptical families already available using quadratic balls; it is significant that both are obtained as special cases of the same quartic family described by eq. 5.1.

Further work, then, would entail developing more robust analysis techniques to determine the unit ball parameters of interesting fields with a greater degree of confidence. The basic method developed here could conceivably be extended to allow for nonlinear GSI; certainly, the application of non-crossing conditions to the unit balls and the resulting limitations on their shapes is necessary for nonlinear generators as well. Additionally, the idea of extending multiplicative cascades to vector and tensor quantities (i.e. Lie cascades [Schertzer and Lovejoy, 1995a]) rather than the scalar quantities currently assumed is an obvious and necessary one, given that the physics underlying these problems is vector and tensor in nature.

A. Appendix: Characterizing the curves.

Rotation dominant case: $a^2 < 0$:

A convenient characterization of the curves is given by the following change of variables:

$$\text{(A.1)} \qquad \begin{aligned} r' \cos\phi &= cy - (e+f)x \\ r' \sin\phi &= |a|y \end{aligned}$$

The lines r' = constant are ellipses with the following major and minor axes:

$$\text{(A.2)} \qquad \frac{1}{\sqrt{(e+f)(e-\sqrt{c^2+f^2})}}, \quad \frac{1}{\sqrt{(e+f)(e+\sqrt{c^2+f^2})}}$$

with axes turned through an angle:

$$\text{(A.3)} \qquad \tan^{-1}\left(\frac{c}{-f+\sqrt{c^2+f^2}}\right)$$

The anisotropy of the ellipses can be characterized by the difference in the squares of the axes divided by the sum of the squares:

$$\text{(A.4)} \qquad \frac{\Delta(\text{squares})}{\sum(\text{squares})} = \frac{\sqrt{c^2+f^2}}{e} = \sqrt{1-\frac{|a|^2}{e^2}}$$

Since $a^2 < 0$, this ratio is bounded by 1, the larger it is, the more elongated the ellipses. For fixed e, anisotropy is a maximum for small $|a|$. In this new elliptical coordinate system, we have the following simplifications:

$$\text{(A.5)} \qquad \begin{aligned} r'^2 &= \underline{x}^T \mathbf{m}_1 \underline{x} \\ r'^2 \cos 2\phi &= \underline{x}^T \mathbf{m}_4 \underline{x} \\ r'^2 \sin 2\phi &= \underline{x}^T \mathbf{m}_5 \underline{x} \end{aligned}$$

so that the basic second order polynomials take the simple form:

$$\text{(A.6)} \qquad \underline{x}^T \mathbf{A}_1 \underline{x} = r'^2(\alpha_1 + \alpha_4 \cos 2\phi + \alpha_5 \sin 2\phi)$$

Taking the case $\alpha_1 = 0$, the curve $f_\lambda(\underline{x}) = 1$ now becomes:

$$\text{(A.7)} \qquad \begin{aligned} r'(\phi)^{-4} &= \frac{|\underline{\alpha}|^2 + |\underline{\beta}|^2}{2}(1 + R'\cos(4\phi - \zeta)) \\ &\lambda^{-4d}\frac{R^2}{2}(1 + R'\cos(4\phi - \zeta)) \end{aligned}$$

where

$$\begin{aligned} R'^2 &= 1 - \left(\frac{2\sin\theta}{r + r^{-1}}\right)^2 \\ R^2 &= \alpha_{4,1}^2 + \alpha_{4,1}^2 + \alpha_{5,1}^2 + \beta_{4,1}^2 + \beta_{5,1}^2 \end{aligned}$$

and;

$$\tan\zeta = \frac{2\cos\theta}{r^{-1} - r}$$

where $r = |\underline{r}_s|/|\underline{r}_4|$ as before, and θ is the angle between $\underline{r}_5$ and $\underline{r}_4$.

Stratification dominance case: $a^2 \geq 0$:

We start with an analogous change of variables in this case:

$$\text{(A.8)} \qquad \begin{aligned} r'\cos\phi &= \frac{1}{\sqrt{2}}(-(e+f)x + (c-a)y) \\ r'\sin\phi &= \frac{1}{\sqrt{2}}(-(e+f)x + (c+a)y) \end{aligned}$$

The curves r' = constant are ellipses similar to those above. The corresponding eigenvalues whose reciprocal root gives the major and minor axes are:

$$c^2+ef+f^2+\sqrt{c^2a^2+e^2(e+f^2)};\, c^2+ef+f^2-\sqrt{c^2a^2+e^2(e+f^2)} \tag{A.9}$$

The corresponding angle is given by

$$\tan^{-1}\left(\frac{c(e+f)}{-c^2+ef+f^2+\sqrt{c^2a^2+e^2(e+f^2)}}\right) \tag{A.10}$$

Similarly:

$$\begin{aligned} r'^2\sin 2\phi &= \underline{x}^T\mathbf{m}_1\underline{x} \\ r'^2(1-\cos 2\phi) &= \underline{x}^T\mathbf{m}_2\underline{x} \\ r'^2(1+\cos 2\phi) &= \underline{x}^T\mathbf{m}_3\underline{x} \end{aligned} \tag{A.11}$$

and

$$\underline{x}^T\mathbf{A}_1\underline{x} = r'^2(\alpha_2+\alpha_3+(\alpha_2-\alpha_3)\cos 2\phi+\alpha_1\sin 2\phi)\ . \tag{A.12}$$

We now obtain:

$$\begin{aligned} r'^{-4}(\phi) &= (\alpha_2^2+\beta_2^2)\Bigg(\frac{3}{2}(r^2+1)+r\cos\theta+2(1-r^2)\cos 2\phi \\ &\qquad +\left(\frac{1}{2}(r^2+1)-r\cos\theta\right)\cos 4\phi\Bigg) \\ &= \lambda^{-4d}R\Bigg(\frac{3}{2}\left(r+\frac{1}{r}\right)+\cos\theta+2\left(\frac{1}{r}-r\right)\cos 2\phi \\ &\qquad +\left(\frac{1}{2}\left(r\frac{1}{r}\right)-\cos\theta\right)\cos 4\phi\Bigg) \end{aligned} \tag{A.13}$$

with

$$R^2=\sqrt{(\alpha_{2,1}^2+\beta_{2,1}^2)(\alpha_{3,1}^2+\beta_{3,1}^2)}$$

where $r = |\underline{r}_3|/|\underline{r}_2|$ and θ is the angle between $\underline{r}_3$ and $\underline{r}_2$. R and θ are invariant under scale transformations; r is given by

$$r=\lambda^{-4a}\sqrt{\frac{(\alpha_3^2+\beta_3^2)}{(\alpha_2^2+\beta_2^2)}} \tag{A.14}$$

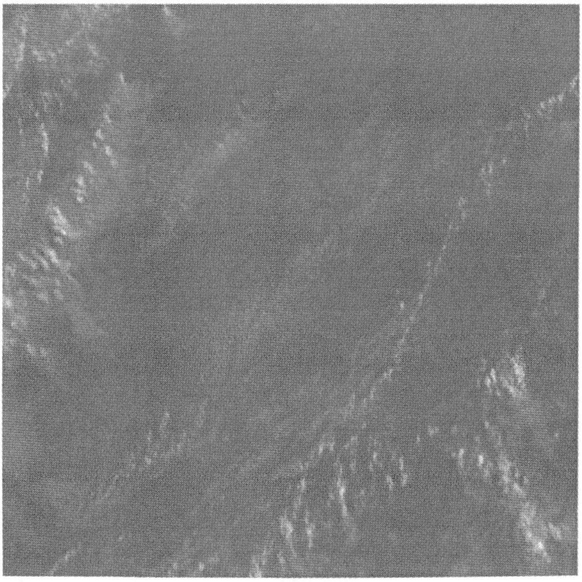

Figure 1.1a) NOAA-9 cloud image, visible spectrum (channel 1)

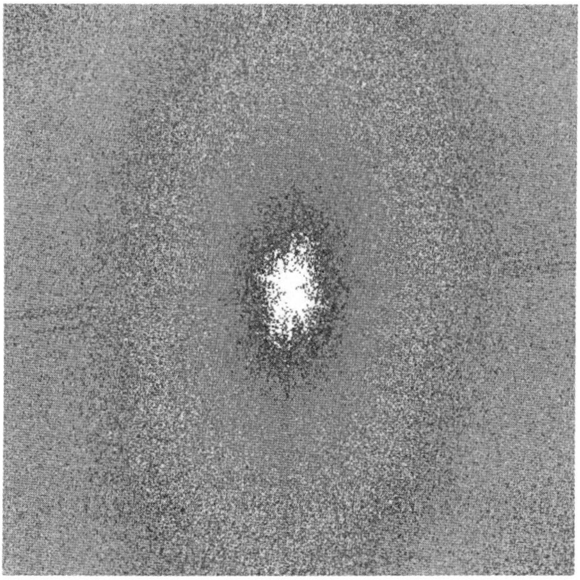

Figure 1.1b) 2-D power spectrum of image a)

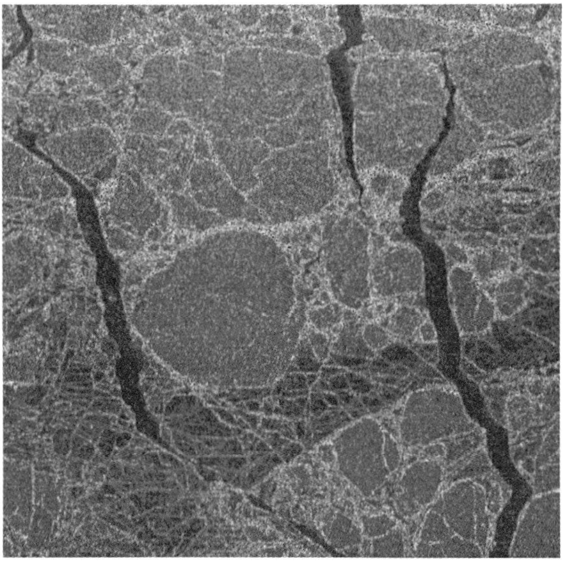

Figure 1.2a) SAR sea ice image

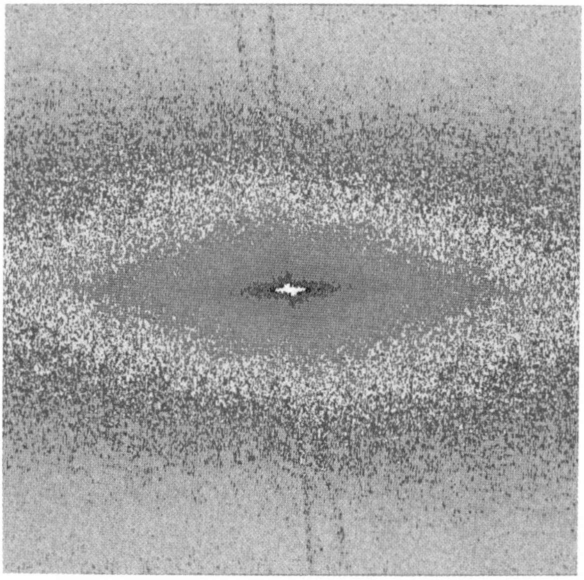

Figure 1.2b) 2-D power spectrum of image a)

Figure 2: Examples of stratification dominant balls

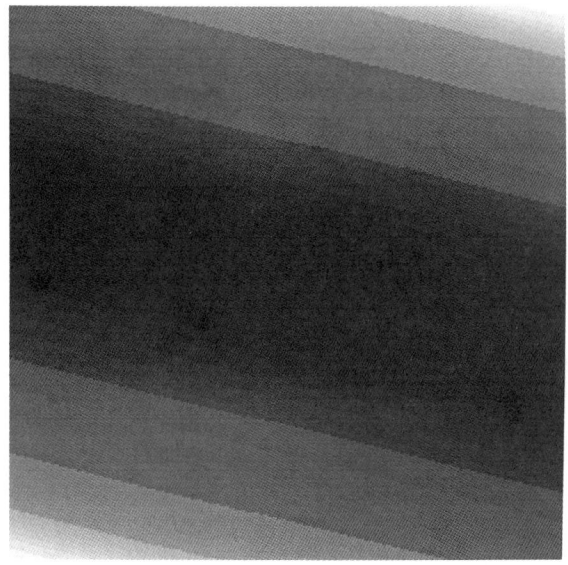

Figure 2.1: c=0.4, f=0.1, e=0.1; r=1.0, $\theta = 0.5$, $R^2 = 1.0$

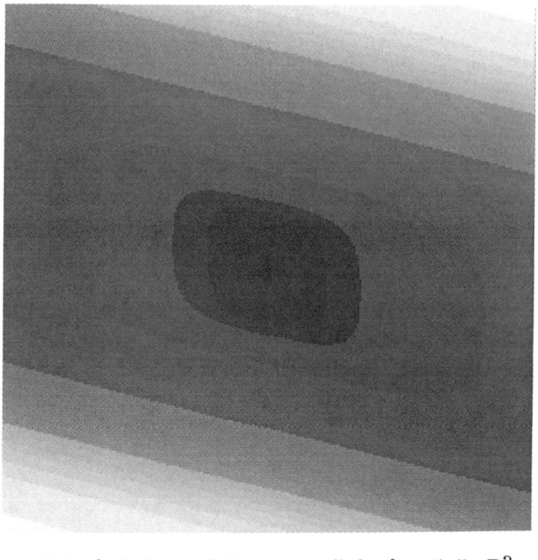

Figure 2.2: c=0.4, f=0.1, e=0.1; r=5.0, $\theta = 1.5$, $R^2 = 1.0 \times 10^{-6}$

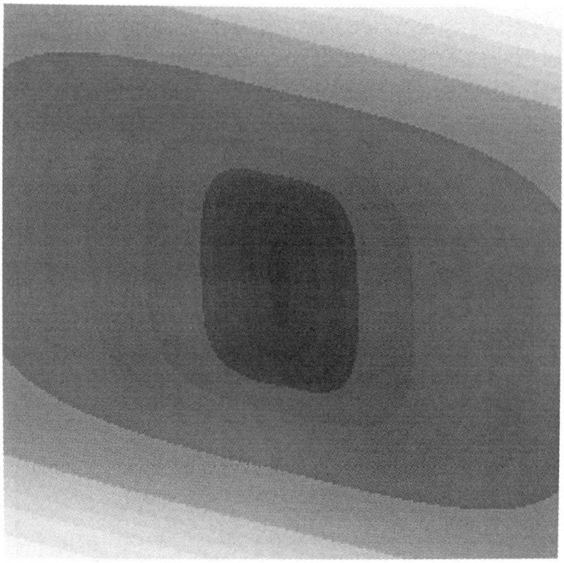

Figure 2.3: c=0.4, f=0.1, e=0.1; r=50.0, $\theta = 1.5$, $R^2 = 1.0 \times 10^{-6}$

Figure 2.4: c=0.1, f=-0.8, e=0.5; r=1.0, $\theta = 1.5$, $R^2 = 1.0 \times 10^{-6}$

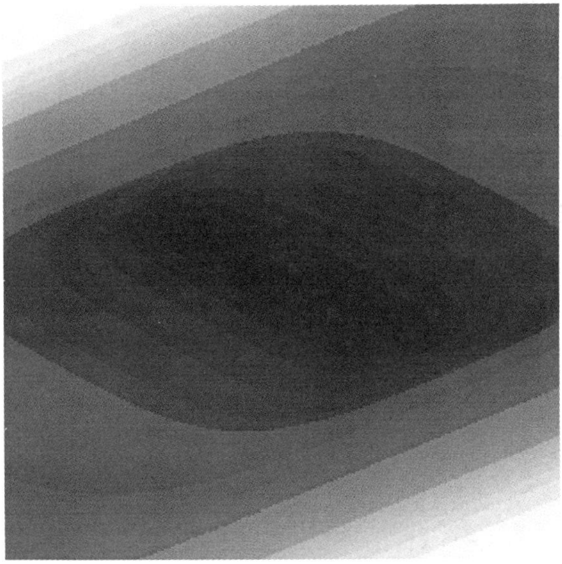

Figure 2.5: c=0.1, f=-0.8, e=0.5; r=50.0, $\theta = 1.5, R^2 = 1.0 \times 10^{-6}$

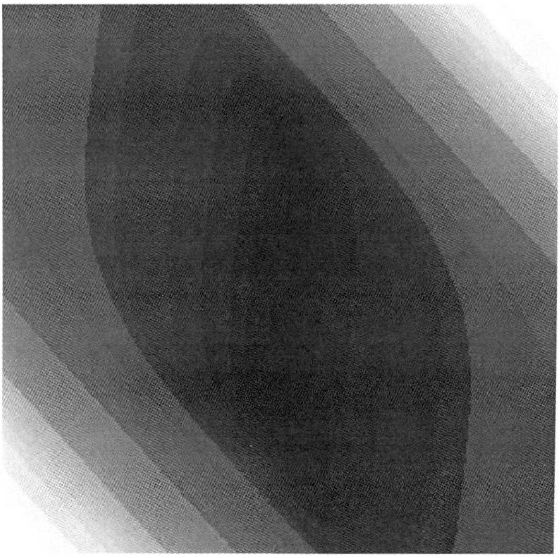

Figure 2.6: c=0.5, f=0.7, e=0.57; r=5.0, $\theta = 0.2, R^2 = 1.0 \times 10^{-6}$

Figure 2.7: c=0.4, f=0.1, e=0.1; r=1.0, $\theta = 1.0, R^2 = 1.0 \times 10^{-6}$

Figure 2.8: c=0.4, f=0.1, e=0.1; r=1.0, $\theta = 0.5, R^2 = 1.0 \times 10^{-6}$

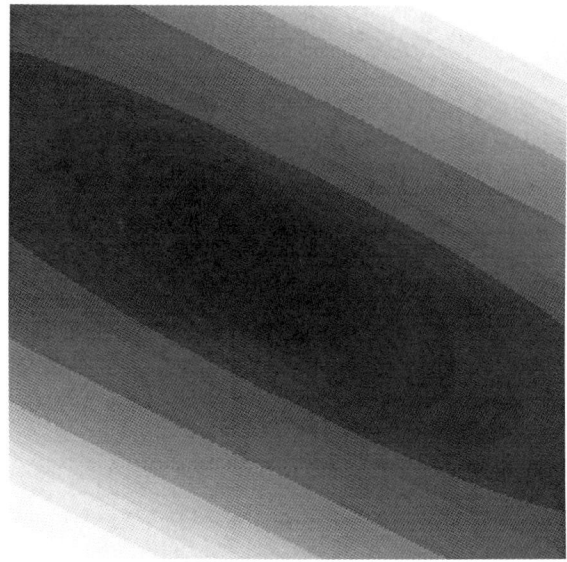

Figures 3: Examples of rotation dominant balls

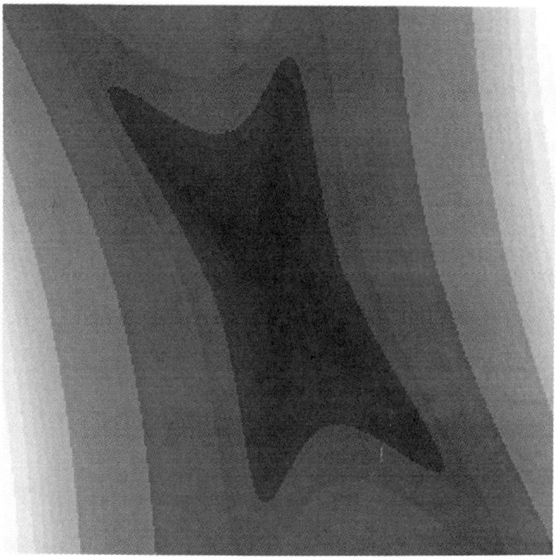

Figure 3.1: c=0.1, f=0.1, e=0.2; r=1.0, $\theta = 0.5$, $R^2 = 1.0$

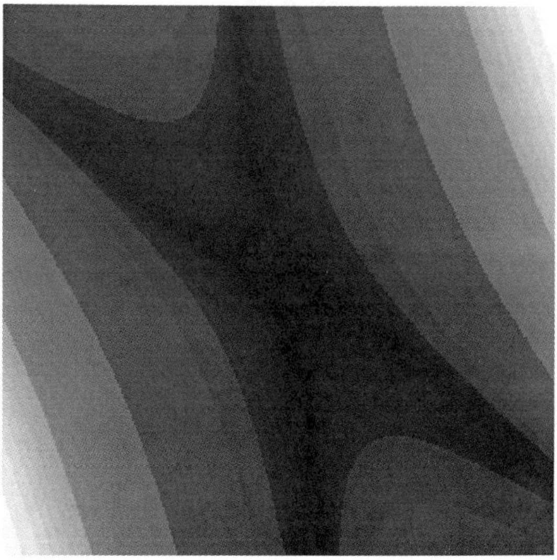

Figure 3.2: c=0.1, f=0.1, e=0.2; r=5.0, $\theta = 0.5$, $R^2 = 1.0$

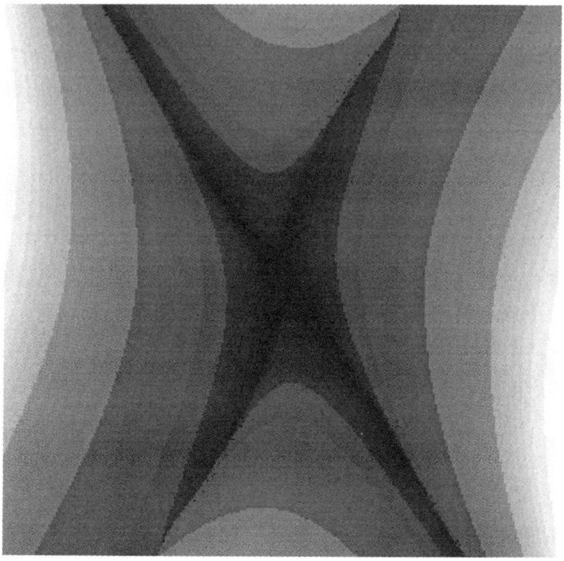

Figure 3.3: c=0.1, f=0.1, e=0.2; r=0.1, $\theta = 0.5, R^2 = 1.0$

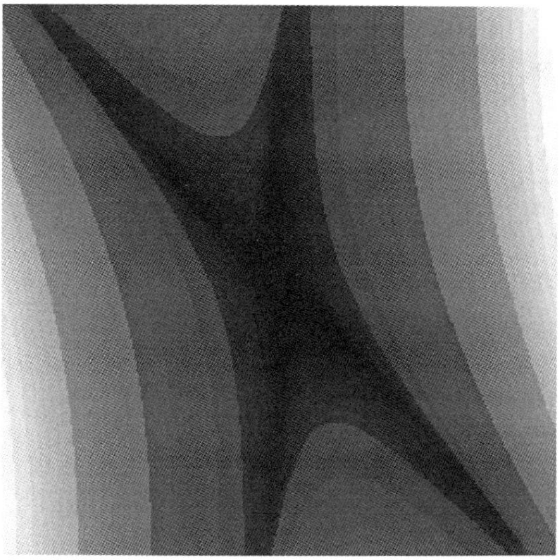

Figure 3.4 c=0.1, f=0.1, e=0.2; r=1.0, $\theta = 0.2, R^2 = 1.0$

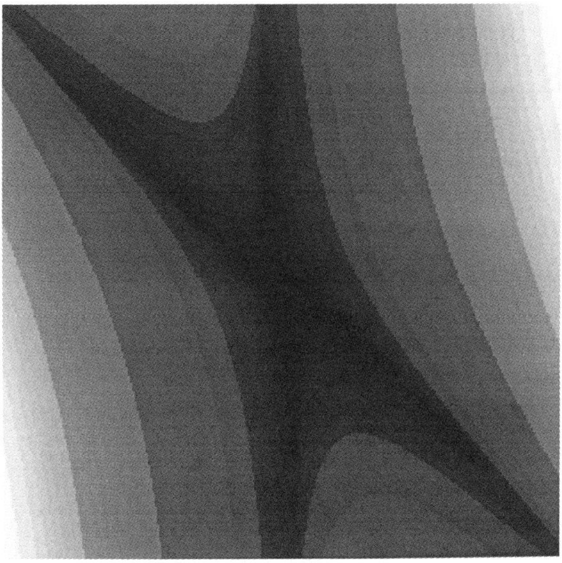

Figure 3.5: c=0.1, f=0.1, e=0.2; r=1.0, $\theta = 0.2, R^2 = 0.05$

Figure 3.6: c=0.1, f=0.1, e=0.2; r=1.0, $\theta = 0.2, R^2 = 20.0$

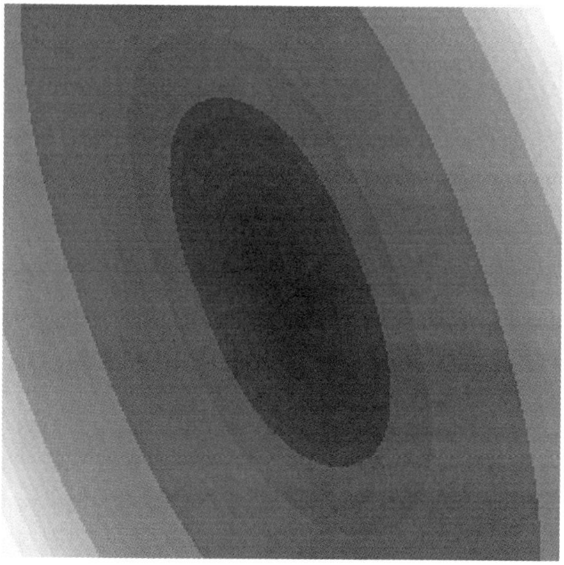

Figure 3.7: c=0.1, f=0.1, e=0.2; r=1.0, $\theta = 1.5$, $R^2 = 1.0$

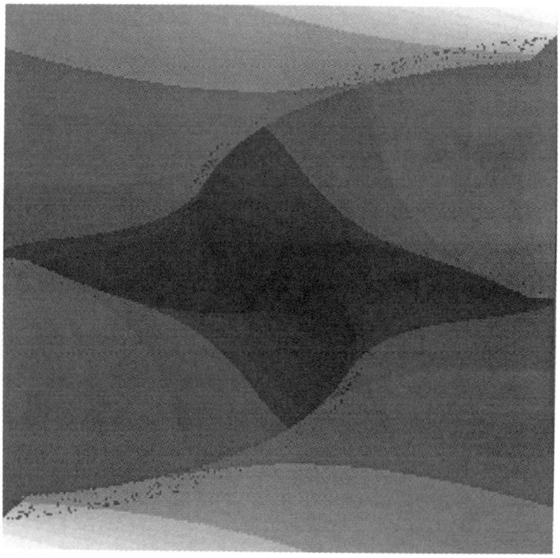

Figure 3.8: c=0.2, f=-0.5, e=1.0; r=1.0, $\theta = 0.5$, $R^2 = 1.0$

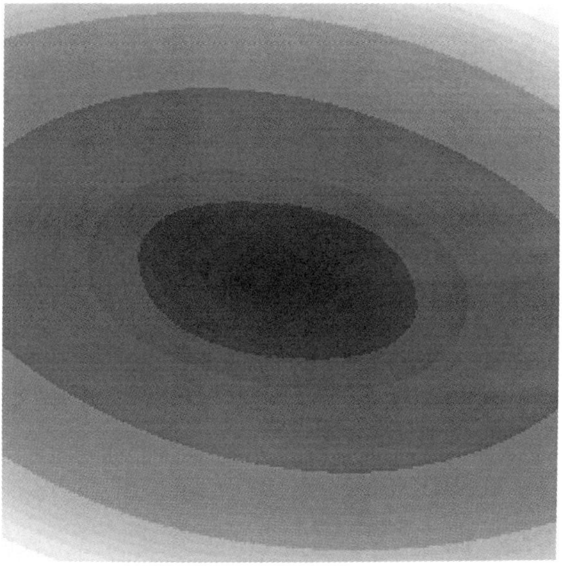

Figure 3.9: c=0.2, f=-0.5, e=1.0; r=1.0, $\theta = 1.5$, $R^2 = 0.05$

Figure 4: Real-space simulated fields corresponding to Fig. 2

Figure 4.1: c=0.4, f=0.1, e=0.1; r=1.0, $\theta = 0.5$, $R^2 = 1.0$

Figure 4.2: c=0.4, f=0.1, e=0.1; r=5.0, $\theta = 1.5, R^2 = 1.0 \times 10^{-6}$

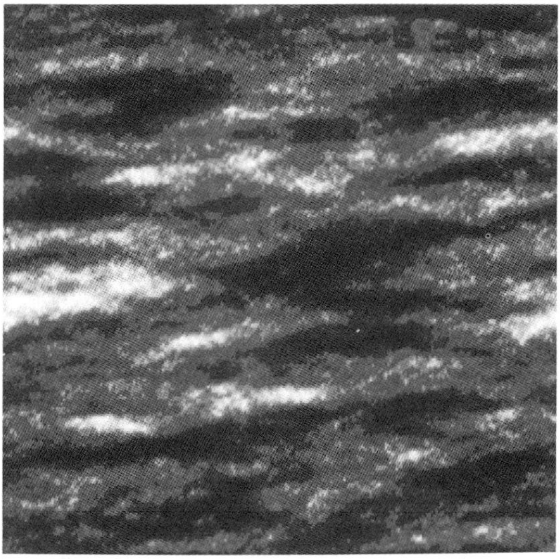

Figure 4.3: c=0.4, f=0.1, e=0.1; r=50.0, $\theta = 1.5, R^2 = 1.0 \times 10^{-6}$

Figure 4.4: c=0.1, f=-0.8, e=0.5; r=1.0, $\theta = 1.5$, $R^2 = 1.0 \times 10^{-6}$

Figure 4.5: c=0.1, f=-0.8, e=0.5; r=50.0, $\theta = 1.5$, $R^2 = 1.0 \times 10^{-6}$

Figure 4.6: c=0.5, f=0.7, e=0.57; r=5.0, $\theta = 0.2, R^2 = 1.0 \times 10^{-6}$

Figure 4.7: c=0.4, f=0.1, e=0.1; r=1.0, $\theta = 1.0, R^2 = 1.0 \times 10^{-6}$

Figure 4.8: c=0.4, f=0.1, e=0.1; r=1.0, $\theta = 0.5$, $R^2 = 1.0 \times 10^{-6}$

Figure 5: Real-space simulated fields corresponding to Fig. 3

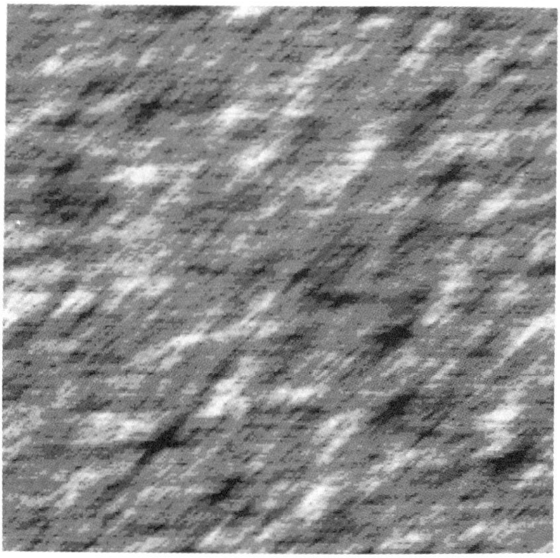

Figure 5.1: c=0.1, f=0.1, e=0.2; r=1.0, $\theta = 0.5$, $R^2 = 1.0$

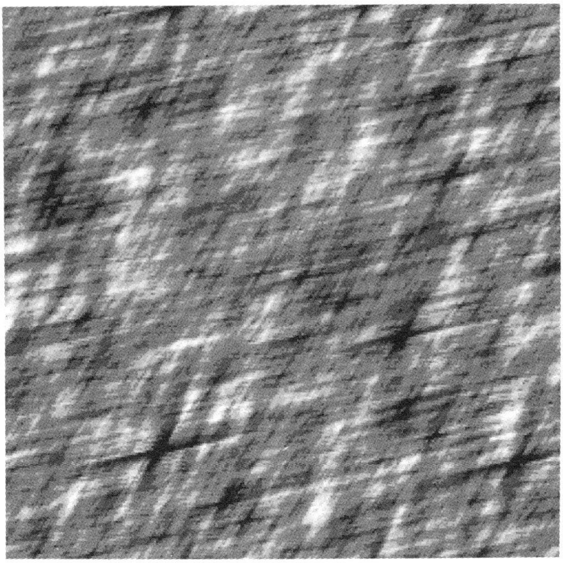

Figure 5.2: c=0.1, f=0.1, e=0.2; r=5.0, $\theta = 0.5$, $R^2 = 1.0$

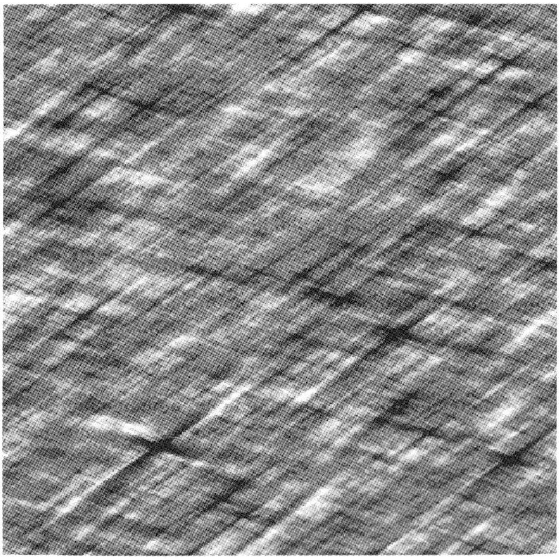

Figure 5.3: c=0.1, f=0.1, e=0.2; r=0.1, $\theta = 0.5$, $R^2 = 1.0$

Figure 5.4: c=0.1, f=0.1, e=0.2; r=1.0, $\theta = 0.2$, $R^2 = 1.0$

Figure 5.5: c=0.1, f=0.1, e=0.2; r=1.0, $\theta = 0.2$, $R^2 = 0.05$

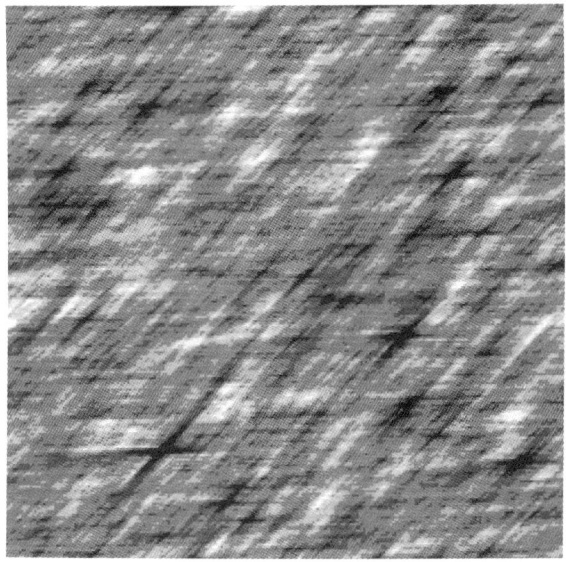

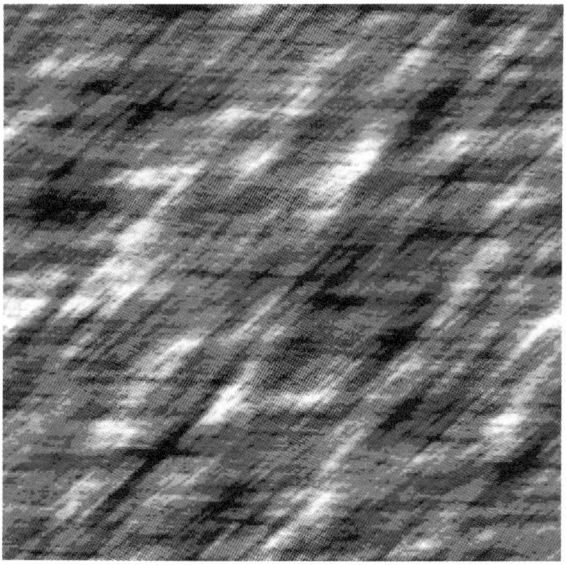

Figure 5.6: c=0.1, f=0.1, e=0.2; r=1.0, $\theta = 0.2, R^2 = 20.0$

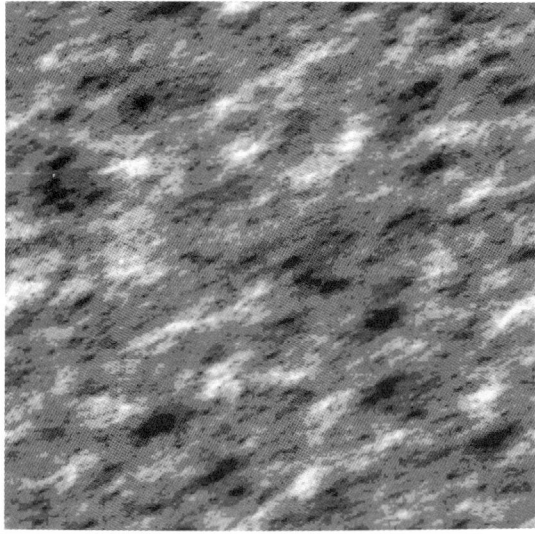

Figure 5.7: c=0.1, f=0.1, e=0.2; r=1.0, $\theta = 1.5, R^2 = 1.0$

Figure 5.8: c=0.2, f=-0.5, e=1.0; r=1.0, $\theta = 0.5$, $R^2 = 1.0$

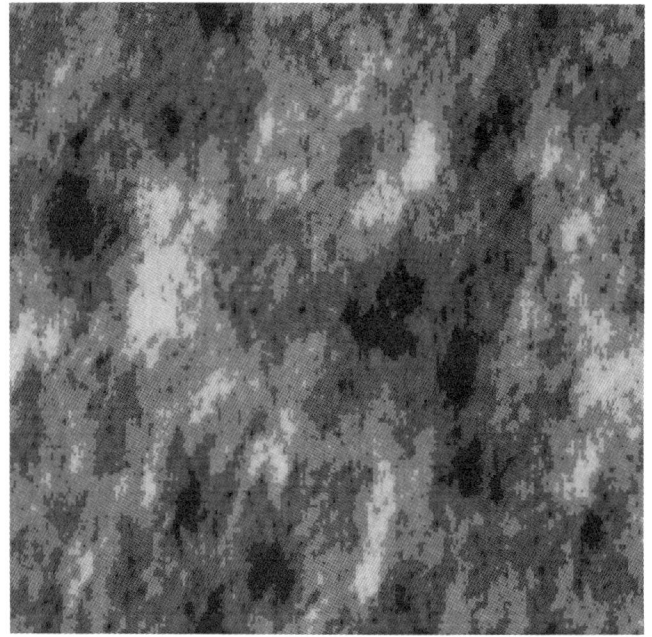

Figure 5.9: c=0.2, f=-0.5, e=1.0; r=1.0, $\theta = 1.5$, $R^2 = 0.05$

REFERENCES

Chigirinskaya, Y., D. Schertzer, S. Lovejoy, A. Lazarev, and A. Ordanovich, *Unified multifractal atmospheric dynamics tested in the tropics Part 1: horizontal scaling and self-organized criticality*, Nonlinear Processes in Geophysics (1), 105–114, 1994.

Davis, A., S. Lovejoy, and D. Schertzer, *Supercomputer simulation of radiative transfer inside multifractal cloud models*, in I.R.S. 92, edited by S. Keevallik and O. Kärner, 112–115, 1992.

Falco, T., F. Francis, S. Lovejoy, D. Schertzer, B. Kerman, and M. Drinkwater, in *IEEE Transactions on Geosciences and Remote Sensing*, 34, 906–914, 1996.

Fox, C.G., and D. Hayes, *Quantitative methods for analyzing the roughness of the seafloor*, Reviews of Geophysics (23), 1–48, 1985.

Frisch, U.P., P.L. Sulem, and M. Nelkin, *A Simple Dynamical Model of Intermittency in Fully Developed Turbulence*, Journal of Fluid Mechanics, 87, 719–24, 1978.

Gabriel, P., S. Lovejoy, D. Schertzer, and G.L. Austin, *Multifractal Analysis of resolution dependence in satellite imagery*, Journal of Geophysical Research, 15, 1373–1376, 1988.

Gupta, V.K., and E. Waymire, *A Statistica l Analysis of Mesoscale Rainfall as a Random Cascade*, Journal of Applied Meteorology, 32, 251–267, 1993.

Hooge, C., *Earthquakes as a Space-Time multifractal Process*, M. Sc. thesis, McGill University, Montreal (Qeubec), Canada, 1993.

Hooge, C., S. Lovejoy, D. Schertzer, S. Pecknold, J.-F. Malouin, and F. Schmitt, *Multifractal Phase Transitions: The Origin of Self-Organized Criticality in Earthquakes*, Non-Linear Processes in Geophysics, 1 (2), 191–197, 1994.

Lavallée, D., S. Lovejoy, D. Schertzer, and P. Ladoy, *Nonlinear variability and landscape topography: analysis and simulation*, in *Fractals in geography*, edited by L. De Cola, and N. Lam, pp. 171–205, Prentice-Hall, 1993.

Lazarev, A., D. Schertzer, S. Lovejoy, and Y. Chigirinskaya, *Unified multifractal atmospheric dynamics tested in the tropics Part II: vertical scaling and generalized scale invariance, Nonlinear Processes in Geophysics* (1), 115–123, 1994.

Lewis, G., S. Lovejoy, and D. Schertzer, *The scale invariant generator technique for parameter estimates in generalized scale invariant systems*, Computers in Geoscience (submitted), 1996.

Lovejoy, S., D. Lavallée, D. Schertzer, and P. Ladoy *The $\ell^{1/2}$ law and multifractal topography: theory and analysis, Nonlinear Processes in Geophysics* (2), 16–22, 1995.

Lovejoy, S., and D. Schertzer, *Generalized Scale Invariance in the Atmosphere and Fractal Models of Rain*, Water Resources Research, 21 (8), 1233–1250, 1985.

Lovejoy, S., and D. Schertzer, *Our multifractal atmosphere: A unique laboratory for non-linear dynamics*, Physics in Canada, 46, 62, 1990.

Lovejoy, S., and D. Schertzer, *How bright is the coast of Brittany?*, in *Fractals in Geoscience and Remote Sensing*, edited by G. Wilkinson, I. Kanellopoulos, and J. Mgier,. 102–151, Office for Official Publications of the European Communities, Luxembourg, 1995.

Lovejoy, S., D. Schertzer, and K. Pflug, *Generalized Scale Invariance and differentially rotation in cloud radiances*, Physica A, 185, 121–127, 1992.

Lovejoy, S., D. Schertzer, and A.A. Tsonis, *Functional Box-Counting and Multiple Dimensions in rain*, Science, 235, 1036–1038, 1987.

Mandelbrot, B.B., *Intermittent turbulence in self-similar cascades: divergence of high moments and dimension of the carrier*, Journal of Fluid Mechanics, 62, 331–350, 1974.

Meneveau, C., K.R. Sreenivasan, P. Kailasnah, and M.S. Fan, *Joint multifractal measures: theory and applications to turbulence*, Physical Review A, 41, 894, 1990.

Novikov, E.A., and R. Stewart, *Intermittency of turbulence and spectrum of fluctuations in energy-dissipation*, Izv. Akad. Nauk. SSSR. Ser. Geofiz., 3, 408–412, 1964.

Pecknold, S., S. Lovejoy, D. Schertzer, C. Hooge, and J.-F. Malouin, *The Simulation of Universal Multifractals*, in *Cellular Automata: Prospects in Astrophysical Applications*, edited by A. Lejeune, and J. Perdang, pp. 228–67, World Scientific, Singapore, 1993.

Pflug, K., S. Lovejoy, and D. Schertzer, *Generalized Scale Invariance, differential rotation and cloud texture*, in *Nonlinear dynamics of Structures*, edited by R.Z. Sagdeev, U. Frisch, A.S. Moiseev, and A. Erokhin, 72–78, World Scientific, 1991.

Pflug, K., S. Lovejoy, and D. Schertzer, *Differential Rotation and Cloud Texture: Analysis Using Generalized Scale Invariance*, Master's of Science Thesis thesis, McGill University, 1992.

Pflug, K., S. Lovejoy, and D. Schertzer, *Generalized Scale Invariance, differential rotation and cloud texture*, Journal of Atmospheric Sciences, 50, 538–553, 1993.

Pilkington, M., and J.P. Todoeschuk, *Fractal Magnetization of Continental Crust*, *Geophysical Research Letter*, 20 (7), 627–630, 1993.

Schertzer, D., and S. Lovejoy, *On the Dimension of Atmospheric motions*, in *Turbulence and Chaotic phenomena in Fluids*, IUTAM, edited by T. Tatsumi, 505–512, Elsevier Science Publishers B. V., 1984.

Schertzer, D., and S. Lovejoy, *The dimension and intermittency of atmospheric dynamics*, in *Turbulent Shear Flows 4*, edited by B. Launder, 7–33, Springer, 1985a.

Schertzer, D., and S. Lovejoy, *Generalised scale invariance in turbulent phenomena*, Physico-Chemical Hydrodynamics Journal, 6, 623–635, 1985b.

Schertzer, D., and S. Lovejoy, *Physical modeling and Analysis of Rain and Clouds by Anysotropic Scaling of Multiplicative Processes*, Journal of Geophysical Research, D 8 (8), 9693–9714, 1987.

Schertzer, D., and S. Lovejoy, *Generalized Scale Invariance and multiplicative processes in the atmosphere*, PAGEOPH, 130, 57–81, 1989.

Schertzer, D., and S. Lovejoy, *Nonlinear geodynamical variability: multiple singularities, universality and observables*, in *Non-linear variability in geophysics: Scaling and Fractals*, edited by D. Schertzer, and S. Lovejoy, 41–82, Kluwer, 1991a.

Schertzer, D., and S. Lovejoy, *Scaling Nonlinear Variability in Geodynamics*, in *Nonlinear Variability in Geophysics: Scaling and Fractals*, edited by D. Schertzer, and S. Lovejoy, 41–82, Kluwer Academic Publishers, Dordrecht, 1991b.

Schertzer, D., and S. Lovejoy, *From scalar cascades to Lie cascades: joint multifractal analysis of rain and cloud processes*, in *Space/time Variability and Interdependance for various hydrological processes*, edited by R.A. Feddes, 153–173, Cambridge University Press, New-York, 1995a.

Schertzer, D., and S. Lovejoy, *The multifractal phase transition route to self-organized criticality*, Physics Reports, in press, 1996.

Schmitt, F., D. Lavallée, D. Schertzer, and S. Lovejoy *Empirical Determination of Universal Multifractal Exponents in Turbulent Velocity Fields*, Physical Review Letter, 68, 305–308, 1992.

Schmitt, F., D. Schertzer, and S. Lovejoy, and Y. Brunet, *Estimation of universal multifractal indices for atmospheric turbulent velocity fields*, Fractals, 3, 568–575.

Tessier, Y., S. Lovejoy, and D. Schertzer, *Universal Multifractals: theory and observations for rain and clouds*, Journal of Applied Meteorology, 32 (2), 223–250, 1993a.

Tessier, Y., S. Lovejoy, D. Schertzer, D. Lavallée, and B. Kerman, *Universal Multifractal Indices For The Ocean Surface At Far Red Wavelengths*, Geophysical Research Letter, 20 (12), 1167–1170, 1993b.

Van Zandt, T.E., S.A. Smith, T. Tsuda, D.C. Fritts, T. Sato, S. Fukao, and S. Kato, *Studies of Velocity Fluctuations in the Lower Atmosphere Using the MU Radar*, Part I: Azimuthal Anisotropy, J. Atmos. Sci. (47), 39–50, 1990.

Veneziano, D., G. Moglen, and R.L. Bras, *Iterate random pulse processes and their spectral properties*, Water Resources (submitted), 1996.

Yaglom, A.M., *The influence on the fluctuation in energy dissipation on the shape of turbulent characteristics in the inertial interval*, Sov. Phys. Dokl., 2, 26–30, 1966.

SHORT-CORRELATION APPROXIMATION IN MODELS OF TURBULENT DIFFUSION*

L. PITERBARG†

Abstract. The problem of stirring a passive scalar by a random velocity field is considered. By the short-correlation approximation one means the assumption that the decorrelation time of the velocity field is infinitely small. There is a widespread misunderstanding that such an approach readily yields the Fokker-Planck equation for the mean field of passive scalar. Indeed, the resulting equation strongly depends on the order of a "hidden" time scale appearing in this problem. This time scale called the turnover time is defined as the ratio of the correlation radius and the mean square velocity fluctuation. We show that the effective diffusivity is different under different assumptions on the order of turnover time. As a consequence we have different physical effects for different forms of the scale separation. Also some new rigorous results related to the regime of superdiffusion are obtained.

Outline.

Introduction. Passive scalar stirring by a random velocity field in the presence of molecular or small scale diffusion is extremely important to the development of fundamental science and practical applications [Csanady, 1973, Okubo, 1980]. For this reason, this field has been an active area of research since the classical papers in the 20-30s by Taylor, Prandtl, Richardson [Monin & Yaglom,1975]. A new impulse for further development of models for turbulent diffusion was given by mathematical works of early 80s [Kozlov, Varadhan & Papanicolaou] concerning homogenization in random media. The motivation for this work is to find a bridge connecting the pioneering works and their extension of the 50-60s (Batchelor, Kraichnan, Roberts), where euristic methods were used, with modern (80-90s) rigorous results in this field.

As a basis for unifying different physical and mathematical approaches

* This work was supported by ONR Grant No. N00014-91-J-1526.

† Center for Applied Mathematical Sciences, University of Southern California, Los-Angeles, CA 90089-1113.

to turbulent diffusion we suggest a "scale" classification. For the sake of simplicity we restrict our consideration to homogeneous velocity fields with only one time scale (correlation time) and only one space scale (correlation radius). Therefore in the simplest case there are four independent time scales: the current time (we consider a Cauchy problem), the molecular diffusion time, the correlation time and the turnover time defined as the ratio of the correlation radius and the mean square velocity fluctuation. If all these scales have the same order of magnitude it is hopeless to make any progress towards an analytical solution. Explicit formulas for the effective diffusivity or explicit equations for the statistical characteristics of the tracer (passive scalar) can be obtained only if one or more from the scales is much smaller (much bigger) than the others. We show that many well known exactly solvable models can be formulated in terms of the order of the mentioned scales. Our focus will be on the so called "short-correlation" approximation meaning that the correlation time of velocity is much smaller than both the current time and the molecular diffusivity time. In this case, under a wide range of conditions, the explicit expression for the turbulent diffusivity can be found, but this expression drastically depends on the relationship between the correlation time and the turnover time. If they are the same order then the turbulent diffusivity depends on the molecular one which can essentially intensify the mixing of the fluid. In contrast if the turnover time is much bigger then the correlation time but still much less than the current time we arrive at the classical Fokker-Planck equation where the effective diffusivity depends on the statistics of the velocity field only. In this case some interesting physical effects follow from the equation describing the evolution of the space correlation function of the passive scalar. We will focus on two of them: the long-term existence of tracer anomalies and a possible sharpening anomalous gradients due to stirring by the homogeneous velocity field.

In summary this paper has two main goals

(i) To enumerate and classify the situations where the turbulent diffusion can be described in an analytical form, using the "scale" language.

(ii) To discuss some physical effects following from exactly solvable models which could appear to be important in practice.

In addition we give a detailed derivation of the mean tracer equation in the homogenization case and some new results concerning the regime of superdiffusion for isotropic turbulence.

As stated above the number of publications related to passive scalar mixing is close to an infinity. For this reason our choice of the reviewed papers is first, quite incomplete and second, subjective. Primarily we give a preference to the works making a loud resonance. The other criterion of choice is the applicability of the results to oceanographic problems. Because of this, we almost pay no attention to models of steady velocity fields. It is note worthy that a detailed discussion of oceanographic applications is beyond the scope of this paper, but we prepare a basis for such a discussion.

Our effort to state recent rigorous results in terms of the time scales serves just this goal. Of course, the present knowledge of the mentioned time scales in the ocean is far from complete. Nevertheless, for some regions and certain conditions we can make reasonable conjectures as to their values. In turn it enables us to have insight into what model of turbulent diffusion may be used in those regions under those conditions.

Finally let us remark that we collected both rigorous and non-rigorous models. For this reason we give specific formulations when we deal with a rigorous model.

This paper is organized as follows. In section 1 the problem statement is given in mathematical form and a simplest exactly solvable model is discussed. Different approaches to the study of random transport are also discussed. In section 2 the main time scales are introduced in a formal way and some examples of exact asymptotics are given. Section 3 is dedicated to the short-correlation approximation. It is divided into four subsections. In subsection 3.1 different cases of this approximation are stated depending on the order of the turnover time. The well studied case of δ-correlated velocity fields is considered in 3.2. The focus is on the equation for the space correlation function of tracer and on two physical effects following from this equation. The first of them is the prolonged lifetime of passive scalar anomalies due to the spatial correlation of the velocity field. The second one consists of sharpening the tracer gradients by a homonenous velocity field. In section 3.3 the homogenization problem is considered. The effect of enhancing the turbulent diffusion by molecular diffusion is discussed. Isotropic velocity fields with an infrared singularity in the spectrum are studied in section 3.4, where, some new results are exhibited and their proof is given in the Appendix.

1. Statement of problem.

1. We study the simplest equation

$$\frac{\partial c}{\partial t} + \boldsymbol{u} \cdot \nabla c = \kappa \nabla^2 c \tag{1.1}$$

governing the evolution of a passive scalar in a random flow. Here $c = c(t, \boldsymbol{x})$ is the tracer concentration at the moment t at the point $\boldsymbol{x}$ in d-dimension Euclidean space E^d. Primarily we consider values $d = 2, 3$ as most important for applications. $\boldsymbol{u} = \boldsymbol{u}(t, \boldsymbol{x})$ is the divergence-free random velocity field, $\nabla \cdot \boldsymbol{u} = 0$, ∇^2 is the Laplacean and κ is the constant non-random coefficient of molecular diffusion.

We assume unless otherwise specified that $\boldsymbol{u}(t, \boldsymbol{x})$ is a homogeneous in $\boldsymbol{x}$, stationary in t Gaussian random field with zero mean, $\langle \boldsymbol{u} \rangle = \mathbf{0}$. Therefore its probabilistic distribution is completely determined by the space-time correlation tensor

$$\boldsymbol{R}_{\boldsymbol{u}}(t, \boldsymbol{x}) = \langle \boldsymbol{u}(s, \boldsymbol{y}) \boldsymbol{u}(s + t, \boldsymbol{y} + \boldsymbol{x})^T \rangle \tag{1.2}$$

or by the spectral tensor $\boldsymbol{E}(\omega, \boldsymbol{k})$ given by

$$\boldsymbol{R}_{\boldsymbol{u}}(t, \boldsymbol{x}) = \int_{E^d} \int_{-\infty}^{\infty} e^{i(\boldsymbol{k}\cdot\boldsymbol{x} - \omega t)} \boldsymbol{E}(\omega, \boldsymbol{k}) d\omega d\boldsymbol{k}. \tag{1.3}$$

Under the additional assumption of isotropy the entries of the correlation matrix are expressed in terms of only the scalar function $R_L(t, r)$

$$R_{ij}(t, \boldsymbol{x}) = \left(R_L(t, r) + \frac{r}{d-1} \frac{\partial R_L(t, r)}{\partial r} \right) \delta_{ij} - \frac{x_i x_j}{r(d-1)} \frac{\partial R_L(t, r)}{\partial r}, \tag{1.4}$$

where $\boldsymbol{x} = (x_1, \ldots, x_d)$, $r = |\boldsymbol{x}|$, δ_{ij} is the Kronecker delta.

In a similar way the spectral tensor can be represented

$$E_{ij}(\omega, \boldsymbol{k}) = E_L(\omega, k) \left(\delta_{ij} - \frac{k_i k_j}{k^2} \right),$$

where $E_L(\omega, k)$ is the longitudinal spectrum.

The general problem we discuss here is to describe the behaviour of the mean field $\langle c(t, \boldsymbol{x}) \rangle$ and the statistics of the fluctuation fields $c'(t, \boldsymbol{x}) = c(t, \boldsymbol{x}) - \langle c(t, \boldsymbol{x}) \rangle$, $\nabla c'(t, \boldsymbol{x})$ under given statistics of the velocity field $\boldsymbol{u}(t, \boldsymbol{x})$. We stress that the statistics of $\boldsymbol{u}$(t,$\boldsymbol{x}$) is not derived from equations of motion, but are determined from the given correlation or spectral tensor of an arbitrary form. On one hand, such an approach enables us to deal with the general form of the velocity spectral or correlation characteristics. But on the other hand the circle of physical problems we can study in that frame is quite restricted and does not include for example the convection problem.

Let us specify the questions we will focus on

(i) Under what conditions does the mean field satisfy the diffusion equation

$$\frac{\partial \langle c \rangle}{\partial t} = \nabla \cdot \boldsymbol{D} \nabla \langle c \rangle + \kappa \nabla^2 \langle c \rangle, \tag{1.5}$$

where $\boldsymbol{D}$ is the effective (turbulent) diffusion tensor?

(ii) How can $\boldsymbol{D}$ be expressed in terms of the velocity correlation or spectral characteristics and κ?

(iii) What are the governing equations for the evolution of the correlation function of the tracer

$$R_c(t, \boldsymbol{x}) = \langle c'(t, \boldsymbol{y}) c'(t, \boldsymbol{y} + \boldsymbol{x}) \rangle$$

and its gradients?

There are many oceanographic works [Okubo, 1980] where equation (1.5) is postulated and formulas for $\boldsymbol{D}$ are found by physical reasoning.

From the mathematical point of view such an approach looks slightly confusing, because the solution of (1.1) is expressible in the formal way as

$$c(t,\boldsymbol{x}) = N_{t,\boldsymbol{x};\kappa}(c_0(\boldsymbol{y}), \boldsymbol{u}(s,\boldsymbol{y}), \boldsymbol{y}\in E^d, 0\le s\le t), \tag{1.6}$$

where the functional on the right hand side of (1.6) which depends on $t, \boldsymbol{x}$ and κ, is linear in the initial field $c_0(\cdot)$ and nonlinear in the velocity field $\boldsymbol{u}(\cdot,\cdot)$. Theoretically speaking we can average this functional over the ensemble $\{\boldsymbol{u}(\cdot,\cdot)\}$, and obtain an expression for $\langle c(t,\boldsymbol{x})\rangle$. Then check the validity of eq. (1.5). Thus we do not need to state (1.5), because it is possible to verify its validity. The trouble is that under general conditions on the velocity this possibility is purely theoretical. Nevertheless, in many important asymptotical cases this averaging can be carried out in an explicit form and consequently eq. (1.5) can be derived rigorously.

We remark that the functional $N\ (\cdot,\cdot)$ can be represented in an explicit Lagrangian form. Namely, let us introduce the statistical ensemble $\{\boldsymbol{w}(\cdot)\}$ of the standard d-dimensional Brownian motion, i.e. $\boldsymbol{w}(\cdot)$ is the vector Gaussian processes with $M\boldsymbol{w}(\cdot) = 0$, and $M(w_i(t)\ w_j(s)) = \delta_{ij}\ t\wedge s$ where $t\wedge s$ is the minimum among t and s, and M denotes averaging with respect to the ensemble $\{\boldsymbol{w}(\cdot)\}$. Define the generalized Lagrangian trajectory $\boldsymbol{\xi}^{(\kappa)}_{t,\boldsymbol{a}}(\cdot)$ passing through the point $(t,\boldsymbol{a})$ as the solution of the following integral equation

$$\boldsymbol{\xi}^{(\kappa)}_{t,\boldsymbol{a}}(s) = \boldsymbol{a} + \int_t^s \boldsymbol{u}\ (s',\ \boldsymbol{\xi}^{(\kappa)}_{t,\boldsymbol{a}}(s'))ds' + \sqrt{2\kappa}\ (\boldsymbol{w}(s) - \boldsymbol{w}(t)). \tag{1.7}$$

Then it can be shown (see for example [Friedman, 1975, p. 147]) that

$$N_{t,\boldsymbol{x},\kappa}(\boldsymbol{u}, c_0) = Mc_0(\boldsymbol{\xi}^{\kappa}_{t,\boldsymbol{x}}(0)). \tag{1.8}$$

Along with $\boldsymbol{\xi}^{\kappa}_{t,\boldsymbol{x}}(\cdot)$ let us introduce the classical Lagrangian trajectory $\boldsymbol{X}(t,\boldsymbol{a})$ as the solution of the integral equation

$$\boldsymbol{X}(t,\boldsymbol{a}) = \boldsymbol{a} + \int_0^t \boldsymbol{u}(s,\ \boldsymbol{X}(s,\boldsymbol{a}))ds. \tag{1.9}$$

Then, in the absence of the molecular diffusion, we have

$$N_{t,\boldsymbol{x},0}(\boldsymbol{u}, c_0) = c_0(\boldsymbol{X}^{-1}(t,\boldsymbol{x})), \tag{1.10}$$

where $\boldsymbol{X}^{-1}(t,\boldsymbol{x})$ satisfies the equation

$$\boldsymbol{X}(t,\ \boldsymbol{X}^{-1}\ (t,\boldsymbol{x})) = \boldsymbol{x}.$$

This solution is unique because $\left|\frac{\partial \boldsymbol{X}}{\partial \boldsymbol{a}}\right| = 1$ due to $\nabla\cdot\boldsymbol{u} = 0$ [Monin, Yaglom, 1975]. From this and (1.10) readily follows that if $c_0(\boldsymbol{x}) = \delta(\boldsymbol{x})$ then

$$\langle c(t,\boldsymbol{x})\rangle = P_{\boldsymbol{X}(t)}(\boldsymbol{x}), \tag{1.11}$$

where $P_{\boldsymbol{X}(t)}(\boldsymbol{x})$ is the probability density of $\boldsymbol{X}(t)$.

Matching (1.6) and (1.8) we have

$$\langle c(t,\boldsymbol{x})\rangle = \left\langle Mc_0\left(\boldsymbol{\xi}^{\kappa}_{t,x}\right)\right\rangle = M\left\langle c_0\left(\boldsymbol{\xi}^{\kappa}_{t,\boldsymbol{x}}\right)\right\rangle, \tag{1.12}$$

where $\boldsymbol{\xi}^{\kappa}_{t,x} = \boldsymbol{\xi}^{\kappa}_{t,x}(0)$.

As mentioned above, in general we cannot compute the double average in (1.12) because we can not find the solution of (1.7) in an explicit form. Moreover, we cannot do this even when $\kappa = 0$. But there is a trivial but useful example in which such a computation can be easily carried out.

Namely, suppose that $\boldsymbol{u} = \boldsymbol{u}(t)$ is the function of time only. Then from (1.7) one can get

$$\boldsymbol{\xi}^{\kappa}_{t,\boldsymbol{a}} = \boldsymbol{a} - \int_0^t \boldsymbol{u}(s)ds + \sqrt{2\kappa}\ \boldsymbol{w}(t). \tag{1.13}$$

Hence from (1.6), (1.8)

$$c(t,\boldsymbol{x}) = Mc_0\left(\boldsymbol{x} - \int_0^t \boldsymbol{u}\ (t-s)ds + \sqrt{2\kappa}\ \boldsymbol{w}(t)\right) = \tag{1.14}$$

$$= \int_{E^d} c_0(\boldsymbol{x} - \boldsymbol{y} - \int_0^t \boldsymbol{u}(s)ds)\ e^{-\frac{y^2}{2t\kappa}}\ \frac{d\boldsymbol{y}}{(2\pi t\kappa)^{d/2}}.$$

To average this functional with respect to $\boldsymbol{u}$ it is convenient use Fourier transform

$$c_0(\boldsymbol{x}) = \int \hat{c}_0(\boldsymbol{k})\ e^{i\boldsymbol{k}\cdot\boldsymbol{x}}d\boldsymbol{x}.$$

Using the well known equality $\langle e^{\xi}\rangle = e^{\frac{1}{2}\langle\xi^2\rangle}$ for Gaussian mean-zero value ξ, one can easily obtain

$$\begin{aligned}\langle c(t,\boldsymbol{x})\rangle &= \int_{E^d}\int_{E^d} \hat{c}_0(\boldsymbol{k})e^{i\boldsymbol{k}\cdot(\boldsymbol{x}+\boldsymbol{y})}\frac{1}{(2\pi t\kappa)^{d/2}}e^{-\frac{y^2}{2t\kappa}}\left\langle e^{-i\boldsymbol{k}\cdot\int_0^t \boldsymbol{u}(s)ds}\right\rangle d\kappa d\boldsymbol{y} = \\ &= \int_{E^d} \hat{c}_0(\boldsymbol{k})\ e^{i\boldsymbol{k}\cdot\boldsymbol{x} - \kappa t\boldsymbol{k}^2 - \frac{1}{2}\boldsymbol{\sigma}^2(t)\boldsymbol{k}\cdot\boldsymbol{k}}d\boldsymbol{k},\end{aligned} \tag{1.15}$$

where

$$\boldsymbol{\sigma}^2(t) = \left\langle \int_0^t \boldsymbol{u}(s)ds\ \left(\int_0^t \boldsymbol{u}(s)ds\right)^T\right\rangle =$$

$$= \int_0^t \int_0^t \boldsymbol{R}_{\boldsymbol{u}}(s_1 - s_2) ds_1 \, ds_2 = \int_{-t}^t (t - |s|) \boldsymbol{R}_{\boldsymbol{u}}(s) ds$$

It follows from (1.15) that $\langle c(t, \boldsymbol{x}) \rangle$ satisfies eq.(1.5) with

$$\boldsymbol{D} = \boldsymbol{D}(t) = \frac{1}{2} \frac{d\boldsymbol{\sigma}^2(t)}{dt} = \int_0^t \boldsymbol{R}_{\boldsymbol{u}}(s) ds. \tag{1.16}$$

In the same manner we can obtain an answer to question (iii).

This simple example shows that in order to get a closed equation for the mean tracer we do not need any additional "closure" conjecture such as e.g. $\langle \boldsymbol{u}' c' \rangle$ is proportional to $\nabla \langle c \rangle$, used in the physical literature [Monin, Yaglom, 1975]. Difficulties arising in averaging the advection-diffusion equation have an analytical origin whereas the problem itself is well posed as a mathematical problem.

There are different kinds of assumptions which enable us to overcome those analytical obstacles. Roughly all such assumptions can be broken down into the three following groups: 1) constraints on scales; 2) constraints on correlations and 3) others.

For example, the well known weak interaction approach of [Hasselman, 1966] and used by Davis in 1982 for the analysis of turbulent diffusion, is based on the assumption that the ratio α of typical particle velocities to typical phase velocities in the Eulerian field, is small, $\alpha \ll 1$. Under this assumption the solution of (1.9) can be expanded in the perturbation series

$$\boldsymbol{X}(t) = \boldsymbol{X}_0(t\alpha^2) + \alpha \boldsymbol{X}_1(t) + \alpha^2 \boldsymbol{X}_2(t) + \ldots. \tag{1.17}$$

As a result, the Lagrangian correlation function is computed in terms of $\boldsymbol{R}_{\boldsymbol{u}}(t, \boldsymbol{x})$ with the accuracy $O(\alpha^4)$. Using relation (1.11) it is not difficult to find the effective diffusivity defined as

$$\widetilde{\boldsymbol{D}} = \frac{\partial}{\partial t} \langle \boldsymbol{X}(t) \boldsymbol{X}(t)^T \rangle. \tag{1.18}$$

We use a different notation for this quantity in order to stress that in the general case, $\widetilde{\boldsymbol{D}}$ does not coincide with $\boldsymbol{D}$ defined by eq.(1.5). Obviously, in the case of the space independent velocity field $\widetilde{\boldsymbol{D}} \equiv \boldsymbol{D}$. The same holds true under the assumption that $\boldsymbol{X}(t)$ is Gaussian .

As an example of correlation constraints let us recall the classical conjecture of Saffman, 1969. Assuming that

$$\langle \boldsymbol{u} \boldsymbol{u} c' \rangle = 0 \tag{1.19}$$

we readily arrive at the relation

$$\boldsymbol{R}_{\boldsymbol{v}}(t) = \langle \boldsymbol{v}(s, \boldsymbol{a}) \boldsymbol{v}(s + t, \boldsymbol{a})^T \rangle = \int_{E^d} \boldsymbol{R}_{\boldsymbol{u}}(t, \boldsymbol{x}) \langle c(t, \boldsymbol{x}) \rangle d\boldsymbol{x}, \tag{1.20}$$

where $\boldsymbol{v}(t,\boldsymbol{a}) = \partial \boldsymbol{X}(t,\boldsymbol{a})/\partial t$ is the Lagrangian velocity, $\boldsymbol{R_v}(t)$ is its correlation tensor and $c(t,\boldsymbol{x})$ is the solution of (1.1) with $\kappa = 0$ and initial condition $c_0(\boldsymbol{x}) = \delta(\boldsymbol{x}-\boldsymbol{a})$. In the spectral form (1.20) can be rewritten as

$$(1.21) \qquad \boldsymbol{R_v}(t) = \int_{-\infty}^{\infty}\int_{E^d} \boldsymbol{E}(\omega,\boldsymbol{k})\, e^{i\omega t}\langle e^{i\boldsymbol{k}\cdot(\boldsymbol{X}(t,\boldsymbol{a})-\boldsymbol{a})}\rangle d\boldsymbol{k}\, d\omega.$$

Let us note that under the Gaussian hypothesis for $\boldsymbol{X}(t,\boldsymbol{a})$ relation (1.21) yields the nonlinear integral equation for the Lagrangian correlation function

$$(1.22) \quad \boldsymbol{R_v}(t) = \int\int e^{i\omega t}\, \boldsymbol{E}(\omega,\boldsymbol{k}) e^{-\frac{1}{2}\left(\int_{-t}^{t}(t-|s|)\boldsymbol{R_v}(s)\boldsymbol{k}\cdot\boldsymbol{k}\, ds\right)} d\omega d\boldsymbol{k}$$

It is likely that eq. (1.22) provides an accurate approximation for $\boldsymbol{R_v}(t)$, but this conclusion should be checked by simulation. On the other hand (1.22) may be viewed as an equation for $\boldsymbol{E}(\omega,\boldsymbol{k})$ and can be used to deduce the Eulerian velocity characteristics from the Lagrangian observations. This problem is very important in oceanography where Lagrangian devices are now used widely [Davis, 1991]. Equality (1.21) was also derived by Roberts in 1961 via the Direct Interaction Approach (DIA) developed by Kraichnan,1959. Basically this approach reduces to correlation constraints as well.

Finally let us mention the method of successive approximation due to [Phythian,1975], as an example from the group "others". The main idea is to look for the solution of (1.9) using the recurrence equation

$$\boldsymbol{X}_{n+1}(t) = \boldsymbol{a} + \int_0^t \boldsymbol{u}(s,\boldsymbol{X}_n(s))ds, \quad \boldsymbol{X}_0(t) \equiv \boldsymbol{a}.$$

It appears that the correlation function for the second approximation $\boldsymbol{X}_2(t)$ can be computed in an explicit form and the approximate formula for the turbulent diffusivity

$$\widetilde{\boldsymbol{D}} \approx \frac{d}{dt}\langle \boldsymbol{X}_2(t)\boldsymbol{X}_2(t)^T\rangle$$

is highly accurate for some models of random flow.

The scale constraints approach is preferable because of two advantages. First, the relation between scales may be stated in a rigorous mathematical form. Later we will illustrate this. Second, scale relations can be verified in many real problems via experimental data. As for correlation constraints the essential disadvantage is that of referring to the unknown field $c'(t,\boldsymbol{x})$ which must be found from the problem. Hence these constraints can be checked only after the problem is solved. For example, in the one dimensional version of the space independent velocity field discussed above we have for the initial field $c_0(x) = \delta(x)$

$$\langle uuc' \rangle = -\frac{\left(\int_0^t R_u(s)ds\right)^2}{2\pi \left(\int_0^t (t-s)R_u(s)ds\right)^{3/2}}$$

at $x = 0$. One can see that this correlation never vanishes, but is small compared with $\langle u^2 \rangle \langle c'^2 \rangle^{1/2}$ for either big t or for small velocity variance $\langle u^2 \rangle$.

2. Time scales. In this chapter, we introduce our definitions for the correlation radius and the correlation time of the underlying velocity field. There are different ways to define these scales. Our definitions reveal the purpose for studying velocity fields which are smooth in space and quickly decorrelating in time. Both of these assumptions are reasonable from a hydrodynamics point of view, due to the dissipation of energy through small space scales and the short memory of developed turbulence [Batchelor, 1982]

Restricting ourselves by the isotropic case we define the correlation radius and time as

$$l_u = \left(\frac{\int\int E_L(\omega, k) k^{d-1} d\omega dk}{\int\int E_L(\omega, k) k^{d+1} d\omega dk} \right)^{1/2} \tag{2.1}$$

and

$$\tau_u = \frac{\int E_L(o, k) k^{d-1} dk}{\int\int E_L(\omega, k) k^{d-1} d\omega dk}. \tag{2.2}$$

The integration with respect to ω is over the entire real line and with respect to k from zero to infinity unless otherwise specified. Let us note that the variance $\sigma_u^2 = \langle |\boldsymbol{u}|^2 \rangle$ is given by

$$\sigma_u^2 = \vartheta_d (d-1) d^{-1} \int\int E_L(\omega, k) k^{d-1} d\omega dk, \tag{2.3}$$

where θ_d is the volume of a unit ball, in part $\vartheta_2 = 2\pi$, $\vartheta_3 = 4\pi$.

Further we will repeatedly turn to the spectrum considered in [Avellaneda, Majda, 1990, 1992]

$$E_L(\omega, k) = \begin{cases} \frac{Cak^{2-d-\alpha+z}}{\omega^2+a^2k^{2z}}, & C_0\bar{k} < k < C_1\bar{k} \\ 0, & \text{otherwise} \end{cases} \tag{2.4}$$

Here $\bar{k}$ is a wave length scale. The constants C_0, C_1 determine the upper and lower cutoff and, C, a, α, z are positive parameters. For the wave spectrum $E_L(k) = \int E_L(\omega, k)d\omega$ we have

$$E_L(k) = C\pi k^{2-d-\alpha} \quad , \quad C_0\bar{k} < k < C_1\bar{k}$$

and hence α determines the slope of spectrum for the above wave number range. The time correlation function of the harmonics due to the wavenumber k decays as $exp\{-k^z t\}$. Thus the parameter z characterizes the rate of decorrelation. Via straightforward calculations one can obtain from (2.1-2.3)

$$l_u = \frac{1}{\bar{k}} \sqrt{\frac{(C_1^{-\alpha+2} - C_0^{-\alpha+2})(4-\alpha)}{(2-\alpha)(C_1^{4-\alpha} - C_0^{4-\alpha})}}, \tag{2.5}$$

$$\tau_u = \frac{1}{a\bar{k}^z} \frac{(2-\alpha)(C_1^{2-\alpha-z} - C_0^{2-\alpha-z})}{\pi(2-\alpha-z)(C_1^{2-\alpha} - C_0^{2-\alpha})}, \tag{2.6}$$

$$\sigma_u^2 = \vartheta_d(d-1) \frac{C\pi\bar{k}^{2-\alpha}}{da(2-\alpha)} (C_1^{2-\alpha} - C_0^{2-\alpha})$$

Let us consider the Cauchy problem for equation (1.1) with an isotropic Gaussian velocity field whose statistics is fully determined by parameters σ_u, τ_u, l_u given an initial field $c_0(\boldsymbol{r})$ determined by a single length scale, l_0. In doing so we assume that $c_0(\boldsymbol{r})$ is either random or deterministic. If for example $c_0(\boldsymbol{r})$ is assumed to be isotropic random field then l_0 can be defined as in (2.1). Thus, we have 6 parameters

$$\{t, \sigma_u, \tau_u, l_u, l_0, \kappa\}, \tag{2.7}$$

which fully determine the mean field $\langle c(t, \boldsymbol{r})\rangle$ and the rest of the tracer statistics.

There are four independent time scales that can be posed in terms of these parameters. Namely the current time t, the Eulerian correlation time

$$\tau_E \equiv \tau_u \tag{2.8}$$

given by (2.2), the turnover time

$$\tau_T = \frac{l_u}{\sigma_u} \tag{2.9}$$

and finally the molecular diffusion time

$$\tau_D = \frac{l_0^2}{\kappa}. \tag{2.10}$$

Therefore τ_E characterizes the correlation time of the time series $\boldsymbol{u}_{\boldsymbol{x}}(t) = \boldsymbol{u}(t, \boldsymbol{x})$ obtained at the arbitrary fixed point $\boldsymbol{x}$, meanwhile τ_T shows how fast a Lagrangian particle trapping by an eddy performs one revolution. In the "eddy" terminology τ_E also can be interpreted as the lifetime of eddy. We call the introduced scales independent since each of them is determined by a distinct subset of the full set (2.7) of parameters. Note that there is another molecular diffusion time scale, $\widetilde{\tau_D} = l_u^2/\kappa$, in this problem, but we cannot consider $\widetilde{\tau_D}$ as independent of τ_T because both include the same parameter l_u.

If all scales have the same order

$$t \sim \tau_E \sim \tau_T \sim \tau_D \tag{2.11}$$

then any effort to find an explicit formula for $\langle c(t, \boldsymbol{x}) \rangle$ fails. To make progress in computing the tracer statistics it is necessary to have separate scales. For instance the separation assumption

$$t \sim \tau_E \sim \tau_D \ll \tau_T \tag{2.12}$$

may be interpreted in mathematical form as a space independent velocity field $\boldsymbol{u}(t, \boldsymbol{x}) = \boldsymbol{u}(t)$ discussed in detail above. Thus under (2.12) all statistics of the passive scalar can be found immediately. Before we give other separation relationships which imply solvability of the turbulent diffusion problem, let us count all possible scale separations. In other words how many different ways exist to place $t, \tau_E, \tau_D, \tau_T$ in the circles on Fig.1, where the arrow means either $\sim$ or $\ll$? If doing so we should take into account that some combinations are equivalent. For example $\tau_E \sim t \sim \tau_D \ll \tau_T$ is equivalent to (2.12). The answer is 74 provided the hopeless configuration (2.11) is excluded. Let us exhibit some examples of the order relation under which an exact asymptotic can be found.

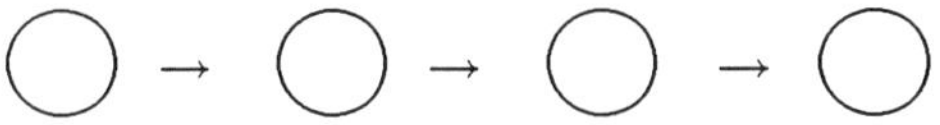

FIG. 1. *To the classification of scale separation.*

First let us address once again the weak interaction approach. The starting point, in terms of the introduced scales is expressible in the form

$$\tau_T \ll \tau_E \sim t \ll \tau_D . \tag{2.13}$$

The assumption $t \ll \tau_D$ means that the molecular diffusion is ignored.

It is not difficult to compute the short time asymptotics. In absence of the molecular diffusion it was first found by Batchelor, 1952 and Roberts, 1961. In our classification this case can be written as

$$t \ll \tau_T \sim \tau_E \ll \tau_D \tag{2.14}$$

Under (2.14) we have

$$P_{\boldsymbol{X}(t,\boldsymbol{a})}(\boldsymbol{x}) \sim \frac{1}{(2\pi t^2\sigma_u^2)^{d/2}} e^{-\frac{(\boldsymbol{X}-\boldsymbol{a})^2}{2t^2\sigma_u^2}} \tag{2.15}$$

and therefore $\boldsymbol{D}(t) = D(t)\boldsymbol{I}$, where $\boldsymbol{I}$ is the identity matrix and

$$D(t) \equiv \widetilde{D}(t) \sim \sigma_u^2 t. \tag{2.16}$$

The adding the molecular diffusion is obvious. If

$$t \ll \tau_T \sim \tau_E \sim \tau_D \tag{2.17}$$

then

$$P_{\boldsymbol{X}(t,\boldsymbol{a})}(\boldsymbol{x}) \sim \frac{1}{2\pi\left[(t^2\sigma_u^2 + \kappa t)\right]^{d/2}} e^{-\frac{(\boldsymbol{x}-\boldsymbol{a})^2}{2(t^2\sigma_u^2+\kappa t)}} \tag{2.18}$$

and hence

$$D(t) \sim \sigma_u^2 t + \kappa. \tag{2.19}$$

Let us note that since formulas (2.18,2.19) do not include in explicit form τ_E and τ_T they cover more than one limiting case in our classification. More exactly they are valid for $t \ll \tau_D \sim \tau_E \ll \tau_T$ and $t \ll \tau_D \sim \tau_T \ll \tau_E$, etc. The same remark is true for some cases discussed below.

A reasonable mathematical model for large Eulerian correlation

$$t \sim \tau_D \sim \tau_T \ll \tau_E \tag{2.20}$$

is the steady (time independent) velocity field

$$\boldsymbol{u}(t, \boldsymbol{x}) = \boldsymbol{u}(\boldsymbol{x}). \tag{2.21}$$

In general form assumption (2.21) does not simplify the problem at all. Thus, the scale separation (2.20) is not sufficient for problem solvability. But the additional order constraint

$$\tau_T \ll t \sim \tau_D \ll \tau_E \tag{2.22}$$

leads to a substantial mathematical theory which will be commented below.

Note that if a scale separation includes two or more symbols $\ll$, then the problem becomes undetermined in the sense that there can be a variety of asymptotical regimes for $\langle c \rangle$. Let us give a simple example concerning the multiscale problem as whole. Imagine that there are three scales m_1, m_2, m_3 determining the behaviour of a variable $z = z(m_1, m_2, m_3)$. The relation

$$m_1 \ll m_2 \ll m_3 \tag{2.23}$$

means that $m_1/m_2 \to 0$, $m_2/m_3 \to 0$. If the variable z is a function of $m_1 m_3/m_2^2$, then assumption (2.23) is not enough to compute the asymptotic of z.

Further, we restrict ourselves to the case where all time scales are rigidly connected and hence will not be a room for an ambiguity. Namely, let $\varepsilon > 0$ be a formal dimensionless small parameter and let us change the coordinates to

$$t' = \alpha(\varepsilon)t, \quad \boldsymbol{r}' = \beta(\varepsilon)\boldsymbol{r},$$

where $\alpha(\varepsilon)$, $\beta(\varepsilon)$ are some dimensionless functions. As a result the equation (1.1) turns to

$$\frac{\partial c}{\partial t'} + \boldsymbol{u}_\varepsilon(t', \boldsymbol{r}') \cdot \nabla c = \kappa_\varepsilon \nabla^2 c$$

where

$$\boldsymbol{u}_\varepsilon(t', \boldsymbol{r}') = \frac{\beta(\varepsilon)}{\alpha(\varepsilon)} \boldsymbol{u} \left(\frac{t'}{\alpha(\varepsilon)}, \frac{\boldsymbol{r}'}{\beta(\varepsilon)} \right), \kappa_\varepsilon = \frac{\beta(\varepsilon)^2}{\alpha(\varepsilon)} \kappa.$$

Such a transformation of variables we call the rescaling or renormalization. It is clear that in this situation all the time scales are functions of ε only and hence any scale separation completely determines the asymptotical behaviour of the mean tracer. For example, (2.2) can be obtained by $\alpha(\varepsilon) = \varepsilon^2$, $\beta(\varepsilon) = \varepsilon$. The latter is equivalent to the following renormalization of the steady velocity field

$$\boldsymbol{u}_\varepsilon(t, \boldsymbol{x}) = \frac{1}{\varepsilon} \boldsymbol{u} \left(\frac{\boldsymbol{x}}{\varepsilon} \right). \tag{2.24}$$

The most important and well-known result in this area is the following homogenization statement. Roughly, if $c_\varepsilon(t, \boldsymbol{x})$ is the solution of

$$\frac{\partial c_\varepsilon}{\partial t} + \boldsymbol{u}_\varepsilon \cdot \nabla c_\varepsilon = \kappa c_\varepsilon$$

due to a square integrable initial condition, then the limiting tracer field

$$\bar{c}(t, \boldsymbol{x}) = \lim_{\varepsilon \to 0} c_\varepsilon(t, \boldsymbol{x})$$

is deterministic and satisfies an equation of form (1.5). This assertion is proved in rigorous mathematical way by different authors [Kozlov, 1980, Varadhan & Papanicolaou, 1982, Avellaneda & Majda, 1990].

3. Short-correlation approximation.

3.1. Classification. We now proceed to the main subject of our consideration, the short-correlation approximation. The basis of this approximation is the following assumption

$$\tau_E \ll t \sim \tau_D \tag{3.1.1}$$

i.e. the Eulerian correlation time is much less than the current time. The diffusion time is allowed to be much bigger than t as well.

Let us note that thus far we have not fixed τ_T in our definition of short-correlation velocity. Suppose that $t \sim 1$ then one can see that there are 5 options for the turnover time (see Fig.2) corresponding to the relations

$$\begin{array}{ll} (A) & \tau_T \gg 1 \\ (B) & \tau_T \sim 1 \\ (C) & \tau_E \ll \tau_T \ll 1 \\ (D) & \tau_T \sim \tau_E \ll 1 \\ (E) & \tau_T \ll \tau_E \ll 1. \end{array} \tag{3.1.2}$$

The assumptions of scale separation $(A), (B), (C), (D)$ can hold for a velocity field with a finite low-frequency energy

$$\int_0^\infty E_L(0,k)k^{d-1}dk < \infty \tag{3.1.3}$$

under the following rescaling

$$\begin{array}{ll} (A) & \boldsymbol{u}_\varepsilon(t,\boldsymbol{x}) = \boldsymbol{u}(t/\varepsilon^2) \\ (B) & \boldsymbol{u}_\varepsilon(t,\boldsymbol{x}) = \boldsymbol{u}(t/\varepsilon^2, \boldsymbol{x}) \\ (C) & \boldsymbol{u}_\varepsilon(t,\boldsymbol{x}) = \frac{1}{\varepsilon}\boldsymbol{u}(t/\varepsilon^2, \boldsymbol{x}) \\ (D) & \boldsymbol{u}_\varepsilon(t,\boldsymbol{x}) = \frac{1}{\varepsilon}\boldsymbol{u}(t/\varepsilon^2, \boldsymbol{x}/\varepsilon). \end{array} \tag{3.1.4}$$

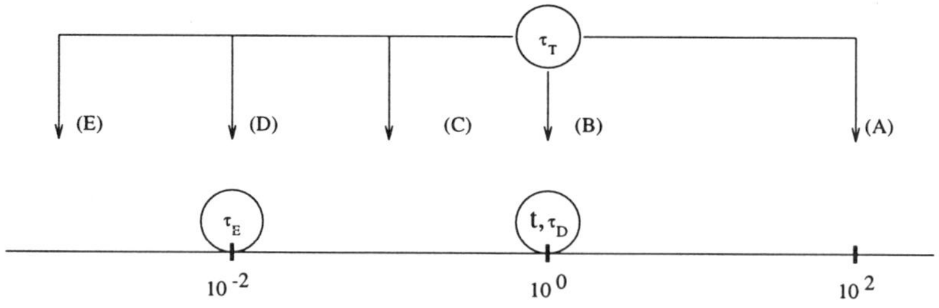

FIG. 2. *Five versions of the short correlation approximation* ($\varepsilon = 10^{-1}$)

Indeed, from condition (3.1.3), definitions (2.1),(2.2) and relation (2.9) it follows for the renormalized velocity field $\boldsymbol{u}_\varepsilon(t,\boldsymbol{x})$ that

$$\tau_E^{(\varepsilon)} \sim \varepsilon^2 \ , \ \tau_D^{(\varepsilon)} \sim 1 \tag{3.1.5}$$

and

$$\text{(3.1.6)}\qquad \begin{array}{ll} \text{(A)} & \tau_T^{(\varepsilon)} = \infty \\ \text{(B)} & \tau_T^{(\varepsilon)} \sim 1 \\ \text{(C)} & \tau_T^{(\varepsilon)} \sim \varepsilon \\ \text{(D)} & \tau_T^{(\varepsilon)} \sim \varepsilon^2. \end{array}$$

The situation with separation (E) from (3.1.2) is much more complicated, nevertheless this case is very important for applications in hydrodynamics and oceanography. Here we only give a particular example of separation (E) and a detailed discussion is deferred to section 3.4 which is devoted to superdiffusion.

Let us consider the spectrum (2.4) with the following restrictions on the parameters

$$\text{(3.1.7)}\qquad \alpha + z > 2\ ,\ \alpha + 2z < 4$$

and assume that the lower boundary $C_0 \equiv C_0\varepsilon$ is small, e.g. in the classical turbulence theory $\varepsilon = (Re)^{-3/4}$ where Re is the Reynolds number [Batchelor, 1982]. Let $c_\varepsilon(t,x)$ be the solution of the Cauchy problem

$$\text{(3.1.8)}\qquad \frac{\partial c_\varepsilon}{\partial t} + \boldsymbol{u}_\varepsilon(t,\boldsymbol{x})\cdot\nabla c_\varepsilon = \kappa_\varepsilon \nabla^2 c_\varepsilon\ ,\ c_0(x) \in L_2(E^d)$$

with the following renormalization

$$\text{(3.1.9)}\qquad \boldsymbol{u}_\varepsilon(t,\boldsymbol{x}) = \frac{\varepsilon}{\rho^2(\varepsilon)}\boldsymbol{u}\left(\frac{t}{\rho^2(\varepsilon)}\ ,\ \frac{\boldsymbol{x}}{\varepsilon}\right),\ \kappa_\varepsilon = \frac{\varepsilon^2}{\rho^2(\varepsilon)}\kappa.$$

It was shown in [Avellaneda, Majda, 1992] that if

$$\text{(3.1.10)}\qquad \rho(\varepsilon) = \varepsilon^{\frac{4-\alpha-z}{2}}$$

then there exist a nondegenerate limit $\lim\limits_{\varepsilon\to 0}\langle c_\varepsilon(t,\boldsymbol{x})\rangle = \bar{c}(t,\boldsymbol{x})$ and the limiting function satisfies an equation of the form (1.5).

The renormalized velocity field (3.1.9) is determined by the same type of spectrum as (2.4), but its parameters are changed by

$$\text{(3.1.11)}\qquad C^{(\varepsilon)} = \frac{C\varepsilon^{4-\alpha}}{\rho^4}\ ,\ C_1^{(\varepsilon)} = C_1/\varepsilon\ ,\ a_\varepsilon = \frac{\varepsilon^z}{\rho^2}a$$

Taking into account conditions (3.1.7) we obtain from the general formulas (2.5,2.6) and definition (2.10) of the diffusion time

$$\text{(3.1.12)}\qquad \tau_E^{(\varepsilon)} \sim \begin{cases} \varepsilon^{6-2\alpha-2z} & \alpha < 2 \\ \varepsilon^{4-\alpha-2z} & \alpha > 2 \end{cases}$$

$$\tau_T^{(\varepsilon)} \sim \varepsilon^{4-\alpha-z}$$

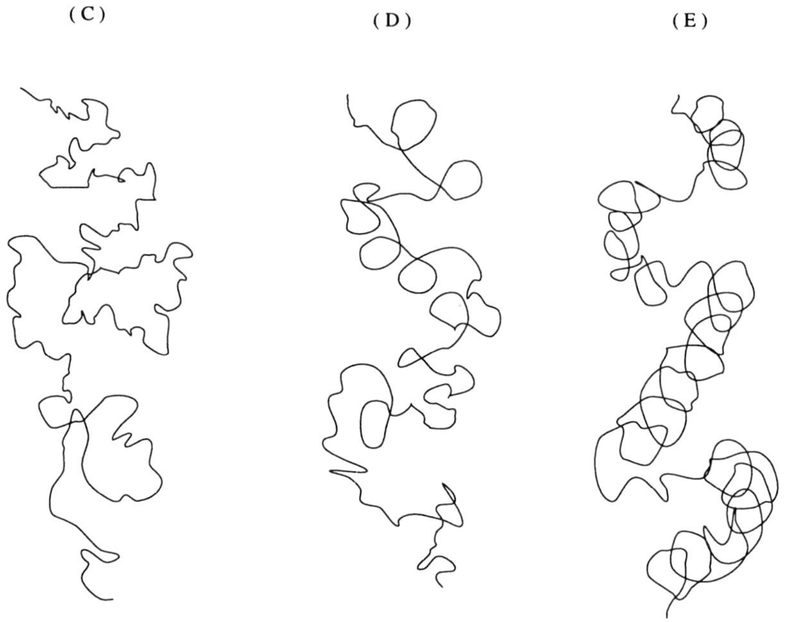

FIG. 3. *Typical Lagrangian trajectories* $\boldsymbol{X}(t)$ *in 2D random flow (C):* $\tau_T \gg \tau_E$ *(D):* $\tau_T \sim \tau_E$ *(E):* $\tau_T \ll \tau_E$

$$\tau_D^{(\varepsilon)} \sim \varepsilon^{2-\alpha-z}$$

From (3.1.12) it follows that in the region (3.1.7) we have the following separation of scales

$$\tau_T \ll \tau_E \ll t \ll \tau_D \tag{3.1.13}$$

To illustrate the difference between cases (C), (D) and (E) we give sketches of typical Lagrangian trajectories for $d = 2$ (Fig.3). In the first case the trajectory does not contain loops, because the eddy is broken earlier than a particle trapped by it completes the turnover. Of course, this case does not preclude self-intersection. In case (D) the particle trapped by an eddy performs a number of full revolutions before the eddy dies. Finally in case (E), the number of complete revolutions can be very large.

From the analytical viewpoint cases (A), (B) are not difficult. For model (A) we have the Fickian equation for the mean concentration with effective diffusivity [Kubo, 1967]

$$\boldsymbol{D} \sim \varepsilon^2 \int_0^\infty \boldsymbol{R_u}(t)dt. \tag{3.1.14}$$

In case (B) we have a similar result

$$D \sim \varepsilon^2 \int_0^\infty R_u(t,0)dt. \tag{3.1.15}$$

In both cases the effective diffusivity is of the order ε^2 and hence is much smaller than molecular diffusivity ($\tau_D \sim 1$). For this reason we do not focus on these cases. The second reason is that (A) and (B) can be deduced from case (C) considered in detail below, by simply multiplying the velocity by ε. Sometimes assumptions (A),(B) are called the quasilinear approximation.

3.2. δ-correlated velocity field. In this section we address the velocity field renormalized as follows

$$u_\varepsilon(t,x) = \frac{1}{\varepsilon}\, u\left(\frac{t}{\varepsilon^2}, x\right) \tag{3.2.1}$$

where ε is small.

From (3.1.5) and (3.1.6, C) it follows that the separation of scales is

$$\tau_E \ll \tau_T \ll 1. \tag{3.2.2}$$

Under certain general conditions the random field $u_\varepsilon(t,x)$ defined by (3.2.1) converges to the white noise process $\dot{w}(t,x)$ in the sense of distributions, i.e.

$$\lim_{\varepsilon\to 0} R_u^{(\varepsilon)}(t,x) = \delta(t)\; B_u(x) \tag{3.2.3}$$

where $R_u^{(\varepsilon)}(t,x) = \frac{1}{\varepsilon^2}\, R_u\left(\frac{t}{\varepsilon^2}, x\right)$ is the correlation function of the field $u_\varepsilon(t,x)$ and

$$B_u(x) = \int_{-\infty}^{\infty} R_u(t,x)dt. \tag{3.2.4}$$

As a result we immediately obtain the Fokker-Planck equation for the mean concentration

$$\frac{\partial\langle c\rangle}{\partial t} = \nabla\cdot D\nabla\langle c\rangle + \kappa\nabla^2\langle c\rangle \tag{3.2.5}$$

with

$$D = \frac{1}{2}B_u(0) = \int_0^\infty R_u(t,0)dt = \pi \int E_u(0,k)dk.$$

This equation is well known and any comments are hardly required about it.

Here we focus on the less known equation for the space correlation function of the tracer

$$R_c(t, \boldsymbol{x}) = \langle c'(t, \boldsymbol{y}) c'(t, \boldsymbol{y} + \boldsymbol{x}) \rangle .$$

In the isotropic case this equation can be written as

$$\frac{\partial R_c(t,r)}{\partial t} = \frac{1}{r^{d-1}} \frac{\partial}{\partial r} r^{d-1} \left(2\kappa + F(r)\right) \frac{\partial R_c(t,r)}{\partial r} \tag{3.2.6}$$

where as before $r = |\boldsymbol{x}|$, $d = 2, 3, \cdots$ is the dimension of space and

$$F(r) = B_L(0) - B_L(r), \; B_L(r) = \int_{-\infty}^{\infty} R_L(t,r) dt. \tag{3.2.7}$$

A similar equation can be derived in the non-isotropic and even in non-homogeneous unstationary case [Molchanov, Piterbarg, 1994; Klyatskin, 1994].

Equation (3.2.6) has a long history. Its predecessor had appeared in [Richardson, 1926] before the notion of correlation function has been widely used. After this, equations of form (3.2.6) were repeatedly recovered in a number of papers [Roberts, 1961, Kraichnan, 1974, Lundgren, 1981, Bennet, 1987 etc.]. Those authors proposed different expressions for the "two-particle" diffusion coefficient $F(r)$ and different assumptions were used. But in a rigorous way, only for the δ-correlated velocity field, can such an equation be obtained [Molchanov, Piterbarg, 1994; Semenov, 1990].

Now we point out two interesting and important physical effects which follow from (3.2.6). The first case addresses the lifetime of tracer anomalies.

Let us consider the following mathematical model. The initial field $c_0(\boldsymbol{x})$ is assumed to be homogeneous and isotropic with zero mean and correlation function

$$R_c^{(0)}(r) = \langle c_0(\boldsymbol{y}) c_0(\boldsymbol{y} + \boldsymbol{x}) \rangle \;, \quad r = |\boldsymbol{x}|. \tag{3.2.8}$$

Thus, the evolution of tracer fluctuations (anomalies) is described by eq. (3.2.6) with initial condition (3.2.8). It can easily be seen that $R_c(t,r)$ decays in time for each r. Let $\sigma_0^2 = R_c^{(0)}(0)$ be the variance of the initial anomalies and $\sigma_t^2 = R_c(t,0)$ be the variance at moment t. It is natural to define the lifetime of anomaly as the time it takes σ_t^2 becomes N times less than σ_0^2, where N can be chosen for practical reasons as 10, 100, e.t.c. Formally, the solution τ_l of the equation

$$R_c(\tau_l, 0) = \frac{1}{N} R_c^{(0)}(0), \tag{3.2.9}$$

is called the anomaly lifetime where $R_c(t,r)$ is the solution of eq. (3.2.6) due to initial condition (3.2.8). The subscript l is for "lifetime".

Here we give an asymptotic formula for τ_l under a large Peclet number $Pe = \sigma_u l_u/\kappa$. A detailed derivation of this formula is given in [Molchanov, Piterbarg, 1994]

$$\tau_l \sim \begin{cases} C_0 \dfrac{l_u^2}{D} \log \dfrac{l_0 (Pe)^{1/2}}{l_u} & Pe^{1/2} \ll \dfrac{l_0}{l_u} \ll 1 \\ C_1 \dfrac{l_u^2}{2D} \log Pe + C_2 \dfrac{l_0^2}{D} & \dfrac{l_0}{l_u} \gg 1 \end{cases} \tag{3.2.10}$$

where the constants C_0, C_1, C_2 depend on N, dimension d and the shape of $R_c^{(0)}(r)$, $l_0 = \sqrt{-R_c^{(0)}(r)/\nabla^2 R_c^{(0)}(0)}$ is the space correlation scale of the initial passive scalar distribution and $D = B_L(0)$ is the eddy diffusivity.

There are two interesting effects following from (3.2.10). First if the initial anomalies have a small size compared to the correlation cell of velocity fields then their lifetime is much longer than it is given by naive estimation

$$\tau_l \sim \frac{l_0^2}{D} \tag{3.2.11}$$

which is based on the analogy between turbulent and molecular diffusion. Such an analogy is only valid when $l_u \gg l_0$. In this case the second part of (3.2.10) yields the classical formula (3.2.11). This effect of prolonging the lifetime by the velocity space correlation is obvious. But the point is that we give a quantitative description of the phenomena.

The second observation being worthy of notice is that τ_N depends on the logarithm of the Peclet number. In some practical situations we know Pe with accuracy only up to 2-3 orders, because of a poor estimates the effective diffusivity. It can be seen that the range of the lifetime variability in this case is much more narrow.

Under the discussed diffusion approximation we can directly study the behaviour of the gradient of the tracer fluctuations also. In particular, the explicit equation for the gradient correlation matrix

$$G(t, \boldsymbol{x}, \boldsymbol{y}) = \langle \nabla c'(t, \boldsymbol{x}) \; \nabla c'(t, \boldsymbol{y})^T \rangle \tag{3.2.12}$$

can be deduced [Molchanov, Piterbarg, 1994, Klyatskin, 1994].

Here we restrict ourselves by the isotropic velocity field and isotropic initial tracer. The question of most interest is whether the gradients are steepened in time? For the molecular diffusion the answer is trivial "no", but for the turbulent diffusion the answer is not quite so obvious. Some examples of increasing gradients by a deterministic velocity field in application to oceanic processes were discussed by Eckart, 1948. But these effects were mostly related to inhomogeniuty of both the velocity and initial tracer distribution. The processes by which gradients increase or decay were called by "stirring" and "mixing" respectively. As for this terminology, questions of our interest are as follows. Is it possible to have stirring under statistically homogeneous conditions? If "yes" how long is the stirring stage? It is clear that mixing stage is obligatory and should follow the stirring because of the molecular diffusion. To answer the stated questions let us analyze the behaviour of the mean square gradient

$$g(t) = \langle |\nabla c'(t, \boldsymbol{x})|^2 \rangle$$

for an arbitrary point $\boldsymbol{x}$. Due to homogeneity, $g(t)$ is independent of $\boldsymbol{x}$. It was shown in [Molchanov, Piterbarg, 1994] that under the condition

$$l_0 > \frac{1}{2} l_u Pe^{-1/2} \tag{3.2.13}$$

the function $g(t)$ increases during the time τ_*, which for large Peclet number is given by

$$\tau_* \sim \frac{l_0^2}{D} \log \frac{l_0 (Pe)^{1/2}}{l_u}. \tag{3.2.14}$$

In other words the relation (3.2.13) gives necessary and sufficient conditions of the stirring stage and (3.2.14) its duration. Later the mean square gradient rapidly decays due to both, turbulent and molecular diffusion. The rate of decay is equal to $[(D+\kappa)t]^{1/2}$. In Fig.4 from [Piterbarg, 1989], typical graphs of $g(t)$ are shown for different values of l_0. To explain the mathematical origin of the obtained conclusion, let us denote the operator in the right hand side of (3.2.6) by

$$H = \frac{1}{r^{d-1}} \frac{\partial}{\partial r} r^{d-1} (2\kappa + F(r)) \frac{\partial}{\partial r}.$$

By applying H to both sides of (3.2.6) we obtain for the function $g(t, r) = -HR_c(t, r)$ the following equation

$$\frac{\partial g}{\partial t} = U(r) g + \widetilde{H} g \tag{3.2.15}$$

where $\widetilde{H} = H + 2 \frac{\partial F}{\partial r} \frac{\partial}{\partial r}$ and

$$U(r) = \frac{d^2 F}{dr^2} + \frac{(d-1)}{r} \frac{dF}{dr} + \frac{1}{F} \left(\frac{dF}{dr} \right)^2,$$

here the function $F(r)$ is given by (3.2.7)

Due to the first term in (3.2.15) and the inequality $U(r) \geq 0$ the solution $g(t, r)$ of (3.2.15) can grow during the initial stage. Noting that

$$g(t) = \nabla^2 R_c(t, r) \Big|_{r=0} = \frac{1}{2\kappa} g(t, r) \Big|_{r=0}$$

we find that $g(t)$ increases under condition (3.2.13). A detailed derivation is given in [Molchanov, Piterbarg, 1994].

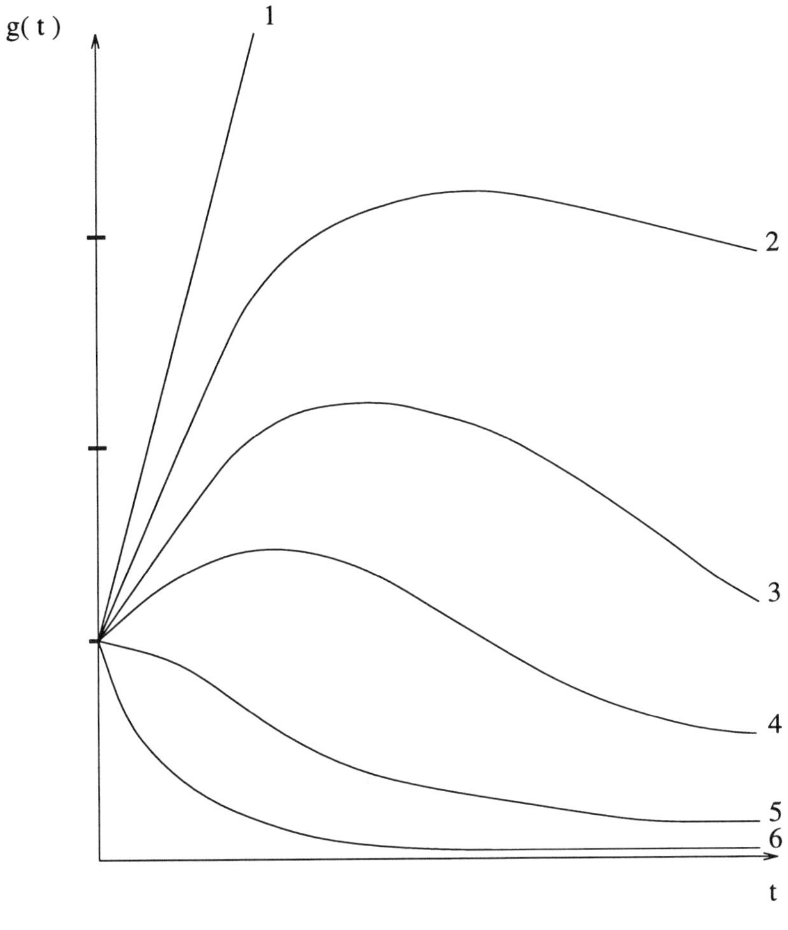

FIG. 4. *Mean square tracer gradient verse time for Pe* = 20 *and different values of* $s \equiv l_0^2 Pe/l_u^2$ *1) s* = 2000; *2) s* = 125; *3) s* = 70; *4) s* = 31; *5) s* = 8; *6) s* = 0.5. *The curves were obtained by numerical solving (3.2.30) with* $R_L(r) = \exp(-r^2/l_u^2)$ *and* $R_c^{(0)}(r) = \exp(-r^2/l_0^2)$.

3.3. Homogenization in unsteady flow. In this section we discuss the behaviour of the solution of the equation

$$\frac{\partial c_\varepsilon}{\partial t} + \boldsymbol{u}_\varepsilon \cdot \nabla c_\varepsilon = \kappa \nabla^2 c_\varepsilon \tag{3.3.1}$$

with the renormalization

$$\boldsymbol{u}_\varepsilon(t, \boldsymbol{x}) = \frac{1}{\varepsilon}\boldsymbol{u}\left(\frac{t}{\varepsilon^2}, \frac{\boldsymbol{x}}{\varepsilon}\right) \tag{3.3.2}$$

where the low frequency energy $\int \boldsymbol{E}(0, \boldsymbol{k})d\boldsymbol{k}$ is finite. In other words we consider the case when $\tau_T \sim \tau_E \ll t \sim \tau_D$. Under certain additional constraints with respect to the velocity field and the condition $c_0(\boldsymbol{x}) \in L_2(E^d)$, the following homogenization statement is true

Proposition 3.1.

$$\lim_{\varepsilon \to 0} c_\varepsilon(t, \boldsymbol{x}) = \bar{c}(t, \boldsymbol{x}) \tag{3.3.3}$$

where $\bar{c}(t, \boldsymbol{x})$ *is the deterministic field satisfying the equation*

$$\frac{\partial \bar{c}}{\partial t} = (\nabla \cdot \boldsymbol{D}\nabla)\bar{c} + \kappa \nabla^2 \bar{c} \tag{3.3.4}$$

with the effective diffusivity given by

$$\boldsymbol{D} = \int\int \boldsymbol{E}(\omega, \boldsymbol{k}) \frac{\boldsymbol{D}_0 \boldsymbol{k} \cdot \boldsymbol{k}}{(\boldsymbol{D}_0 \boldsymbol{k} \cdot \boldsymbol{k})^2 + \omega^2} d\omega d\boldsymbol{k} \tag{3.3.5}$$

where $\boldsymbol{D}_0 = \kappa \boldsymbol{I} + \pi \int \boldsymbol{E}(0, \boldsymbol{k})d\boldsymbol{k}$.

This statement was rigorously proved by Avellaneda & Majda, 1990 for the 2D shear flow with the velocity component spectrum of type (2.2).

There are some strong evidences that (3.3.3) holds for general homogeneous velocity field fastly decorrelating in time and ergodic in space [Molchanov, Piterbarg, 1994], although a rigorous proof is not found yet. In the same paper, formula (3.3.5) is obtained via “rigorous” summation of diagrams. In the present paper some gaps in that derivation are covered up (see Appendix).

Thus, in the considered case, the fluctuations of passive scalar appear to be of a small order compared to the mean field. The other important consequence of the considerable scale separation is a dependence of the turbulent diffusivity on the molecular one. This case sharply differs from the δ-correlation approximation. From (3.3.5) it follows that the molecular diffusion is able to affect the turbulent diffusion. Let us give an example where this effect is drastic.

Let us consider the 2D shear flow $\boldsymbol{u}(t,\boldsymbol{x}) = (u_1(t,x_2),0)$, where the wave frequency spectrum of u_1 is

$$(3.3.6) \qquad E_1(\omega,\boldsymbol{k}) = \begin{cases} \frac{3\sqrt{2}\sigma_u^2 q^{3/2}|\omega|k_2^2}{4\pi\omega_0^{3/2}(\omega^2+q^2k_2^4)}\delta(k_1), & |\omega| < \omega_0 \\ 0, & |\omega| > \omega_0 \end{cases}$$

where ω_0, σ_u, q are the parameters such that

$$\int \int_{-\omega_0}^{\omega_0} E_1(\omega,\boldsymbol{k})d\omega d\boldsymbol{k} = \sigma_u^2.$$

Assume that the molecular diffusion acts in the direction x_2 only. In agreement with the homogenization property the limiting equation for the passive scalar is as follows

$$(3.3.7) \qquad \frac{\partial \bar{c}}{\partial t} = D_1(\kappa)\frac{\partial^2 \bar{c}}{\partial x_1^2} + \kappa\frac{\partial^2 \bar{c}}{\partial x_2^2}$$

where the effective diffusivity in x_1-direction is easily computed from the general formula (3.3.5)

$$(3.3.8) \qquad D_1(\kappa) = \frac{3\sigma_u^2}{\omega_0}\,\frac{q\sqrt{\kappa}}{(\kappa+q)(\sqrt{\kappa}+\sqrt{q})}.$$

From (3.3.8) one can see that $D_1(0) = 0$, but if $\kappa \neq 0$ we can choose q in such a way that $D_1(\kappa) = 3\sigma_u^2/4\omega_0$. To do this we should set $q = \kappa$. For example if $\kappa = 10^{-10}\ cm^2/s$ and $\sigma_u^2/\omega_0 = 10^{10}\ cm^2/s$ we can find a spectrum in the form of (3.3.6) such that the turbulent diffusion in x_1-direction increases from 0 to $0.75 \times 10^{10}\ cm^2/s$. Hence, at least theoretically the molecular diffusion can enhance the turbulent one dramatically. The physical mechanism, clearly, is quite simple. If $\kappa = 0$ there is no relation between streamlines and since $E_1(\omega)\,\big|_{\omega=0} = 0$, the turbulent diffusivity in x_1-direction is zero. Indeed, it can be shown that before turning on the molecular diffusion, the displacement of a Lagrangian particle in x_1-direction satisfies $\langle X_1(t)^2\rangle \sim t^{1/2}$. In the terminology of the review [Isichenko, 1992] we have a subdiffusion. When the molecular diffusion in x_2-direction is turned on, generalized Lagrangian particles have the possibility of jumping from one stream line to another. Due to this interaction the particles can travel far along x_1-direction. More exactly, after switching on the molecular diffusion one will have $M\langle \xi_1^2(t)\rangle \sim D_1(\kappa)t$, where the generalized Lagrangian trajectory $\xi_1(t)$ is found from the stochastic equation $\dot{\xi}_1 = u_1(t,\sqrt{2\kappa}w_2(t))$, where $w_2(t)$ is the Wiener process, (Fig. 5).

Let us note that for big κ (or small τ_D) formula (3.3.8) gives the following asymptotic

$$D_1(\kappa) \sim \frac{3\sigma_u^2 q}{\omega_0 \kappa}$$

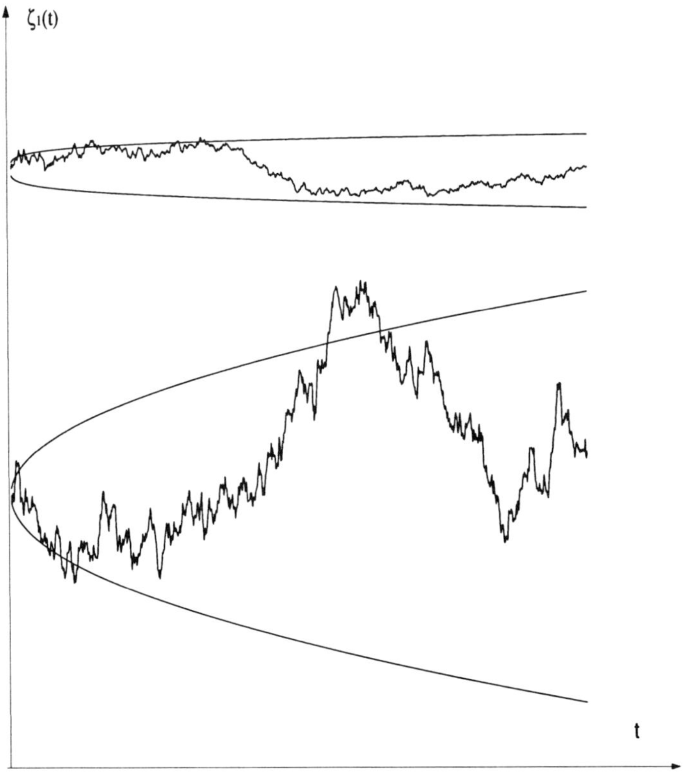

FIG. 5. *The typical generalized lagrangian trajectories for $\kappa = 0$ (top) and $\kappa > 0$ (bottom)*

i.e. the effective diffusivity is inversely proportional to the molecular one. A similar result was obtained by Taylor, 1953 for a deterministic flow through a tube. Let us note that the Taylor's reasoning fails for the small molecular diffusion and does not concern enhancing the turbulent diffusion by the molecular one.

3.4. Superdiffusion. The choice of the rescaling factor ε^{-2} for t in (3.32) is due to the fact that, under the condition

$$0 < \int \|\boldsymbol{E}(0, \boldsymbol{k})\| \boldsymbol{k} < \infty, \tag{3.4.1}$$

where $\|\cdot\|$ is a matrix norm, the mean-square particle displacement satisfies the scaling $\langle \boldsymbol{X}^2(t)\rangle \sim t$.

In section 3.1 we discussed an example of superdiffusion. Here we formulate a general result related to the same spectrum (2.4) with $C_0 = C_0\varepsilon$.

Its proof is given in the Appendix. First let us introduce some notation. Let us consider three regions in the quarterplane $\{(\alpha, z),\ \alpha \geq 0,\ z \geq 0\}$

$$\begin{aligned} G_1 &= \{\alpha + z > 2, \quad \alpha + 2z < 4\} \\ G_2 &= \{\alpha + 2z > 4, \quad 2z(1-\alpha) < (2-\alpha)^2, \\ &\qquad (6 - 2\alpha - z)(\alpha + 2z - 2) > 2z\} \\ G_3 &= \{2z(1-\alpha) > (2-\alpha)^2, \quad z < 4 - \alpha\} \end{aligned} \tag{3.4.2}$$

(see. Fig.6)

Proposition 3.2. *Let $\rho = \rho_j$ be the rescaling function in region G_j, $j = 1, 2, 3$ determined by*

$$\rho_1 = \varepsilon^{\frac{4-\alpha-z}{2}}, \quad \rho_2 = \varepsilon^{2-\frac{\alpha}{2}-\beta}, \quad \rho_3 = \varepsilon,$$

where $\beta = \frac{z}{\alpha+2z-2}$, then there exists the limit

$$\bar{c}(t, \boldsymbol{x}) = \lim_{\varepsilon \to 0} \langle c_\varepsilon(t, \boldsymbol{x}) \rangle$$

for the solution of the equation

$$\frac{\partial}{\partial t} c_\varepsilon + \boldsymbol{u}_\varepsilon \cdot \nabla c_\varepsilon = \kappa_\varepsilon \nabla^2 c_\varepsilon \tag{3.4.3}$$

where $\boldsymbol{u}_\varepsilon = \frac{\varepsilon}{\rho^2}\, \boldsymbol{u}(t/\rho^2, \boldsymbol{x}/\varepsilon)$, $\kappa_\varepsilon = \frac{\varepsilon^2}{\rho^2}\kappa$, and $\bar{c}(t, \boldsymbol{x})$ satisfies the equation

$$\frac{\partial \bar{c}}{\partial t} = D_i \nabla^2 \bar{c}$$

for $(\alpha, z) \in G_i$ with the effective diffusivity given by

$$\begin{aligned} D_1 &= \tfrac{C}{a(\alpha+z-2)} (C_0 \bar{k})^{2-\alpha-z} \\ D_2 &= q_{\alpha,z} (C_0 \bar{k})^{2-\alpha-2\beta} \\ D_3 &= q_{\alpha,z} (C_1 \bar{k})^{2-\alpha-2\beta} + \kappa \end{aligned} \tag{3.4.4}$$

where $q_{\alpha,z} = \frac{A^{-\beta}\Gamma(\beta)\beta}{|2-\alpha-2\beta|}$,

$$A = C\pi\vartheta_d(d-1)a^{\frac{\alpha+z-2}{2}} \frac{1}{dz} \int_0^\infty x^{-1/\beta} \left(1 - \frac{1}{x} + \frac{1}{x} e^{-x}\right) dx. \tag{3.4.5}$$

Let us make sure that we are staying in the frame of short-correlation approximation. For region G_1, the result was shown in section 3.1. For G_2 and G_3 we have from (2.6) and (3.1.11)

$$\tau_E \sim \rho^2 \varepsilon^{2-\alpha-z}$$

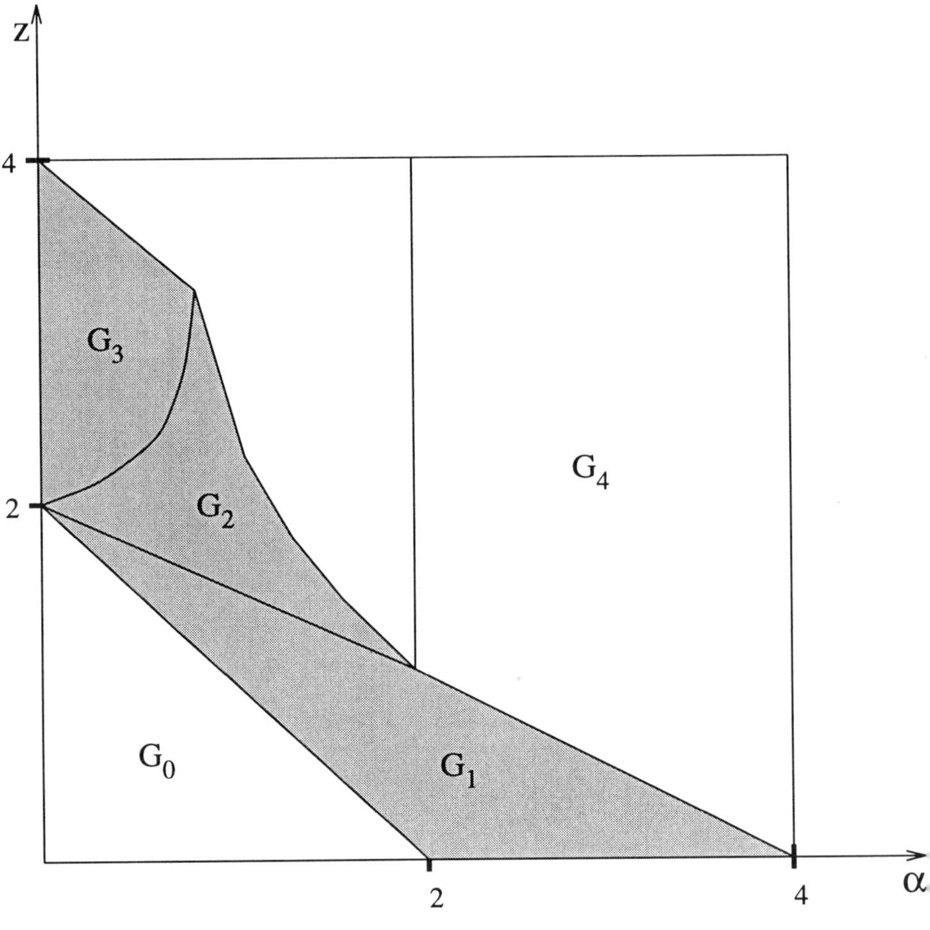

FIG. 6. *Phase diagram in (α, z) plane for isotropic velocity field. In the dashed area $\tau_T \ll \tau_E \ll 1$, in G_0 : $\tau_T \sim \tau_E \ll 1$ and in G_4 : $\tau_T \ll 1 \ll \tau_E$*

and hence we must require that in

$$\begin{array}{ll} G_2: & 4-\alpha-\frac{2z}{\alpha+2z-2}+2-\alpha-z>0 \\ G_3: & 4-\alpha-z>0. \end{array}$$

Obviously this follows from definition (3.4.2) and therefore the formulated proposition gives a description of turbulent diffusion in the case

$$\tau_T \ll \tau_E \ll t.$$

Let us note that for G_1 and G_2 we have $t \ll \tau_D$, but for G_3 the molecular diffusion has the order one. The next remark is that the region $G_0: \ \alpha+z<2$ in Fig.6 gives an example of the separation

$$\tau_T \sim \tau_E \ll t \sim \tau_D$$

considered in the general case of the previous section. Also let us mention that the turbulent diffusion in region G_4 (Fig.6) can be described as well at least for $z<1$ [Avellaneda, Majda, 1992]. In this region the relation

$$\tau_T \ll t \ll \tau_D \ll \tau_E \tag{3.4.6}$$

holds. One can check this knowing that $\rho=\varepsilon^{1-\alpha/4}$ in G_4.

The separation (3.4.6) is not covered by the short-correlation approximation as well as we believe all the quarter-plane (α, z) except regions G_0, G_1, G_2, G_3. Some evidence of this assertion can be found in the reasoning given in the Appendix.

Conclusion. The main point we would like to emphasize is that the short-correlation approximation $\tau_E \ll t$ does not necessarily lead to the Fokker-Planck equation. This equation describes only some particular cases of such a scale separations. The other essential issue is that there are a variety of diffusion regimes even for the simple isotropic velocity field under short-correlation assumption. Some efforts have been made to overcome the short-correlation assumption and to clear the limits of its applicability [Davis, 1987, Zambianchi & Griffa,1994 Klyatskin, 1994]. This circle of problems is beyond of our work and we refer the reader to the last papers.

Appendix. Here we give a proof of Proposition 3.2 from section 3.4 and formula (3.3.5).

Let $P=P(t,\boldsymbol{x};s,\boldsymbol{y})$ be the fundamental solution (Green's function) of (1.1), i.e. P satisfies (1.1) in $t,\boldsymbol{x}$. That is,

$$\partial P/\partial t+\boldsymbol{u}\cdot\nabla_{\boldsymbol{x}}P=\kappa\nabla_{\boldsymbol{x}}^2 P \tag{A.1}$$

with the initial condition

$$P|_{t=s}=\delta(\boldsymbol{x}-\boldsymbol{y}), \tag{A.2}$$

where the subscript indicates the variable on which the operator acts.

Together with (1.1) consider the equation with "fixed" space coordinate and denote by $P^0(t, \boldsymbol{x}; s, \boldsymbol{y})$ its Green's function, i.e.

$$\partial P^0/\partial t + \boldsymbol{u}(t, \boldsymbol{y}) \cdot \nabla_{\boldsymbol{x}} P^0 = \kappa \nabla_{\boldsymbol{x}}^2 P^0, \tag{A.3}$$

$$P^0|_{t=s} = \delta(\boldsymbol{x} - \boldsymbol{y}). \tag{A.4}$$

Then

$$P = P^0 + RP \tag{A.5}$$

where operator R is given by

$$(Rf)(t, \boldsymbol{x}) = -\int_s^t \int_E \nabla_{\boldsymbol{z}} P^0(t, \boldsymbol{x}; \sigma, \boldsymbol{z}) \cdot (\boldsymbol{u}(\sigma, \boldsymbol{y}) - \boldsymbol{u}(\sigma, \boldsymbol{z})) f(\sigma, \boldsymbol{z}) d\boldsymbol{z} d\sigma, \tag{A.6}$$

$$E \equiv E^d \text{ and } P^0(t, \boldsymbol{x}; s, \boldsymbol{y}) = \frac{1}{[4\pi\kappa(t-s)]^{d/2}} \exp\left\{ -\frac{(\boldsymbol{x} - \boldsymbol{y} - \int_s^t \boldsymbol{u}(\sigma, \boldsymbol{y}) d\sigma)^2}{4\kappa(t-s)} \right\}.$$

From (A.5) we obtain the following expansion for P

$$P = P^0 + RP^0 + R^2 P^0 + \cdots. \tag{A.7}$$

Convergence of series (A.7) is proved in [Molchanov & Piterbarg, 1994]. Using the Fourier transformation of P^0 in space coordinates, the m-th term of the series in (A.7) may be expressed in the form

$$\begin{aligned} \xi_m \equiv R^m P^0 = \frac{i^m}{(2\pi)^{dm}} \int_{G_m} \int_{E^m} \int_{E^{m+1}} \Bigg(\prod_{j=1}^m (\boldsymbol{u}(s_j, \boldsymbol{z}_j) - \\ -\boldsymbol{u}(s_j, \boldsymbol{z}_{j-1})) \cdot \boldsymbol{k}_j \Bigg) \exp\Bigg\{ \sum_{j=0}^m [i\boldsymbol{k}_{j+1} \cdot (\boldsymbol{z}_{j+1} - \boldsymbol{z}_j - \\ \int_{s_j}^{s_{j+1}} \boldsymbol{u}(\sigma, \boldsymbol{z}_j) d\sigma) - \boldsymbol{k}_{j+1}^2 \kappa (s_{j+1} - s_j)] \Bigg\} (d\boldsymbol{k})^{m+1} \times \\ \times (d\boldsymbol{z})^m (ds)^m \end{aligned} \tag{A.8}$$

where $G_m = \{s_1, \cdots, s_m : t \geq s_m \geq s_{m-1} \geq \cdots \geq s_1\}$,

$$E^m = \underbrace{E \times \cdots \times E}_{m \text{ times}}, \quad s_0 \equiv 0, \quad s_{m+1} \equiv t, \quad \boldsymbol{z}_0 \equiv \boldsymbol{y},$$

$$z_{m+1} \equiv x, \quad k_{m+1} \equiv k, \quad (dk)^{m+1} = dk_1 dk_2 \cdots dk_m dk,$$

$$(ds)^m = ds_1 ds_2 \cdots ds_m, \quad (dz)^m = dz_1 dz_2 \cdots dz_m.$$

Now we replace u and κ in (A.8) by u_ε and κ_ε respectively and obtain m-th term $\xi_{m,\varepsilon}$ of the expansion of the Green function for equation (3.4.3). The expectation of $\xi_{m,\varepsilon}$ may be written as follows

$$\langle \xi_{m,\varepsilon} \rangle = \langle R_\varepsilon^m P_\varepsilon^0 \rangle = i^m \int_{G_m} \int_{E^m} \int_{E^{m+1}} \langle \xi_1 \cdots \xi_m e^{-i\eta} \rangle \times$$

$$\text{(A.9)} \quad \times (2\pi)^{-dm} \exp\left\{ \sum_{j=0}^{m} [ik_{j+1}, \cdot(z_{j+1} - z_j) - \kappa_\varepsilon k_{j+1}^2 (s_{j+1} - s_j) \right\} \times$$

$$\times (dk)^{m+1} (dz)^m (ds)^m,$$

where

$$\text{(A.10)} \qquad \xi_j = \big(u_\varepsilon(s_j, z_j) - u_\varepsilon(s_j, z_{j-1})\big) \cdot k_j, \quad j = 1, \cdots, m$$

$$\text{(A.11)} \qquad \eta = \sum_{j=0}^{m} \int_{s_j}^{s_{j+1}} u_\varepsilon(\sigma, z_j) d\sigma \cdot k_{j+1}$$

are the Gaussian random values with zero mean. Note that

$$\langle \xi_1 \cdots \xi_m e^{i\eta} \rangle = e^{-\frac{1}{2}\langle \eta^2 \rangle} \times$$

$$\sum_{\substack{i, \cdots, i_k \\ j_1, \cdots, j_{m-k}}} \times \langle \xi_{i_1} \cdots \xi_{i_k} \rangle \langle \eta \xi_{j_1} \rangle \cdots \langle \eta \xi_{j m-k} \rangle$$

where the sum is taken over all $k = 0, \cdots, m$ and all sets of indices $I = (i_1, \cdots, i_k)_1$ and $J = (j_1, \cdots, j_{m-k})$ such that $I \cap J = \emptyset$, $I \cup J = (1, \cdots, m)$ and if $k = m$ or $k = 0$, then the corresponding sets J or I are considered to be empty. It can easily be seen that $\langle \xi_{\varepsilon,m} \rangle = 0$ for odd m. For even m the main contribution into integral (A.9) is due to term

$$\langle \xi_1 \xi_2 \rangle \cdots \langle \xi_{m-1} \xi_m \rangle e^{-\frac{1}{2}\langle \eta^2 \rangle}$$

because of the short correlation assumption.

Let us demonstrate this in the homogenization case (3.4.1), when $\rho(\varepsilon) = \varepsilon$. First, estimate the contribution to $\langle \xi_{m,\varepsilon} \rangle$ in (A.9) due to the term

$$\langle \xi_{i_1} \xi_{j_1} \rangle \cdots \langle \xi_{i_p} \xi_{j_p} \rangle e^{-\frac{1}{2}\langle \eta^2 \rangle},$$

where $(i_1, j_1, \ldots, i_p, j_p)$, $i_k < j_k$, is a permutation of $(1, 2, \ldots, 2p-1, 2p)$ such that either

$$\text{(A.12)} \qquad \begin{array}{ll} (i_1, \ldots, i_p) \neq (1, 3, \ldots, 2p-1) & \text{or} \\ (j_1, \ldots, j_p) \neq (2, 4, \ldots, 2p). & \end{array}$$

Let $\boldsymbol{s} = (s_1, \ldots, s_m)$ be a fixed point from G_m and $S = \cup_{k=1}^p [s_{i_k}, s_{j_k}]$ be the set in the line. This set, of course, can be represented as a union of q, $(q \leq p)$ non-overlapped segments. For example, if $(i_1, \ldots, i_p) = (1, 3, \ldots, 2p-1)$ and $(j_1, \ldots, j_p) = (2, 4, \ldots, 2p)$ then $q = p$, if $i_1 = 1$, $j_p = 2p$ then $q = 1$, etc.

Let $\delta = \delta(\varepsilon)$ tends to zero along with ε and

$$M_\delta = \{\boldsymbol{s} : \quad |s_{i_k} - s_{j_k}| \leq \delta, \quad k = 1, \ldots, p\}.$$

Lemma A.1. *For any bounded function $f_\varepsilon(\boldsymbol{s})$*

$$\int_{G_m \cap M_\delta} f_\varepsilon(\boldsymbol{s}) d\boldsymbol{s} \leq C\delta^{m-q}.$$

Proof. Let

$$S = [s_1, s_{k_1}] \cup [s_{k_1+1}, s_{k_2}] \cup \ldots \cup [s_{k_{q-1}+1}, s_{2_p}]$$

be a decomposition of S into non-overlapped segments. If $\boldsymbol{s} \in G_m \cap M_\delta$ then $|s_{k_l} - s_{k_{l-1}+1}| \leq C_l \delta$, since all original segments including in $[s_{k_{l-1}+1}, s_{k_l}]$ are chained, where $l = 1, \ldots, q$ and for the sake of convenience we suppose $k_0 = 0$, $k_q = 2p$. Thus,

$$|G_m \cap M_\delta|_m \leq |V_1|_{k_1} |V_2|_{k_2 k_1} \cdots |V_q|_{m-k_{q-1}},$$

where $|V|_k$ is the volume of V in the k-dimensional Euclidean space and $V_l = \{|s_{k_l} - s_{k_{l-1}+1}| \leq C_l \delta\}$. Obviously

$$|V_l|_{k_l - k_{l-1}} \leq c_l \delta^{k_l - k_{l-1} - 1}.$$

From the last two inequalities it follows that

$$|G_m \cap M_\delta|_m \leq C_l \delta^{m-q}$$

The statement of the lemma readily follows from the latter.

Let us set

$$b_{m,\varepsilon} = i^m \int\limits_{G_m}\int\limits_{E^m}\int\limits_{E^{m+1}} \langle\xi_{i_1}\xi_{j_1}\rangle\cdots\langle\xi_{i_p}\xi_{j_p}\rangle e^{-\frac{1}{2}\langle\eta^2\rangle}$$

(A.13)
$$\times(2\pi)^{-dm}\exp\left\{\sum_{j=0}^{m}[i(\boldsymbol{k}_{j+1},\boldsymbol{z}_{j+1}-\boldsymbol{z}_j)-\kappa\boldsymbol{k}_{j+1}^2(s_{j+1}-s_j)\right\}\times$$

$$\times(d\boldsymbol{k})^{m+1}(d\boldsymbol{z})^m(ds)^m.$$

One can compute the correlation $\langle\xi_{i_k}\xi_{j_k}\rangle$ using the definition (A.10) of ξ_i. As a result $b_{m,\varepsilon}$ is represented in the following form

$$b_{m,\varepsilon} = \varepsilon^{-2p}\int\limits_{G_m} F\left(\frac{s_{i_1}-s_{j_1}}{\varepsilon^2},\ldots,\frac{s_{i_p}-s_{j_p}}{\varepsilon^2};\boldsymbol{x},\boldsymbol{y}_0\right)(ds)^m,$$

where F is a bounded function. Using Lemma A.1 with $\delta=\varepsilon^2$ we get

$$b_{m,\varepsilon} \le C\varepsilon^{-2p}\int\limits_{G_m\cap M_{\varepsilon^2}} F\left(\frac{s_{i_1}-s_{j_1}}{\varepsilon^2},\ldots,\frac{s_{i_p}-s_{j_p}}{\varepsilon^2};\boldsymbol{x},\boldsymbol{y}_0\right)(ds)^m \le$$

$$\le C_1\varepsilon^{2(p-q)},$$

From the latter it can be seen that only if $q=p$, i.e. if $(i_1,\ldots,i_p)=(1,3,\ldots,2p-1)$ and $(j_1,\ldots,j_p)=(2,4,\ldots,2p)$ then $b_{m,\varepsilon}$ does not tend to zero when $\varepsilon\to 0$. If (A.12) holds then necessarily $\lim_{\varepsilon\to 0} b_{m,\varepsilon}=0$. In the case of superdiffusion we should use Lemma A.1 with $\delta=\rho^2\varepsilon^{2-\alpha-z}$ in regions G_2 and G_3 and with the corresponding Eulerian scale correlation in G_2.

Now we proceed to the terms in expansion (A.9) containing correlation such as $\langle\xi_j\eta\rangle$. For the sake of definiteness let us consider the following term

$$d_{m,\varepsilon} = i^m \int\limits_{G_m}\int\limits_{E^m}\int\limits_{E^{m+1}} \langle\xi_1\xi_2\rangle\cdots\langle\xi_{2p-3}\xi_{2p-2}\rangle\langle\xi_{2p-1}\eta\rangle\langle\xi_{2p}\eta\rangle e^{-\frac{1}{2}\langle\eta^2\rangle}$$

(A.14)
$$\times(2\pi)^{-dm}\exp\left\{\sum_{j=0}^{m}[i\boldsymbol{k}_{j+1}\cdot(\boldsymbol{z}_{j+1}-\boldsymbol{z}_j)-\kappa\boldsymbol{k}_{j+1}^2(s_{j+1}-s_j)\right\}\times$$

$$\times(d\boldsymbol{k})^{m+1}(d\boldsymbol{z})^m(ds)^m.$$

The correlation $r_\varepsilon(\boldsymbol{z}_j-\boldsymbol{z}_{j-1},\boldsymbol{k}_j,\boldsymbol{k}_{j+1})\equiv\langle\xi_j\eta\rangle$ can be estimated from (A.10, A.11) as follows

$$r_\varepsilon(\boldsymbol{z}_j-\boldsymbol{z}_{j-1},\boldsymbol{k}_j,\boldsymbol{k}_{j+1}) \sim \int\limits_0^\infty \left[\boldsymbol{R_u}(\sigma,\boldsymbol{0})-\boldsymbol{R_u}\left(\frac{\boldsymbol{z}_{j-1}-\boldsymbol{z}_j}{\varepsilon}\right)\right]\boldsymbol{k}_j\cdot\boldsymbol{k}_{j+1}d\sigma -$$

$$\int_{-\infty}^{0} \left[\boldsymbol{R}_{\boldsymbol{u}}(\sigma, 0) - \boldsymbol{Ru}\left(\frac{\boldsymbol{z}_j - \boldsymbol{z}_{j-1}}{\varepsilon} \right) \right] \boldsymbol{k}_j \cdot \boldsymbol{k}_j d\sigma.$$

Let us change variables in (A.14) by $\boldsymbol{v}_j = \boldsymbol{z}_j - \boldsymbol{z}_{j-1}$, $j = 1, \ldots, m$. As the result (A.14) can be rewritten as follows

$$d_{m,\varepsilon} = \int_{E^m} \int_{E^{m+1}} \exp \left\{ \sum_{j=0}^{m-1} i(\boldsymbol{k}_{j+1} - \boldsymbol{k}) \cdot \boldsymbol{v}_{j+1} \right\} r_\varepsilon(\boldsymbol{v}_{2_p-1}, \boldsymbol{k}_{2_p-1}, \boldsymbol{k}_{2_p}) \times$$
$$\times r_\varepsilon(\boldsymbol{v}_{2_p}, \boldsymbol{k}_{2_p}, \boldsymbol{k}) F(\boldsymbol{v}_1, \boldsymbol{v}_2, \ldots, \boldsymbol{v}_{2_p-2}; \boldsymbol{k}_1, \ldots, \boldsymbol{k}_{2_p}, \boldsymbol{k}; \boldsymbol{x}, \boldsymbol{y}_0)(d\boldsymbol{k})^{m+1}(d\boldsymbol{v})^m,$$

where F is a bounded function fastly decaying when $|\boldsymbol{k}_j| \to \infty$. Integrating with respect to $\boldsymbol{v}_{2_p}$ yields

$$d_{m,\varepsilon} \sim \varepsilon^d \int_{E^{m-1}} \int_{E^{m+1}} \exp \left\{ \sum_{j=0}^{2_p-1} i(\boldsymbol{k}_{j+1} - \boldsymbol{k}) \cdot \boldsymbol{v}_{j+1} \right\} r_\varepsilon(\boldsymbol{v}_{2_p-1}, \boldsymbol{k}_{2_p-1}, \boldsymbol{k}_{2_p}) \times$$
$$\left[\int_{-\infty}^{0} \hat{\boldsymbol{R}}_{\boldsymbol{u}}(\sigma, \boldsymbol{k}_j) \boldsymbol{k}_j \cdot \boldsymbol{k} d\sigma - \int_{0}^{\infty} \hat{\boldsymbol{R}}\boldsymbol{u}(\sigma, -\boldsymbol{k}_j) \boldsymbol{k}_j \cdot \boldsymbol{k}_j d\sigma \right] \times$$
$$F(\boldsymbol{v}_1, \boldsymbol{v}_2, \ldots, \boldsymbol{v}_{2_p-2}; \boldsymbol{k}_1, \ldots, \boldsymbol{k}_{2_p}, \boldsymbol{k}; \boldsymbol{x}, \boldsymbol{y}_0)(d\boldsymbol{k})^{m+1}(d\boldsymbol{v})^m,$$

We used the property $\boldsymbol{R}_{\boldsymbol{u}}(\sigma, 0) = \boldsymbol{R}_{\boldsymbol{u}}(-\sigma, 0)$. From the latter it follows that $d_{m,\varepsilon}$ disappears when $\varepsilon \to 0$. We believe that in the general case quantities $d_{m,\varepsilon}$ are also neglectible. Thus

$$\langle \xi_{m,\varepsilon} \rangle \cong (-1)^p \int_{G_m} \int_{E^m} \int_{E^{m+1}} \langle \xi_1, \xi_2 \rangle \cdots \langle \xi_{m-1} \xi_m \rangle e^{-\frac{1}{2}\langle \eta^2 \rangle} \times$$
$$\times (2\pi)^{-dm} \exp \left\{ \sum_{j=0}^{m} [i(\boldsymbol{k}_{j+1}, \boldsymbol{z}_{j+1} - \boldsymbol{z}_j) - \kappa \boldsymbol{k}_{j+1}^2 (s_{j+1} - s_j) \right\} \times$$
$$\times (d\boldsymbol{k})^{m+1} (d\boldsymbol{z})^m (ds)^m.$$

Since the short-correlation assumption the terms in the sum (A.11) are asymptotically independent. Hence we can conclude that for even $m = 2p$ and $\varepsilon \to 0$

$$\langle \xi_{m,\varepsilon} \rangle \cong (-1)^p \int_{G_m} \int_{E^m} \int_{E^{m+1}} \prod_{j=1}^{p} (\boldsymbol{Q}(s_{2j} - s_{2j-1}, \boldsymbol{z}_{2j} -$$
$$-\boldsymbol{z}_{2j-1}, \boldsymbol{z}_{2j-1} - \boldsymbol{z}_{2j-2}) \boldsymbol{k}_{2j} \cdot \boldsymbol{k}_{2j-1}) \times$$
$$\times \exp \left\{ \sum_{j=0}^{m} \left[i\boldsymbol{k}_{j+1} \cdot (\boldsymbol{z}_{j+1} - \boldsymbol{z}_j) - \kappa_\varepsilon \boldsymbol{k}_{j+1}^2 (s_{j+1} - s_j) \right) \right.$$

$$-\boldsymbol{D}_\varepsilon(s_{j+1} - -s_j)\boldsymbol{k}_j \cdot \boldsymbol{k}_j]\Bigg\}(d\boldsymbol{k})^{m+1}(d\boldsymbol{z})^m(ds)^m,$$

where

$$\boldsymbol{Q}_\varepsilon(t,\boldsymbol{z},\boldsymbol{v}) = \langle \boldsymbol{u}_\varepsilon(t,\boldsymbol{z})\boldsymbol{u}_\varepsilon^T(o,\boldsymbol{o})\rangle - \langle \boldsymbol{u}_\varepsilon(t,\boldsymbol{o})\boldsymbol{u}_\varepsilon^T(o,\boldsymbol{o})\rangle +$$

$$+\langle \boldsymbol{u}_\varepsilon(t,\boldsymbol{u})\boldsymbol{u}_\varepsilon^T(o,\boldsymbol{o})\rangle - \langle \boldsymbol{u}_\varepsilon(t,\boldsymbol{z}+\boldsymbol{v})\boldsymbol{u}_\varepsilon^T(o,\boldsymbol{o})\rangle,$$

$$\boldsymbol{D}_\varepsilon(t) = \frac{1}{2}\int_o^t\int_o^t \langle \boldsymbol{u}_\varepsilon(s_1,\boldsymbol{z})\boldsymbol{u}_\varepsilon^T(s_2,\boldsymbol{z})\rangle ds_1 ds_2.$$

After changing and renumbering the variables of integration according to

$$s_j' = \frac{s_{2j} - s_{2j-1}}{\rho^2}, \quad \boldsymbol{z}'_j = \frac{\boldsymbol{z}_{2j} - \boldsymbol{z}_{2j-1}}{\varepsilon}, \quad \boldsymbol{k}_j' = \varepsilon \boldsymbol{k}_{2j}$$

$$s_j'' = s_{2j-1}, \quad \boldsymbol{z}_j'' = \boldsymbol{z}_{2j-1}, \quad \boldsymbol{k}_j'' = \boldsymbol{k}_{2j-1}$$

$$j = 1, \cdots, p$$

we obtain

$$\langle \xi_{2p,\varepsilon}\rangle \cong (-1)^p \rho^{2p} \int_E \int_{G_p'} \int_{E^p} \int_{E^p} \prod_{j=1}^p (\boldsymbol{Q}_\varepsilon(s_j', \boldsymbol{z}_j')\boldsymbol{k}\cdot\boldsymbol{k})\times$$

$$\times \exp\{i(\boldsymbol{k}_j'\cdot\boldsymbol{z}_j') - \varepsilon^{-2}(\boldsymbol{D}_\varepsilon(\rho^2 s_j')\boldsymbol{k}_j'\cdot\boldsymbol{k}_j') + \kappa_\varepsilon \rho^2 s_j' \boldsymbol{k}_j'^2]\}\times$$

$$\text{(A.15)} \qquad \times(d\boldsymbol{k}')^p(d\boldsymbol{z}')^p)\exp\Bigg\{-\sum_{j=1}^{p+1}[(\boldsymbol{D}_\varepsilon(s_j'' - s_{j-1}'')\boldsymbol{k}\cdot\boldsymbol{k}) +$$

$$+\kappa_\varepsilon(s_j'' - s_{j-1}'')\boldsymbol{k}^2] + i\boldsymbol{k}\cdot(\boldsymbol{x}-\boldsymbol{y})\Bigg\} \times (ds_1')^p(ds_2'')^p d\boldsymbol{k},$$

where the time integration region is given by

$$G_p' = \Bigg\{0 \le s_1'' \le \ldots \le s_p'' < t,\ 0 \le s_1' \le \frac{s_2'' - s_1''}{\rho^2}, \ldots,$$

$$0 \le s_{p-1}' \le \frac{s_p'' - s_{p-1}''}{\rho^2},\ 0 \le s_p' \le \frac{t - s_p''}{\rho^2}\Bigg\},$$

the auxiliary function $\boldsymbol{Q}_\varepsilon$ is defined by

$$\boldsymbol{Q}_\varepsilon(s,\boldsymbol{z}) = \frac{\varepsilon^2}{\rho^4}(\boldsymbol{R}_{\boldsymbol{u}}(s,\boldsymbol{z}) - \boldsymbol{R}_{\boldsymbol{u}}(s,\boldsymbol{0}))$$

and the approximation $s_{2j-1} - s_{2j-2} = s''_j - s''_{j-1} - \rho^2 s'_{j-1} \approx s''_j - s''_{j-1}$ was used at the second exponent of (A.15).

Setting

$$G_\varepsilon(t, \boldsymbol{k}) = \exp\{-\boldsymbol{D}_\varepsilon(t)\boldsymbol{k}\cdot\boldsymbol{k} - \kappa_\varepsilon t\boldsymbol{k}^2\}$$

and

$$H_\varepsilon(t, \boldsymbol{k}) = \frac{\rho^2}{(2\pi)^d} G_\varepsilon(t, \boldsymbol{k}) \int_E \int_o^{t/\rho^2} (\hat{\boldsymbol{Q}}(s, \boldsymbol{k}')\boldsymbol{k}\cdot\boldsymbol{k}) G_\varepsilon(\rho s, \varepsilon^{-1}\boldsymbol{k}') ds d\boldsymbol{k}',$$

where $\hat{\boldsymbol{Q}}_\varepsilon(s, \boldsymbol{k}') = \int_E e^{i\boldsymbol{k}'\cdot\boldsymbol{z})}\boldsymbol{Q}_\varepsilon(s, \boldsymbol{z})d\boldsymbol{z}$, we can rewrite (A.15) as follows

$$\langle \xi_{2p,\varepsilon}\rangle \cong (-1)^p \int_E G_\varepsilon * (H_\varepsilon)^{p*} e^{i\boldsymbol{k}\cdot(\boldsymbol{x}-\boldsymbol{y})} d\boldsymbol{k}, \tag{A.16}$$

where the star means convolution with respect to time variable and $f^{p*} = \underbrace{f * \ldots * f}_{p}$.

Let us introduce the Laplace transform

$$f^*(\lambda) = \int_0^\infty e^{-\lambda t} f(t) dt.$$

and denote $\hat{P}^*_\varepsilon(\lambda, \boldsymbol{k})$ the Laplace transform in t and Fourier transform in $\boldsymbol{x} - \boldsymbol{y}$ of the Green function $P_\varepsilon(t, \boldsymbol{x}; s, \boldsymbol{y})$ determined by (A.1),(A.2) with $\boldsymbol{u}_\varepsilon(t, \boldsymbol{x}) = \frac{\varepsilon}{\rho^2}\boldsymbol{u}\left(\frac{t}{\rho^2}, \frac{\boldsymbol{x}}{\varepsilon}\right)$ and $\kappa_\varepsilon = \frac{\varepsilon^2}{\rho^2}\kappa$. Then from (A.7),(A.8) and (A.16) it follows that for small ε, $\rho(\varepsilon)$

$$\hat{P}^*_\varepsilon(\lambda, \boldsymbol{k}) \cong \frac{G^*_\varepsilon(\lambda, \boldsymbol{k})}{1 + H^*_\varepsilon(\lambda, \boldsymbol{k})}. \tag{A.17}$$

Noting that

$$\frac{\partial G_\varepsilon(t, \boldsymbol{k})}{\partial t} = -G_\varepsilon(t, \boldsymbol{k}) \left(\kappa_\varepsilon \boldsymbol{k}^2 + \frac{\varepsilon^2}{\rho^2} \int_0^t \boldsymbol{R_u}\left(\frac{s}{\rho^2}, 0\right) \boldsymbol{k}\cdot\boldsymbol{k}\, ds\right)$$

we have

$$H^*_\varepsilon(\lambda, \boldsymbol{k}) = L^*_\varepsilon(\lambda, \boldsymbol{k}) + \lambda G^*_\varepsilon(\lambda, \boldsymbol{k}) - 1 \tag{A.18}$$

where

$$L_\varepsilon(t, \boldsymbol{k}) = G_\varepsilon(t, \boldsymbol{k}) S_\varepsilon(t, \boldsymbol{k}), \tag{A.19}$$

and

$$S_\varepsilon(t,\boldsymbol{k})=\frac{\varepsilon^2}{(2\pi)^d\rho^2}\int\limits_E\int\limits_E\int\limits_0^{t/\rho^2}\left(e^{i\boldsymbol{k}'\cdot\boldsymbol{z}}\boldsymbol{R}_{\boldsymbol{u}}(s,\boldsymbol{z})\boldsymbol{k}\cdot\boldsymbol{k}\right)G_\varepsilon(\rho s,\varepsilon^{-1}\boldsymbol{k}')dsd\boldsymbol{z}d\boldsymbol{k}'+\kappa_\varepsilon\boldsymbol{k}^2.$$

(A.20)

Substituting (A.18) to (A.17) we obtain

$$\hat{P}^*_\varepsilon(\lambda,\boldsymbol{k})=\frac{1}{\lambda+L^*_\varepsilon(\lambda,\boldsymbol{k})/G^*_\varepsilon(\lambda,\boldsymbol{k})}. \tag{A.21}$$

Comparing (A.19) and (A.21) one can see that if $\lim_{\varepsilon\to0}S_\varepsilon(t,\boldsymbol{k})=\boldsymbol{Dk}\cdot\boldsymbol{k}$, where $\boldsymbol{D}$ is a matrix independent of t then

$$\lim_{\varepsilon\to0}\hat{P}^*_\varepsilon(\lambda,k)=\frac{1}{\lambda+\boldsymbol{Dk}\cdot\boldsymbol{k}}$$

and hence the limiting equation is

$$\frac{\partial\bar{c}}{\partial t}=\nabla\cdot\boldsymbol{D}\nabla\bar{c}$$

From this remark formula (3.3.5) follows readily because under finiteness of $\int\boldsymbol{E}(0,\boldsymbol{k})d\boldsymbol{k}$ we should set $\rho=\varepsilon$ and then

$$\lim_{\varepsilon\to0}S_\varepsilon(t,\boldsymbol{k})=\boldsymbol{Dk}\cdot\boldsymbol{k}+\kappa\boldsymbol{k}^2,$$

where $\boldsymbol{D}$ is given by (3.3.5).

In the isotropic case with spectrum (2.4) straightforward computations yield

$$G_\varepsilon(t,\boldsymbol{k})=e^{-\boldsymbol{k}^2(D_\varepsilon(t)+\kappa_\varepsilon t)},$$

$$S_\varepsilon(t,\boldsymbol{k})=S_\varepsilon(t)\boldsymbol{k}^2,\qquad\text{where}$$

$$D_\varepsilon(t)=\frac{B\varepsilon^2}{\rho^2}\int\limits_{k_0\varepsilon}^{k_1}k^{1-z-\alpha}\left(t-\frac{\rho^2}{ak^z}+\frac{\rho^2}{ak^z}e^{-\frac{ak^zt}{\rho^2}}\right)dk, \tag{A.22}$$

$$S_\varepsilon(t)=\frac{B\varepsilon^2}{\rho^4}\int\limits_o^t\int\limits_{\varepsilon k_o}^{k_1}k^{1-\alpha}e^{-\frac{ask^z}{\rho^2}-\frac{k^2}{\varepsilon^2}(D_\varepsilon(s)+s\kappa_\varepsilon)}dkds+\kappa_\varepsilon, \tag{A.23}$$

and $k_0=C_0\bar{k},\ k_1=C_1\bar{k},\ B=C\pi\vartheta_d(d-1)d^{-1}$.

When the infrared catastrophe occurs, i.e. $\int_{\varepsilon k_0}^{k_1}E_L(0,k)k^{d-1}$ indefinitely increases ($\varepsilon\to0$) then the choice of ρ and the limiting equation for $\bar{c}$

are determined by the asymptotical behavior of $D_\varepsilon(t)$ and $S_\varepsilon(t)$ which is described in the following statement.

Lemma A.2.
1) *If* $(\alpha, z) \in G_1$ *and* $\rho = \rho_1$ *then*

$$\lim_{\varepsilon \to o} D_\varepsilon(t) = D_1 t, \quad \lim_{\varepsilon \to o} S_\varepsilon(t) = D_1$$

where D_1 *is given by (3.4.4)*
2) *If* $(\alpha, z) \in G_2 \cup G_3$ *and* $\rho = \varepsilon^m$ *such that* $\varepsilon^z/\rho^2 \to 0$ *then*

$$\lim_{\varepsilon \to o} \frac{D_\varepsilon(t)}{\varepsilon^\gamma} = At^{1/\beta},$$

whereA is given by (3.4.5) and

$$\gamma = 2 - \frac{2m}{\beta} < 0$$

and if $\rho = \rho_2$ *or* ρ_3 *then*

$$\lim_{\varepsilon \to 0} S_\varepsilon(t) = \lim_{\varepsilon \to 0} \left(\frac{B\varepsilon^2}{\rho^4} \int_o^t \int_{k_o\varepsilon}^{k_1} k^{1-\alpha} e^{-\frac{Ak^2 s^{1/\beta}}{\varepsilon^{2-\gamma}}} dk ds + \kappa_\varepsilon \right) =$$

$$= \begin{cases} D_2 & (\alpha, z) \in G_2 \\ D_3 + \kappa & (\alpha, z) \in G_3 \end{cases}$$

where D_2, D_3 *are given by (3.4.4).*

From this Lemma and (A.19,A.21) Proposition 3.2 follows immediately.

Proof of Lemma A.2. After substituting $\rho = \varepsilon^m$, $0 < m \leq 1$ to $D_\varepsilon(t)$ given in (A.22) we obtain

$$D_\varepsilon(t) = Bz^{-1} a^{(z+\alpha-2)/2} t^{1/\beta} \varepsilon^{2-2m+2m(2-\alpha-z)/z} \times$$
$$\times \int_{b_1(\varepsilon)}^{b_2(\varepsilon)} x^{-1/\beta} \left(1 - \frac{1}{x} + \frac{1}{x} e^{-x}\right) dx,$$

where

$$b_1(\varepsilon) = ak_0^z t\varepsilon^{z-2m}, \quad b_2(\varepsilon) = ak_0^z t\varepsilon^{-2m} \to \infty.$$

Note that the function under the integral looks like $x^{(2-\alpha-z)/z}$ at zero and like $x^{-1+(2-\alpha-z)/z}$ at infinity. Thus, the integral is always convergent at infinity.

If

$$2m < z, \tag{A.24}$$

then $b_1(\varepsilon) \to 0$ and we have two cases: first, $\alpha < 2$, in which the integral converges and hence

$$D_\varepsilon(t) \sim Bz^{-1}a^{(z+\alpha-2)/2}t^{1/\beta}\varepsilon^{2-2m+2m(2-\alpha-z)/z} \times \int_0^\infty x^{-1/\beta}\left(1-\frac{1}{x}+\frac{1}{x}e^{-x}\right)dx, \tag{A.25}$$

and second, $\alpha > 2$, in which the integral diverges and hence

$$D_\varepsilon(t) \sim B(\alpha-2)^{-1}k_0^{2-\alpha}t^2\varepsilon^{4-4m-\alpha}, \tag{A.26}$$

Finally, if

$$2m > z, \tag{A.27}$$

then $b_1(\varepsilon) \to \infty$ and we arrive at

$$D_\varepsilon(t) \sim B(\alpha+z-2)^{-1}k_0^{2-\alpha-z}t\varepsilon^{4-2m-\alpha-z}. \tag{A.28}$$

In summary, we have

$$D_\varepsilon(t) \sim Ct^p\varepsilon^q, \tag{A.29}$$

where p, q are defined in (A.25), (A.26), (A.28) for the different cases. After substituting (A.29) to formula (A.23 for $S_\varepsilon(t)$ and changing the variable of integration we get

$$S_\varepsilon(t) = B\varepsilon^{4-4m-\alpha}\int_o^t\int_{k_o}^{k_1/\varepsilon} k^{1-\alpha}\exp\{-ask^z\varepsilon^{z-2m} - Ck^2\varepsilon^p s^q - s\kappa\varepsilon^{2-2m}\}dkds + \kappa_\varepsilon. \tag{A.30}$$

If $(\alpha, z) \in G_1$, i.e. $\alpha + z > 2$ and $\alpha + 2z < 4$, and $m = (4-\alpha-z)/2$ then (A.27) holds and hence (A.28) is true. After substituting the corresponding p, q to (A.30) and integrating with respect to s we have

$$S_\varepsilon(t) \sim B\varepsilon^{4-4m-\alpha}\int_{k_o}^{k_1/\varepsilon}\frac{k^{1-\alpha}}{ak^z\varepsilon^{z-2m} + Ck^2\varepsilon^p} \times$$
$$\times(1 - \exp\{-t(ak^z\varepsilon^{z-2m} + Ck^2\varepsilon^p\})dk,$$

where the terms with κ_ε vanish because of $m < 1$. Since $\varepsilon^{z-2m} \to \infty$ we have

$$\lim_{\varepsilon\to 0} S_\varepsilon(t) = B\lim_{\varepsilon\to 0}\varepsilon^{4-4m-\alpha}\int_{k_o}^{k_1/\varepsilon}\frac{k^{1-\alpha}}{ak^z\varepsilon^{z-2m}}dk = \frac{B}{a(\alpha+z-2)}k_0^{2-\alpha-z},$$

whereby 1) is proven.

Note, that if $(\alpha, z) \in G_2 \cup G_3$ and

$$m = \begin{cases} 2 - \alpha/2 - \beta, & (\alpha, z) \in G_2, \\ 1, & (\alpha, z) \in G_3. \end{cases} \tag{A.31}$$

then (A.24) holds. Indeed, in the first case $z - 2m = (\alpha + 2z - 4)(\alpha + z - 2)/(\alpha + 2z - 2)$ and in the second case $z - 2m = z - 2$. Since in both cases $\alpha < 2$ we should use asymptotic (A.25). It can easily be seen that for $p = 2 - 2m + m(2 - \alpha - z)$ prescribed by (A.25) quantity $Ck^2\varepsilon^p s^q$, where C, p, q come from (A.25), surpresses other terms in the exponent in (A.30) since its value at $k = k_1/\varepsilon$ is much bigger then that of for other terms. Taking this into account we obtain that for both regions

$$\begin{aligned} S_\varepsilon(t) &\sim B\varepsilon^{4-4m-\alpha} \int_0^t \int_{k_0}^{k_1/\varepsilon} k^{1-\alpha} \exp\{-Ck^2\varepsilon^p s^q\} dk ds + \kappa_\varepsilon \sim \\ &\sim q_{\alpha,z}|2 - \alpha - 2\beta|\varepsilon^{4-4m-\alpha-p\beta} \int_{k_o}^{k_1/\varepsilon} k^{1-\alpha-2\beta} dk + \kappa_\varepsilon, \end{aligned} \tag{A.32}$$

The behaviour of the latter integral depends on the sign of $v \equiv 2 - \alpha - 2\beta = (2z(1-\alpha) - (\alpha - 2)^2)/(\alpha + 2z - 2)$. If $(\alpha, z) \in G_2$ and m is given by (A.31) then $v < 0$ and this integral converges when $\varepsilon \to 0$. Thus

$$S_\varepsilon(t) \sim q_{\alpha,z}\varepsilon^{4-4m-\alpha p\beta} k_0^{2-\alpha 2\beta}.$$

By substituting $m = 2 - \alpha/2 - \beta$ and $p = 2 - 2m/\beta$ from (A.25) to the last equality we obtain that the exponent of ε is zero.

Thus, if m is given by (A.31) then $S_\varepsilon(t)$ has the finite limit given in Proposition 3.2.

If $(\alpha, z) \in G_3$ and $m = 1$ then $v > 0$ and the integral on the right-hand side of (A.32) diverges when $\varepsilon \to 0$. Thus

$$S_\varepsilon(t) \sim q_{\alpha,z}\varepsilon^{2-4m+(2-p)\beta} k_1^{2-\alpha-2\beta} + \kappa.$$

Hence if $m = 1$ then $S_\varepsilon(t)$ has the finite limit given in Proposition 3.2. Lemma A.2 is proven.

REFERENCES

[1] Avellaneda M., Majda A.J., *Mathematical models with exact renormalization for turbulent transport*, Commun. Math. Phys., v. 131, pp. 181–429 (1990).

[2] Avellaneda M., Majda A.J., *Renormalization theory for eddy diffusivity in turbulent transport*, Physical Review Letters, v. 68, N20, pp. 3028–3031 (1992).

[3] Batchelor, G.K., *Diffusion field of homogeneous turbulence II.* The relative motion of particles, Proc. Camb. Phil. Soc., 48, pp. 345–363 (1952).
[4] Batchelor, G.K., *The Theory of homogeneous turbulence*, Cambridge University, Cambridge (1982).
[5] Bennet, A.F., *A Lagrangian analysis of turbulent diffusion*, Reviews of Geophysics, v. 25, pp. 799–822 (1987).
[6] Csanady, G.T., *Turbulent Diffusion in the Environment, Geophysics and Astrophysics Monographs*, (Reidel, Dordrecht) (1973).
[7] Davis, R.E., *On relating Eulerian and Lagrangian velocity statistics: single particles in homogeneous flows*, J. Fluid Mechanics, v. 114, pp. 1–26 (1982).
[8] Davis, R.E., *Modelling eddy transport of passive tracers*, Journal of Marine Research, v. 45, pp. 635–666 (1987).
[9] Davis, R.E., *Observing the general circulation with floats*, Deep-Sea Research, v. 38, pp. 5531–5571 (1991).
[10] Eckart C., *An analysis of the stirring and mixing process in incompressible fluids*, Journal of Marine Research, v. 7, N3, pp. 265–325 (1948).
[11] Hasselman, K., *Feynman diagrams and interaction rules of wave-wave scattering processes*, Rev. Geophys. Space Phys, v. 4, pp. 1–32 (1966).
[12] Isichenko, M.B., *Percolation, statistical topography, and transport in random media*, Reviews of Modern Physics, v. 64, No.4, pp. 961–1043 (1992).
[13] Klyatskin, V.I., *Statistical description of diffusing tracers in random velocity fields*, Uspekhi Fizicheskikh Nauk, Russian Academy of Science, v. 164, n5, pp. 531–544 (1994).
[14] Kozlov, S.M., *Averaging of random operators*, Transactions of Moscow Math. Soc., v. 37, pp. 167–180 (1980).
[15] Kraichnan, R.M., *The structure of isotropic turbulence at very high Reynolds numbers*, J. Fluid Mech., v. 5, pp. 497–543, (1959).
[16] Kraichnan, R.M., *Convection of a passive scalar by a quasi-uniform random straining field*, J. Fluid Mech., v. 64, pp. 737–762 (1974).
[17] Kubo, R., *Stochastic Liouville equation*, J. Math. Phys., v. 4, pp. 174–183 (1963).
[18] Lundgren, T.S., *Trubulent pair dispersion and scalar diffusion*, J. Fluid Mech., v. 111, pp. 27–57 (1981).
[19] Molchanov, S.A. and Piterbarg, L.I., *Heat propagation in random flows*, Russian Journal of Mathematical Physics, v. 1, 3, pp. 353–376, Wiley & Sons, Inc., (1994).
[20] Monin, A.S. and Yaglom, A.M., *Statistical fluid mechanics; Mechanics of turbulence*, Cambridge, MA (1975).
[21] Okubo, A., *Diffusion and Ecological problems: mathematical models*, Springer-Verlag, New York (1980).
[22] Papanicolaou, G. and Varadhan, S.R.S., *Boundary value problems with rapidly oscillating coefficients* In: *Random Fields, colloq. Math.*, Soc. Janos Bolyai, 27. Fritz, J., Kebowitz, J.L., Szasz, D.(eds.) pp. 835–875, Amsterdam: North Holland (1982).
[23] Phythian, R., *Dispersion by random velocity fields*, J. Fluid Mech., v. 67, pp. 145–153 (1975).
[24] Piterbarg L., *The Dynamics and Prediction of the Large-Scale SST Anomalies (Statistical Approach)*, Leningrad, Gidrometeoizdat (in Russian) (1989).
[25] Richardson, L.F., *Atmospheric diffusion shown on a distance-neighbor graph*, Proc. Roy. Soc., v. 110, pp. 709–727 (1926).
[26] Roberts, P.H., *Analytical theory of turbulent diffusion*, J. Fluid Mech., v. 11, pp. 257–283 (1961).
[27] Saffman, P.G., *An application of the Weiner-Hermite expansion to diffusion of a passive scalar in a homogeneous turbulent flow*, Phys. Fluids, v. 12, pp. 1786–1798 (1969).
[28] Semenov, D.V., *Averaging of differential equations of parabolic type with random coefficients*, Math. Notes, v. 45, N5, pp. 123–126 (1989).

[29] Taylor, G.I., *Dispersion of soluble matter in solvent flowing slowly through a tube*, Proc. Roy. Soc. A, v. 219, pp. 186–203 (1953).

[30] Zambianchi, E. and Griffa, A., *Effects of finite scales of turbulence on dispersion estimates*, Journal of Marine Research, v. 52, pp. 129–148 (1994).

COMMENTS ON ESTIMATION AND PREDICTION FOR AUTOREGRESSIVE AND MOVING AVERAGE NONGAUSSIAN SEQUENCES*

MURRAY ROSENBLATT†

Autoregressive and moving average models have been studied for a long period of time, particularly in the Gaussian case, with respect to the problems of prediction and that of estimation of the coefficients of the models. It is only in recent years that especial attention has been paid to the case of nonGaussian models where it has been realized that the corresponding problems may have a more complicated but richer structure. A discrete time autoregressive moving average model is a solution x_t of the system of equations

$$\sum_{j=0}^{p} a_j x_{t-j} = \sum_{k=0}^{q} b_k \xi_{t-k} \tag{1.1}$$

where the sequence ξ_t is a sequence of independent, identically distributed random variables with $E\xi_t \equiv 0, E\xi_t^2 = \sigma^2 > 0(\sigma^2 < \infty)$. The coefficients a_j, b_k are real and it is conventional to set $a_0 = b_0 = 1$. There is a stationary solution to the system if and only if the polynomial

$$a(z) = \sum_{j=0}^{p} a_j z^j$$

has no zeros of absolute value one and this solution is uniquely determined. The estimation problem is that of estimating the coefficients a_j, b_k given the sequence of observations $x_1, \ldots, x_n$. The stationary solution x_t is causal, if the polynomial $a(z)$ has all its zeros of absolute value greater than one, in the sense that there is a one-sided representation of x_t in terms of the present and past of the ξ sequence

$$x_t = \sum_{j=0}^{\infty} \alpha_j \xi_{t-j}$$

with the coefficients α_j decreasing to zero exponentially fast as $j \to \infty$. Let

$$b(z) = \sum_{k=0}^{q} b_k z^k.$$

* This research was supported in part by Office of Naval Research Grant N00014-90-J1372.

† Department of Mathematics, University of California-San Diego, 9500 Gilman Drive, La Jolla CA 92093-0112.

The condition that both $a(z)$ and $b(z)$ have all their zeros of absolute value greater than one is sometimes called a minimum phase condition. In the case of Gaussian sequences x_t (the ξ's are then Gaussian) one cannot distinguish between roots inside and outside the unit circle since the whole probability structure of the x sequence is determined by its spectral density

$$f(\lambda) = \frac{\sigma^2}{2\pi}|b(e^{-i\lambda})/a(e^{-i\lambda})|^2,$$

a function determined simply by the modulus of $a(z), b(z)$ on the boundary of the unit circle. In the Gaussian case it is conventional to assume that the roots are outside the unit circle since the estimation problem is then identifiable. In the nonGaussian case one can actually determine the location of the roots of $a(z)$ and $b(z)$ in contrast to the situation in the Gaussian case. A detailed discussion of methods of estimation in the Gaussian case can be found in Brockwell and Davis [3]. In the nonGaussian minimum phase case estimates based on a formal Gaussian likelihood can still be used and are consistent and asymptotically normal. However, they can be improved upon if there is knowledge available of the distribution of the ξ random variables. If the ξ random variables have a smooth positive density improved estimation of the parameters is discussed in the paper of Kreiss [6]. Estimation in the nonGaussian nonminimum phase context is taken up in Breidt, *et al.* [1] and Lii and Rosenblatt [7] for the autoregressive and the moving average models. It should be noted that if the ξ distribution is discrete a formal likelihood may not exist due to singularity of the marginal distributions but still effective estimates of the parameters with even stronger properties may exist (see Davis and Rosenblatt [5]).

However, our main concern in this short note is the prediction problem. In Rosenblatt [8] the first order autoregressive stationary scheme

$$x_t = \beta x_{t-1} + \xi_t, \quad 0 < |\beta| < 1, \quad 0 < \mathrm{var}(\xi_t) = \sigma^2 < \infty$$

is considered. The best predictor of x_t given the past in mean square is linear, namely $x_t^* = \beta x_{t-1} + E\xi_t$. However, there one shows that the best predictor of x_t given the future in mean square is linear if and only if the sequence is Gaussian. In Breidt, *et al.* (1992) this is shown to hold for backward prediction in the case of a minimum phase autoregressive scheme under stronger moment conditions. As a simple example consider the case in which $0 < \beta < \frac{1}{2}$ and

$$\xi_t = \begin{cases} 0 & \text{with probability } \frac{1}{2} \\ 1 & \text{with probability } \frac{1}{2} \end{cases}$$

In that case one can predict perfectly backwards and

$$x_t = \begin{cases} \frac{1}{\beta} x_{t+1} & \text{if } x_{t+1} < 1 \\ \frac{1}{\beta}(x_{t+1} - 1) & \text{if } x_{t+1} \geq 1. \end{cases}$$

If G is the distribution function of ξ_t and F that of x_t, $0 < \beta < 1$,

$$F(\cdot) = G(\cdot) * F(\beta^{-1}\cdot)$$

where $*$ denotes the convolution operation. For convenience let $E\xi_t = 0$ and set

$$\overline{G}(x) = \int_{-\infty}^{x} u dG(u).$$

It is then shown that the best predictor of x_t given the future is

$$E(x_t|x_{t+1} = x) = \frac{1}{\beta}x - \frac{1}{\beta}\{\overline{G} * F(\beta^{-1}\cdot)\}(dx)/F(dx).$$

A number of explicit examples have been discussed by Chiu [4]. Let $\psi(\cdot)$ be the characteristic function of $\xi_t, \eta(\cdot)$ that of x_t and $\phi(\tau_1, \tau_2)$ the joint characteristic function of x_t and x_{t+1}. If ξ_t has the following mixture of a discrete distribution and density function as specifying its distribution

$$\beta\delta_0 + (1 - \beta)e^{-x}dx$$

then

$$\psi(\tau) = \tfrac{1-\beta i\tau}{1-i\tau},$$
$$\eta(\tau) = \textstyle\prod_{j=0}^{\infty} \psi(\beta^j\tau) = \tfrac{1}{1-i\tau}$$

while

$$\phi(\tau_1, \tau_2) = \psi(\tau_2)\eta(\beta\tau_2 + \tau_1) = \frac{1 - \beta i\tau_2}{1 - i\tau_2}\frac{1}{1 - i(\beta\tau_2 + \tau_1)}.$$

The distribution of x is clearly exponential. The joint distribution of x_t and x_{t+1} is given by

$$P(dx, dy) = \beta e^{-x}\delta_{y-\beta x}dx + (1 - \beta)e^{-y}e^{-x(1-\beta)}dxdy \quad \text{for } y \geq \beta x, x \geq 0$$

and zero otherwise. The best predictor of x_t given x_{t+1} is

$$E[x_t|x_{t+1}] = \frac{1}{1 - \beta} - \frac{1}{1 - \beta}e^{-x_{t+1}(1-\beta)/\beta}$$

for $x_{t+1} \geq 0$ and zero otherwise. The claim is that the prediction error for the best predictor is

$$2\frac{1 - \beta}{2 - \beta}$$

while that for the best linear predictor is

$$1 - \beta^2$$

Chiu also carries out the computations for x_t having the density

$$x^p e^{-x}, \quad (p \text{ an integer})$$

for $x > 0$. In this case the ξ_t characteristic function is

$$\psi(z) = \left(\frac{i - i\beta\tau}{1 - i\tau} \right)^p .$$

Similar computations have also been carried out for the case in which x_t has a symmetric exponential distribution.

Stationary autoregressive schemes of order p, whether minimum phase or not, are shown to be Markov sequences of order p in [9]. If all moments are finite and zeros of $a(z)$ are simple, the best predictor in mean square is shown to be linear if and only if the sequence is minimum phase or Gaussian in the same paper.

We give a short argument for a proof of a result for first order moving averages comparable to the result cited above for first order autoregressive sequences. Let

$$x_t = \xi_t + b\xi_{t-1}, \quad E\xi_t \equiv 0$$

with the ξ_t's independent and identically distributed with $0 < E\xi_t^2 = \sigma^2 < \infty$. The sequence is invertible if $|b| < 1$, that is, there is a one-sided representation of ξ_t in terms of the present and past of the x_t sequence. The sequence is noninvertible if $|b| > 1$. The joint characteristic function of $x_{t-k}, k = 0, 1, \ldots$ is

$$(1.2) \quad \phi(\tau_0, \tau_1, \ldots) = E\left\{ \exp\left(i \sum_{k=0}^{\infty} \tau_k x_{t-k} \right) \right\} = \psi(\tau_0) \prod_{k=1}^{\infty} \psi(\tau_k + b\tau_{k-1}).$$

Now

$$\begin{aligned} \frac{\partial}{\partial \tau_0} \phi(\tau_0, \tau, \ldots)|_{\tau_0 = 0} &= \int i x_t \exp\left(i \sum_{k=1}^{\infty} \tau_k x_{t-k} \right) dF(x_t, x_{t-1}, \ldots) \\ (1.3) \qquad &= \int \exp\left(i \sum_{k=1}^{\infty} \tau_k x_{t-k} \right) iE(x_t | x_{t-1}, \ldots) \\ &\qquad dF(x_{t-1}, x_{t-2} \ldots) \\ &= b\psi'(\tau_1) \prod_{k=2}^{\infty} \psi(\tau_k + b\tau_{k-1}) \end{aligned}$$

where F is the joint distribution of $x_{t-k}, k = 0, 1, \ldots$ However if the best predictor in mean square is linear when $|b| > 1$

$$(1.4) \qquad E(x_t | x_{t-1}, \ldots) = \sum_{k=1}^{\infty} a_k x_{t-k}$$

it is clear that

$$a_k = (-1)^{k+1} b^{-k},\ k = 1, 2, \ldots \tag{1.5}$$

Then expression (1.3) could equally well be written as

$$\begin{aligned}
&\sum_{k=1}^{\infty} a_k \frac{\partial}{\partial \tau_k} \psi(\tau_0) \prod_{k=1}^{\infty} \psi(\tau_k + b\tau_{k-1})|_{\tau_0=0} \\
&= \sum_{k=1}^{\infty} [\psi'(\tau_k + b\tau_{k-1})\psi(\tau_{k+1} + b\tau_k) + b\psi(\tau_k + b\tau_{k-1})\psi'(\tau_{k+1} + b\tau_k)] \\
&\quad a_k \psi(\tau_0) \prod_{s \neq k, k+1} \psi(\tau_s + b\tau_{s-1})|_{\tau_0=0}.
\end{aligned} \tag{1.6}$$

If the two expressions (1.3) and (1.5), which are equal under the assumption of a linear expression for the regression (1.4), are divided by

$$\psi(\tau_1) \prod_{k=2}^{\infty} \psi(\tau_k + b\tau_{k-1})$$

and one sets $\tau_2 = \tau_3 = \cdots = 0$, one obtains the equality

$$b\frac{\psi'(\tau_1)}{\psi(\tau_1)} = \frac{1}{b}\left[\frac{\psi'(\tau_1)}{\psi(\tau_1)} + b\frac{\psi'(\tau_1)}{\psi(\tau_1)}\right] - \frac{1}{b^2}\frac{\psi'(b\tau_1)}{\psi(\tau_1)} \tag{1.7}$$

for $|\tau_1| < T$ for some $T > 0$. Let

$$\varphi(\tau_1) = \frac{\psi'(\tau_1)}{\psi(\tau_1)} - \sigma^2 \tau_1 .$$

We then obtain the equation (1.7) with $\varphi(\tau_1)$ replacing $\psi'(\tau_1)/\psi(\tau)$ or

$$a\varphi(t) = \varphi(at)$$

for $|t| < T$. We know that $\varphi(t) = o(t)$. An argument like that in [8] implies that $\varphi(t) \equiv 0$. The assumption of a linear regression when $|b| > 1$ implies that the moving average is Gaussian.

REFERENCES

[1] F. Breidt, R. Davis, K.S. Lii and M. Rosenblatt, *Maximum Likelihood estimation for noncausal autoregressive processes*, J. Mult. Anal. 36, (1991) pp. 175–198.

[2] F. Breidt, R. Davis and W. Dunsmuir, *On backcasting in linear time series models*, New Directions in Time Series Analysis Part 1, Springer-Verlag, (1992) pp. 25-40.

[3] P. Brockwell and R. Davis, *Time Series: Theory and Methods*, Springer-Verlag, (1991).

[4] Y. Chiu, *Topics on prediction and representation of stationary processes*, thesis at University of California, San Diego, (1994).

[5] R. Davis and M. Rosenblatt, *Parameter estimation for some time series models without contiguity*, Statistics and Probability Letters 11, (1991) pp. 515–521.

[6] J. Kreiss, *On adaptive estimation in stationary ARMA processes*, Ann. Statist. 15, (1987) pp. 112–133.

[7] K.S. Lii and M. Rosenblatt, *An approximate maximum likelihood estimation for non-Gaussian non-minimum phase moving average processes*, J. Mult. Anal. 43, (1992) pp. 272–299.

[8] M. Rosenblatt, *A note on prediction and an autoregressive sequences*, in *Stochastic Processes* (editors: Cambanis, Ghosh, Karandikar, Sen) Springer-Verlag, (1993) pp. 291–295.

[9] M. Rosenblatt, *Prediction and nonGaussian autoregressive stationary sequences*, Annals of Applied Probability, vol. 5, no. 1, (1995) pp. 239–247.

PROBABILITY DISTRIBUTIONS OF PASSIVE TRACERS IN RANDOMLY MOVING MEDIA

A.I. SAICHEV* AND W.A. WOYCZYNSKI†

Abstract. Statistical properties of fluctuations of the density of passive tracer and related random fields in a randomly moving medium are discussed. Diffusion approximation and a Gaussian velocity field model is used. We find probability distributions of density fields and of Jacobians in a chaotically compressible medium. Formulas connecting statistical characteristics of random fields in Lagrangian and in Eulerian coordinates are provided. For an incompressible medium, we analyze statistical properties of the passive scalar field's gradient, and also statistics of the total gradient and the length of a contour carried in the chaotic flow of an incompressible fluid.

1. Introduction. A study of evolution of statistical characteristics of random fields of passive tracers in chaotically moving media has been of a continuing interest for a long time (see e.g. Batchelor [2], Kraichnan [16], Csanady [5], Avellaneda, Majda [1], Majda [18], Piterbarg [20]). The interest was motivated by the fact that the knowledge of those characteristics is important for solution of a number of applied problems, including the environmental problems of pollution of atmosphere and the oceans. Most of the publications devoted to the study of transport processes and diffusion of the passive tracer restrict themselves to efforts to derive and analyze equations for the mean values of the passive tracer density, or to an analysis of the diffusion of a single particle of the tracer (see e.g. Davis [6], Lipscomb, Frenkel, ter Haar [16], Careta et al. [3]). However, these characteristics are not able to describe various subtle peculiarities of the density fields of passive tracers, such as creation of complex, layered structures of the passive tracer field, density thereof rapidly changes with a slight displacement of the point of observation. These, and other phenomena are more adequately described by probability distributions, and their analysis occupies the significant portion of this paper.

In experimental studies of behavior of passive tracers in turbulent flows two, basically different, types of methods of measurement are utilized. The first type uses fixed, immobile (or moving along a determined in advance path) sensors which measure Eulerian characteristics of the chaotically moving medium. The second type uses atmospheric probes, meteorological balloons, buoys and floating devices that move with the medium, " frozen" into the flow, which measure Lagrangian characteristics of such fields such as velocity and density fields. For that reason, in the present paper we pay great attention to a comparison of statistical properties of random fields in Eulerian and Lagrangian representations, and also to finding connections between analogous Lagrangian and Eulerian statistical characteristics of

* Radio Physics Department, Nizhny Novgorod University, Nizhny Novgorod, Russia.
† Department of Statistics and Center for Stochastic and Chaotic Processes in Science and Technology, Case Western Reserve University, Cleveland, Ohio 44106.

the randomly moving medium itself. The recovery of such connections is important not only for the sake of comparison of results of the above mentioned two types of experiments. These connecting formulas are also useful in the theoretical analysis of a randomly moving medium, when it may be more convenient to conduct analysis in one (say Lagrangian) representation, but the final results are needed in the other (say, Eulerian) representation.

The main goal of the present paper is to investigate probabilistic and spatio-temporal characteristics of the passive tracer in a randomly moving medium with a random velocity field $\boldsymbol{v}(\boldsymbol{x},t)$. Both, the case of compressible, and the case of incompressible ($\boldsymbol{\nabla}\cdot\boldsymbol{v}\equiv 0$) media will be considered. The full program of such investigation should include a prior recovery of the statistical properties of the velocity field $\boldsymbol{v}(\boldsymbol{x},t)$ satisfying nonlinear equations of the hydrodynamic type, for example the Navier-Stokes equation for an incompressible fluid. However, this is an extraordinarily difficult problem which is not completely solved until this day (see e.g. Avellaneda, Majda [1], Majda [18]), and also Saichev, Woyczynski [21,22]) for simplified Burgers' turbulence models).

For that reason we will work within the framework followed by many authors (see references quoted in the Bibliography section) and assume that the statistical properties of the random field $\boldsymbol{v}(\boldsymbol{x},t)$ are given in advance. Namely, we will assume that $\boldsymbol{v}(\boldsymbol{x},t)$ *is a zero-mean, smooth isotropic and homogeneous in space, and stationary in time, Gaussian random field with a known covariance matrix*

$$\langle v_i(\boldsymbol{x},t)v_j(\boldsymbol{x}+\boldsymbol{s},t+\tau)\rangle = b_{ij}(\boldsymbol{s},\tau). \tag{1.1a}$$

Recall, that the covariance matrix of a statistically isotropic and homogeneous three-dimensional velocity field can be represented in the form

$$b_{ij}(\boldsymbol{s},\tau)=\kappa\left[F_\perp(s,\tau)\delta_{ij}+\frac{s_i s_j}{s}\frac{d}{ds}F_\perp(s,\tau)\right] \tag{1.1b}$$

$$+(1-\kappa)\left[G_\parallel(s,\tau)\delta_{ij}+\left(\delta_{ij}-\frac{s_i s_j}{s^2}\right)\frac{s}{2}\frac{d}{ds}G_\parallel(s,\tau)\right],$$

where $F_\perp(s,\tau)$ corresponds to the potential component of the velocity field, and $G_\parallel(s,\tau)$—to the rotational component. Coefficient $0\le\kappa\le 1$ takes into account the relative contribution of the potential component, and of the rotational component, to the dispersion of velocity fluctuations. For $\kappa=1$, the velocity field is purely potential, and for $\kappa=0$, it is purely rotational.

Functions $F_\perp$ and $G_\parallel$ have a transparent physical meaning: $F_\perp$ is the transversal correlation of the potential component of the velocity field, and $G_\parallel$ is the longitudinal correlation of its rotational component. The longitudinal correlation of the potential component is expressed in terms of the transversal correlation by the formula

$$F_{\|} = \frac{d}{ds}(sF_{\perp}),$$

and the transversal correlation of the rotational component is computable from its longitudinal correlation through the Karman relation

$$G_{\perp} = \frac{1}{2s}\frac{d}{ds}(s^2 G_{\|}).$$

In what follows, we will often use the simplest "Gaussian" model for functions $F_{\perp}$ and $G_{\|}$:

$$F_{\perp}(s,\tau) = G_{\|}(s,\tau) = \frac{1}{3}\sigma_v^2 \exp\left(-\frac{s^2}{2l_v^2} - \frac{\tau^2}{2\tau_v^2}\right). \tag{1.1c}$$

There are reasons to expect that even such a crude model of fluctuations of the velocity field permits adequate description of numerous qualitative and quantitative features of evolution of the passive tracer in flows of randomly moving media.

As is well known, in the Eulerian coordinate system, the density field $\rho(\boldsymbol{x}, t)$ of the passive tracer satisfies the continuity equation

$$\frac{\partial \rho}{\partial t} + \boldsymbol{\nabla} \cdot \boldsymbol{v}\rho = \kappa \Delta \rho, \tag{1.2}$$

$$\rho(\boldsymbol{x}, t = 0) = \rho_0(\boldsymbol{x}), \tag{1.3}$$

where $\rho_0(\boldsymbol{x})$ is the initial density field of the passive tracer and κ is the coefficient of molecular diffusion. Throughout this paper we will restrict ourselves to the case of the "massive enough" tracer particles which, practically, are not affected by collisions with the particles of the surrounding medium, and at the same time are "light enough" to be carried without resistance by the flow of the medium. In such a case, one can neglect the molecular diffusion term in (1.2) and use a more simple "standard" continuity equation

$$\frac{\partial \rho}{\partial t} + \boldsymbol{\nabla} \cdot \boldsymbol{v}\rho = 0. \tag{1.4}$$

The density of the passive tracer is directly connected with the laws of motion of individual particles. To recall this connection, let us write a random vector differential equation

$$\frac{d\boldsymbol{X}}{dt} = \boldsymbol{v}(\boldsymbol{X}, t), \tag{1.5}$$

$$\boldsymbol{X}(\boldsymbol{y}, t=0)=\boldsymbol{y},$$

describing the random position $\boldsymbol{X}(\boldsymbol{y},t)$ of a particle with initial (Lagrangian) coordinates $\boldsymbol{y}$. It is equivalent to (1.4) (but not to (1.2)). Whereas, continuity equations (1.2) and (1.4) are the basic equation describing an evolution of the passive tracer density in the Eulerian system of coordinates, equation (1.5) constitutes the basic equation of motion in the Lagrangian coordinate system. If one knows the laws of motion of a tracer particle with arbitrary Lagrangian coordinates $\boldsymbol{y}$, one can write the solution of the continuity equation (1.4) in the form

$$\rho(\boldsymbol{x},t)=\int \rho_0(\boldsymbol{y})\delta\Big(\boldsymbol{x}-\boldsymbol{X}(\boldsymbol{y},t)\Big)d\boldsymbol{y}. \tag{1.6}$$

Notice, that under relatively weak conditions imposed on the velocity field $\boldsymbol{v}(\boldsymbol{x},t)$, the vector function

$$\boldsymbol{x}=\boldsymbol{X}(\boldsymbol{y},t) \tag{1.7}$$

gives a one-to-one smooth mapping $\boldsymbol{x}\leftrightarrow\boldsymbol{y}$. In addition, the formula

$$\delta\Big(\boldsymbol{x}-\boldsymbol{X}(\boldsymbol{y},t)\Big)=\frac{1}{J(\boldsymbol{y},t)}\delta\Big(\boldsymbol{y}-\boldsymbol{Y}(\boldsymbol{x},t)\Big) \tag{1.8}$$

holds true, where the vector function

$$\boldsymbol{y}=\boldsymbol{Y}(\boldsymbol{x},t), \tag{1.9}$$

is the inverse function to function (1.7), and

$$J(\boldsymbol{y},t)=\left|\frac{\partial \boldsymbol{X}(\boldsymbol{y},t)}{\partial \boldsymbol{y}}\right| \tag{1.10}$$

is the Jacobian of the transformation from Lagrangian coordinates $\boldsymbol{y}$ to Eulerian coordinates $\boldsymbol{x}$. The Jacobian satisfies equation

$$\frac{dJ}{dt}=u(\boldsymbol{X},t)J,\quad J(\boldsymbol{y},t=0)=1, \tag{1.11}$$

where the auxiliary scalar field

$$u(\boldsymbol{x},t)=\boldsymbol{\nabla}\cdot\boldsymbol{v}(\boldsymbol{x},t), \tag{1.12}$$

is identically equal to 0 in an incompressible fluid.

Substituting (1.8) into (1.6), we obtain another useful expression for the density of passive tracer:

$$\rho(\boldsymbol{x},t)=\frac{\rho_0(\boldsymbol{Y}(\boldsymbol{x},t))}{J(\boldsymbol{Y}(\boldsymbol{x},t))}. \tag{1.13}$$

2. Statistical description of Lagrangian fields in a randomly moving compressible flow.

2.1. General framework. We begin with a survey of some key problems of the statistical description of random fields in the Lagrangian coordinate system and, in particular, of the fluctuations of the Lagrangian field of a passive tracer in a randomly moving compressible medium.

First of all, observe that in the physical situations the passive tracer can subject to compression and rarefaction even in the case when the fluid carrying the tracer is itself incompressible. It is true for instance for floating tracers (such as oil slicks and other pollutants) which are carried on the water surface in a quasi-2D motions of the ocean surface. Indeed, the divergence of the velocity field

$$\nabla \cdot \boldsymbol{v} = \frac{\partial v_1}{\partial x_1} + \frac{\partial v_2}{\partial x_2} + \frac{\partial v_3}{\partial x_3}$$

in the depth of the ocean is practically equal to 0. Let x_3 be the vertical coordinate and (x_1, x_2) be horizontal coordinates in the plane in which the passive tracer is constrained. The divergence of the horizontal component of the velocity field, which determines the behavior of a floating passive tracer is not equal to zero and, in view of the previous observation,

$$\mathrm{div}_h\ \boldsymbol{v} = \frac{\partial v_1}{\partial x_1} + \frac{\partial v_2}{\partial x_2} = -\frac{\partial v_3}{\partial x_3} \neq 0,$$

so that a floating passive tracer is subject to the laws of motion of a passive tracer in a 2D *compressible* flow. The situation is similar in the case of a channel flow where water, despite vertical mixing, has a horizontal motion only in the direction of channel walls. Motion of a floating tracer on its surface can then be described as the motion of a passive tracer in a 1D compressible medium. The above comments also apply to the analysis of heavy, bottom hugging tracer particles which are mixed by the fluid motions but remain constrained to the bottom layer. A simple example of such a situation can be observed in a stirred tea pot with tea leaves floating just above its bottom.

An analytic description of statistical fluctuations of tracer density is simplest in the Lagrangian coordinate system, in which its behavior is determined by a system of ordinary differential equations (1.5) and (1.11). Statistical properties of a fixed, physically infinitesimal volume of the passive tracer are determined by the joint probability distribution

$$f_L(\boldsymbol{x}, j; \boldsymbol{y}, t) = \Big\langle \delta(\boldsymbol{x} - \boldsymbol{X}(\boldsymbol{y}, t))\delta(j - J(\boldsymbol{y}, t)) \Big\rangle \tag{2.1}$$

of its coordinate $\boldsymbol{X}$ and Jacobian J, which quantitatively describes the effects of compression and expansion of the volume of the passive tracer. Subscript L signifies that the probability distribution of the random field is considered in the Lagrangian representation. In what follows we shall simply call similar statistical characteristics and probability distributions—Lagrangian.

Our problem has been already significantly simplified by an assumption that the velocity field $\boldsymbol{v}(\boldsymbol{x}, t)$ is Gaussian, with a known correlation tensor (1.1). However, even in this case, for arbitrary spatial and temporal characteristic scales of the velocity field, the problem of statistical analysis of evolution of a passive tracer field remains quite intractable. For that reason, we will make one more simplifying assumption in our model, and suppose that

$$\sigma_v \tau_v \ll l_v, \tag{2.2}$$

where $\sigma_v = \sqrt{\langle v^2 \rangle}$ is the standard deviation of fluctuations of the velocity, τ_v is the characteristic time scale, and l_v is the characteristic length scale of the velocity field $\boldsymbol{v}(\boldsymbol{x}, t)$. Inequality (2.2) has a clear physical interpretation. It requires that the displacement $\delta X \sim \sigma_v \tau_v$ during the characteristic time τ_v of changes in its velocity, be much smaller than the length scale l_v of the velocity field.

Inequality (2.2) is a condition of applicability of the diffusion approximation which permits a passage from the stochastic equations (1.5) and (1.11) to a closed equation for the probability distribution (2.1). A detailed and quite complex derivation of that equation will appear later in this section. Another approach, using a powerful apparatus of the Furutsu-Novikov-Donsker formula can be found in Klyatskin, Woyczynski, Gurarie [14] and Klyatskin, Woyczynski [15].

The diffusion approximation takes into account, fundamental for the description of processes of turbulent diffusion, differences between Eulerian and Largangian statistics of the velocity field. We will illustrate this by comparing statistical means of the divergence of velocity field (1.12). The Eulerian mean

$$\langle u(\boldsymbol{x}, t) \rangle = 0$$

(see (1.12)) in view of the general assumptions imposed in Section 1 on the velocity field $\boldsymbol{v}$. Next, let us calculate the mean of field u in the diffusion approximation and in the Lagrangian coordinate system, where it has the form

$$U(\boldsymbol{y}, t) = u(\boldsymbol{X}(\boldsymbol{y}, t), t). \tag{2.3}$$

For this purpose, we shall split solution of the stochastic equation (1.5) into a sum of two vector fields

$$\boldsymbol{X}(\boldsymbol{y}, t) = \boldsymbol{X}_0(\boldsymbol{y}, t) + \boldsymbol{Z}(t_0, t), \quad (t_0 > 0), \tag{2.4}$$

where the first field satisfies a stochastic differential equation

$$\frac{d\boldsymbol{X}_0}{dt} = \boldsymbol{v}(\boldsymbol{X}_0, t)\chi(t_0 - t), \tag{2.5}$$

$$\boldsymbol{X}_0(\boldsymbol{y}, t=0) = \boldsymbol{y},$$

and where $\chi(t_0 - t)$ is the Heaviside function equal 1 to the left of t_0 and equal 0 to the right of t_0. The inclusion of the latter eliminates the influence on $\boldsymbol{X}_0(\boldsymbol{y}, t)$ of fluctuations of the velocity field in the time interval $[t_0, t]$. That influence on coordinates of a fixed tracer particle is taken into account by the second term in (2.4), which satisfies equation

$$\frac{d\boldsymbol{Z}_0}{dt} = \boldsymbol{v}\Big(\boldsymbol{X}_0 + \boldsymbol{Z}(t_0, t), t\Big), \tag{2.6}$$

with the initial condition $\boldsymbol{Z}_0(t_0, t = t_0) = 0$ and notation $\boldsymbol{X}_0(\boldsymbol{y}, t) = \boldsymbol{X}_0$.

In the first stage, the diffusion approximation replaces the exact solution of equation (2.6) by its Born approximation

$$\boldsymbol{Z}_1 = \int_{t_0}^{t} \boldsymbol{v}(\boldsymbol{X}_0, t')dt'. \tag{2.7}$$

Roughly speaking, the replacement of $\boldsymbol{Z}$ by $\boldsymbol{Z}_1$ is justified for

$$|\boldsymbol{Z}| \ll l_v, \quad (|\boldsymbol{Z}_1| \ll l_v). \tag{2.8}$$

Replacing $\boldsymbol{Z}$ in (2.4) by $\boldsymbol{Z}_1$, and then substituting (2.4) into (2.3), we arrive at an approximate equality

$$U(\boldsymbol{y}, t) \approx u(\boldsymbol{X}_0 + \boldsymbol{Z}_1, t).$$

Assumption (2.8) also justifies the second stage of the diffusion approximation, where the field u appearing on the right-hand side of the above approximate equality is expanded into the Taylor series in powers of $\boldsymbol{Z}_1$, and only terms of order ≤ 1 are retained. This gives

$$U(\boldsymbol{y}, t) \approx u(\boldsymbol{X}_0, t) + (\boldsymbol{Z}_1 \cdot \boldsymbol{\nabla}_{\boldsymbol{X}_0})u(\boldsymbol{X}_0, t). \tag{2.9}$$

Suppose that the velocity field $\boldsymbol{v}(\boldsymbol{x}, t)$ and, consequently, field $u(\boldsymbol{x}, t)$ have a finite temporal range τ_v of statistical dependence. This means that, for arbitrary $\boldsymbol{x}_1$ and $\boldsymbol{x}_2$, random vectors $\boldsymbol{v}(\boldsymbol{x}_1, t_1)$ and $\boldsymbol{v}(\boldsymbol{x}_2, t_2)$ are statistically independent whenever $|t_1 - t_2| \geq \tau_v$. Let us select the length of the time interval $[t_0, t]$ considered above to be

$$t - t_0 = 2\tau_v. \tag{2.10}$$

Then, the statistical mean of the first term on the right-hand side of (2.9) is equal to 0 in view of the statistical independence of $u(\boldsymbol{x}, t)$ and $\boldsymbol{X}_0(\boldsymbol{y}, t)$. Hence, in view of (2.9), (2.7), and the statistical homogeneity of the velocity field, the mean of the Lagrangian field $U(\boldsymbol{y}, t)$ is approximately equal to

$$\langle U(\boldsymbol{y}, t)\rangle \approx -\int_{t-2\tau_v}^{t} \langle u(\boldsymbol{X}_0, t')u(\boldsymbol{X}_0, t)\rangle dt'. \tag{2.11}$$

The approximation also reflects the fact that, in our case, $\boldsymbol{X}_0(\boldsymbol{y},t') = \boldsymbol{X}_0(\boldsymbol{y},t) = \boldsymbol{X}_0$.

Next, consider the conditional mean

$$\langle u(\boldsymbol{X}_0,t')u(\boldsymbol{X}_0,t)\rangle_{\boldsymbol{x}}, \tag{2.12}$$

given the condition that the random vector $\boldsymbol{X}_0$ takes a given value $\boldsymbol{x}$. Then the means under the integral sign in (2.11) will be obtained by an addition averaging over $\boldsymbol{X}_0$. Notice, that in the right half of the interval of integration $(t-\tau_v,t)$, in view of the statistical independence of $\boldsymbol{X}_0$ from $u(\boldsymbol{x},t)$ and $u(\boldsymbol{x},t')$, the conditional mean (2.12) is equal to the unconditional mean

$$\langle u(\boldsymbol{x},t')u(\boldsymbol{x},t)\rangle = b_u(0,t-t'), \tag{2.13}$$

where

$$b_u(s,\tau) = -\sum_{i,j}\frac{\partial^2 b_{ij}(\boldsymbol{s},\tau)}{\partial s_i \partial s_j}, \tag{2.14a}$$

is the covariance function of the Gaussian random field $u(\boldsymbol{x},t)$ and $b_{ij}(\boldsymbol{s},\tau)$ are as in (1.1). In particular, substituting expression (1.1b) into (2.14) we get that

$$b_u(s,\tau) = -\kappa\left(\frac{8}{s}\frac{dF_\perp}{ds} + 7\frac{d^2F_\perp}{ds^2} + s\frac{d^3F_\perp}{ds^3}\right). \tag{2.14b}$$

In the left half of the interval of integration $t' \in (t-2\tau_v, t-\tau_v)$, field $u(\boldsymbol{x},t)$ is statistically independent from both $\boldsymbol{X}_0$ and $u(\boldsymbol{x},t')$, and the conditional mean (2.12) is split into a product of means

$$\langle u(\boldsymbol{X}_0,t')u(\boldsymbol{X}_0,t)\rangle_{\boldsymbol{x}} = \langle u(\boldsymbol{X}_0,t')\rangle_{\boldsymbol{x}}\ \langle u(\boldsymbol{X}_0,t)\rangle = 0.$$

For $t' < t-\tau_v$, the covariance function (2.14) of the field $u(\boldsymbol{x},t)$ vanishes as well. Therefore, in the whole domain of integration $[t-2\tau_v,t]$ one can replace the conditional mean (2.12), and with it the means under the integral sign in (2.11), by the covariance function (2.13). Respectively, the lower limit of integration in (2.11) can be replaced by $-\infty$. As a result, the diffusion approximation gives that

$$\langle U(\boldsymbol{y},t)\rangle = -B, \tag{2.15}$$

where $B = B(0)$, and

$$B(s) = \int_{-\infty}^{0} b_u(s,\tau)d\tau. \tag{2.16}$$

Notice that, in the case of Gaussian velocity fields, inequality (2.2), and conditions (2.8) and (2.10), can be viewed as equivalent. It should be also mentioned that there exist approximations more sophisticated and exact than the diffusion approximation sketched above, which permit calculation of similar Lagrangian means (see. e.g. Davis [6], Fradkin [7]). However, the diffusion approximation, as simple as it is, permits an adequate description of many universal characteristic features of fluctuations of the density field in both, compressible and incompressible randomly moving media.

2.2. Equation for Lagrangian probability distribution in the diffusion approximation. The above subsection provided a detailed illustration of the diffusion approximation ideas on the simple example of calculation of the statistical mean $\langle U(\boldsymbol{y},t)\rangle$ (see (2.15)). Now, the same ideas will be used to derive a closed equation for the probability distribution (2.1) itself.

To begin with, consider a smooth function $\phi(\boldsymbol{x},j)$. Differentiating the composed function

$$\phi(\boldsymbol{X}(\boldsymbol{y},t),J(\boldsymbol{y},t))$$

with respect to time, using equalities (1.5) and (1.11), and then averaging the obtained identity, we get that

$$\frac{d}{dt}\langle\phi(\boldsymbol{X},J)\rangle \tag{2.17}$$

$$=\Big\langle\Big(\boldsymbol{v}(\boldsymbol{X},t)\boldsymbol{\nabla}_{\boldsymbol{X}}\Big)\phi(\boldsymbol{X},J)\Big\rangle+\Big\langle u(\boldsymbol{X},t)J\frac{\partial}{\partial J}\phi(\boldsymbol{X},J)\Big\rangle.$$

Let us compute, in the diffusion approximation, the means on the right-hand side of (2.17). We start with representing the random field $J(\boldsymbol{y},t)$ in the form analogous to (2.4), namely

$$J(\boldsymbol{y},t)=J_0(\boldsymbol{y},t)+Q(t_0,t), \tag{2.18}$$

where $J_0(\boldsymbol{y},t)$ satisfies equation

$$\frac{dJ_0}{dt}=u(\boldsymbol{X},t)J\chi(t_0-t),\quad J_0(\boldsymbol{y},t=0)=1, \tag{2.19}$$

in which, in analogy with equation (2.5), we have excluded the influence of the velocity field $\boldsymbol{v}(\boldsymbol{x},t)$ on the behavior of field $J_0(\boldsymbol{y},t)$ for $t>t_0$. The second summand in (2.18) satisfies an equation analogous to (2.6), namely

$$\frac{dQ}{dt}=u(\boldsymbol{X}_0+\boldsymbol{Z},t)(J_0+Q),\quad Q(t_0,t=t_0)=0. \tag{2.20}$$

Suppose, that all the assumptions imposed on field $\boldsymbol{v}(\boldsymbol{x},t)$ in derivation of relation (2.15) are still in force. Then, as in (2.7), one can restrict oneself to the Born approximation of the solution of equation (2.20), that is

$$Q_1(t_0,t)=J_0(\boldsymbol{y},t)\int_{t_0}^{t}u(\boldsymbol{X}_0,t')\,dt'. \tag{2.21}$$

Clearly, we have taken into account the fact that, for $t > t_0$, neither $\boldsymbol{X}_0$ nor J_0 depend on t'.

Now, let us proceed directly to the calculation of the last mean in equation (2.17). For that purpose, let us rewrite it in the form

$$\left\langle u(\boldsymbol{X},t)J\frac{\partial}{\partial J}\phi(\boldsymbol{X},J)\right\rangle \tag{2.22}$$

$$= \left\langle u(\boldsymbol{X}_0+\boldsymbol{Z},t)(J_0+Q)\frac{\partial}{\partial J_0}\phi(\boldsymbol{X}_0+\boldsymbol{Z},J_0+Q)\right\rangle .$$

Assume that $\phi(\boldsymbol{x},j)$ is a sufficiently smooth function of its arguments so that the expression inside the mean brackets on the right-hand side of equality (2.22) can be expanded in a Taylor series in powers of $\boldsymbol{Z}$ and Q, and then sufficiently accurately represented by the part of that expansion that contains terms of degree at most 1. In addition, we replace $\boldsymbol{Z}$ and Q by their Born approximations $\boldsymbol{Z}_1$ and Q_1. All of these give that

$$\left\langle u(\boldsymbol{X},t)J\frac{\partial}{\partial J}\phi(\boldsymbol{X},J)\right\rangle \approx \left\langle u(\boldsymbol{X}_0,t)J_0\frac{\partial}{\partial J_0}\phi(\boldsymbol{X}_0,J_0)\right\rangle \tag{2.23}$$

$$+\left\langle J_0\frac{\partial}{\partial J_0}\phi(\boldsymbol{X}_0,J_0)(\boldsymbol{Z}_1\nabla_{\boldsymbol{X}_0})u(\boldsymbol{X}_0,t)\right\rangle$$

$$+\left\langle u(\boldsymbol{X}_0,t)Q_1(t_0,t)\frac{\partial}{\partial J_0}J_0\frac{\partial}{\partial J_0}\phi(\boldsymbol{X}_0,J_0)\right\rangle$$

$$+\left\langle u(\boldsymbol{X}_0,t)(\boldsymbol{Z}_1(t_0,t)\boldsymbol{\nabla}_{\boldsymbol{X}_0})J_0\frac{\partial}{\partial J_0}\phi(\boldsymbol{X}_0,J_0)\right\rangle .$$

If equation (2.10) is satisfied, then the first mean on the right-hand side of the above equation splits into the product of means and

$$\left\langle u(\boldsymbol{X}_0,t)J_0\frac{\partial}{\partial J_0}\phi(\boldsymbol{X}_0,J_0)\right\rangle = \langle u(\boldsymbol{X}_0,t)\rangle\left\langle J_0\frac{\partial}{\partial J_0}\phi(\boldsymbol{X}_0,J_0)\right\rangle = 0, \tag{2.24}$$

in view of the statistical independence of the field $u(\boldsymbol{x},t)$ from the values of fields $J_0(\boldsymbol{y},t)$ and $\boldsymbol{X}_0(\boldsymbol{y},t)$, and the fact that $\langle u(\boldsymbol{x},t)\rangle = 0$.

Let us turn to calculation of the second mean on the right-hand side of equality (2.23) which we will write in the form

$$\left\langle J_0\frac{\partial}{\partial J_0}\phi(\boldsymbol{X}_0,J_0)(\boldsymbol{Z}_1\boldsymbol{\nabla}_{\boldsymbol{X}_0})u(\boldsymbol{X}_0,t)\right\rangle \tag{2.25}$$

$$= \left\langle J_0 \frac{\partial}{\partial J_0} \phi(\boldsymbol{X}_0, J_0) \left\langle (\boldsymbol{Z}_1 \boldsymbol{\nabla}_{\boldsymbol{X}_0}) u(\boldsymbol{X}_0, t) \right\rangle_{X_0, J_0} \right\rangle ,$$

where $\langle\ .\ \rangle_{X_0, j_0}$ signifies averaging under condition that values of $\boldsymbol{X}_0$ and J_0 are given, and the outside brackets signify averaging with respect to random quantities $\boldsymbol{X}_0$ and J_0. Substituting the explicit expression (2.7) for $\boldsymbol{Z}_1$ into the conditional mean, that average can be written in the form

$$\left\langle (\boldsymbol{Z}_1 \boldsymbol{\nabla}_{\boldsymbol{X}_0}) u(\boldsymbol{X}_0, t) \right\rangle_{X_0, J_0}$$

$$= \int_{t-2\tau_v}^{t} \left\langle \left(\boldsymbol{v}(\boldsymbol{X}_0, t') \boldsymbol{\nabla}_{\boldsymbol{X}_0} \right) u(\boldsymbol{X}_0, t) \right\rangle_{\boldsymbol{X}_0, J_0} dt'.$$

Repeating considerations that permitted us to pass from equality (2.11) to equality (2.15), we obtain that the above conditional mean is independent of the conditions and, in the diffusion approximation, is equal to

$$\left\langle (\boldsymbol{Z}_1 \boldsymbol{\nabla}_{\boldsymbol{X}_0}) u(\boldsymbol{X}_0, t) \right\rangle_{\boldsymbol{X}_0, J_0} = -B.$$

As a result, equality (2.25) assumes the form

(2.26)
$$\left\langle J_0 \frac{\partial}{\partial J_0} \phi(\boldsymbol{X}_0, J_0) (\boldsymbol{Z}_1 \boldsymbol{\nabla}_{\boldsymbol{X}_0}) u(\boldsymbol{X}_0, t) \right\rangle = -B \left\langle J_0 \frac{\partial}{\partial J_0} \phi(\boldsymbol{X}_0, J_0) \right\rangle .$$

In an analogous fashion one can also show that the two remaining means on the right-hand side of (2.23) are, in the diffusion approximation, as follows:

$$\left\langle u(\boldsymbol{X}_0, t) Q_1(t_0, t) \frac{\partial}{\partial J_0} J_0 \frac{\partial}{\partial J_0} \phi(\boldsymbol{X}_0, J_0) \right\rangle = B \left\langle \frac{\partial}{\partial J_0} J_0 \frac{\partial}{\partial J_0} \phi(\boldsymbol{X}_0, J_0) \right\rangle$$

$$\left\langle u(\boldsymbol{X}_0, t) (\boldsymbol{Z}_1(t_0, t) \boldsymbol{\nabla}_{\boldsymbol{X}_0}) J_0 \frac{\partial}{\partial J_0} \phi(\boldsymbol{X}_0, J_0) \right\rangle = 0.$$

Consequently, equality (2.23) takes the form

$$\left\langle u(\boldsymbol{X}, t) J \frac{\partial}{\partial J} \phi(\boldsymbol{X}, J) \right\rangle$$

$$= -B \left\langle J_0 \frac{\partial}{\partial J_0} \phi(\boldsymbol{X}_0, J_0) \right\rangle + B \left\langle \frac{\partial}{\partial J_0} J_0 \frac{\partial}{\partial J_0} \phi(\boldsymbol{X}_0, J_0) \right\rangle ,$$

or, after simplifications,

(2.27)
$$\left\langle u(\boldsymbol{X}, t) J \frac{\partial}{\partial J} \phi(\boldsymbol{X}, J) \right\rangle = B \left\langle J_0^2 \frac{\partial^2 \phi(\boldsymbol{X}_0, J_0)}{\partial J_0^2} \right\rangle .$$

With the help of similar arguments, in the diffusion approximation, we arrive at the following expression for the first mean on the right-hand side of equality (2.17):

$$\Big\langle \boldsymbol{v}(\boldsymbol{X},t)\boldsymbol{\nabla}_{\boldsymbol{X}}\phi(\boldsymbol{X},J)\Big\rangle = D\Big\langle \Delta_{\boldsymbol{X}_0}\phi(\boldsymbol{X}_0,J_0)\Big\rangle, \tag{2.28}$$

where the diffusion coefficient $D = D(0)$ with

$$D(s) = \frac{1}{N}\int_{-\infty}^{0} b_{ii}(\boldsymbol{s},\tau)\,d\tau, \tag{2.29}$$

and N is the dimension of space. Notice, that in the three-dimensional case, in view of (1.1b), the integrand in (2.29) is equal to

$$b_{ii} = \frac{1}{s^2}\frac{d}{ds}s^3[\kappa F + (1-\kappa)G].$$

Utilizing relations (2.27), (2.28) and (2.29), we can rewrite equality (2.17) in the form

$$\frac{d}{dt}\langle\phi(\boldsymbol{X},J)\rangle = D\Big\langle \Delta_{\boldsymbol{X}_0}\phi(\boldsymbol{X}_0,J_0)\Big\rangle + B\left\langle J_0^2\frac{\partial^2\phi(\boldsymbol{X}_0,J_0)}{\partial J_0^2}\right\rangle. \tag{2.30}$$

We will utilize once more the discussed above smoothness of function $\phi(\boldsymbol{X},J)$. More exactly, we will assume the function to be sufficiently nice so that, without going outside the boundaries of accuracy of the approximate equality (2.30), we can replace on its right-hand side, $\boldsymbol{X}_0$ and J_0 by $\boldsymbol{X}$ and J. In this situation, equation (2.30) takes the form

$$\frac{d}{dt}\langle\phi(\boldsymbol{X},J)\rangle - D\Big\langle \Delta_{\boldsymbol{X}}\phi(\boldsymbol{X},J)\Big\rangle - B\left\langle J^2\frac{\partial^2\phi(\boldsymbol{X},J)}{\partial J^2}\right\rangle = 0. \tag{2.31}$$

Observe, that to calculate the statistical means entering in this equation only knowledge of probability distribution $f_L(\boldsymbol{x},j;\boldsymbol{y},t)$ (see (2.1)) is needed. Putting together all the summands in (2.31), and writing the statistical averages explicitly, we obtain that

$$\int d\boldsymbol{x}\int dj\left[\phi(\boldsymbol{x},j)\frac{\partial f_L}{\partial t} - Df_L\Delta_{\boldsymbol{x}}\phi(\boldsymbol{x},j) - Bf_Lj^2\frac{\partial^2\phi(\boldsymbol{x},j)}{\partial j^2}\right] = 0. \tag{2.32}$$

We will require that $\phi(\boldsymbol{x},j)$, together with its partial derivatives, converges to zero as $|x|\to\infty$, and $|j|\to\infty$. In such a case, after integration by parts, we can pass from (2.32) to an equivalent equation

$$\int d\boldsymbol{x}\int dj\phi(\boldsymbol{x},j)\left[\frac{\partial f_L}{\partial t} - D\Delta_{\boldsymbol{x}}f_L - B\frac{\partial^2}{\partial j^2}(j^2 f_L)\right] = 0. \tag{2.32}$$

The above equation, given that the supply of functions $\phi(\boldsymbol{x}, j)$ satisfying conditions imposed above is sufficient, implies that the probability distribution $f_L(\boldsymbol{x}, j; \boldsymbol{y}, t)$ fulfills a closed, Fokker-Planck-Kolmogorov type equation

$$\frac{\partial f_L}{\partial t} = D\Delta_{\boldsymbol{x}} f_L + B\frac{\partial^2}{\partial j^2}(j^2 f_L), \tag{2.33}$$

that is well known in the theory of Markov processes. It should be considered together with the initial condition

$$f_L(\boldsymbol{x}, j; \boldsymbol{y}, t = 0) = \delta(\boldsymbol{x} - \boldsymbol{y})\delta(j - 1). \tag{2.34}$$

Without getting bogged down by mathematical details, we would like to comment that the way we used smoothness conditions on function $\phi(\boldsymbol{x}, j)$ means that the smoother probability densities $f_L(\boldsymbol{x}, j; \boldsymbol{y}, t)$ vary as functions of variables $\boldsymbol{x}$ and j (or, alternatively, the slower f_L varies as a function of time), the more accurate the approximate equation (2.33) is. From the physical viewpoint, the above comment means that equation (2.33) is valid in the "large time scale", with the characteristic scale of variability τ_f, much larger than the correlation time $\tau_{\boldsymbol{v}}$ of the random velocity field $\boldsymbol{v}(\boldsymbol{x}, t)$.

Also, notice that in the case of the "Gaussian" form (1.1c) of functions $F_\perp$ and $G_\parallel$, coefficients D and B appearing in the equation can be expressed through the variance σ_v^2 of the velocity field, and its characteristic spatial and temporal scales l_v and τ_v, as follows:

$$D = \sqrt{2\pi}\sigma_v^2\tau_v/3,$$

$$B = 5\kappa\sqrt{2\pi}\sigma_v^2\tau_v/l_v^2.$$

Finally, it is useful to observe that the Lagrangian mean (2.15) which is part of the structure of the last term in equation (2.33), determines numerous characteristic features of the evolution of the Jacobian field $J(\boldsymbol{y}, t)$, which will be studied in detail in the following sections.

3. Probabilistic properties of realizations of Lagrangian Jacobian and density fields.

3.1. Auxiliary Markov processes. In this section we will provide a detailed analysis of implications of equation (2.33) for Lagrangian probability distributions of fields $\boldsymbol{X}(\boldsymbol{y}, t)$ and $J(\boldsymbol{y}, t)$. Its solution, in view of attendant initial conditions, splits into a product of two factors

$$f_L(\boldsymbol{x}, j; \boldsymbol{y}, t) = f_X(\boldsymbol{x}; \boldsymbol{y}, t) f_J(j; t), \tag{3.1}$$

each of which has a clear cut physical meaning. The first factor represents the probability distribution for the coordinates of a fixed particle at time t, that is

$$f_X(\boldsymbol{x}; \boldsymbol{y}, t) = \langle \delta(\boldsymbol{x} - \boldsymbol{X}(\boldsymbol{y}, t)) \rangle, \tag{3.2}$$

and it satisfies the standard diffusion equation

$$\frac{\partial f_X}{\partial t} = D\Delta f_X, \tag{3.3}$$

$$f_X(\boldsymbol{x}; \boldsymbol{y}, t=0) = \delta(\boldsymbol{x} - \boldsymbol{y}).$$

In particular, the standard law of diffusion of the passive tracer

$$\langle (\boldsymbol{X}(\boldsymbol{y}, t) - \boldsymbol{y})^2 \rangle = ND = \int_{-\infty}^{0} b_{ii}(0, \tau)\, d\tau$$

follows form (3.3).

The second factor on the right-hand side of (3.1) gives a quantitative description of probabilistic properties of compressions and expansions of physically infinitesimal volumes of the passive tracers and satisfies equation

$$\frac{\partial f_J}{\partial t} = B\frac{\partial^2}{\partial j^2}(j^2 f_J), \tag{3.4}$$

$$f_J(j; t=0) = \delta(j-1).$$

From now on, our analysis of evolution of physical fields will actively utilize the notion of statistically equivalent random processes, that is processes with identical finite-dimensional distributions, and in particular with identical equations for their joint probability distributions. For Markov processes, which are determined by their infinitesimal generators, the above two conditions are equivalent.

More precisely, we will discuss, at the physical level of rigor, properties of simpler, but statistically equivalent auxiliary processes, and then attribute these properties to corresponding random fields which are of principal interest in this paper. This will be done for fixed space points and only time dependence results will be discussed. Although such an approach has a different degree of validity in different concrete situations, it is always heuristically useful and helps a deeper understanding of the fine structure of realizations of a random field. We leave the justification of mathematical validity of such an approach for a different publication.

Equation (3.4) can be interpreted as a Fokker-Planck-Kolmogorov equation for transition probabilities of an auxiliary Markov process $J(t)$, which satisfies a (Stratonovich) stochastic differential equation

$$\frac{dJ}{dt} + BJ = \eta(t)J, \tag{3.5}$$

$$J(t=0) = 1,$$

where $\eta(t)$ is a Gaussian white noise with covariance function

$$\langle \eta(t)\eta(t+\tau)\rangle = 2B\delta(\tau). \tag{3.6}$$

The solution of stochastic equation (3.5) is of the form

$$J(t) = \exp[-Bt + w(t)], \tag{3.7}$$

where $w(t)$ is the Brownian motion process such that

$$\langle w(t)\rangle = 0, \quad \text{and} \quad \langle w^2(t)\rangle = 2Bt. \tag{3.8}$$

Formulas (3.7) and (3.4) imply that the probability density for the process $J(t)$, and thus also field $J(\boldsymbol{y}, t)$ in a neighborhood of a fixed tracer particle with Lagrangian coordinate $\boldsymbol{y}$, is lognormal and

$$f_J(j;t) = \frac{1}{2j\sqrt{\pi Bt}} \exp\left[-\frac{\ln^2(je^{Bt})}{4Bt}\right], \quad j > 0, \tag{3.9}$$

with the cumulative distribution function

$$F_J(j;t) = \int_0^j f_J(j;t)dj = \Phi\left(\frac{\ln(je^{Bt})}{2\sqrt{Bt}}\right), \tag{3.10}$$

where $\Phi(z) = (1/\sqrt{\pi})\int_{-\infty}^{z} \exp(-y^2)dy$ is the standard error function.

Observe, that the realizations of process (3.7) (and thus also those of $J(\boldsymbol{y}, t)$ for a fixed $\boldsymbol{y}$) have a complex fine structure which apparently contradicts their statistical properties. For example, consider the time behavior of statistical moments of process $J(t)$. It follows from (3.4) that the n-th moment $\langle J^n(t)\rangle$ satisfies a closed equation

$$\frac{d\langle J^n\rangle}{dt} = Bn(n-1)\langle J^n\rangle,$$

$$\langle J^n(0)\rangle = 1.$$

Its solution being

$$\langle J^n\rangle = \exp[n(n-1)Bt], \tag{3.11}$$

it follows that the mean Jacobian is conserved, that is

$$\langle J\rangle = \langle J(\boldsymbol{y}, t)\rangle = 1. \tag{3.12}$$

This result for field $J(\boldsymbol{y}, t)$ is not difficult to justify as a consequence of the physically obvious law of conservation of the fluid volume. According to such conservation law, even in compressible medium, statistically homogeneous flows of the medium behave in such a way that the mean volume occupied by an arbitrary fixed portion of the medium remains constant. For an incompressible medium, this conservation law becomes a trivial dynamical identity $J(\boldsymbol{y}, t) \equiv 1$.

Formula (3.11) also implies that all the higher moments of J, beginning with $n = 2$, grow exponentially with the passage of time. For example, for $n = 2$,

$$\langle J^2(t) \rangle = e^{2Bt}. \tag{3.13}$$

This property seems to imply an appearance, as t increases, of very high peaks in the realizations of $J(t)$, which would explain the exponential growth of higher moments and of the variance

$$\sigma_J(t) = \langle J^2 \rangle - \langle J \rangle^2 = e^{2Bt} - 1. \tag{3.14}$$

Geometrically, an appearance of peaks where $J(\boldsymbol{y}, t) \gg 1$ signifies the sojourn of a fixed particle in a strongly rarefied region. However, the picture of a particle finding itself at large times in an increasingly rarefied region is contradicted by the behavior of probabilistic characteristics (3.9-10) of the process $J(t)$ and field $J(\boldsymbol{y}, t)$. In particular, it follows from (3.10) that the probability that the random field $J(\boldsymbol{y}, t)$ exceeds the mean level $\langle J \rangle \equiv 1$ is equal to

$$P\Big(J(\boldsymbol{y},t) > 1\Big) = \Phi\left(-\frac{\sqrt{Bt}}{2}\right) \sim \frac{2}{\sqrt{\pi Bt}} \exp\left(-\frac{Bt}{4}\right), \quad (t \to \infty) \tag{3.15}$$

and for $Bt \gg 1$, when the moments of J grow exponentially, it converges exponentially to 0.

The geometric meaning of equality (3.15) becomes clear if we recall that probability (3.15) can be written as a statistical mean

$$P(J > 1) = \langle \chi(J(\boldsymbol{y}, t) - 1) \rangle, \tag{3.16}$$

where $\chi(z)$ is the Heaviside function equal to 1 for $z \geq 0$ and equal to 0 for $z < 0$. Integrating equality (3.16) over a fixed domain Ω in the Lagrangian coordinate system, and taking into account the statistical homogeneity of the random field $J(\boldsymbol{y}, t)$, we arrive at equality

$$P(J > 1) = \frac{\langle V(J > 1) \rangle}{V},$$

where V is the volume of the domain of integration Ω, and $V(J > 1)$—volume of the region where $J(\boldsymbol{y}, t) > 1$. Equation (3.15) shows that, on the average, the ratio of volumes of these domains exponentially decays as t increases, which means that, as time grows, the overwhelming majority of particles find themselves in the strongly compressed regions where $J \ll 1$.

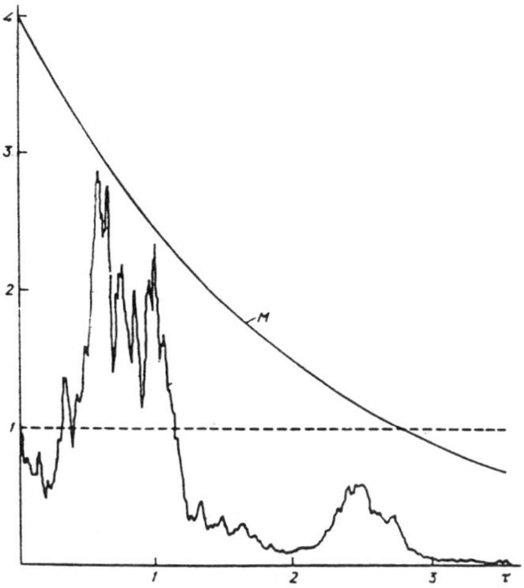

FIG. 1. *Dominating curve for one half of the realizations of the Jacobian random field $J(\boldsymbol{y}, t)$.*

3.2. Extremal properties of realizations of Jacobian and density fields. The above argument about concentration of particles in the areas of high compression can be even more convincingly illustrated by the extremal properties of realizations of the process $J(t)$, which is statistically equivalent to random field $J(\boldsymbol{y}, t)$. Recall, that in view of the well known extremal property of the process $w(t) - Bt$ (Brownian motion with a regular drift), it is easy to show (see e.g. Klyatskin, Saichev [13]) that, with probability

$$Q = 1 - A^{(r-1)}, \tag{3.17}$$

realizations of process $J(t)$, for all t, are dominated by an exponentially decreasing (for $r > 0$) curve

$$M(t) = Ae^{-rBt}, \tag{3.18}$$

where $A > 1$ and $0 < r < 1$. In particular, one half of all realizations of $J(t)$ are located beneath the curve (see Fig. 1)

$$M(t) = 4\exp(-Bt/2).$$

In other words, beginning with time

$$t_0 = \frac{4}{B}\ln 2,$$

one half of the particles of the medium find themselves in the compressed region of the space where

$$J(t) < 4\exp\left(-\frac{Bt}{2}\right) < 1,$$

whenever $t > t_0$.

The above result on "trapping" of particles in the compressed regions is also supported by another result of the theory of Markov processes (see e.g. Klyatskin and Saichev [13]) according to which the random area

$$S = \int_0^\infty J(t)dt, \tag{3.19}$$

under the realizations of process $J(t)$ has a nondegenerate probability distribution

$$P(S < \theta) = \exp\left(-\frac{1}{B\theta}\right). \tag{3.20}$$

This implies that the realizations of $J(t)$ converge to 0 (as $t \to \infty$) sufficiently fast for the improper integral (3.19) to be finite with probability 1. More exactly, for any $p < 1$,

$$P(S < -1/B \ln p) = p.$$

It is interesting to observe that although, as expected, all moments of the distribution (3.20) are infinite, and in particular the mean

$$\langle S \rangle = \int_0^\infty \langle J \rangle dt = \int_0^\infty dt = \infty,$$

the inverse moments, for example

$$\langle 1/S \rangle = B$$

are finite. This means that the area under realizations of $J(t)$ may not be too small.

Let us summarize the basic facts concerning statistics of the Jacobian $J(\boldsymbol{y}, t)$, and of the auxiliary, statistically equivalent process $J(t)$. The main conclusion is that the exponential growth of moments of Jacobian (3.11) and (3.13) does not mean that the sensor moving with the medium, which measures the degree of compression or expansion of the medium, will in course of time fall into more and more rarefied region where $J \gg 1$. Just the opposite happens, as is clear from the behavior of the probability distribution of the Jacobian (3.9-10), of the dominating curves (3.18), and of the probability distribution (3.20) of the area (3.19) under the realizations of $J(t)$. The sensor will eventually fall in compressed ($J \ll 1$) regions, where the degree of compression increases with time. Hence, the time dependence of the Jacobian's moments $\langle J^n \rangle$ does not reflect the true behavior of the realizations of the Jacobian J. The exponential growth of the Jacobian's moments is caused by the so-called ensemble effect. In our particular case it means that some (but far from all) realizations of $J(\boldsymbol{y}, t)$ display large peaks. The characteristic height level of these peaks increases with time.

However, the relative portion of realizations of $J(\boldsymbol{y},t)$ which develop such peaks $J \gg 1$, decreases. As a result, although $\langle J^2 \rangle$—the ensemble mean of the square of the Jacobian—increases exponentially in time because of high peaks in separate realizations, practically all the experiments will register a more or less monotone decay of $J(t)$, so that, beginning with a certain instant of time the inequality $J < 1$ is fulfilled.

Physical phenomena involving compression and expansion of infinitesimally small volumes of the passive tracer are usually described in terms of density. For that reason we will rephrase the above results in the language of density fluctuations.

It is clear from (1.13) that the density field $R(\boldsymbol{y},t)$ of the passive tracer in the Lagrangian coordinate system (as different from $\rho(\boldsymbol{x},t)$ —the density field in the Eulerian coordinate system) is described by a simple expression

$$R(\boldsymbol{y},t) = \frac{\rho_0(\boldsymbol{y})}{J(\boldsymbol{y},t)}, \tag{3.21}$$

where $\rho_0(\boldsymbol{y})$ is the initial given density field. For a fixed Lagrangian coordinate $\boldsymbol{y}$, statistical properties of realizations of the random Lagrangian density field are more conveniently studied with the help of an auxiliary process

$$R(t) = \rho_0 / J(t). \tag{3.22}$$

It follows from (3.6) that R satisfies a stochastic differential equation

$$\frac{dR}{dt} = BR - \eta(t)R,$$

with the initial condition $R(0) = \rho_0$, the solution thereof is of the form

$$R(t) = \rho_0 \exp[Bt - w(t)].$$

This (or the solution of corresponding Fokker-Planck equation) gives the cumulative distribution function for the density:

$$P\Big(R(\boldsymbol{y},t) < \rho\Big) = \Phi\left(\frac{\ln((\rho/\rho_0)e^{-Bt})}{2\sqrt{Bt}}\right).$$

The median curve

$$\bar{\rho} = \rho_0 e^{Bt}$$

corresponding to equality

$$P\Big(R(\boldsymbol{y},t) < \rho\Big) = \frac{1}{2}$$

grows exponentially with time, as do curves

$$m(t) = \frac{\rho_0}{A} e^{rBt},$$

which bound from below, with probability Q (see (3.17)), the realizations of the density field for $t > 0$. In other words, as time progresses, practically every particle of the tracer falls in the compressed region with an exponentially growing density. In this case, the moments of the density field

$$\langle R^n \rangle = \rho_0^n \exp[n(n+1)Bt] \tag{3.23}$$

and, in particular, the mean density of the passive tracer is described by the formula

$$\langle R(\boldsymbol{y}, t) \rangle = \rho_0(\boldsymbol{y}) e^{2Bt},$$

which adequately reflects evolution of the Lagrangian density field at a point with fixed Lagrangian coordinate $\boldsymbol{y}$.

4. Lagrangian vs. Eulerian statistical characteristics of randomly moving media. The above analysis of evolution of the passive tracer was carried out in the Lagrangian coordinate system. However, fixed sensors are also used to study fluctuations of the passive tracer fields, and they register behavior of these fields in the Eulerian coordinate system. Eulerian statistical characteristics are, in general, not only quantitatively but also qualitatively different from the corresponding Lagrangian statistical characteristics. Nevertheless, both types of characteristics and, in particular, characteristics of density fields, are connected by universal relationships (see e.g. Gurbatov, Malakhov, Saichev [8] and Gurbatov, Saichev [9]), and with their help one can recover Eulerian statistics of a random field from known Lagrangian statistics of the same field. Some of such formulas will be established in the present section.

To begin, consider the joint Eulerian probability distribution

$$f_E(\boldsymbol{y}, j; \boldsymbol{x}, t) = \Big\langle \delta(\boldsymbol{y} - \boldsymbol{Y}(\boldsymbol{x}, t))\delta(j - j(\boldsymbol{x}, t)) \Big\rangle \tag{4.1}$$

of the Lagrangian coordinates $\boldsymbol{Y}(\boldsymbol{x}, t)$ (see (1.9)) of a particle which is located at a point with Eulerian coordinates $\boldsymbol{x}$, and of the Eulerian Jacobian field

$$j(\boldsymbol{x}, t) = J(\boldsymbol{Y}(\boldsymbol{x}, t), t),$$

which is obtained from the corresponding Lagrangian field $J(\boldsymbol{y}, t)$ by replacing $\boldsymbol{y}$ with $\boldsymbol{Y}(\boldsymbol{x}, t)$. We transform the right-hand side of (4.1) using formula (1.8) to get that

$$f_E(\boldsymbol{y}, j; \boldsymbol{x}, t) = j f_L(\boldsymbol{x}, j; \boldsymbol{y}, t), \tag{4.2}$$

where f_L is the Lagrangian probability distribution (2.1) encountered before. Relation (4.2) is the first in a series of promised formulas connecting Lagrangian and Eulerian characteristics.

A closer look at (4.2) shows that, in an incompressible medium, where

$$j(\boldsymbol{x},t) = J(\boldsymbol{y},t) = 1,$$

it is equivalent with the identity

$$f_{\boldsymbol{Y}}(\boldsymbol{y};\boldsymbol{x},t) = f_{\boldsymbol{X}}(\boldsymbol{x};\boldsymbol{y},t), \tag{4.3}$$

where

$$f_{\boldsymbol{Y}}(\boldsymbol{y};\boldsymbol{x},t) = \langle\delta(\boldsymbol{y} - \boldsymbol{Y}(\boldsymbol{x},t))\rangle. \tag{4.4}$$

Hence, in an incompressible medium, the probability distribution (3.2) of the Eulerian coordinates of a fixed particle, and that of the Lagrangian coordinates of the particle which is located at point $\boldsymbol{x}$ at time t satisfies a reciprocity condition (4.3). In the general case of a compressible medium that relationship may fail.

In the general case of a compressible medium, if the random field $\boldsymbol{v}(\boldsymbol{x},t)$ of velocities is statistically homogeneous, then another useful formula connecting Lagrangian and Eulerian probability distributions of the Jacobian follows from (4.2). Indeed, in this case, the probability distributions entering in (4.2) depend, apart from j, only on the difference of the spatial coordinates $\boldsymbol{x}$ and $\boldsymbol{y}$, that is

$$f_E(\boldsymbol{y} - \boldsymbol{x}, j; t) = j f_L(\boldsymbol{x} - \boldsymbol{y}, j; t). \tag{4.5}$$

Integrating both sides with respect to either $\boldsymbol{x}$ or $\boldsymbol{y}$, we get that

$$f_j(j;t) = j f_J(j;t). \tag{4.6}$$

This equality connects the probability distribution of the Eulerian Jacobian field

$$f_j(j;t) = \langle\delta(j - j(\boldsymbol{x},t))\rangle \tag{4.7}$$

with the probability distribution

$$f_J(j;t) = \langle\delta(j - J(\boldsymbol{y},t))\rangle \tag{4.8}$$

of the corresponding Lagrangian field, which satisfies, in the diffusion approximation, equation (3.4). Factor j in (4.6), by which these two distributions differ, takes into account the increase in the proportion of expanded particles of the fluid for which $j > 1$ in the statistical ensemble of Eulerian fields, as compared to the statistical ensemble of Lagrangian fields.

Since both probability distributions in (4.6) have to satisfy the normalization condition, we automatically get conservation laws

$$\langle J(\boldsymbol{y},t)\rangle = 1, \quad \langle 1/j(\boldsymbol{x},t)\rangle = 1. \tag{4.9}$$

Both of them have a transparent physical meaning. The first coincides with equality (3.12), which was obtained earlier in the diffusion approximation. At that time we noticed its universal character, and its relationship with the law of conservation of the volume occupied by the medium. The meaning of the second conservation law in (4.9) also becomes clear if we assume that, at the initial instant of time $t = 0$, the density of the medium, say ρ_0, is identical at all points of space. Multiplying the second equality in (4.9) by ρ_0, and noticing that

$$\rho_0/j(\boldsymbol{x},t) = \rho(\boldsymbol{x},t) \tag{4.10}$$

is the Eulerian density field of the medium, we get that

$$\langle \rho(\boldsymbol{x},t)\rangle = \rho_0. \tag{4.11}$$

So, the second equation in (4.9) expresses the conservation law for the mean Eulerian density of the medium. The latter, in turn, is a consequence of the dynamical mass conservation law for the medium.

Substituting in (4.6) the Lagrangian probability distribution (3.9) of the Jacobian found in the diffusion approximation, we obtain, also in the diffusion approximation, an explicit expression for the Eulerian probability distribution of the Jacobian:

$$f_j(j;t) = \frac{1}{2j\sqrt{\pi Bt}}\exp\left[-\frac{\ln^2(je^{-Bt})}{4Bt}\right], \quad j > 0. \tag{4.12}$$

However, in this case, the equation satisfied by the probability distribution is more enlightening than the probability distribution itself. Substituting $f_j = f_j/j$ in (3.4), we arrive at an equation

$$\frac{\partial f_j}{\partial t} = B\frac{\partial}{\partial j}\left(j^2\frac{\partial f_j}{\partial j}\right), \tag{4.13}$$

$$f(j;t=0) = \delta(j-1).$$

This equation can be also treated as a Fokker-Planck-Kolmogorov equation for the transition probabilities of an auxiliary Markov process $j(t)$, which is statistically equivalent with the Jacobian field $j(\boldsymbol{x},t)$ at a fixed point $\boldsymbol{x}$. This auxiliary process satisfies a stochastic differential equation

$$\frac{dj}{dt} = Bj - \eta(t)j, \tag{4.14}$$

$$j(t=0) = 1.$$

Notice, that this equation can be formally obtained from equation (3.5) by substituting $J(t) = 1/j(t)$. Therefore, in the case under consideration, a sort of inversion principle is satisfied: the Eulerian Jacobian j (the Jacobian at a fixed point $\boldsymbol{x}$ of the space) behaves as a function of time as an inverse Lagrangian Jacobian J^{-1} (where J is the Jacobian in the neighborhood of a fixed particle of the medium with the Lagrangian coordinate $\boldsymbol{y}$.)

In particular, if—with a given probability Q—one can find an exponentially decreasing dominating curves $M(t)$ (see (3.18)) under which lie graphs of $Q \cdot 100$ percent of realizations of $J(t)$, then the same percentage of realization of $j(t)$ will, for any $t > 0$, lie above the exponentially growing minorant curves $1/M(t)$.

In other words, if a given number of particles of the medium fall (as time t grows) once and forever in the region with exponentially growing compression (where $J \ll 1$), then the same number of fixed immobile sensors, uniformly distributed in the whole space, will at the same time register a strong expansion of the medium (that is $j \gg 1$) in their neighborhood. The observed, mutually inverse behavior of the auxiliary processes $J(t)$ and $j(t)$, which are (in different representations) statistically equivalent with corresponding physical Jacobian field, is not a contradiction. It explains the nature of the spatial structure of the distribution of the passive tracer in chaotic flows of a compressible medium: As time grows, tracer particles clump up into relatively compact clusters of high density which are surrounded by large regions, where the density of passive tracer is much smaller than the average.

Formula (4.2) connects the Lagrangian and Eulerian probability distribution of the Jacobian field. From the view point of physical applications, it is desirable to obtain a similar formula which would connect probability distributions of the random density fields of the passive tracer. Its derivation is analogous to the derivation of formula (4.2), and gives the final result of the form

$$\rho\varphi_E(\boldsymbol{y}, \rho; \boldsymbol{x}, t) = \rho_0(\boldsymbol{y})\varphi_L(\boldsymbol{x}, \rho; \boldsymbol{y}, t), \tag{4.15}$$

where

$$\varphi_L(\boldsymbol{x}, \rho; \boldsymbol{y}, t) = \Big\langle \delta(\boldsymbol{x} - \boldsymbol{X}(\boldsymbol{y}, t))\delta(\rho - R(\boldsymbol{y}, t)) \Big\rangle \tag{4.16}$$

is the joint Lagrangian probability distribution of random fields $\boldsymbol{X}(\boldsymbol{y}, t)$ and $R(\boldsymbol{y}, t)$, and

$$\varphi_E(\boldsymbol{y}, \rho; \boldsymbol{x}, t) = \Big\langle \delta(\boldsymbol{y} - \boldsymbol{Y}(\boldsymbol{x}, t))\delta(\rho - \rho(\boldsymbol{x}, t)) \Big\rangle \tag{4.17}$$

is the Eulerian joint probability distribution of the Lagrangian coordinate $\boldsymbol{Y}(\boldsymbol{x}, t)$ and of the density $\rho(\boldsymbol{x}, t)$. In (4.15), as in the rest of this article, the initial density field $\rho_0(\boldsymbol{x})$ is assumed to be deterministic. Integrating

(4.15) over all $\boldsymbol{y}$ we arrive at a formula

$$\rho\varphi_\rho(\rho;\boldsymbol{x},t)=\int\rho_0(\boldsymbol{y})\varphi_L(\boldsymbol{x},\rho;\boldsymbol{y},t)d\boldsymbol{y}, \tag{4.18}$$

which expresses the one-point probability distribution of the Eulerian density through a joint Lagrangian distribution of $\boldsymbol{X}$ and R defined in (4.16). Now, integrating both sides of (4.18) with respect to ρ we arrive at the well known (see e.g. Csanady [5]) statement of the basic theorem of turbulent diffusion:

$$\langle\rho(\boldsymbol{x},t)\rangle=\int\rho_0(\boldsymbol{y})f_{\boldsymbol{X}}(\boldsymbol{x};\boldsymbol{y},t)d\boldsymbol{y}. \tag{4.19}$$

It expresses the mean density of the passive tracer through the probability distribution (3.2) of coordinates of fixed particles.

We shall also indicate another, sometimes more convenient, modification of formula (4.18). The problem is that the Lagrangian probability distribution entering on the right-hand side of equality (4.18) depends not only on the "objective" properties of the chaotically moving medium, but also on the "subjective", often created by the experimenter, initial density $\rho_0(\boldsymbol{y})$ of the tracer. In order to clearly separate the impact of "objective" and "subjective" factors, we can express φ_L (see (4.16)) through a "completely objective" Lagrangian probability distribution f_L (see (2.1)). As a result, after simple transformations, we get that

$$\varphi_\rho(\rho;\boldsymbol{x},t)=\frac{1}{\rho^3}\int\rho_0^2(\boldsymbol{y})f_L\left(\boldsymbol{x},\rho_0(\boldsymbol{y})/\rho;\boldsymbol{y},t\right)d\boldsymbol{y}. \tag{4.20}$$

Naturally, the related formulas (4.2), (4.15), (4.18) and (4.20), which connect Lagrangian and Eulerian probability distributions, can be generalized to the case of multipoint probability distributions describing joint statistical properties of Lagrangian and Eulerian fields at different points of space. As an example we shall quote two formulas. The first is

$$f_E(j_i,j_2;\boldsymbol{x}_1,\boldsymbol{x}_2,t_1,t_2)= \tag{4.21}$$

$$j_1j_2\int\int f_L(\boldsymbol{x}_1,\boldsymbol{x}_2,j_1,j_2;\boldsymbol{y}_1,\boldsymbol{y}_2,t_1,t_2)d\boldsymbol{y}_1d\boldsymbol{y}_2,$$

where the joint, two-point and two-time, probability distribution

$$f_E(j_i,j_2;\boldsymbol{x}_1,\boldsymbol{x}_2,t_1,t_2)=\Big\langle\delta(j_1-j(\boldsymbol{x}_1,t_1))\delta(j_2-j(\boldsymbol{x}_2,t_2))\Big\rangle, \tag{4.22}$$

of the field $j(\boldsymbol{x},t)$ appears on the left-hand, and the Lagrangian probability distribution

$$f_L(\boldsymbol{x}_1,\boldsymbol{x}_2,j_1,j_2;\boldsymbol{y}_1,\boldsymbol{y}_2,t_1,t_2)= \tag{4.23}$$

$$\left\langle \delta(\boldsymbol{x}_1 - \boldsymbol{X}(\boldsymbol{y}_1, t_1))\delta(\boldsymbol{x}_2 - \boldsymbol{X}(\boldsymbol{y}_2, t_2))\delta(j_1 - J(\boldsymbol{y}_1, t_1))\delta(j_2 - J(\boldsymbol{y}_2, t_2))\right\rangle$$

appears on the right-hand side.

The second assumes that the velocity field of the medium is statistically homogeneous in space and that $t_1 = t_2 = t$. Then, formula (4.21) takes a simpler form

$$f_E(j_1, j_2; \boldsymbol{s}, t) = j_1 j_2 \int\int f_L(\boldsymbol{s}, j_1, j_2; \boldsymbol{s}_0, t) d\boldsymbol{s}_0, \tag{4.24}$$

where $\boldsymbol{s} = \boldsymbol{x}_1 - \boldsymbol{x}_2$,

$$\begin{aligned} &f_L(\boldsymbol{s}, j_1, j_2; \boldsymbol{s}_0, t) \\ &= \left\langle \delta(\Delta \boldsymbol{X}(\boldsymbol{y}, \boldsymbol{s}_0, t) - \boldsymbol{s})\delta(j_1 - J(\boldsymbol{y} + \boldsymbol{s}_0, t))\delta(j_2 - J(\boldsymbol{y}, t))\right\rangle, \end{aligned} \tag{4.25}$$

and

$$\Delta \boldsymbol{X} = \boldsymbol{X}(\boldsymbol{y} + \boldsymbol{s}_0, t) - \boldsymbol{X}(\boldsymbol{y}, t).$$

5. Eulerian statistics of the passive tracer density field. In this section we will recover the one-point Eulerian probability distributions of the passive tracer density field. Recall, that they are expressed (see formula (4.20)) via the Lagrangian joint probability distribution of the coordinates and of the Jacobian. Given that, in the diffusion approximation in (4.20), the Lagrangian probability distribution splits into a product (3.1), we obtain that

$$\varphi_\rho(\rho; \boldsymbol{x}, t) = \left\langle J(t)\varphi(\rho; \boldsymbol{x}, t|\rho_0/J(t))\right\rangle, \tag{5.1}$$

where $J(t)$ is an auxiliary process satisfying the stochastic differential equation (3.5), and

$$\varphi(\rho; \boldsymbol{x}, t|\rho_0(\boldsymbol{x})) = \frac{1}{\rho}\int \rho_0(\boldsymbol{y}) f_{\boldsymbol{Y}}(\boldsymbol{y}; \boldsymbol{x}, t)\delta(\rho - \rho_0(\boldsymbol{y}))d\boldsymbol{y} \tag{5.2}$$

is the probability distribution of passive tracer density in an incompressible medium. In particular, (5.1), (5.2) and (4.3) imply that the statistical moments of the density field are equal to

$$\langle \rho^n(\boldsymbol{x}, t)\rangle = \langle J^{1-n}(t)\rangle \int \rho_0^n(\boldsymbol{y}) f_{\boldsymbol{X}}(\boldsymbol{x}; \boldsymbol{y}, t)d\boldsymbol{y}. \tag{5.3}$$

They differ only by the factor of

$$\langle J^{1-n}(t)\rangle = e^{n(n-1)Bt} \tag{5.4}$$

from the moments of density in an incompressible medium.

Observe the following interesting fact. It is clear from equation (5.4) and (3.11) that

$$\langle J^{1-n}(t)\rangle = \langle J^n(t)\rangle.$$

This equality, together with (5.3), means that the probability distribution (5.1) of the passive tracer density field coincides with the probability distribution of the auxiliary process

$$\rho_{\text{aux}}(\boldsymbol{x},t) = \rho_0(\boldsymbol{Y}(\boldsymbol{x},t))J(t). \tag{5.5}$$

Let us consider properties of this process in some detail, in the case when the initial passive tracer density is identical at all points of space, that is when

$$\rho_0(\boldsymbol{x}) = \rho_0 = const. \tag{5.6}$$

In this situation

$$\rho_{\text{aux}}(\boldsymbol{x},t) = \rho(t) = \rho_0 J(t). \tag{5.7}$$

This equality looks somewhat paradoxical, since it equates the density and the Jacobian which is inverse to it in the arithmetical sense. A similar inversion phenomenon was already encountered when we considered the time evolution of Lagrangian and Eulerian Jacobians J and j. The paradox is explained by the fact that passing from Lagrangian to Eulerian statistics of the same field (say, the density fields $\rho(\boldsymbol{x},t)$ and $R(\boldsymbol{y},t)$) the probability measures undergo a deformation: The compression of a region of space has a smaller statistical weight in the Eulerian ensemble than in the Lagrangian ensemble (see Kraichnan [16]), Gurbatov, Malakhov, Saichev [8], Gurbatov, Saichev [9]). Based on equation (5.7), we can repeat most of what has been said about the contradictory statistical properties of $J(t)$.

In particular, notice that although density moments

$$\langle \rho^n(\boldsymbol{x},t)\rangle = \rho_0^n e^{n(n-1)Bt} \tag{5.8}$$

grow exponentially with time in view of (5.7) and (3.11), the probability that random field $\rho(\boldsymbol{x},t)$ exceeds the mean level $\langle\rho\rangle = \rho_0$ is given by expression (3.15), and converges exponentially to 0 as $t \gg 1/B$.

Extending an analogy between the time evolutions of $\rho(\boldsymbol{x},t)$ and of $J(t)$, we can presume that their realizations behave in a similar fashion. Hence, realizations of $\rho(\boldsymbol{x},t)$ lie underneath curves $\rho_0 M(t)$ (see (3.18)) and eventually decay to 0.The exponential growth of moments of the Eulerian density (5.8) can be explained by an appearance in some (but far from all) realizations of large peaks of density $\rho \gg \rho_0$. In the physical language it means that a majority of fixed sensors finds itself eventually in the regions of low tracer concentration (where $\rho \ll \rho_0$), and only a few of them will be lucky enough to lie inside a "macroparticle" — a compact region of high density, and register a high ($\rho \gg \rho_0$) values of the passive tracer density (see Fig. 2).

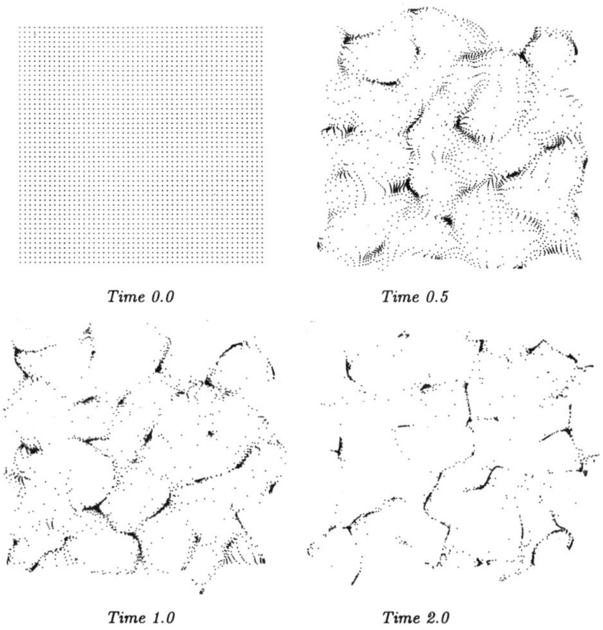

FIG. 2. *Evolution of clusters of high tracer density in a compressible flow.*

6. Fluctuations of the passive tracer density gradient in a randomly moving incompressible medium. Although compressions and expansions, leading to density fluctuations in a compressible medium, are absent in an incompressible medium, the random motions of the latter can transform, initially smooth and simple curves of constant passive tracer concentration into more and more complex objects. This is a result of initially close particles being separated and initially far particles being brought together. Consequently, regions with vastly different tracer densities can be located next to each other, separated by clear cut boundaries in the neighborhood thereof the density changes rapidly.

In this section we give a quantitative description of the increasing spatial complexity of density fields in a randomly moving incompressible medium. For the sake of simplicity, we restrict ourselves to an analysis of a 2D case, where components of the velocity field can be expressed by a stream function $\Psi(\boldsymbol{x}, t)$ via formulas

$$v_1 = \frac{\partial \Psi}{\partial x_2}, \quad v_2 = -\frac{\partial \Psi}{\partial x_1}. \tag{6.1}$$

The statistical properties of the stream function are assumed to be known.

The spatial complexity of density field realizations can be characterized, for example, by the gradient $\nabla_{\boldsymbol{x}} \rho(\boldsymbol{x}, t)$ of density field ρ and we will study its statistical properties. To begin, let us recall that, in an incompressible medium, the density field is given by expression

$$\rho(\boldsymbol{x},t)=\rho_0(\boldsymbol{Y}(\boldsymbol{x},t)), \tag{6.2}$$

where $\rho_0(\boldsymbol{x})$ is the initial deterministic density profile, and $\boldsymbol{y}\mapsto\boldsymbol{Y}(\boldsymbol{x},t)$ is the mapping of Eulerian coordinates into Lagrangian coordinates. As a result, we get that

$$\boldsymbol{\nabla}_{\boldsymbol{x}}\rho=\boldsymbol{e}_1\left(\frac{\partial\rho_0}{\partial y_1}j_{11}+\frac{\partial\rho_0}{\partial y_2}j_{21}\right)+\boldsymbol{e}_2\left(\frac{\partial\rho_0}{\partial y_1}j_{12}+\frac{\partial\rho_0}{\partial y_2}j_{22}\right), \tag{6.3}$$

where $\boldsymbol{e}_1,\boldsymbol{e}_2$ are the unit coordinate vectors in Eulerian coordinates, and j_{lm} are components of the transition matrix from Eulerian to Lagrangian coordinates, that is

$$j_{lm}=\frac{\partial Y_l(\boldsymbol{x},t)}{\partial x_m},\qquad l,m=1,2. \tag{6.4}$$

The statistical properties of gradient (6.3) are more conveniently studied in the Lagrangian coordinate system. For that reason, we shall express fields $j_{lm}(\boldsymbol{x},t)$ entering into (6.3) through components

$$J_{lm}(\boldsymbol{y},t)=\frac{\partial X_l(\boldsymbol{y},t)}{\partial y_m} \tag{6.5}$$

of the transition matrix from Lagrangian to Eulerian coordinates. After differentiation of the obvious vector identity

$$\boldsymbol{x}=\boldsymbol{X}(\boldsymbol{Y}(\boldsymbol{x},t),t)$$

with respect to x_1 and x_2, and after solving obtained equations with respect to j_{lm}, we get that

$$j_{11}=J_{22}/J,\quad j_{22}=J_{11}/J,\quad j_{12}=-J_{12}/J,\quad J_{21}=-J_{21}/J. \tag{6.6}$$

Substituting (6.6) into (6.3), and taking into account that in the case of an incompressible medium $J\equiv 1$, we find that

$$\boldsymbol{\nabla}_{\boldsymbol{x}}\rho=|\boldsymbol{\nabla}_{\boldsymbol{y}}\rho_0|(\boldsymbol{e}_1J_1+\boldsymbol{e}_2J_2), \tag{6.7}$$

where $|\boldsymbol{\nabla}_{\boldsymbol{y}}\rho_0|$ is the initial norm of the density gradient at point $\boldsymbol{y}$,

$$J_1(\boldsymbol{y},t,\theta_0)=J_{22}\cos\theta_0-J_{21}\sin\theta_0, \tag{6.8a}$$

$$J_2(\boldsymbol{y},t,\theta_0)=J_{11}\sin\theta_0-J_{12}\cos\theta_0, \tag{6.8b}$$

and θ_0 is the angle between the y_1-axis and the initial direction of the gradient vector.

It is not hard to prove that, in the Lagrangian coordinate system, random fields J_1, J_2 of (6.8) satisfy the following system of stochastic differential equations

$$\frac{dJ_2}{dt} = \delta J_2 - \beta J_1, \quad \frac{dJ_1}{dt} = -\delta J_1 + \gamma J_2, \tag{6.9}$$

$$J_1(\boldsymbol{y}, t = 0, \theta_0) = \cos\theta_0, \quad J_2(\boldsymbol{y}, t = 0, \theta_0) = \sin\theta_0,$$

where

$$\delta(\boldsymbol{x}, t) = \frac{\partial^2 \Psi(\boldsymbol{x}, t)}{\partial x_1 \partial x_2}, \quad \beta(\boldsymbol{x}, t) = \frac{\partial^2 \Psi(\boldsymbol{x}, t)}{\partial x_2^2}, \quad \gamma(\boldsymbol{x}, t) = \frac{\partial^2 \Psi(\boldsymbol{x}, t)}{\partial x_1^2}.$$

Solutions of equations (6.9) can, in turn, be represented in the form

$$J_1 = e^{\chi} \cos\theta, \quad J_2 = e^{\chi} \sin\theta. \tag{6.10}$$

Thus, expression for the gradient (6.7) assumes a particularly transparent form

$$\boldsymbol{\nabla}_{\boldsymbol{x}} \rho = |\boldsymbol{\nabla}_{\boldsymbol{y}} \rho_0| g \boldsymbol{n}, \tag{6.11}$$

where

$$g(\boldsymbol{y}, t, \theta_0) = e^{\chi} = |\boldsymbol{\nabla}_{\boldsymbol{x}} \rho| / |\boldsymbol{\nabla}_{\boldsymbol{y}} \rho_0| \tag{6.12}$$

is the relative change in the magnitude of the gradient, and

$$\boldsymbol{n} = \boldsymbol{e}_1 \cos\theta + \boldsymbol{e}_2 \sin\theta \tag{6.13}$$

is the unit vector describing fluctuations of the gradient's direction vector as a function of time.

It follows from (6.9-10) that χ and θ satisfy stochastic differential equations

$$\frac{d\chi}{dt} = -\delta \cos 2\theta + \mu \sin 2\theta, \quad \chi(\boldsymbol{y}, t = 0, \theta_0) = 0, \tag{6.14a}$$

$$\frac{d\theta}{dt} = \delta \sin 2\theta + \mu \cos 2\theta - \nu, \quad \theta(\boldsymbol{y}, t = 0, \theta_0) = \theta_0, \tag{6.14b}$$

where

$$\nu = (\gamma + \beta)/2, \quad \mu = (\gamma - \beta)/2. \tag{6.15}$$

In the subsequent statistical analysis of fields χ and θ we shall assume that the velocity field potential $\Psi(\boldsymbol{x}, t)$ is a Gaussian, statistically isotropic in space and stationary in time random field with a known covariance function. In the diffusion approximation, which we will employ again, for statistical characteristics of the above fields we will need parameter B, defined by equality

$$\int_{-\infty}^{0} \langle \Psi(\boldsymbol{x}, t)\Psi(\boldsymbol{x}, +\boldsymbol{s}, t+\tau)\rangle d\tau = \ldots + \frac{B}{8}(s_1^2 + s_2^2)^2 - \ldots,$$

where s_1, s_2 are components of vector $\boldsymbol{s}$.

Let us consider the joint probability distribution

$$f(\chi, \theta; t, \theta_0) = \Big\langle \delta(\chi(\boldsymbol{y}, t, \theta_0) - \chi)\delta(\theta(\boldsymbol{y}, t, \theta_0) - \theta)\Big\rangle \tag{6.16}$$

of random fields χ and θ. One can show that, in the diffusion approximation, it satisfies equation

$$\frac{\partial f}{\partial t} = B\left(\frac{\partial^2 f}{\partial \chi^2} - 2\frac{\partial f}{\partial \chi} + 3\frac{\partial^2 f}{\partial \theta^2}\right), \tag{6.17}$$

$$f(\chi, \theta; t = 0, \theta_0) = \delta(\chi)\delta(\theta - \theta_0),$$

which implies that random fields $\chi(\boldsymbol{y}, t, \theta_0)$ and $\theta(\boldsymbol{y}, t, \theta_0)$, and the field of the relative change of the gradient of $g(\boldsymbol{y}, t, \theta_0)$ (the latter being of our main interest), are statistically equivalent with the following stochastic processes:

$$\chi(t) = \omega(t) + 2Bt, \tag{6.18a}$$

$$\theta(t, \theta_0) = \theta_0 + 3\kappa(t), \tag{6.18b}$$

and

$$g(t) = \exp[\omega(t) + 2Bt], \tag{6.18c}$$

where $\omega(t)$ and $\kappa(t)$ are statistically independent Brownian motions with identical variances

$$\langle \omega^2(t)\rangle = \langle \kappa^2(t)\rangle = 2Bt.$$

In particular, it follows from (6.18) that the gradient's mean value

$$\langle g(\boldsymbol{y}, t, \theta_0)\rangle = \exp(3Bt),$$

increases exponentially, but that, at the same time, the mean value of the unit vector $\boldsymbol{n}$ decays exponentially as

$$\langle \boldsymbol{n}(\boldsymbol{y}, t, \theta_0)\rangle = \boldsymbol{n}_0 \exp(-3Bt),$$

where

$$\boldsymbol{n}_0 = \boldsymbol{e}_{1_0} \cos\theta_0 + \boldsymbol{e}_2^0 \sin\theta_0, \tag{6.19}$$

and $\boldsymbol{e}_1^0$ and $\boldsymbol{e}_2^0$ are the unit vectors in the Lagrangian coordinate system. Consequently, the tracer density's mean gradient, in the vicinity of a fixed tracer particle with given Lagrangian coordinate $\boldsymbol{y}$, remains unchanged so that

$$\langle \boldsymbol{\nabla}_{\boldsymbol{x}} \rho \rangle = \boldsymbol{\nabla}_{\boldsymbol{y}} \rho_0.$$

The corresponding Eulerian density gradient's mean, at a point with fixed coordinate $\boldsymbol{x}$, can be obtained by averaging the above equality with respect to random Lagrangian coordinates $\boldsymbol{Y}(\boldsymbol{x}, t)$. This gives

$$\langle \boldsymbol{\nabla}_{\boldsymbol{x}} \rho(\boldsymbol{x}, t) \rangle = \int f_{\boldsymbol{Y}}(\boldsymbol{y}; \boldsymbol{x}, t) \boldsymbol{\nabla}_{\boldsymbol{y}} \rho_0(\boldsymbol{y})\, d\boldsymbol{y}.$$

In contrast to the mean with fixed Lagrangian coordinates, the Eulerian mean, in general, depends on time. The probability distribution $f_{\boldsymbol{Y}}(\boldsymbol{y}; \boldsymbol{x}, t)$ (see (4.4)) appearing in the above formula is, in the diffusion approximation, of the form

$$f_{\boldsymbol{Y}}(\boldsymbol{y}; \boldsymbol{x}, t) = \left(\frac{1}{4\pi D t}\right)^{N/2} \exp\left(-\frac{(\boldsymbol{y} - \boldsymbol{x})^2}{4Dt}\right),$$

where N is the dimension of space (in our case $N = 2$).

Now, we turn to a discussion of statistical properties of the relative rate of change of the gradient's magnitude g. Statistically equivalent process $g(t)$ (see 6.18c) enjoys properties that are similar to those of process $R(t)$ in (3.22). Thus, process $g(t)$ has a lognormal distribution with the cumulative distribution function

$$P\Big(g(t) < g\Big) = \Phi\left(\frac{\ln(g e^{-2Bt})}{2\sqrt{Bt}}\right),$$

and the probability that it assumes values smaller than the initial value $g(0) = 1$ quickly decays to 0 as time increases. More exactly,

$$P\Big(g(t) < 1\Big) = \Phi(-\sqrt{Bt}).$$

The moments of g behave in a fashion similar to those of R in (3.23), and grow exponentially with time:

$$\langle g^n \rangle = \exp[n(n+2)Bt]$$

for $n > 0$.

Exponentially increasing minorant curves for process $g(t)$, which bound it from below with arbitrarily close to 1 probability $\bar{P} < 1$, can also be found.

In summary, process $R(t)$ models the density of a randomly moving compressible medium at a point with Lagrangian coordinate $\boldsymbol{y}$, and its statistical properties are drastically different from the properties of process $\rho(t)$, modulating its density at a point with Eulerian coordinate $\boldsymbol{x}$. In the case considered in this section, process $g(t)$ models a relative rate of change of the density gradient in both Lagrangian and Eulerian coordinates. To be more precise one has to remember that $g(t)$ models the rate of change of the gradient at Eulerian point $\boldsymbol{x}$, relative to the initial magnitude of the gradient in the neighborhood of a particle which, at time instant t, is located at point $\boldsymbol{x}$.

7. Fluctuations of parameters of contours carried by a randomly moving incompressible medium. In this section we will study the evolution of the length $l(t)$ of a contour $\mathcal{L}$ of a constant tracer density—another, important for applications, geometric characteristic of a randomly moving incompressible medium. It can be expressed by a formula

$$l(t) = \int \delta\Big(\rho_0(\boldsymbol{Y}(\boldsymbol{x},t)) - \rho\Big)|\boldsymbol{\nabla}_{\boldsymbol{x}}\rho| d^2x. \tag{7.1}$$

Another related characteristic of the spatial evolution of the density field is the integral

$$A(t) = \int \delta\Big(\rho_0(\boldsymbol{Y}(\boldsymbol{x},t)) - \rho\Big)(\boldsymbol{\nabla}_{\boldsymbol{x}}\rho)^2 d^2x, \tag{7.2}$$

which will be called here the *total gradient of the density field.* Its physical meaning becomes clear if we observe that the double integral (7.2) is equal to the line integral

$$A(t) = \oint |\boldsymbol{\nabla}_{\boldsymbol{x}}\rho|\, dl$$

of the magnitude of the density gradient along the curve $\mathcal{L}$ of constant density. As in the case of the density gradient studied in the previous section, the statistical properties of integrals of type (7.1-2) are easier to investigate by switching integrals from Eulerian to Lagrangian variables of integration. We shall show this on the example of the total gradient (7.2). In that case, after changing variables, we get that

$$A(t) = \int \delta(\rho_0(\boldsymbol{y}) - \rho)(\boldsymbol{\nabla}_{\boldsymbol{x}}\rho)^2 d^2y.$$

We shall assume that the initial density is given by a deterministic, radially symmetric function $\rho_0(\boldsymbol{y}) = \rho_0(|\boldsymbol{y}|)$, which monotonically decreases with

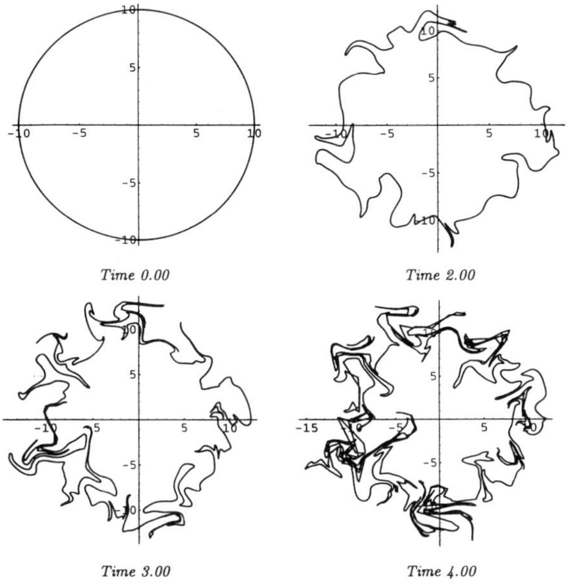

FIG. 3. *Contour carried by an incompressible random flow.*

the distance from the origin of the Lagrangian coordinate system. The initial contour in this case and the computer simulation of the corresponding contours $\mathcal{L}$, for selected instants $t > 0$, is shown of Fig. 3.

Under these circumstances, the above integral takes the form

$$A(t) = \frac{r_0(\rho)}{|\nabla_{\boldsymbol{y}}\rho_0|} \int_{-\pi}^{\pi} (\nabla_{\boldsymbol{x}}\rho)^2 \Big|_{\boldsymbol{y}=r_0\boldsymbol{n}_0} d\theta_0, \tag{7.3}$$

where $r_0(\rho)$ is the solution of equation $\rho_0(r) = \rho$, and $\boldsymbol{n}_0$ is determined by (6.19). Substituting in (7.3) the explicit expression (6.11) for the density gradient, we get that

$$A(t) = A_0 G,$$

where $A_0 = l_0|\nabla_{\boldsymbol{y}}\rho_0|$ is the initial total gradient, $l_0 = 2\pi r_0$ is the initial length of the contour, and the dimensionless quantity

$$G(t) = \frac{1}{2\pi} \int_{-\pi}^{\pi} g^2(\boldsymbol{y}, t, \theta_0) \Big|_{\boldsymbol{y}=r_0\boldsymbol{n}_0} d\theta_0 \tag{7.4}$$

characterizes the relative rate of change in time of the total gradient of the selected contour.

It is also easy to see that, similarly, the relative rate of change of length of that contour, can be expressed by the integral

$$\Omega(t) = l(t)/l_0 = \frac{1}{2\pi} \int_{-\pi}^{\pi} g(\boldsymbol{y}, t, \theta_0) \Big|_{\boldsymbol{y}=r_0\boldsymbol{n}_0} d\theta_0.$$

The statistical means of the above introduced relative quantities are thus expressed by formulas

$$\langle G(t)\rangle = \langle g^2(t)\rangle = e^{8Bt}, \quad \langle \Omega(t)\rangle = \langle g(t)\rangle = e^{3Bt}.$$

We should stress, however, that only the statistical means of processes G and g^2, and Ω and g coincide. Their other statistical characteristics differ, both quantitatively and qualitatively. We shall demonstrate this on the example of the relative rate of change of the total gradient G from (7.4). In our argument we will assume, for simplicity, that our contour of integration is small enough so that we can take $\boldsymbol{y} = 0$ in (7.4).

Let us substitute in (7.4) $g^2 = J_1^2 + J_2^2$, utilize formulas (6.8), and take into account the fact that matrix elements J_{lm} do not depend on angle θ_0. Then, after integration, we get that

$$G = \frac{1}{2}(J_{11}^2 + J_{22}^2 + J_{12}^2 + J_{21}^2).$$

In view of the condition of incompressibility

$$J_{11}J_{22} - J_{12}J_{21} \equiv 1$$

the previous equation can be rewritten in the form

$$G = 1 + \frac{1}{2}[(J_{11} - J_{22})^2 + (J_{12} + J_{21})^2],$$

wherefrom it follows that inequality $G(t) \geq 1$ is always satisfied, even as random field g^2 appearing in (7.4) takes values smaller than 1 with positive probability.

Finally, we turn to the statistical properties of G. Introduce auxiliary functions

$$C = \frac{1}{2\pi}\int_{-\pi}^{\pi} g^2 \cos 2\theta d\theta_0, \quad S = \frac{1}{2\pi}\int_{-\pi}^{\pi} g^2 \sin 2\theta d\theta_0,$$

and pass from equations (6.14) to a system of stochastic differential equations for G, C, and S:

$$\frac{1}{2}\frac{dG}{dt} = \mu S - \delta C, \quad G(0) = 1,$$

$$\frac{1}{2}\frac{dC}{dt} = \nu S - \delta G, \quad C(0) = 0,$$

$$\frac{1}{2}\frac{dS}{dt} = \mu G - \nu C, \quad S(0) = 0.$$

If we introduce polar coordinates A and φ for C and S by means of equations

$$C = A\cos\varphi, \quad S = A\sin\varphi,$$

then the above system gives three stochastic differential equations

$$\frac{1}{2}\frac{dG}{dt} = (\mu\sin\varphi - \delta\cos\varphi)A, \quad G(0) = 1, \tag{7.5a}$$

$$\frac{1}{2}\frac{dA}{dt} = (\mu\sin\varphi - \delta\cos\varphi)G, \quad A(0) = 0, \tag{7.5b}$$

$$A\frac{1}{2}\frac{d\varphi}{dt} = (\mu\cos\varphi + \delta\sin\varphi)G - \nu A, \quad \varphi(0) = 0 \tag{7.5c}$$

for G, A, and φ. Hence, it is clear that equations (7.5a) and (7.5b) have solutions

$$G = \cosh Z(t), \quad A = \sinh Z(t), \tag{7.6}$$

where auxiliary process $Z(t)$ and $\varphi(t)$ satisfy a closed system of two equations

$$\frac{dZ}{dt} = 2(\mu\sin\varphi - \delta\cos\varphi), \quad Z(0) = 0,$$

$$\frac{d\varphi}{dt} = 2(\mu\cos\varphi + \delta\sin\varphi)\coth Z - 2\nu.$$

After rather straightforward but tedious calculations one can show that the one-point probability distribution

$$w(z;t) = \langle\delta(Z(t) - z)\rangle$$

of random process $Z(t)$ satisfies, in the diffusion approximation, equation

$$\frac{\partial w}{\partial\tau} + \frac{\partial}{\partial z}(\coth zw) = \frac{\partial^2 w}{\partial z^2}, \tag{7.7}$$

$$w(z;0) = \delta(z),$$

where $\tau = 4Bt$ is the dimensionless time. It means, that the statistically equivalent process $Z(\tau)$satisfies a stochastic differential equation

$$\frac{dZ}{d\tau} = \coth Z + \xi(\tau).$$

Consequently, according to (7.11), the corresponding statistically equivalent process $G(\tau)$ satisfies a Stratonovich stochastic differential equation

$$\frac{dG}{d\tau} = G + \sqrt{G^2 - 1}\xi(\tau), \quad G(0) = 1,$$

where $\xi(\tau)$ is a Gaussian white noise with covariance function

$$\langle \xi(\tau)\xi(\tau')\rangle = 2\delta(\tau - \tau').$$

Hence, the probability distribution

$$f(u;t) = \langle \delta(G(t) - u)\rangle = \langle \delta(\cosh Z - u)\rangle,$$

of G satisfies a Fokker-Planck-Kolmogorov equation

$$\frac{\partial f}{\partial \tau} = \frac{\partial}{\partial u}(u^2 - 1)\frac{\partial f}{\partial u}, \tag{7.8}$$

$$f(u;\tau = 0) = \delta(u - 1).$$

In particular, it follows from (7.8) that the moment functions of G satisfy a chain of equations

$$\frac{d\langle G^n\rangle}{d\tau} = n(n+1)\langle G^n\rangle - n(n-1)\langle G^{n-2}\rangle,$$

so that, for example, the mean square of G and its third moment are equal to

$$\langle G^2\rangle = \frac{1}{3}(1 + 2e^{6\tau}),$$

$$\langle G^3\rangle = \frac{1}{5}(2e^{12\tau} + 3e^{2\tau}).$$

Luckily, the above equation (7.8) can be solved explicitly in terms of the Legendre function of the first kind $P_{-1/2+i\mu}(u)$ ($\mu \geq 0$) (see Papanicolaou [19]), Kesten, Papanicolaou [11]), which is the solution of the Legendre equation

$$\frac{d}{du}(u^2 - 1)\frac{d}{du}P_{-1/2+i\mu}(u) = -(\mu^2 + 1/4)P_{-1/2+i\mu}(u), \tag{7.9}$$

and which has various integral representations in terms of elementary function, such as

$$P_{-1/2+i\mu}(u) = \frac{2\cosh \mu\pi}{\pi}\int_0^\infty \frac{\cos \mu\theta}{\sqrt{2(u + \cosh \theta\tau)}}d\theta, \quad \alpha \geq 0. \tag{7.10}$$

Multiplying both sides of equation (7.8) by the Legendre function, and integrating them twice by parts with respect to u, in the limits from 1 to ∞, we get an easy ordinary differential equation

$$\frac{d}{d\tau}\tilde{f}(\mu;\tau) = -(\mu^2 + 1/4)\tilde{f}(\mu;\tau) \tag{7.11}$$

for the Fourier-Legendre transform

$$\tilde{f}(\mu;\tau) = \int_1^\infty du\, f(u,\tau)P_{-1/2+i\mu}(u)$$

of function f. With the initial condition, corresponding to the initial condition in equation (7.8), its solution is

$$\tilde{f}(\mu;\tau) = \exp[-(\mu^2 + 1/4)\tau].$$

Taking its inverse Fourier-Legendre transform we get a solution of the initial problem (7.8)

$$f(u;\tau) = \int_0^\infty d\mu\, \mu \tanh(\pi\mu) \exp[-(\mu^2 + 1/4)\tau] P_{-1/2+i\mu}(u),$$

which, using the integral representation (7.10), can be written in a more explicit form

$$f(u;\tau) = \frac{\exp[-\tau/4 + \pi^2/4\tau]}{2\sqrt{2\pi\tau}} \int_0^\infty \frac{\exp[-\theta^2/4\tau]\sinh\theta \sin(\pi\theta/2\tau)d\theta}{\sqrt{(\cosh\theta + u)^3}}. \tag{7.12}$$

For $\tau = 1$, the graph of the probability distribution $f(u,\tau)$ of the

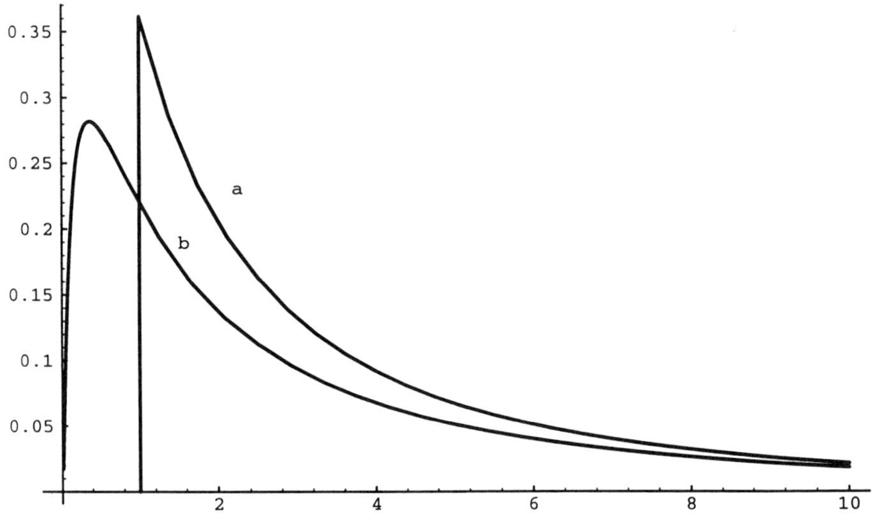

FIG. 4. *Comparison of probability densities (a) $f(u;1)$, and (b) $f_{g^2}(u;1)$ of, respectively, relative rate G of change of the total gradient, and of the square of gradient's magnitude g^2.*

relative rate G of change of the total gradient is provided on Fig. 4 together with the graph of the lognormal density

$$f_{g^2}(u;\tau) = \frac{1}{2u\sqrt{\pi\tau}} \exp\left[-\frac{\ln^2(ue^{-\tau})}{4\tau}\right],$$

of the quantity g^2, where g is the gradient's magnitude.

Finally, we provide an expression for the moment generating function

$$\Theta(\lambda;\tau) = \int_1^\infty f(u;\tau)e^{-\lambda u}\,du$$

of the quantity G. In our case,

$$\Theta(\lambda;\tau) = e^{-t} - \sqrt{\frac{\lambda}{2\tau}} \exp\left(-\lambda - \frac{\tau}{4} + \frac{\pi^2}{4\tau}\right) \times$$

$$\int_0^\infty \exp\left(-\frac{\theta^2}{4\tau} + 2\lambda\cosh^2\frac{\theta}{2}\right) \sin\frac{\pi\theta}{2\tau} \sinh\theta \left[1 - F\left(\sqrt{2\lambda}\cosh\frac{\theta}{2}\right)\right] d\theta,$$

where

$$F(z) = \frac{2}{\sqrt{\pi}} \int_0^z e^{-t^2}\,dt.$$

Since statistical moments of the stochastic process $G(\tau)$ serve as coefficients in the Taylor expansion

$$\Theta(\lambda;\tau) = \sum_{n=0}^{\infty} \frac{(-1)^n \lambda^n}{n!} \langle G^n \rangle,$$

using the formula

$$\frac{1}{\sqrt{\pi\tau}} \int_0^{\infty} \exp\left(-\frac{\theta^2}{4\tau}\right) \sin\frac{\pi\theta}{2\tau} \sinh(\beta\theta)\, d\theta = \exp\left(\beta^2\tau - \frac{\pi^2}{4\tau}\right) \sin(\beta\pi),$$

borrowed from the tables of integrals, one can also obtain the expressions for $\langle G \rangle$, $\langle G^2 \rangle$, and $\langle G^3 \rangle$, which were provided above.

8. Conclusions. In this paper, we studied statistical properties of fluctuations of passive tracer density in a randomly moving medium with Gaussian random field of velocities. Our attention was focused on analysis of full probability distributions arising in this context. By comparison, majority of papers in this area restrict themselves to a derivation of the diffusion equation for the mean of the passive tracer density.

Even in the framework of a relatively simple diffusion approximation, we managed to find a sufficiently rich cartoon of time evolution of the density and field realizations. In particular, we found out that in the compressible case, the tracer will concentrate in compact regions of high density, surrounded by large expanses of space with density significantly below average. The differences in the qualitative time behavior of density realizations in Lagrangian and in Eulerian coordinates have been explained. We noted that the growth with time of high moments of tracer fluctuations in a randomly moving compressible medium does not give an adequate picture of the actual behavior of realizations of the fluctuating density. For that reason, a more appropriate and adequate probabilistic analysis of fluctuations is of fundamental importance.

We have also shown, that fluctuations of the norm of the density gradient in an incompressible medium has a, characteristic for processes in media with random multiplicative parameters, lognormal distribution, and the mean length of a contour carried by the medium, grows exponentially with time. In real physical situations, this exponential growth will be, of course, cut off by the molecular diffusion effects. Other quantities, related to the flow of the passive tracer, also deserve a closer investigation with the help of tools discussed in this article. One attractive candidate here is the Ertel's potential vorticity $(\nabla \times \boldsymbol{v}) \cdot \nabla\rho$ (see e.g. Herring, Kerr, Rotunno [10]), and we intend to return to it in a separate paper.

Acknowledgment. This work was partially supported by the U.S. Office of Naval Research, and by grant R8Q000 from the International Science Foundation. We also thank Dr. Yiming Hu and ONR HSAP participants for their help in preparation of illustrations.

REFERENCES

[1] AVELLANEDA M., MAJDA A. (1992), Approximate and exact renormalization theories for a model for turbulent transport, *Phys. Fluids* A 4, 41-57.

[2] BATCHELOR G.K. (1959), Small-scale variation of convected quantities like temperature in turbulent fluid, *J. Fluid Mech.* 5, 113.

[3] CARETA A., SAGUES F., RAMIREZ-PISCINA L., SANCHO J.M. (1993), Effective diffusion in a stochastic velocity field, *J. Stat. Phys.* 71, 235-313.

[4] CRISANTI A., VULPIANI A. (1993), On the effects of noise and drift on diffusion in fluids, *J. Stat. Phys.* 70, 197-211.

[5] CSANADY G.T. (1980), *Turbulent Diffusion in the Environment*, Reidel, Boston.

[6] DAVIS R.E. (1982), On relating Eulerian and Lagrangian velocity statistics: single particles in homogeneous flow, *J. Fluid Mech.* 74, 1-26.

[7] FRADKIN L. (1991), Comparison of Lagrangian and Eulerian approaches to turbulent diffusion, *Plasma Physics and Controll. Fusion*, 685.

[8] GURBATOV S.N., MALAKHOV A., SAICHEV A.I. (1991), *Non-linear Random Waves and Turbulence in Non-dipersive Media: Waves, Rays and Particles*, Manchester U Press.

[9] GURBATOV S.N., SAICHEV A.I. (1993), Inertial nonlinearity and chaotic motion of particle fluxes, *Chaos 3*, 333-358.

[10] HERRING J.R., KERR R.M., ROTUNNO R. (1994), Ertel's potential vorticity in unstratified turbulence, *J. Atmospheric Sciences* 51, 35-47.

[11] KESTEN H., PAPANICOLAOU G.C. (1979), A limit theorem for turbulent diffusion, *Comm. Math. Phys.* 65, 97-128.

[12] KLYATSKIN V. (1986), *Method of Imbedding in the Theory of Wave Propagation*, Moscow, Nauka.

[13] KLYATSKIN V., SAICHEV A.I., (1992), Statistical and dynamical localization of plave waves in randomly layered media, *Soviet Physics Usp.* 35 (3), 231-247.

[14] KLYATSKIN V., WOYCZYNSKI W.A., D. GURARIE (1996), Diffusing passive tracers in random incompressible flows: statistical topography aspects, *J. Stat. Phys.* 84, 797-836.

[15] KLYATSKIN V., WOYCZYNSKI W.A. (1994), Dynamical and statistical characteristics of geophysical fields and waves, and related boundary-value problems, this volume.

[16] KRAICHNAN R.H. (1970), Diffusion by a random velocity field, *Phys. Fluids* 13, 22-31.

[17] LIPSCOMB T.C., FRENKEL A.L., TER HAAR D. (1970), On the convection of a passive scalar by a turbulent Gaussian velocity field, *J. Stat. Phys.* 63, 305-313.

[18] MAJDA A. (1993), Explicit inertial range renormalization theory in a model for turbulent diffusion, *J. Stat. Physics* 73, 515-542.

[19] PAPANICOLAOU G.C. (1971), Wave propagation in one-dimensional random medium, , *SIAM J. Appl. Math. 21*, 13-18.

[20] PITERBARG L. (1994), Short-correlation approximation in models of turbulent diffusion, this volume.

[21] SAICHEV A.I., WOYCZYNSKI W.A. (1995a), Model description of passive tracer density fields in the framework of Burgers' turbulence, in *Non-*

linear Stochastic PDE's: Burgers Turbulence and Hydrodynamic Limit, IMA Volume 77, Springer-Verlag, 167-192.

[22] Saichev A.I., Woyczynski W.A. (1996), Density fields in Burgers' and KdV-Burgers' turbulence, *SIAM J. Appl. Math.* 56(1), 1-36.

[23] Zirbel C.L. (1993), Stochastic flows: dispersion of a mass distribution and Lagrangian observations of a random field, *Princeton U Ph.D. Dissertation*, 197-211.

THREE-DIMENSIONAL BURGERS' EQUATION AS A MODEL FOR THE LARGE-SCALE STRUCTURE FORMATION IN THE UNIVERSE

SERGEI F. SHANDARIN*

Abstract. As galaxy redshift surveys probe deeper into the universe, they uncover ever more dramatic structures in the large-scale distribution of galaxies. In particular, the CfA2 and SSRS2 surveys to an apparent magnitude limit of 15.5 exhibit an impressive complex of sheets, filaments, and clusters. The formation of the large-scale structure in the universe results from the gravitational amplification of the primordial small perturbations of density. The primordial density perturbations are thought to be random fields originated as quantum fluctuations at the very early stage. Thus the understanding of the formation of the large-scale structure may reveal important information about the early universe and the laws of fundamental physics. One of the major obstacles to understanding the formation of the large-scale structure is the complexity of the evolution of the density inhomogeneities at the nonlinear stage when the observable structures form. One way of addressing this problem is to run three-dimensional numerical simulations. Here we review another approach based on the approximate analytic model of the nonlinear gravitational instability utilizing Burgers' equation of the nonlinear diffusion.

1. Introduction. The term large-scale structure in the universe is referred to a distribution of galaxies on the scales roughly from 1 Mpc to 100 Mpc [17], where 1 $Mpc = 10^6\ pc \approx 3 \cdot 10^{24}\ cm$ is a unit of length commonly used in cosmology. Galaxies can not probe much smaller scales because of discreteness, and on larger scales the galaxy distribution becomes almost homogeneous. The redshift surveys reveal spectacular abundance of structures often described as filamentary, network, or bubble structure [17], [6], [4] (see Figure 1). The origin of the large-scale structure is one of the most important problems in modern cosmology. Many fundamental issues in physics, cosmology and astronomy ranging from speculations on the physical nature of dark matter, to the measurement of angular anisotropy of the microwave background radiation and determination of the epoch of galaxy formation join together here [21].

The most popular and best developed class of theories of structure formation is based on the assumption that it started from primeval small amplitude density perturbations which grew by gravitational instability. Primeval perturbations are assumed to arise as vacuum fluctuations during the very early stage when the universe was expanding exponentially (inflationary universe) [13]. Afterwards, the density perturbations had a long history before they become galaxies, clusters of galaxies, superclusters and voids. The formation of galaxies is a very difficult problem itself. Many complex physical processes like star formation and supernova explosions are very important for understanding the galaxy formation. We shall discuss the mass distribution assuming that galaxies are fairly good

* Department of Physics and Astronomy, University of Kansas, Lawrence, KS 66045.

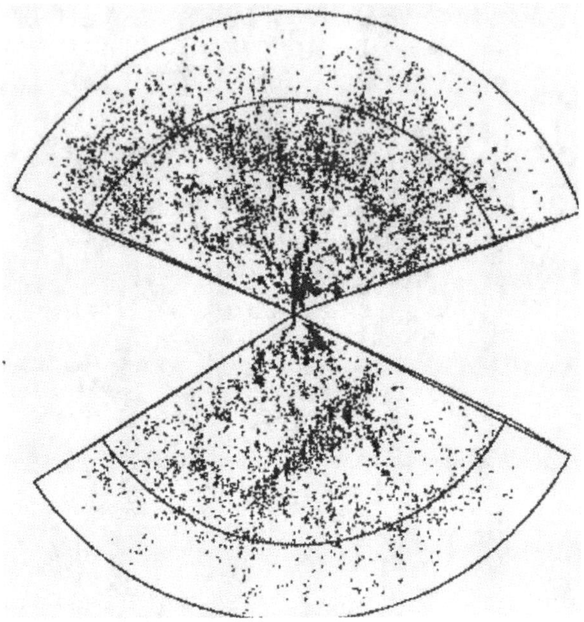

FIG. 1. *This is a low resolution version of Fig.2 from [4]. It displays galaxies from two redshift catalogs to an apparent magnitude limit of 15.5 and distances less than* $120\ h^{-1}$ *Mpc: CfA2 (north; upper portion) and SSRS2 (south; lower portion). For CfA2 the box shows the declination limits* $8^\circ.5 \le \delta \le 44^\circ.5$ *and the right ascension limits* $8^h \le \alpha \le 17^h$. *In the south the box limits are* $-40^\circ \le \delta \le -2^\circ.5$ *and the right ascension limits are* $20^h.8 \le \alpha \le 4^h$. *There are 9325 galaxies in the original image.*

(though not perfect) tracers of mass on large scales.

As long as the density perturbations are small by amplitude their evolution is described by the linear theory of gravitational instability (see e.g. [28]; [18]; [22]. The linear theory is very well understood; in particular, it predicts the rate of growth of the perturbations at different stages of the evolution of the universe. Unfortunately, it is not applicable to the nonlinear stage when the amplitude of the density perturbations becomes large and the observable structures (sheets, filaments, and clusters of galaxies) form.

One of the most difficult obstacles arising in understanding models is the analysis of the evolution of perturbations at the nonlinear stage, when the typical amplitudes of density inhomogeneities become larger than mean density in the universe: $\delta\rho/\rho \ge 1$. The most straightforward way to address the problem of the nonlinear evolution is to do gravitational three-dimensional N-body simulations. Usually in simulations of this type the medium is assumed to consist of collisionless particles, in agreement with the popular hypothesis that the most of the mass in the universe is in the form of weakly interacting particles such as massive neutrinos or axions. The trajectory of each particle is calculated in the gravitational field generated by all the particles. Boundary conditions are commonly assumed to be periodic.

Here we review an alternative approach to the problem of the large-scale structure in the universe. We present an approximate analytic technique to solve the nonlinear gravitational instability based on Burgers' equation. Other analytic and semi-analytic methods are discussed in an excellent review by Sahni and Coles [19].

The initial condition is the result of the linear theory applied to the earlier stages. In the linear regime the density fluctuations are assumed to be a Gaussian random field specified by the spectrum and the amplitude. The current measurements of the angular fluctuations in the temperature of the microwave background radiation by COBE (Cosmic Background Explorer) and other experiments put strong constraints on the initial fluctuations. In particular, the amplitude of the temperature fluctuations suggests that the scale where the density fluctuations have recently reached nonlinearity is about $5 - 10$ Mpc which is in a good agreement with the observations of the large-scale galaxy distribution (see e.g. [26]).

According to the prediction of cosmic inflation, the geometry of the universe is assumed to be flat (see e.g.[13]. For simplicity we also assume that the cosmological term $\Lambda = 0$. Such a model is specified by only two parameters: the Hubble constant H_0 describing the current rate of expansion of the universe and the fractional density of baryons Ω_b. The astronomical measurement of the Hubble constant is not very accurate: $50 < H_0 < 100$ $km\ s^{-1}\ Mpc^{-1}$; this uncertainty is usually expressed in terms of a parameter h: $H_0 = 100\, h\ km\ s^{-1}\ Mpc^{-1}$ with $0.5 < h < 1$. The density of baryons as well as the density of other components is convenient to measure in the units of the critical density: $\Omega_b = \bar{\rho}_b/\rho_{cr}$, where the critical density $\rho_{cr} \equiv 3H_0^2/8\pi G \approx 2\cdot 10^{-29} g\ cm^{-3}$ is a parameter separating the closed cosmological models $\Omega_{tot} > 1$ from the open ones $\Omega_{tot} < 1$. An open universe has negative spatial curvature and expands forever, and a closed universe has positive curvature and eventually collapses. If $\Omega_{tot} = 1$ the spatial geometry of the universe is flat and it expands forever.

We shall assume that about 90% or more of mass in the universe is in a form of weakly interacting collisionless particles (dark matter) and the remaining few percent of the mass is in baryons; $\Omega_{tot} = \Omega_{dm} + \Omega_b = 1$ [21]. Although all luminous objects (e.g. stars) are made from baryons the baryon component dynamically is not very important on large scales. Therefore we shall study the evolution of density fluctuations in the collisionless medium neglecting the baryonic component.

2. Basic equations. The evolution of density inhomogeneities can be described by a system of three partial differential equations comprising

the continuity, Euler, and the Poisson equation (see e.g. [28], [18], [22]). It is convenient to exclude the homogeneous expansion of the universe from the analysis by introducing new variables. Instead of the coordinates $\mathbf{r}$ the comoving coordinates $\mathbf{x} = \mathbf{r}/a(t)$ are commonly used, here $a(t)$ is the scale factor describing the homogeneous expansion of the universe; in a flat universe $a(t) \propto t^{2/3}$. Instead of the velocity $\mathbf{u}$ the so called peculiar velocity $\mathbf{u}_p = \mathbf{u} - \dot{a}/a \cdot \mathbf{r}$ is used; if $\mathbf{u}_p = 0$ the velocity field on the scales smaller than the cosmological horizon ($R_H \approx 3,000\ h^{-1}\ Mpc$) is described by the Hubble law: $\mathbf{u} = \frac{\dot{a}}{a} \cdot \mathbf{r} \equiv H(t) \cdot \mathbf{r}$). The velocities and gravitational field in the process we discuss are nonrelativistic ($v \ll c$, $\varphi \ll c^2$) thus the evolution of density inhomogeneities can be described by the equations of classic hydrodynamics and gravity. In terms of the comoving coordinates and peculiar velocities the equations are as follows:
the continuity equation

$$\frac{\partial \rho}{\partial t} + \frac{1}{a}\nabla \cdot (\rho \mathbf{u}_p) = -3\frac{\dot{a}}{a}\rho, \tag{2.1}$$

the Euler equation

$$\frac{\partial \mathbf{u}_p}{\partial t} + \frac{1}{a}(\mathbf{u}_p \cdot \nabla)\mathbf{u}_p = -\frac{1}{a}\nabla\phi - \frac{\dot{a}}{a}\mathbf{u}_p, \tag{2.2}$$

and the Poisson equation

$$\frac{1}{a^2}\nabla^2\phi = 4\pi G(\rho - \overline{\rho}), \tag{2.3}$$

where $\dot{a} \equiv da/dt$; ρ and $\overline{\rho}$ are respectively the density and mean density of mass; ϕ is the perturbation of the gravitational potential due to the inhomogeneities of density; G is the gravitational constant.

The pressure is neglected since we study the medium interacting only gravitationally. Additional terms on the right hand side of the continuity and Euler equations are due to the homogeneous expansion of the universe and the factor $1/a$ is due to differentiation with respect to the comoving coordinates $\mathbf{x}$: $\nabla \equiv \partial/\partial x_i \equiv a \cdot \partial/\partial r_i$. We assume that the initial condition is small density and smooth velocity perturbations imposed on a homogeneous density distribution.

As long as the amplitude of the density perturbations remains small their evolution can be analyzed in the linear approximation obtained by the linearization of the above equations. The exact solution of the linearized system has one growing mode which is the major object of our analysis. The velocity in the growing mode is a potential vector field. In the linear regime the spatial structure of the perturbations (in the comoving coordinates) remains unchanged and its amplitude is proportional to the growing solution b_g of the differential equation

$$a\frac{d^2b}{dt^2} + 2\dot{a}\frac{db}{dt} + 3\ddot{a}b = 0. \tag{2.4}$$

The scale factor a is assumed to be a known function of time and in a flat matter dominated universe $b_g(t) \propto a(t) \propto t^{2/3}$. It is convenient to make yet another transformation of the variables:

$$\eta(\mathbf{x}, b_g) = a^3 \rho(\mathbf{x}, t), \tag{2.5}$$

$$\mathbf{v}(\mathbf{x}, b_g) \equiv \frac{d\mathbf{x}}{db_g} = \frac{1}{a\dot{b}_g} \mathbf{v}_p(\mathbf{x}, t), \tag{2.6}$$

$$\varphi(\mathbf{x}, b_g) = (3\ddot{a}ab_g)^{-1} \phi(\mathbf{x}, t), \tag{2.7}$$

and also to reparametrize the time coordinate by the monotonic function of time $b_g(t)$ describing the growth of the perturbations. Finally, after explicit use of the function $b_g = a$ (which assumes $\Omega_{tot} = 1$) and introducing the velocity potential Φ: $\mathbf{v} = -\nabla\Phi$ the equations take the form

$$\frac{\partial \eta}{\partial a} + \nabla \cdot (\eta \mathbf{v}) = 0, \tag{2.8}$$

$$\frac{\partial \mathbf{v}}{\partial a} + (\mathbf{v} \cdot \nabla)\mathbf{v} = -\frac{3}{2a}\nabla(\varphi - \Phi), \tag{2.9}$$

$$\nabla^2 \varphi = \frac{1}{a}\frac{\eta - \bar{\eta}}{\bar{\eta}}. \tag{2.10}$$

In the linear regime both the gravitational potential φ and velocity potential Φ remain constant and equal to each other, and the right hand side of eq.2.9 vanishes. In 1970 Zel'dovich [27] suggested to extrapolate this condition well into the nonlinear regime; the corresponding solution is known in cosmology as the Zel'dovich approximation.

3. Zel'dovich approximation. The Zel'dovich approximation is convenient to formulate as a mapping from the Lagrangian space $L\{q\}$ into Eulerian space $E\{x\}$

$$\mathbf{x}(\mathbf{q}, a) = \mathbf{q} + a \cdot \mathbf{v}_0(\mathbf{q}), \tag{3.1}$$

which obviously follows from eq.2.9 assuming that $\varphi \approx \Phi$. The initial velocity potential $\Phi_0(\mathbf{q})$ ($v_{0i}(\mathbf{q}) = -\partial\Phi_0/\partial q_i$) is assumed to be a smooth random Gaussian field specified by the spectrum $P_\Phi(k)$. In cosmology the initial condition is usually characterized by the spectrum $P_\delta(k)$ of the linear density fluctuations $\delta \equiv (\rho - \bar{\rho})/\bar{\rho} \equiv (\eta - \bar{\eta})/\bar{\eta}$ which is obviously related to the spectrum of the potential

$$P_\Phi(k) = k^{-4} P_\delta(k). \tag{3.2}$$

Utilizing the law of mass conservation one finds the density as a function of the time coordinate a and the Lagrangian coordinate $\mathbf{q}$

$$\eta(\mathbf{q}, a) = \frac{\bar{\eta}}{[1 - a\,\lambda_1(\mathbf{q})][1 - a\,\lambda_2(\mathbf{q})][1 - a\,\lambda_3(\mathbf{q})]}, \tag{3.3}$$

where $\lambda_1(\mathbf{q})$, $\lambda_2(\mathbf{q})$ and $\lambda_3(\mathbf{q})$ are the eigenvalues of the deformation tensor $d_{ij} = \partial^2\Phi_0/\partial q_i \partial q_j$. Combining eqs.3.1 and 3.3 one can find the density distribution in the Eulerian space. It follows from eq.3.1 and eq.3.3 that the first objects form at the maxima of the largest eigenvalue and have very oblate shapes. After Zel'dovich they are known in cosmology as "pancakes". Recent three-dimensional gravitational N-body simulations are in a perfect agreement with this conclusion [23]. The pancakes originate as the three-stream flow regions bounded by caustics, the surfaces of formally infinite density. The shape and other characteristics of the pancakes are determined by catastrophe theory [1]. At the later stages the Zel'dovich solution predicts several different types of singularities which are classified in [1].

We use this opportunity to remark that the Zel'dovich approximation (in two-dimensional space) is very similar to the equations describing the propagation of light in geometric optics [29].

The Zel'dovich approximation proved to be very good until orbit crossing when caustics form and the multi-stream flows occur (see e.g. [22] and references therein). (At this stage the original equations must be obviously modified in order to incorporate the multi-stream flows.) However, the Zel'dovich approximation predicts the multi-stream flow regions to broaden very fast which contradicts to the results of the N-body simulations [5], [9]. Numerical studies of the orbits in the multi-stream flow regions show that the velocity component orthogonal to the pancakes randomizes very quickly [12]. As a result the pancakes observed in the N-body simulations remain quite thin. This result has become a physical basis for the adhesion approximation.

4. Adhesion model. The general idea of the adhesion model is very simple. We wish to use the Zel'dovich solution everywhere except the regions of multi-stream flows. By adding a diffusion term into the Euler equation one can suppress the formation of the multi-stream flow regions. Assuming that the gravitational potential is approximately equal to the velocity potential $\varphi \approx \Phi$ and adding a viscosity term $\nu\nabla^2\mathbf{v}$ in eq.2.9 one obtains the equation of the nonlinear diffusion [7], [8]

$$\frac{\partial \mathbf{v}}{\partial a} + (\mathbf{v}\cdot\nabla)\mathbf{v} = \nu\nabla^2\mathbf{v}. \tag{4.1}$$

Generally speaking the viscosity term need not to be in the form of eq.4.1 but choosing this particular form one obtains Burgers' equation that has an exact analytic solution [3]. For potential motion $\mathbf{v} = -\nabla\Phi$ eq.4.1 can be solved by performing the Hopf-Cole substitution $\Phi(\mathbf{x}, a) = -2\nu \log U(\mathbf{x}, a)$. As a result eq.4.1 translates into the familiar linear diffusion equation

$$\frac{\partial U}{\partial a} = \nu\nabla^2 U. \tag{4.2}$$

Solving eq.4.2 for the velocity we obtain

$$\mathbf{v}(\mathbf{x},a) = \frac{\int d^3q \, \left(\frac{\mathbf{x}-\mathbf{q}}{a}\right) \, \exp[-S(\mathbf{x},a;\mathbf{q})/2\nu]}{\int d^3q \, \exp[-S(\mathbf{x},a;\mathbf{q})/2\nu]}, \tag{4.3}$$

where the "action"

$$S(\mathbf{x},a;\mathbf{q}) = -\Phi_0(\mathbf{q}) + \frac{(\mathbf{x}-\mathbf{q})^2}{2a}. \tag{4.4}$$

In cosmology the adhesion model has been used in two forms: one assumes a small but finite value of the viscosity parameter ν and the other assumes it is infinitesimal: $\nu \to 0$ [19].

For finite ν the trajectory of a particle can be determined by solving the integral equation [24], [25], [16], [15]

$$\mathbf{x}(\mathbf{q},a) = \mathbf{q} + \int_0^a da' \; \mathbf{v}[\mathbf{x}(\mathbf{q},a'),a'], \tag{4.5}$$

and the resulting density can be determined from the continuity equation

$$\eta(\mathbf{x},a) = \bar{\eta}/det(\frac{\partial x_i}{\partial q_j}). \tag{4.6}$$

For an infinitesimal value of the viscosity parameter $\nu \longrightarrow 0$, the integrals in eq.4.3 can be evaluated using the method of steepest descents [7], [8], [10], [11]. In this case

$$\mathbf{v}(\mathbf{x},a) = \frac{\mathbf{x}-\mathbf{q}(\mathbf{x},a)}{a}, \tag{4.7}$$

where $\mathbf{q}(\mathbf{x},a)$ is the coordinate of the absolute minimum of the action $S(\mathbf{x},a;\mathbf{q})$ at given $\mathbf{x}$ and a. The points $\mathbf{q}$ that minimize the action obviously satisfy the Zel'dovich equation 3.1.

This solution (eq.4.7) has an interesting geometrical interpretation (see Figure 2). One can find the Eulerian coordinate $\mathbf{x}$ of the particle with initial Lagrangian coordinate $\mathbf{q}$ at a chosen time simply by projecting the apex of the paraboloid

$$P(\mathbf{x},a;\mathbf{q}) = \frac{(\mathbf{x}-\mathbf{q})^2}{a} + C, \tag{4.8}$$

assuming that there is a constant C satisfying simultaneously two conditions: 1) the paraboloid P is tangent to the initial velocity potential Φ_0 at $\mathbf{q}$ and 2) it does not cross Φ_0 at any point. At early times a is small and the curvature of the paraboloid is greater than that of Φ_0 and the above two conditions can be easily fulfilled. The Zel'dovich approximation is universally valid at this stage.

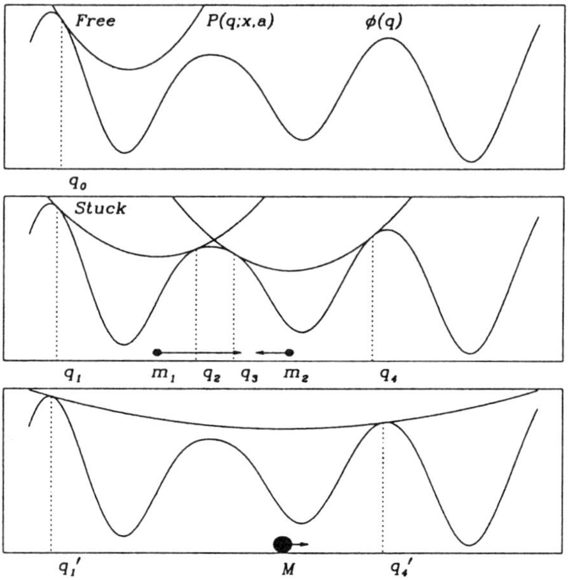

FIG. 2. *Geometrical prescription of descending a paraboloid onto the initial velocity potential in order to find the Eulerian positions of particles and knots in one-dimensional case. The particle having Lagrangian coordinate q_0 in the uppermost panel has the Eulerian coordinate of the paraboloid apex. In the middle panel corresponding to a later stage it is stuck into the knot m_1 which velocity is shown by the arrow. The knot m_2 is determined by the second paraboloid in the middle panel. The lower panel shows the knot M formed as a result of merging m_1 and m_2. Adapted from [20].*

As time passes the curvature of the paraboloid decreases. As a result the points $\mathbf{q}$ appear that do not satisfy the above two conditions. Such a point has been stuck into a surface due to orbit crossing. Again, such surfaces can be found by projecting the apex of the paraboloid but this time the paraboloid is tangent to the initial velocity potential in two points simultaneously. These surfaces correspond to the pancakes; their thickness is proportional to the value of the viscosity parameter and therefore is infinitesimal as $\nu \to 0$.

Later two more types of structures form: the filaments and knots. The filaments correspond to the case when the paraboloid touches the initial potential in three points simultaneously and the knots when it touches the potential in four points. The set of surfaces, filaments and knots make a cellular structure: large regions of low density are separated by the surfaces, the filaments are at the intersections of the surfaces, and the knots are at the intersections of the filaments. This geometrical construction can be viewed as the skeleton of the real structure. In one-dimensional case illustrated by Figure 2 obviously there are no surfaces nor filaments.

5. Accuracy of the adhesion model. Apart from the regions of high density where the viscosity term (eq.4.1) plays a significant role the adhesion model is exact in one-dimensional case. It means that the velocity field outside the high density regions is predicted exactly. Also the motion of the clumps is described very accurately. This is because the Zel'dovich solution is exact outside the multi-stream flow regions in one-dimensional case [22].

In more interesting two- and especially three-dimensional case the adhesion model is only an approximation. Therefore the question arises about the accuracy of the model. The both variants of the adhesion model have been thoroughly tested against the gravitational N-body simulations in two and three dimensions. Both the N-body simulation and the adhesion model used the identical initial conditions and were compared at several stages of the evolution.

The geometrical version of the adhesion model was tested against the two-dimensional N-body simulations with the initial power law spectra $P_\delta(k) \propto k^n$ with spectral indices $n = 2, 0$ and -2 and various cutoffs [11]

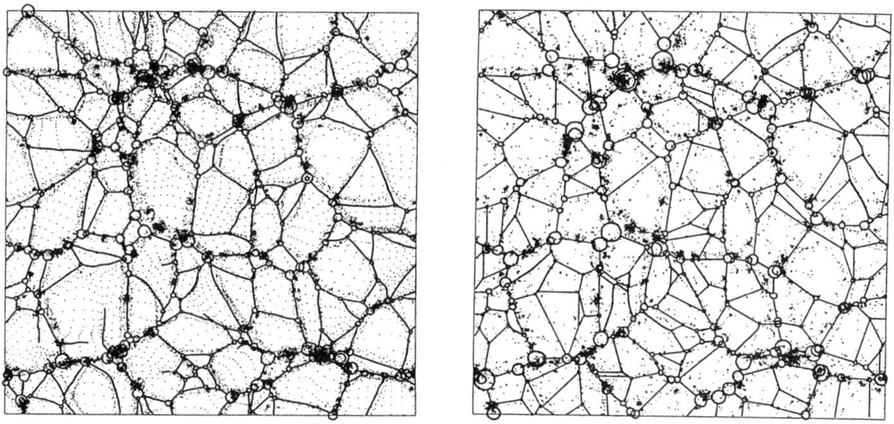

FIG. 3. *Composite picture of the results of the two-dimensional N-body simulation and the adhesion model for the flat initial spectrum $P_\delta(k) \propto k^0$. The left hand side panel shows the model with the cutoff of the initial spectrum at $k_c = 32\ k_f$, here k_f is the fundamental frequency corresponding to the box size: $L_{box} = 2\pi/k_f$; the right hand side panel shows the model without a cutoff (except at the Nyquist frequency $k_{Ny} = 256$). Not all the particles can be shown, but this is a fair sample). Solid lines and circles represent the skeleton of the structure constructed by the paraboloid technique. The area of circles is proportional of the mass of knots. Adapted from [11].*

The N-body simulations used the particle-mesh code with 512^2 particles on equal mesh and periodic boundary conditions (for the details see [2]). The code constructing the skeleton of the structure is described in [10]. It has been found that the skeleton reproduces the density distribution extremely well (see Figure 3) for all choices for the parameters of the initial spectra until the stage when the scale of the nonlinearity k_{nl}^{-1} defined by the equation

$$a^2 \int_0^{k_{nl}} P_\delta(k)\, d^D k = 1 \tag{5.1}$$

reaches the characteristic scale of the initial velocity potential Φ_0: $k_{nl}^{-1} \leq R_{\Phi_0}$, here D is the dimensions of the space. The scale of the potential is defined from the expansion of the correlation function $\xi_{\Phi_0}(r) = \xi_{\Phi_0}(0)(1 - r^2/2R_{\Phi_0}^2 + \cdots)$:

$$R_{\Phi_0} = (2D)^{1/2}\, \frac{\sigma_0}{\sigma_1}, \tag{5.2}$$

where σ_0 and σ_1 are the dispersions of the potential and its gradient, respectively. In the N-body simulations we deal with finite ranges and therefore R_{Φ_0} always exists. At the later stages $k_{nl}^{-1} \geq R_{\Phi_0}$ the adhesion model remain qualitatively correct though its accuracy somewhat deteriorates.

The version of the adhesion model utilizing a finite viscosity parameter ν has been quantitatively compared to fully nonlinear, numerical three-dimensional gravitational N-body simulations [15]. The initial perturbations were again random Gaussian fields with power-law spectra with the spectral indices $n = -2, -1, 0, +1$. The particle-mesh N-body and adhesion simulations both used 128^3 particles on a 128^3 mesh and periodic boundary conditions. In these simulations of the adhesion model, the smallest value of the viscosity parameter ν that did not produce numerical overflows has been used. For further discussion of the N-body and adhesion simulations see [14] and [25] respectively. The both codes calculated the particle positions therefore the corresponding density distributions could be easily generated (Figure 4). The primary tool of the quantitative comparison was the cross-correlation coefficient for the density fields obtained from the gravitational N-body simulation and the adhesion model with the identical initial conditions. Also the density distribution functions and the power spectra were compared. The adhesion model produces an excessively filamentary distribution due to smoothing effects in high density regions. As a result the density distribution function in the adhesion model is lower than that of the N-body simulation at high densities $\rho/\bar{\rho} \geq 10-20$ depending on the initial spectrum and the power spectrum of the nonlinear distribution falls off steeper on small scales $k \geq k_{nl}$.

6. Summary. The adhesion model based on three-dimensional Burgers' equation of the nonlinear diffusion has been found to work very well in explaining the large-scale features of the structure of the universe. Comparison with gravitational two- and three-dimensional N-body simulations has shown remarkable agreement till very late nonlinear times. The main drawback of the adhesion approximation is probably the fact that it cannot describe accurately the density distribution within pancakes and filaments.

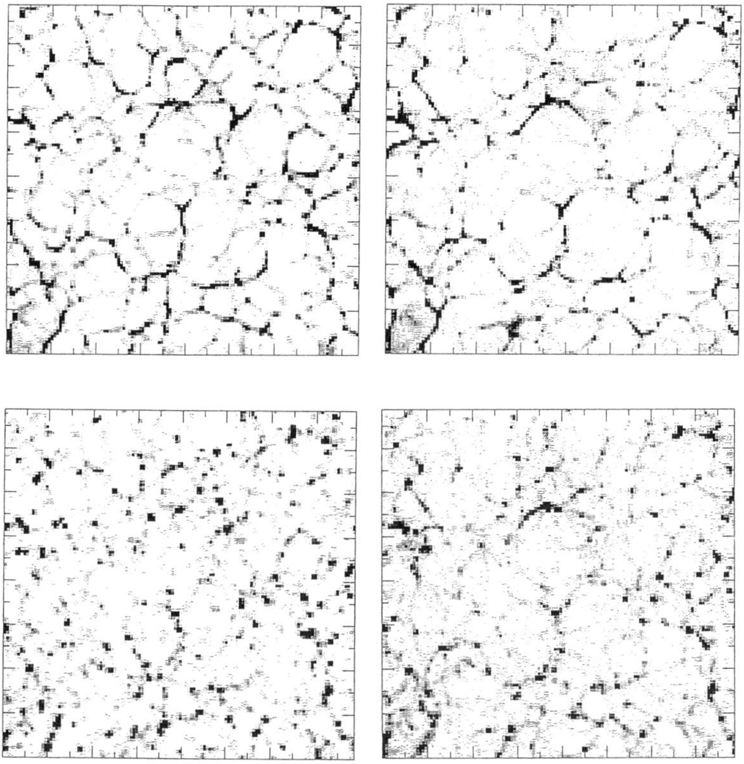

FIG. 4. *A gray scale plot of thin ($L_{box}/128$) slices through the simulation cubes for $n = 0$ (left hand side panels) and $n = -1$ (right hand side panels) initial spectra, at the stages when $k_{nl} = 8k_f$. The gravitational three-dimensional N-body simulations are shown in the bottom panels and the adhesion model simulations are shown in the top panels. Adapted from [15].*

The adhesion model provides a natural qualitative explanation of the origin of the large-scale coherent structures such as superpancakes and superfilaments, as a result of coherent motion of clumps due to large-scale inhomogeneities in the initial gravitational potential (compare Figure 1) and 5). The formation and evolution of large-scale structure is described by the adhesion model as a two stage process [20]. During the first stage matter falls into pancakes and then moves along them towards filaments and then along filaments to collect finally in knots. At the end of the first stage the formation of the skeleton of the large scale structure is complete and virtually all of the matter in the universe is located in one of three structural units: pancakes, filaments or knots. The second stage sees the deformation of the large-scale structure skeleton due to the dynamical motion of pancakes, filaments and especially knots. At this stage knots merge into larger knots and small voids disappear giving space to growth of the

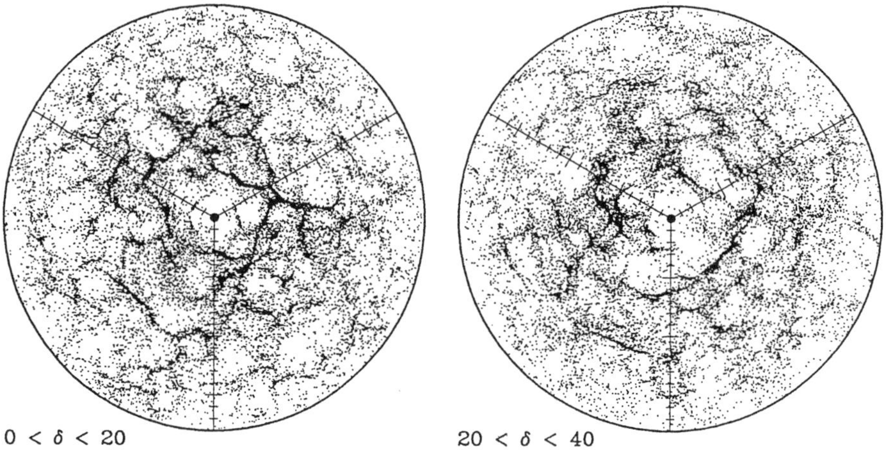

$0 < \delta < 20$
$m < 16.5$
32803 galaxies

$20 < \delta < 40$
$m < 16.5$
27068 galaxies

FIG. 5. *Simulated galaxy distributions are drawn from the adhesion model simulation based on the biased Cold Dark Matter cosmological model. These redshift-angle projections show all "galaxies" with apparent magnitude less than 16.5 and distance less than* 180 $h^{-1}Mpc$. *Adapted from [24].*

larger ones. Eventually almost all the mass concentrates in knots. Depending on the initial spectrum the knots may move coherently in such a manner that they concentrate to superpancakes and superfilaments. The superpancakes and superfilaments can be identified by applying the adhesion model to *smoothed* initial potential [11].

Acknowledgment. I am grateful to my collaborators on the adhesion model S.Gurbatov, L.Kofman, D.Pogosyan, V.Sahni, A.Saichev, and B.Sathyaprakash for innumerous discussions. I acknowledge NSF grant AST-9021414, NSF EPSCoR grant OSR-9255223, and NASA grant NAGW-2923.

REFERENCES

[1] Arnol'd, V.I., Shandarin, S.F., Zel'dovich, Ya.B. 1982, *Geophys. Astrophys. Fluid Dyn.*, **20**, 111.

[2] Beacom, J.F., Dominik, K.G., Melott, A.L., Perkins, S., and Shandarin, S.F. 1991, *Astrophys. J.*, **372**, 351.

[3] Burgers, J.M. 1974, *The Non-linear Diffusion Equation* (Dordrecht: Reidel).

[4] da Costa, L.N., Geller, M.J., Pellegrini, P.S., Latham, D.W., Fairall, A.P., Marzke, R.O., Willmer, C.N.A., Huchra, J.P., Calderon, J.H., and Kurtz, M.J. 1994, *Astrophys. J. Lett.*, **424**, L1.

[5] Doroshkevich, A.G., Kotok, E.V., Novikov, I.D., Polyudov, A.N., Shandarin, S.F.,

and Sigov Yu.S. 1980, *Mon. Not. R. astr. Soc.*, **192** 321.
[6] Geller, M., and Huchra, J. 1989, *Science*, **246**, 897.
[7] Gurbatov, S.N., Saichev, A.I., and Shandarin, S.F. 1985. *Sov. Phys. Doklady*, **30**, 921.
[8] Gurbatov, S.N., Saichev A.I., and Shandarin, S.F. 1989, *Mon. Not. R. astr. Soc.*, **236** 385.
[9] Klypin, A.A., and Shandarin, S.F. 1983, *Mon. Not. R. astr. Soc.*,, **204**, 891.
[10] Kofman, L.A., Pogosyan, D.Yu., and Shandarin S.F. 1990, *Mon. Not. R. astr. Soc.*, **242**, 200.
[11] Kofman, L., Pogosyan, D., Shandarin, S.F., and Melott, A.L. 1992, *Astrophys. J.*, **393**, 437.
[12] Kotok, E.V., and Shandarin, S.F. 1987, *Sov. Astr.*, **64**, 1144.
[13] Linde, A.D. 1990, *Particle Physics and Inflationary Cosmology* (Harwood, New York).
[14] Melott, A.L., and Shandarin, S.F. 1993, *Astrophys. J.*, **410**, 469.
[15] Melott, A.L., Shandarin, S.F., and Weinberg, D.H. 1994, *Astrophys. J.*, **428**, 28.
[16] Nusser, A., and Dekel, A. 1990, *Astrophys. J.*, **362**, 14.
[17] Oort, J. 1983, *Ann. Rev. Astron. Astrophys.*, **21**, 373.
[18] Peebles, P.J.E. 1980, *The Large-scale Structure of the Universe* (Princeton: Princeton University Press).
[19] Sahni, V., and Coles, P. 1995, *Physics Reports*, 262, 1.
[20] Sahni, V., Sathyaprakash, B.S., and Shandarin, S.F. 1994, *Astrophys. J.*, **431**, 20.
[21] Shandarin, S.F., Doroshkevich, A.G., and Zel'dovich, Ya.B. 1983, *Sov. Phys. Usp.*, **26**, 46.
[22] Shandarin S.F., and Zel'dovich, Ya.B. 1989, *Rev. Mod. Phys.*, **61**, 185.
[23] Shandarin, S.F., Melott, A.L., McDavitt, K., Pauls, J., and Tinker, J. 1995, *Phys. Rev. Lett.*, 75, 7.
[24] Weinberg, D., and Gunn, J. 1989, *Astrophys. J. Lett.*, **352**, L25.
[25] Weinberg, D., and Gunn, J. 1990, *Mon. Not. R. astr. Soc.*, **247**, 260.
[26] White, M., Scott, D., and Silk, J. 1994 *Ann. Rev. Astron. Astrophys.*, **32**, 319.
[27] Zel'dovich, Ya.B. 1970, *Astron. Astrophys.*, 5 84.
[28] Zel'dovich, Ya.B., and Novikov, I.D. 1983, *The Structure and Evolution of the Universe* (Chicago/London: University of Chicago Press).
[29] Zel'dovich, Ya.B., Mamaev, A.V., and Shandarin, S.F. 1983 *Sov. Phys. Usp.*, **26**, 77.

NON-MEAN FIELD APPROACH TO SELF-ORGANIZATION OF LANDFORMS VIA STOCHASTIC MERGER

HUBERT SHEN*

Abstract. The stochastic merging model of Werner et al. (1993) for self-organization in landforms such as wind ripples is extended to the fully-statistical, non-mean field case, using functional methods recently developed for the closure of turbulence, kinetic equation and nonlinear two-timescale problems. The role of fluctuations and spatial correlations in the dynamics is clarified and the regime of validity of the mean-field, single-particle distribution function approximation is determined. This approach lends new support to the notion that growth and stabilization of macroscopic geophysical structures may be understood on a deeper level via a fundamentally probabilistic model.

1. Introduction. The emergence of macroscopic geophysical structures of a definite length scale, from an initially uniform state, has traditionally been viewed as treatable within a deterministic, continuum mechanics approximation (e.g., Anderson 1990). More recently (e.g., Forrest & Haff 1992, Werner & Gillespie 1993), such self-organization has been modelled as an essentially probabilistic or discrete stochastic process, involving repeated collisional events in which discrete microscopic amounts of material are exchanged between structures, resulting in fluctuations in their sizes. The distribution of structure sizes is determined by competing processes, whereby smaller structures either grow at the expense of larger structures or are partially absorbed by larger structures, complete absorption being viewed as a merger between the two structures. Such a stochastic model has been written down in the context of describing the formation of ripples in sand by wind (ibid.) and has been solved in a mean-field approximation. Our goal here is to extend the solution of Werner et al. to the non-mean field case, examine the evolution of the probability distribution itself rather than just its mean, and compare with mean-field results.

Explicitly, recall the stochastic merging model of Werner et al. for self-organization of a ring of N links which is partitioned into contiguous segments ("worms") of varying length. Each worm of length n corresponds to a ripple of sand of roughly triangular cross-section and linear dimension n; motion of the head of a worm corresponds to translation of a ripple by removal of a thin layer of sand from its upwind side and deposition of the sand on the downwind side. Ripple speed is inversely proportional to its size, since it is the sum of the distances travelled by grains in the ripple per unit time (proportional to the surface area of the ripple exposed to impacts by wind-driven grains) divided by the total number of grains in the ripple (proportional to the volume of the ripple.)

Let $P(n)$ denote the probability of finding a worm of length n and

* Department of Physics, University of Toronto, Toronto, Ontario M5S 1A7 Canada.

$P(n, n')$ denote the joint probability of finding a worm of length n followed by one of length n'. Both are understood to depend on time, in general. Let the transition probability per unit time that the ith worm increases its length w_i by one link, i.e., that the head of the ith worm moves forward by one link, while the head of the $(i-1)$th worm remains stationary, be given by

$$W(w_i \to w_i + 1) = \frac{\beta}{w_i}\left(1 - \frac{\beta}{w_{i-1}}\right) \approx \frac{\beta}{w_i} \tag{1.1}$$

in the small-β approximation. Then the evolution of the 1-worm probability is given within this approximation by

$$\frac{1}{\beta}\frac{\partial P(n)}{\partial t} = \sum_{j=1}^{N}\left[\frac{P(j, n-1)}{n-1} + \frac{P(n+1, j)}{j} - \frac{P(j, n)}{n} - \frac{P(n, j)}{j}\right] \tag{1.2}$$

Note that in general the joint probability will be asymmetric

$$P(n, n') \neq P(n', n) \tag{1.3}$$

because of the unidirectional nature of the motion (physically, sand ripples do not move backwards against the prevailing wind).

Statistically, the faster-moving smaller worms are favored to overtake the more slowly-moving larger worms and hence grow at the expense of the larger worms. However, it has been shown that the statistically unlikely fluctuations in which smaller worms are overtaken and become yet smaller, with the smallest worms (those of unit length) swallowed by overtaking neighbors (so-called merging events), are important. These processes are captured in the third term of the evolution equation, with transition rate

$$W(1 \to 0) = \beta \sum_{j=1}^{N} \frac{P(1, j)}{j} \tag{1.4}$$

and lead to a net irreversible decrease in the total number of worms. (The model contains no mechanism for spontaneous generation of elementary worms of unit length.) These merging events have been shown (ibid.) in a mean-field solution to give rise to logarithmic growth in time of the mean worm length

$$\langle n \rangle = \sum_{n=1}^{N} nP(n) \tag{1.5}$$

This growth of mean worm size to a macroscopic length scale, accompanied by a tapering-off of the growth rate, appears to capture salient features of the observed self-organization process in sand dunes and other landforms. Given the importance of against-the-odds processes such as mergers, one would like (for self-consistency or completeness) to be able to examine more closely the regime of validity of the mean field approximation and to evaluate the importance of fluctuations and correlations. In order to do this, we must go beyond the mean-field and single-worm distribution approximations in which worms are assumed to be followed by "average" worms of length $\langle n \rangle$, with joint probability

$$P(n, n') = P(n)\delta_{n',\langle n \rangle} \tag{1.6}$$

Instead, we examine the full statistical evolution problem (1.2), a closure problem which has been considered intractable until now.

2. Generating function solution. We are motivated here by exact closure methods developed recently for *continuous* variable problems, e.g., the problem of determining joint probability distributions for *(i)* particle velocities in an interacting gas (Shen 1994a), *(ii)* the streamfunction in rotating stratified flow (ibid., Shen 1993), or for *(iii)* the fast and slow mode amplitudes in a Fourier-truncated model such as the Lorenz model (Shen 1994b). Accordingly, we define a function of dummy variables a, b:

$$Q(a,b) \equiv \sum_{n,n'=1}^{N} e^{i(-an+bn')} \frac{P(n,n')}{n'} - ib \tag{2.1}$$

This is a moment-generating function in the sense that contributions from terms of order $(a)^0(b)^1$ and higher in the Taylor series expansion of the exponential reduce to moments of the worm length in the limit $a, b \to 0$. This generating function may be inverse Fourier-transformed to recover the joint distribution

$$\frac{P(n,n')}{n'} = \int_C da\, db\, e^{-i(-an+bn')} [Q(a,b) + ib] \tag{2.2}$$

where

$$\int_C da\, db \equiv \left(\frac{1}{2\pi}\right)^2 \int_0^{2\pi} da \int_0^{2\pi} db \tag{2.3}$$

The "ib" term in (2.1) is retained for probability normalization purposes; in particular, it allows us to write the normalization condition

$$1 = \sum_{n=1}^{N} P(n) = \sum_{n,n'=1}^{N} P(n,n') \tag{2.4a}$$

as

$$\frac{\partial}{\partial(ib)} Q(a,b)\Big|_{a=b=0} = 0 \tag{2.4b}$$

The contribution of the "ib" term to (2.2) vanishes because n and n' are nonzero. N is assumed to be large enough that, to a good approximation,

$$\sum_{n=1}^{N} e^{i(a-a')n} \approx \delta(a-a') \tag{2.4c}$$

such that (2.2) will satisfy (2.1), as well as (2.1) satisfying (2.2).

In terms of Q, the reduction condition may be written, using the Kronecker delta function $\delta(b,0)$ to liberate the second index n', as

$$\begin{aligned} P(n) &= \sum_{n'=1}^{N} P(n,n') \\ &= \int_C da\, db\, e^{-i(-an+bn')} \delta(b,0) \frac{\partial}{\partial(ib)} Q(a,b) \end{aligned} \tag{2.5a}$$

In addition, of course, we must have

$$\begin{aligned} P(n) &= \sum_{n'=1}^{N} P(n',n) \\ &= \int_C da\, db\, e^{-i(-an'+bn)} \delta(a,0) \frac{\partial}{\partial(ib)} Q(a,b) \end{aligned} \tag{2.5b}$$

Because of the Kronecker delta in (2.5ab), the "n'" in the exponents may be replaced by n in each equation. Hence (2.5ab) may be combined to obtain a more general form of the reduction condition, which we will use shortly,

$$\begin{aligned} P(n) &= \\ &\int_C da\, db\, e^{-i(-an+bn)} \left[\alpha\delta(a,0) + (1-\alpha)\delta(b,0)\right] \frac{\partial}{\partial(ib)} Q(a,b) \end{aligned} \tag{2.5c}$$

where the dimensionless constant α will turn out to parameterize the time scale for growth.

The first term in the evolution equation (1.2) is similar to the expression in the reduction condition (2.5b) (but without the dummy derivative) and is hence given by

$$\sum_{j=1}^{N} \frac{P(j, n-1)}{n-1} = \int_C da\, db\, e^{-i[-an'+b(n-1)]}\delta(a,0)Q(a,b) \tag{2.6a}$$

while the second term in the evolution equation follows by similar arguments

$$\sum_{j=1}^{N} \frac{P(n+1, j)}{j} = \int_C da\, db\, e^{-i[-a(n+1)+bn']}\delta(b,0)Q(a,b) \tag{2.6b}$$

Similarly, the remaining terms in the evolution equations become

$$-\sum_{j=1}^{N} \frac{P(j, n)}{n} = -\int_C da\, db\, e^{-i[-an'+bn]}\delta(a,0)Q(a,b) \tag{2.6c}$$

$$-\sum_{j=1}^{N} \frac{P(n, j)}{j} = -\int_C da\, db\, e^{-i[-an+bn']}\delta(b,0)Q(a,b) \tag{2.6d}$$

As before, because of the Kronecker delta in (2.6a-d), the "n'" in the exponents of those equations may be replaced by n. Then, combining (2.5c, 2.6a-d), the evolution equation may be written as

$$\begin{aligned} \int_C da\, db\, e^{-i(-an+bn)}\Big\{ &[\alpha\delta(a,0) + (1-\alpha)\delta(b,0)] \frac{\partial}{\partial t}\frac{\partial}{\partial(ib)} Q(a,b) \\ &- \Big[\left(e^{ib}-1\right)\delta(a,0) + \left(e^{ia}-1\right)\delta(b,0)\Big] Q(a,b)\Big\} = 0 \end{aligned} \tag{2.7}$$

This equation will be satisfied if the integrand vanishes, i.e., if

$$\begin{aligned} &[\alpha\delta(a,0) + (1-\alpha)\delta(b,0)] \frac{\partial}{\partial t}\frac{\partial}{\partial(ib)} Q(a,b) = \\ &\Big[\left(e^{ib}-1\right)\delta(a,0) + \left(e^{ia}-1\right)\delta(b,0)\Big] Q(a,b) \end{aligned} \tag{2.8}$$

Because the factors on the right-hand side multiplying $\delta(b,0)$ or $\delta(a,0)$ vanish when a or b vanish, respectively, a sufficient condition for (2.8) to be satisfied is that

$$\frac{1}{\beta}\frac{\partial}{\partial t}\frac{\partial}{\partial(ib)} Q(a,b) = \left[\frac{e^{ib}-1}{\alpha} + \frac{e^{ia}-1}{1-\alpha}\right] Q(a,b) \tag{2.9}$$

along the lines $a = 0$ and $b = 0$ in the ab plane, excluding the dummy origin $a = b = 0$, where (2.9) is automatically satisfied if normalization (2.4b) holds. Equation (2.9) will be satisfied if

$$\frac{1}{\beta}\frac{\partial}{\partial t}\frac{\partial}{\partial G}Q(a,b) = Q(a,b) \tag{2.10}$$

where

$$G(a,b) \equiv i\int_0^b db' \left[\frac{e^{ib'}-1}{\alpha} + \frac{e^{ia}-1}{1-\alpha}\right] + G_1(ia)$$
$$= \frac{e^{ib}-1}{\alpha} + ib\frac{e^{ia}-1}{1-\alpha} - \frac{ib}{\alpha} + G_1(ia) \tag{2.11}$$

where $G_1(ia)$ is a constant of integration, to be determined. The most general solution of (2.9) is then given by

$$Q(a,b) = -\int_0^{+\infty} dk\; kA^2(k)\sinh\left[\frac{G(a,b)}{k} + 2k\beta t\right] \tag{2.12}$$

where $A(0) = 0$. This solution has been chosen to satisfy the realizability condition that mean worm length is positive:

$$\langle n(t)\rangle = \frac{\partial^2}{\partial(-ia)\partial(ib)}Q(a,b)\Big|_{a=b=0} \tag{2.13a}$$

$$= \int_0^{+\infty} dk\; A^2(k)\cosh\left[\frac{G_1(0)}{k} + 2k\beta t\right] \tag{2.13b}$$

The correlation function between adjacent worms

$$\langle n(t)n'(t)\rangle = \frac{\partial^3}{\partial(-ia)\partial^2(ib)}Q(a,b)\Big|_{a=b=0} \tag{2.13c}$$

$$= G_1'(0)\int_0^{+\infty} dk\; \frac{A^2(k)}{k}\sinh\left[\frac{G_1(0)}{k} + 2k\beta t\right] \tag{2.13d}$$

will be positive if $G_1(0), G_1'(0)$ are both positive.

Note that the correlation function in general is nonzero and does not equal the square of the mean, affirming the importance of correlations between adjacent worms. Another way to see this is to note that, because $\partial Q(a,b)/\partial b$ is in general not symmetric under exchange of (a,b), the joint probability is in general asymmetric (1.3) under exchange of adjacent worms, as expected. This rules out solutions in which adjacent worms are statistically independent, i.e., in general we obtain

$$P(n,n') \neq P(n)P(n') \tag{2.14}$$

Normalization (2.4b) is satisfied by the solution (2.12) since, by (2.11),

$$\frac{\partial}{\partial(ib)}Q(a,b)\Big|_{a=b=0} \propto \frac{\partial}{\partial(ib)}G(a,b)\Big|_{a=b=0} = 0 \tag{2.15}$$

Unlike the mean field case, the mean worm length as a function of time is determined by the coefficients $A(k)$, which in turn depend upon the values of *all* moments of the length at $t=0$, as contained in the moment-generating function (2.12). Actually, time integration of the evolution equation (2.8) is only required along the lines $\{a=0, b=0\}$, so it is sufficient to match our solution (2.12) to given initial values along those lines. Explicitly,

$$Q(0,b,t=0) = \sum_{n,n'=1}^{N} e^{ibn'}\frac{P(n,n',t=0)}{n'} - ib \tag{2.16a}$$

$$= -\int_0^{+\infty} dk\; kA^2(k)\sinh\left[\frac{e^{ib}-1-ib+G_1(0)}{k}\right] \tag{2.16b}$$

which fixes $A(k)$, while

$$Q(a,0,t=0) = \sum_{n,n'=1}^{N} e^{-ian}\frac{P(n,n',t=0)}{n'} \tag{2.17a}$$

$$= -\int_0^{+\infty} dk\; kA^2(k)\sinh\left[\frac{G_1(ia)}{k}\right] \tag{2.17b}$$

which fixes $G_1(ia)$. The inversion of (2.16ab, 2.17ab) to obtain the coefficients $A(k)$ for a given initial distribution $P(n,n')$ is under investigation. For a, b and k discrete with M values each, we see that (2.16ab) reduces to a system of M linear equations in M unknowns (the $A^2(k_j)$), while (2.17ab) reduces to M independent equations, each of which may be solved for its corresponding $G_1(a_j)$.

For the purposes of obtaining $\langle n(t)\rangle$, it would suffice to solve just (2.16ab). It is not clear at this point how to guarantee that the resulting $A^2(k)$ are positive, however. Alternatively, one may simply *choose* $A(k), G_1(a)$; mean worm length $\langle n(t)\rangle$ would then be determined from (2.13b), as well as the initial 1-worm probability (from Fourier inverting (2.16ab)) and the initial worm-shortening transition rates (from Fourier inverting (2.17ab)).

3. Mean field limit. For comparison purposes with earlier mean-field results (op cit.), let us substitute the mean field joint probability (1.6) into the definition of the generating function (2.1), yielding

$$Q_{MF}(a,b) = \frac{e^{ib\langle n\rangle}}{\langle n\rangle} \sum_{n=1}^{N} e^{-ian} P(n) - ib \tag{3.1}$$

Then the evolution equation (2.9) simplifies to

$$\frac{1}{\beta}\frac{\partial}{\partial t}\Big\{\langle n\rangle[Q_{MF}(a,b)+ib]\Big\} = \left[\frac{e^{ib}-1}{\alpha} + \frac{e^{ia}-1}{1-\alpha}\right] Q_{MF}(a,b) \tag{3.2}$$

This may be integrated to obtain the implicit solution

$$\begin{aligned} \langle n(t)\rangle &= \langle n(0)\rangle \cdot \\ &\cdot \frac{Q_{MF}(a,b,0)+ib}{Q_{MF}(a,b,t)+ib} \exp\left[\beta\left(\frac{e^{ib}-1}{\alpha} + \frac{e^{ia}-1}{1-\alpha}\right)\int_0^t \frac{dt'}{\langle n(t')\rangle}\right] \\ &- \frac{ib}{Q_{MF}(a,b,t)+ib}\left[\frac{e^{ib}-1}{\alpha} + \frac{e^{ia}-1}{1-\alpha}\right]\beta t \end{aligned} \tag{3.3}$$

At the dummy origin, the secular term vanishes and this solution reduces to

$$\langle n(t)\rangle = \langle n(0)\rangle \frac{Q_{MF}(0,0,0)}{Q_{MF}(0,0,t)} \tag{3.4}$$

which, however, is merely a tautology (using (3.1)) and does not impart any new information about time dependence.

In order to recover information about dynamics, we allow a to have an infinitesimal imaginary part $i\epsilon$; the generating function (2.1) and inverse Fourier transform (2.2) will remain well-defined for finite (but large) N as long as

$$\epsilon \ll \frac{1}{N} \tag{3.5}$$

Then the generating function (3.1) becomes

$$Q_{MF}(i\epsilon, 0) = \frac{1}{\langle n\rangle}\left(1 + \epsilon\langle n\rangle + \frac{\epsilon^2}{2}\langle n^2\rangle + \cdots\right) \tag{3.6}$$

and the mean field equation of motion (3.2) becomes

$$\frac{\partial}{\partial t}\left\{\epsilon\langle n\rangle + \frac{\epsilon^2}{2}\langle n^2\rangle\right\} = \tag{3.7}$$
$$\frac{\beta}{1-\alpha}\left(-\epsilon + \frac{\epsilon^2}{2}\right)\frac{1}{\langle n\rangle}\left(1 + \epsilon\langle n\rangle + \frac{\epsilon^2}{2}\langle n^2\rangle\right)$$

To lowest order in ϵ, this becomes

$$\epsilon\frac{\partial\langle n\rangle}{\partial t} = \frac{\epsilon}{\alpha - 1}\frac{\beta}{\langle n\rangle} \tag{3.8}$$

which may readily be integrated, yielding

$$\langle n\rangle = \left[\langle n(0)\rangle^2 + \frac{2\beta}{\alpha - 1}t\right]^{1/2} \tag{3.9}$$

Note that this solution is independent of ϵ as desired. This solution is expected to remain valid for times such that $\epsilon\langle n\rangle \ll 1$, or, using (3.5), for times up until approximately

$$t_{\text{max}} \equiv (\alpha - 1)\frac{N^2 - \langle n(0)\rangle^2}{2\beta} \tag{3.10}$$

The constant α is chosen to be greater than 1 and is seen to parameterize the time scale for growth. (Imposing realizability allows us to discard the case of $\alpha < 1$.) This time scale is of order $(1/N^2)$ of the time scale t_{max} for worm "longevity". Hence there should be a time regime on the order of N^2 growth times during which fully-developed structures with essentially stabilized size should be evident.

4. Growth of fluctuations. In order to check the self-consistency of the mean field approximation, we examine the next higher-order terms in the generating function (3.6) and evolution equations (3.7). (Note that fluctuations are not contained in $Q_{MF}(0, b)$), so that introducing an infinitesimal imaginary part into b, rather than a, would not provide the systematic self-consistent prescription for the evolution of higher moments which we obtain here.) The $o(\epsilon^2)$ terms in (3.7) yield

$$\frac{\partial}{\partial t}\langle n^2\rangle = \frac{2\beta}{\alpha - 1}\left(1 - \frac{1}{2\langle n\rangle}\right) \tag{4.1}$$

The right hand side represents the driving of the fluctuations by the mean evolution and is already known from the mean field solution (3.9). Substituting yields

(4.2) $$\langle n^2(t)\rangle = \frac{2\beta}{\alpha-1}t - \langle n(t)\rangle + \langle n^2(0)\rangle + \langle n(0)\rangle$$

The realizability condition that

(4.3) $$\langle n^2(t)\rangle > 0$$

will be satisfied if initially

(4.4) $$\langle n^2(0)\rangle > 0$$

because

(4.5a) $$\langle n(0)\rangle \geq 1$$

which implies by (3.9) that

(4.5b) $$\langle n(t)\rangle > 1$$

which implies by (4.1) that $\langle n^2(t)\rangle$ increases with time. The further requirement that

(4.6) $$\sigma^2 \equiv \frac{\langle n^2\rangle - \langle n\rangle^2}{\langle n\rangle^2} > 0$$

may then be shown to be always satisfied if it is satisfied initially, since (4.2) may be written using (3.9) as

(4.7a) $$\langle n^2(t)\rangle - \langle n(t)\rangle^2 + \langle n(t)\rangle = \langle n^2(0)\rangle - \langle n(0)\rangle^2 + \langle n(0)\rangle$$

or

(4.7b) $$\langle n^2(t)\rangle - \langle n(t)\rangle^2 + \langle n(t)\rangle = \text{conserved}$$

This would at first appear to suggest that the variance decreases in time as the mean increases, thereby validating the mean field approximation at long times. In fact, however, the boundedness of the *initial* mean and variance causes the growth of the mean worm size to be fluctuation-limited, in the sense that the time-dependent variance cannot decrease below zero. Hence the mean field approximation must break down after some finite time T_b, given by the solution of

(4.8) $$\langle n(T_b)\rangle = \langle n^2(0)\rangle - \langle n(0)\rangle^2 + \langle n(0)\rangle$$

Initially-large fluctuations, as might be expected for sand ripples growing from an initially uniform state, would also invalidate the mean field approximation (1.6). In either case, the mean field solution (3.9, 4.2) would have to be supplanted by the more general solution (2.13b, 2.13d).

5. Discussion. Using functional methods, we have extracted non-mean field results for the stochastic merging model of Werner et al., previously thought to be an intractable problem. We have taken into account *(i)* correlation and statistical dependence between adjacent worms, and *(ii)* fluctuations in worm length, and shown both effects, neglected in the mean field approximation, to be potentially significant ((1.3, 2.13a-d, 2.14) and (4.8), respectively). Resulting predictions for growth of mean worm length are qualitatively not inconsistent with data but require more detailed comparison. Initial worm distributions and subsequent realizable evolution may be generated which satisfy the full master equations, but inversion of arbitrary initial conditions in a manner which guarantees realizability remains to be clarified. Extension to modelling self-organization in other geomorphological structures is under investigation.

Acknowledgements.
The author would like to thank Ted Shepherd for support from the Natural Sciences and Engineering Research Council and the Atmospheric Environment Service of Canada.

REFERENCES

[1] R.S. Anderson 1990 Earth Sci. Rev. **29**, 77.
[2] S. Forrest and P.K. Haff 1992 Science **255**, 1240.
[3] B.T. Werner and D.T. Gillespie 1993 Phys. Rev. Lett. **71**, 3230.
[4] B.T. Werner 1992, in *Proceedings of the Geophysical Fluid Dynamics Summer School* (Woods Hole, Massachusetts), p. 167.
[5] H.H. Shen 1993, in *Nonlinear Waves and Weak Turbulence, with Applications to Oceanography and Condensed Matter Physics*, ed. N. Fitzmaurice, D. Gurarie, F. McCaughan, W. Woyczynski, Birkhäuser, Boston.
[6] H.H. Shen 1994ab, submitted.

ASYMPTOTICS OF SOLUTIONS OF BURGERS' EQUATION WITH RANDOM PIECEWISE CONSTANT DATA

DONATAS SURGAILIS*

Abstract. We discuss large-time asymptotics of a suitably rescaled solution of Burgers' equation with a random stationary piecewice constant initial potential, in the situations when (1) the viscosity $\mu > 0$ is constant, or (2) $\mu \sim 1/t \to 0$. In the first case, the rescaled solution of Burgers' equation converges in distribution to a random solution of the heat equation with a stable noise initial data, or the logarithmic derivative of the latter, depending on the distribution of the random heights and random constancy intervals of the potential. In the second case, the asymptotics is determined by high fluctuations of the potential described by a Poisson statistics, and the limit (velocity) field consists of pure shock waves traveling with random speed and coalescing at collisions. This extends the recent result of Molchanov, Surgailis, Woyczynski [MSW95], obtained for a Gaussian initial potential.

Key words. Burgers'equation, statistical solutions, renewal model, scaling limits, stable distribution, Poisson convergence of extremes

1. Introduction and the main results. The investigation of various asymptotics of random (statistical) solutions of Burgers' equation

$$u_t + uu_x = \frac{1}{2}\mu u_{xx} \tag{1.1}$$

has gained considerable attention in recent years; see e.g. Albeverio, Molchanov, Surgailis [AMS94]; Gurbatov, Malakhov, Saichev [GMaSa91]; Avellaneda, E. [AvE93]; Sinai [Si92] and the references herein. In the simplest case, the randomness of $u(t, x)$ is due to a strictly stationary random initial potential $\xi(x) = -\int^x u(0, y)dy, x \in \mathbf{R}$, and one investigates the large-time behavior of the rescaled solution

$$u_T(t, x) = B_T u(T^2 t, Tx)$$

as $T \to \infty$, assuming the viscosity parameter $\mu > 0$ constant. Rigorous results on the convergence of $u_T(t, x)$ in this context were obtained for several models of $\xi(x)$, including the Gaussian model and the shot-noise model (Bulinskii, Molchanov [BuM91], Funaki, Surgailis, Woyczynski [FSW95] etc.).

In this paper, we consider the so-called *renewal model*, which is a piecewise constant process

$$\xi(x) = \xi_j, \quad x_j \le x < x_{j+1} \tag{1.2}$$

taking random constant value ξ_j on random interval $[x_j, x_{j+1})$, $j \in \mathbf{Z}$. We assume that $\xi_j, j \in \mathbf{Z}$ and the interval lengths $\Delta x_j = x_{j+1} - x_j, j \in \mathbf{Z}$

* Institute of Mathematics and Informatics, Lithuanian Academy of Sciences, ul. Akademijos 4, Vilnius, 2600, Lithuania.

are all independent, the distribution functions $P[\xi_j \le a]$, $P[\Delta x_j \le b]$ do not depend on j, and the average interval length $E\Delta x_j = 1$. We also assume that the renewal process $\{x_j\}$ is *stationary*, which guarantees the stationarity of $\xi(x)$ (1.2). In particular, the marginal distribution function $P[\xi(x) \le a] = P[\xi(0) \le a] = P[\xi_0 \le a]$, for any $x \in \mathbf{R}$.

Sect. 2-4 deal with the case of a constant viscosity ($\mu = 1$). In particular, we are interested in the situation when the second exponential moment $Ee^{2\xi_0}$ and/or the second moment $E\Delta x_0^2$ are infinite but the corresponding tail distributions decay hyperbolically at infinity. In such a case, the limit $lim_{T\to\infty} u_T(t,x)$ is not Gaussian, as one usually expects, but is described by a $\alpha-$ stable distribution ($0 < \alpha < 2$). The exact form of the limit depends on the asymptotics of the tail probabilities. The results can be summarized in the following theorem, where we assume $\mu = 1$. Put $\eta_i = e^{\xi_i}$,

$$p(t,x,y) = (2\pi\mu t)^{-1/2} \exp\{-(x-y)^2/2\mu t\}.$$

Write $f(a) \sim g(a)$ if $\lim f(a)/g(a) \in (0,\infty)$ exists, and denote $\Rightarrow$ the weak convergence of (finite dimensional) probability distributions.

THEOREM 1.1. (i). *Let* $E\eta_0^2 < \infty$, $E\Delta x_0^2 < \infty$. *Then*

$$T^{3/2} u(T^2 t, Tx) \Rightarrow \text{const.} \int_{\mathbf{R}} p_x(t,x,y) Z_2(dy). \tag{1.3}$$

(ii). *Let* $P[\eta_0 > a] \sim a^{-\alpha}$ $(a \to \infty)$, $E\Delta x_0^\beta < \infty$, *and* $1 < \alpha < 2$, $\alpha < \beta$. *Then*

$$T^{2-1/\alpha} u(T^2 t, Tx) \Rightarrow \text{const.} \int_{\mathbf{R}} p_x(t,x,y) Z_\alpha(dy). \tag{1.4}$$

(iii). *Let* $E\eta_0^\alpha < \infty$, $P[\Delta x_0 > b] \sim b^{-\beta}$ $(b \to \infty)$, *and* $1 < \beta < 2$, $\beta < \alpha$. *Then*

$$T^{2-1/\beta} u(T^2 t, Tx) \Rightarrow \text{const.} \int_{\mathbf{R}} p_x(t,x,y) Z_\beta(dy). \tag{1.5}$$

(iv). *Let* $P[\eta_0 > a] \sim a^{-\alpha}$ $(a \to \infty)$, $E\Delta x_0^\beta < \infty$, *and* $0 < \alpha < 1$, $\beta > 1$. *Then*

$$Tu(T^2, Tx) \Rightarrow \text{const.} \left(\log \int_{\mathbf{R}} p(t,x,y) Z_\alpha(dy) \right)_x. \tag{1.6}$$

In (1.3)-(1.6), Z_α, $0 < \alpha \le 2$, is an $\alpha-$*stable noise*, see Sect. 3. (ii) - (iv) can be generalized in the sense that the power-law asymptotics of tail probabilities may include a slowly varying factor (see Lemma 3.1 below). The restriction $\beta > 1$, or $E\Delta x_0 < \infty$, is due to the fact that the last condition is necessary and sufficient in order that a stationary renewal process $\{x_j\}$ with the given distribution of interval lengths Δx_j exists (see e.g. Daley, Vere-Jones [DVJ88]).

The case of vanishing viscosity ("large Reynolds numbers") was considered, at the physical level of rigor, by a number of authors (we refer, in particular, to [GMaSa91] and Fournier, Frisch [FF83], which contain an extensive discussion of physical aspects of the problem, and of Burgers' turbulence in general). A rigorous approach can be found in Molchanov, Surgailis, Woyczynski [MSW95], who considered the case of a stationary Gaussian initial potential $\xi(x)$, with zero mean and variance 1. There, it was proved that, under general conditions on the covariance of the Gaussian process and with $\mu^{-1} = T^2(2\log T)^{1/2} \to \infty$, one has the convergence

$$T(\log T)^{1/2} u(T^2(2\log T)^{1/2}t, Tx) \Rightarrow v(t,x) \tag{1.7}$$

to a limit (velocity) field $v(t,x)$, with "saw-tooth" trajectories, and defined via a Poisson process $\{(y_j, u_j)\}$ in the plane having intensity $e^{-u}dydu$, according to the formula

$$v(t,x) = (x - y_{j^*(t,x)})/t. \tag{1.8}$$

$(y_{j^*(t,x)}, u_{j^*(t,x)})$ is the point of the Poisson process which maximizes the difference $u_j - (x-y_j)^2/2t$, i.e.

$$u_{j^*(t,x)} - (x - y_{j^*(t,x)})^2/2t = \max_j \left[u_j - (x-y_j)^2/2t\right]. \tag{1.9}$$

Here, we extend this result to the renewal potential $\xi(x)$ of (1.2). Similarly as in [MSW95], the study is based on the existence of the (Poisson) limit of high fluctuations of the potential, under a suitable scaling. This imposes certain conditions on the distribution function $F(a) := P[\xi(0) \le a]$ as $a \to +\infty$, which are well-studied in the case of weakly dependent stationary processes with discrete time (Leadbetter, Lindgren, Rootzén [LLR83]).

In Theorem 1.2, we assume that $H(a) = 1 - F(a)$ is continuous, strictly monotone and strictly positive for all sufficiently large a. Its inverse $H^{-1}(u)$ is well-defined for sufficiently small $u > 0$ and $H^{-1}(0+) = +\infty$.

THEOREM 1.2. *Let the following conditions be satisfied.*

(j). *There are normalizing constants $A_T = H^{-1}(1/T)$ and $B_T > 0$ such that for any $a \in \mathbf{R}$ there exists the limit*

$$\lim_{T\to\infty} TH(A_T + a/B_T) = G(a) \in [0, +\infty]. \tag{1.10}$$

(jj). *For any $a_0 > G_- := \inf\{a \in \mathbf{R} : G(u) < \infty\}$, there exist $T_0 > 0$ and a continuous monotone function $\bar{G}_0(a)$ on $[a_0, \infty)$ such that for all $a \ge a_0$*

$$\sup_{T > T_0} TH(A_T + a/B_T) \le \bar{G}_0(a), \tag{1.11}$$

and

$$(1.12) \qquad \int_{\mathbf{R}} \bar{G}_0(a_0 + y^2)dy < \infty.$$

$$\text{(jjj).} \qquad E \log_+(1/\Delta x_0) < \infty.$$

Then, with $\mu = (T^2 B_T)^{-1} \to 0$, *one has the convergence*

$$(1.13) \qquad T B_T u(T^2 B_T t, Tx) \Rightarrow v(t, x).$$

The limit random field $v(t, x)$ *is defined by (1.8), (1.9), with the difference that here* $\{(y_j, u_j)\}$ *is a Poisson process on* $\mathbf{R} \times (G_-, +\infty)$, *having intensity measure* $-dydG(u)$.

Condition (j) implies that $F(a)$ belongs to the domain of attraction of a *max-stable* distribution ([LLR83], [BGT87]). Recall that a distribution function $F(a)$ is said asymptotically max-stable if for any $a \in \mathbf{R}$ there exists the (nontrivial) limit

$$\lim_{n \to \infty} F^n(A_n + a/B_n) = e^{-G(a)},$$

where A_n, $B_n > 0$ are some normalizing constants. The limit function $e^{-G(a)}$ has one of the three well-known parametric forms (Type I, II or III extreme value distributions) [LLR83]. In particular, under the assumptions of Theorem 1.2, there are only two possibilities: either

$$(1.14) \qquad G(a) = e^{-ca},$$

$a \in \mathbf{R}$, with $c > 0$ (Type I distribution), or

$$(1.15) \qquad G(a) = 1/(1 + ca)^\gamma,$$

if $a > -1/c, G(a) = +\infty$, if $a < -1/c$ (Type II distribution), where $c > 0$ and $\gamma > 1/2$, according to condition (jj). Moreover, in the latter case, $H(a)$ is necessarily regularly varying at infinity with the exponent $-\gamma$ [LLR83]. The class of asymptotically max-stable distribution functions attracted to a Type I distribution, includes many familiar distributions such as normal, exponential, fractional exponential etc. For example, $H(a) = (2\pi)^{-1/2} \int_a^\infty e^{-y^2/2} dy$ satisfies (j)-(jj) with $A_T = (2 \log T)^{1/2} - \frac{1}{2}(2 \log T)^{-1/2}(\log \log T + \log 4\pi)$, $B_T = (2 \log T)^{1/2}$, $G(a) = e^{-a}$, $G_- = -\infty$, and $\bar{G}_0(a) = 2e^{-a}$, $T_0 = e^{2((-a_0)_+)^2}$; $x_+ := \max(x, 0)$. For the exponential distribution $H(a) = e^{-a}$, one has $A_T = \log T$, $B_T = 1$, $G(a) = e^{-a}$, $G_- = -\infty$, and $\bar{G}_0(a) = e^{-a}$, $T_0 = e^{(-a_0)_+}$. An open question remains whether Theorem 1.2 can be extended to Type III asymptotically max-stable distributions, corresponding to *bounded* random variables ξ_j. Note that, under conditions of Theorem 1.2, $Ee^{\alpha\xi(0)}$ may be infinite for any $\alpha > 0$. Condition (jjj) guarantees that high fluctuations of $\xi(x)$ are not very narrow.

Similarly as most other studies of Burgers' equation, the present work relies heavily on the Hopf-Cole formula

$$u(t,x) = -\mu \frac{\int_{\mathbf{R}} p_x(t,x,y)e^{\xi(y)/\mu}dy}{\int_{\mathbf{R}} p(t,x,y)e^{\xi(y)/\mu}dy} = -\mu\left(\log\int_{\mathbf{R}} p(t,x,y)e^{\xi(y)/\mu}dy\right)_x \tag{1.16}$$

for the solution of Burgers' equation (1.1), where $\xi(x)$ is the the initial potential. In particular, the rescaled solution (corresponding to $\mu = 1$) can be written as

$$u_T(t,x) \equiv B_T u(T^2t, Tx) = -\frac{B_T\langle\eta(T\cdot), p_x(t,x,\cdot)\rangle}{T\langle\eta(T\cdot), p(t,x,\cdot)\rangle}, \tag{1.17}$$

which reduces the proof of Theorem 1.1 to the study of the (scaling) limits of $\langle\eta(T\cdot),\phi\rangle$ and of $\langle\eta(T\cdot) - E\eta_0,\phi\rangle$, $\phi \in \mathcal{S}$ (= the Schwartz space of test functions), given in Sect. 2-4 below; c.f. Surgailis, Woyczynski [SW93]. Sect. 5 contains the proof of Theorem 1.2.

2. Stable and related probability distributions. Let us recall some known facts about stable distributions and their domains of attraction. Write $A_1(T) \simeq A_2(T)\,(T\to\infty)$ if $A_1(T) = A_2(T)h(T)$, where $h(T)$ varies slowly as $T\to\infty$. A random variable Z is said α–stable ($0<\alpha<2$) if its characteristic function

$$Ee^{i\theta Z} = \exp\{im\theta - p|\theta|^\alpha(1 + iq\omega(\theta,\alpha))\}, \quad \theta\in\mathbf{R}, \tag{2.1}$$

where $m \in \mathbf{R}$, $p > 0$, $-1 \le q \le 1$ are parameters, and $\omega(\theta,\alpha) = \tan(\pi\alpha/2)$ if $\alpha \neq 1, = (2/\pi)\log|\theta|$ if $\alpha = 1$. Introduce the family $\mathcal{D}(\alpha)$ $(0<\alpha<2)$ of random variables ζ such that there exist a function $h(a) = h(a;\zeta), a > 0$, slowly varying as $a\to\infty$, and two constants $c_\pm = c_\pm(\zeta) \ge 0$, $i = 1,2$, $c_+ + c_- > 0$, such that

$$P[\zeta > a] = (c_+ + o(1))h(a)/a^\alpha, \tag{2.2}$$

$$P[\zeta < -a] = (c_- + o(1))h(a)/a^\alpha, \tag{2.3}$$

$a > 0$. $\mathcal{D}(\alpha)$ is the domain of attraction of the α–stable law with parameters $m = 0$ and

$$p = \rho(\alpha)(c_+ + c_-), \qquad q = (c_+ - c_-)/(c_+ + c_-), \tag{2.4}$$

where $\rho(\alpha) > 0$ depends only on α, see e.g. Ibragimov, Linnik [IL71]. In other words, if $\zeta \in \mathcal{D}(\alpha)$ and $\zeta_1, \zeta_2, ...$ are independent copies of ζ, then for any $a\in\mathbf{R}$

$$\lim_{N\to\infty} P[(\zeta_1 + ... + \zeta_N - A_N)/C_N \le a] = P[Z\le a],$$

where Z is an $\alpha-$stable variable with the parameters (2.4), and $A_N = A_N(\zeta), C_N = C_N(\zeta) \simeq T^{1/\alpha}$ are some normalizing constants. In particular, C_N can be chosen from the relation

$$h(C_N)C_N^{-\alpha} = (p + o(1))/N. \tag{2.5}$$

In terms of characteristic functions, $\zeta \in \mathcal{D}(\alpha)$ is equivalent to the representation

$$Ee^{i\theta\zeta} = \exp\{im\theta - p|\theta|^\alpha h(1/\theta)(1 + iq\frac{\theta}{|\theta|}\omega(\theta, \alpha)(1 + o(1))\} \tag{2.6}$$

where $m \in \mathbf{R}$ is arbitrary, while p, q are related by (2.4) to c_+, c_- in (2.2), (2.3).

Write $\zeta \in \mathcal{E}(\alpha)\,(\alpha > 0)$ if $E|\zeta|^\alpha < \infty$. Put $\mathcal{D}_+(\alpha) = \{\zeta \in \mathcal{D}(\alpha) : \zeta > 0\}, \mathcal{E}_+(\alpha) = \{\zeta \in \mathcal{E}(\alpha) : \zeta > 0\}$. Using the definition of $\mathcal{D}_+(\alpha)$ and simple properties of slowly varying functions, one can easily verify the following facts which we use in the proof of the convergence (1.3)-(1.6). Let $\zeta_1 \in \mathcal{D}_+(\alpha_1)$, $\zeta_2 \in \mathcal{E}_+(\alpha_2), 0 < \alpha_1 < \alpha_2 < 1$. Then $\zeta := \zeta_1\zeta_2 \in \mathcal{D}_+(\alpha_1)$, with $h(a;\zeta) = h(a;\zeta_1)$ and $c_+(\zeta) = c_+(\zeta_1)E\zeta_2^{\alpha_1}$. Similarly, in the case $1 < \alpha_1 < \alpha_2 < 2$, the product $\zeta := (\zeta_1 - E\zeta_1)\zeta_2 \in \mathcal{D}(\alpha_1)$, with $h(a;\zeta) = h(a;\zeta_1)$ and $c_+(\zeta) = c_+(\zeta_1)E\zeta_2^{\alpha_2}$, $c_-(\zeta) = 0$. Finally, if $\zeta_1 \in \mathcal{E}_+(\alpha_1)$, $\zeta_2 \in \mathcal{D}_+(\alpha_2)$ and $1 < \alpha_2 < \alpha_1 < 2$, then $\zeta := (\zeta_1 - E\zeta_1)\zeta_2 \in \mathcal{D}(\alpha_2)$, with $h(a;\zeta) = h(a;\zeta_2)$ and $c_+(\zeta) = c_+(\zeta_2)E(\zeta_1 - E\zeta_1)_+^{\alpha_2}$, $c_-(\zeta) = c_+(\zeta_2)E(E\zeta_1 - \zeta_1)_+^{\alpha_2}$.

3. Convergence to stable stochastic integrals. An $\alpha-$stable noise Z_α $(0 < \alpha < 2)$ is a generalized process with independent values ([GV64]), whose characteristic functional is given by

$$Ee^{i\langle Z_\alpha,\phi\rangle} = \exp\left\{im\langle 1,\phi\rangle - p\int_{\mathbf{R}} |\phi(y)|^\alpha(1+iq\frac{\phi(y)}{|\phi(y)|}\omega(\phi(y),\alpha))dy\right\}, \phi \in \mathcal{S}, \tag{3.1}$$

where m, p, q are the same as in (2.1), and $\mathcal{S}$ is the Schwartz space of rapidly decreasing $C^\infty-$functions. In what follows, we consider the case $m = 0$ only. It is well-known that $\langle Z_\alpha, \phi\rangle$ can be extended to arbitrary $\phi \in L^\alpha$, which is usually written as the stochastic integral $\int \phi(y)Z_\alpha(dy) \equiv \int \phi dZ_\alpha$ [KwW92]. For $0 < \alpha < 1$, $q = 1$ the random measure $Z_\alpha(dy) \geq 0$. Also, let us introduce a Gaussian white noise Z_2 having the characteristic functional

$$Ee^{i\langle Z_2,\phi\rangle} = \exp\left\{-\frac{1}{2}\int_{\mathbf{R}} \phi^2(y)dy\right\}, \quad \phi \in L^2.$$

LEMMA 3.1. *Let $\eta(x) = e^{\xi(x)}$, where $\xi(x)$ is the stationary process of (1.2), and let $\phi \in \mathcal{S}$ be any test function.*

(i). *Let* $\eta_0 \in \mathcal{E}_+(2)$, $\Delta x_0 \in \mathcal{E}_+(2)$; $\sigma^2 := E(\eta_0 - E\eta_0)^2 E\Delta x_0^2$. *Then*

$$T^{1/2}\langle \eta(T\cdot) - E\eta_0, \phi\rangle \Rightarrow \sigma\langle Z_2, \phi\rangle, \tag{3.2}$$

$$\langle \eta(T\cdot), \phi\rangle \Rightarrow E\eta_0\langle 1, \phi\rangle. \tag{3.3}$$

(ii). *Let* $\eta_0 \in \mathcal{D}_+(\alpha)$, $\Delta x_0 \in \mathcal{E}_+(\beta)$, *and* $1 < \alpha < \beta < 2$. *Then*

$$TC_T^{-1}\langle \eta(T\cdot) - E\eta_0, \phi\rangle \Rightarrow \langle Z_\alpha, \phi\rangle, \tag{3.4}$$

$$\langle \eta(T\cdot), \phi\rangle \Rightarrow E\eta_0\langle 1, \phi\rangle, \tag{3.5}$$

where $C_T = C_T(\zeta) \simeq T^{1/\alpha}$ *are the normalizing constants corresponding to* $\zeta := (\eta_0 - E\eta_0)\Delta x_0 \in \mathcal{D}(\alpha)$, *and* Z_α *is a completely antisymmetric* $\alpha-$*stable noise with parameters* $p, q\,(= 1)$ *given by (2.4), with* $c_+ = c_+(\zeta) = c_+(\eta_0)E\Delta x_0^\alpha$, $c_- = c_-(\zeta) = 0$.

(iii). *Let* $\eta_0 \in \mathcal{E}_+(\alpha)$, $\Delta x_0 \in \mathcal{D}_+(\beta)$ *and* $1 < \beta < \alpha < 2$. *Then*

$$TC_T^{-1}\langle \eta(T\cdot) - E\eta_0, \phi\rangle \Rightarrow \langle Z_\beta, \phi\rangle, \tag{3.6}$$

$$\langle \eta(T\cdot), \phi\rangle \Rightarrow E\eta_0\langle 1, \phi\rangle, \tag{3.7}$$

where $C_T = C_T(\zeta) \simeq T^{1/\beta}$ *are the normalizing constants corresponding to* $\zeta := (\eta_0 - E\eta_0)\Delta x_0 \in \mathcal{D}(\beta)$, *and* Z_β *is a* $\beta-$*stable noise with parameters* $p, q(< 1)$ *given by (2.4), with* $c_+ = c_+(\zeta) = c_+(\Delta x_0)E(\eta_0 - E\eta_0)_+^\beta > 0$, $c_- = c_-(\zeta) = c_+(\Delta x_0)E(E\eta_0 - \eta_0)_+^\beta > 0$.

(iv). *Let* $\eta_0 \in \mathcal{D}_+(\alpha)$, $\Delta x_0 \in \mathcal{E}_+(\beta)$ *and* $0 < \alpha < 1, \beta > 1$. *Then*

$$TC_T^{-1}\langle \eta(T\cdot), \phi\rangle \Rightarrow \langle Z_\alpha, \phi\rangle, \tag{3.8}$$

where $C_T = C_T(\zeta) \simeq T^{1/\alpha}$ *are the normalizing constants corresponding to* $\zeta := \eta_0\Delta x_0 \in \mathcal{D}_+(\alpha)$, *and* Z_α *is a* $\alpha-$*stable noise with parameters* $p, q\,(= 1)$ *given by (2.4), with* $c_+ = c_+(\zeta) = c_+(\eta_0)E\Delta x_0^\alpha$, $c_- = c_-(\zeta) = 0$.

The proof of Lemma 3.1 uses an approximation of the corresponding integrals by sums of independent random variables, and a standard probabilistic argument of characteristic functions. The details being tedious and basically similar in all four cases (i)-(iv), we present the proof of (iii) only.

4. Proof of Lemma 3.1 (iii). Write U_T for the left hand side of (3.6). Then

$$U_T = C_T^{-1}\sum_j \bar{\eta}_j \Delta x_j \phi(\tilde{x}_{j,T}),$$

where $\bar{\eta}_j := \eta_j - E\eta_j$, and $\tilde{x}_{j,T} \in (x_j/T, x_{j+1}/T)$. For $0 < K < \infty$, define

$$U_{T,K}^- := C_T^{-1}\sum_{|j| \le KT} \bar{\eta}_j \Delta x_j \phi(\tilde{x}_{j,T}),$$

$$U^+_{T,K} := C_T^{-1} \sum_{|j|>KT} \bar\eta_j \Delta x_j \phi(\tilde x_{j,T}),$$
$$V^-_{T,K} := C_T^{-1} \sum_{|j|\le KT} \bar\eta_j \Delta x_j \phi(j/T),$$
$$V^+_{T,K} := C_T^{-1} \sum_{|j|>KT} \bar\eta_j \Delta x_j \phi(j/T).$$

To prove (3.6), it suffices to show that, uniformly in $T > 1$,

$$U^+_{T,K} = o_P(1), \tag{4.1}$$
$$V^+_{T,K} = o_P(1) \tag{4.2}$$

as $K \to \infty$, and that, for any $K < \infty$ fixed,

$$U^-_{T,K} - V^-_{T,K} = o_P(1), \tag{4.3}$$
$$V^-_{T,K} \Rightarrow \int_{-K}^{K} \phi(y) Z_\beta(dy) \tag{4.4}$$

as $T \to \infty$.

To prove (4.1)-(4.4), we use the independence of the sequences $\{\eta_j\}$ and $\{x_j\}$ together with the asymptotics of the characteristic function of $\Delta x_j \in \mathcal{D}_+(\beta)$ (see (2.6)), and the fact that $\bar\eta_j \in \mathcal{E}(\alpha)\,(\alpha > 1)$, $E\bar\eta_j = 0$ imply

$$\log E e^{i\theta\bar\eta_j} = O(|\theta|^\alpha) \quad (\theta \to 0). \tag{4.5}$$

Moreover, we use the law of large numbers for the renewal sequence $\{x_j\}$, namely

$$x_j/j \to 1 \quad (|j| \to \infty) \qquad \text{a.s.} \tag{4.6}$$

Let us show (4.1) ((4.2) is similar). By (4.6), $|\tilde x_{j,T}| > \frac{1}{2}|j|/T$ hence $|\phi(\tilde x_{j,T})| <$ const. $|\phi(j/2T)|$ for all sufficiently large $|j|$, and any $T > 1$. Note that

$$\sup_{|j|>KT} C_T^{-1} \Delta x_j |\phi(j/2T)| = o_P(1) \tag{4.7}$$

as $K \to \infty$ uniformly in $T > 1$. Indeed, using the definition of C_T, see (2.5), for any $\epsilon > 0$ one obtains

$$P[\sup_{|j|>TK} C_T^{-1} \Delta x_j |\phi(j/2T)| > \epsilon]$$
$$\le \sum_{|j|>TK} P[\Delta x_j > \epsilon\, C_T/|\phi(j/2T)|]$$

$$= O\Big(C_T^{-\beta} \sum_{|j|>TK} h(\epsilon C_T/|\phi(j/2T)|; \Delta x_0)|\phi(j/2T)|^\beta\Big)$$

$$= O\left(T^{-1} \sum_{|j|>TK} |\phi(j/2T)|^\beta\right)$$

$$= O\left(\int_{|y|>K/2} |\phi(y)|^\beta \, dy\right) = o(1)$$

as $K \to \infty$, uniformly in $T > 1$, which proves (4.7).

By (4.5), (4.7), for any $\theta \in \mathbf{R}$,

$$Ee^{i\theta U^+_{T,K}} = EE[e^{i\theta U^+_{T,K}}|\{x_j\}]$$

$$= E\exp\left\{-O\left(|\theta/C_T|^\alpha \sum_{|j|>KT} \Delta x_j^\alpha |\phi(j/2T)|^\alpha\right)\right\} + o(1)$$

$$= \exp\left\{-O\left(|\theta/C_T|^\beta \sum_{|j|>KT} h(|\theta\phi(j/2T)/C_T; \Delta x_0)|\phi(j/2T)|^\beta\right)\right\} + o(1)$$

$$= \exp\left\{-O\left(|\theta|^\beta \int_{|y|>K/2} |\phi(y)|^\beta dy\right)\right\} + o(1) = 1 + o(1)$$

(4.8)

as $K \to \infty$, uniformly in $T > 1$, where we have used the fact that the random variables $\Delta x_j^\alpha \in \mathcal{D}_+(\beta/\alpha)$, and and the corresponding asymptotics of their characteristic function. This proves (4.1).

Consider (4.3). It suffices to show that for any $\theta \in \mathbf{R}$

$$Ee^{i\theta(U^-_{T,K} - V^-_{T,K})} \to 1 \quad (T \to \infty). \tag{4.9}$$

Note first that for any $K < \infty$

$$\sup_{|j|\le TK} C_T^{-1} \Delta x_j = O_P(1) \quad (T \to \infty), \tag{4.10}$$

as the left hand side converges in distribution to a nondegenerate max-stable random variable ([LLR83], [BGT87]). As $\sup_{T>1} |\phi(\tilde{x}_{j,T}) - \phi(j/T)|$-$\to 0\,(|j| \to \infty)$ a.s. because of (4.6) and the uniform continuity of $\phi \in \mathcal{S}$, from (4.10) one obtains

$$\sup_{|j|\le TK} C_T^{-1} \Delta x_j |\phi(\tilde{x}_{j,T} - \phi(j/T)| = o_P(1) \quad (T \to \infty).$$

Therefore, using similar argument as in (4.8),

$$Ee^{i\theta(U^-_{T,K} - V^-_{T,K})} = EE[e^{i\theta(U^-_{T,K} - V^-_{T,K})}|\{x_j\}]$$

$$= E \exp\left\{ -O \left(|\theta/C_T|^\alpha \sum_{|j|\le TK} \Delta x_j^\alpha |\phi(\tilde{x}_{j,T}) - \phi(j/T)|^\alpha \right) \right\} + o(1)$$

$$= E \exp\left\{ -o \left(|\theta/C_T|^\alpha \sum_{|j|\le TK} \Delta x_j^\alpha \right) \right\} + o(1)$$

$$= \exp\{-|\theta|^\beta o(1)\} + o(1) = 1 + o(1) \quad (T \to \infty),$$

which proves (4.9) and hence (4.3).

Convergence (4.4) follows from a standard probabilistic argument, as $V^-_{T,K}$ is the (weighted) sum of *independent* random variables $\bar{\eta}_j \Delta x_j \in \mathcal{D}(\beta)$; see Feller [Fe66], or [IL71]. Finally, (3.7) is a simple consequence of the law of large numbers.

5. Proof of Theorem 1.2. Denote the left hand side of (1.13) by $v_T(t,x)$. Using the Hopf-Cole formula and the mean value theorem, it can be rewritten as

$$(5.1) \quad v_T(t,x) = -\frac{\sum_j \exp\{T^2(u_{j,T} - (y_{j,T} - x)^2/2t)\}\Delta x_j d_{j,T}(t,x)}{\sum_j \exp\{T^2(u_{j,T} - (y_{j,T} - x)^2/2t)\}\Delta x_j},$$

where

$$(5.2) \qquad u_{j,T} = B_T(\xi_j - A_T), \quad y_{j,T} \in (x_j/T, x_{j+1}/T),$$

$$(5.3) \qquad d_{j,T}(t,x) = d_{j,T}(t,x;x_j,x_{j+1}) = \frac{T}{t\Delta x_j} \int_{x_j/T}^{x_{j+1}/T} (x-y) e^{T^2((x-y_{j,T})^2 - (x-y)^2)/2t} dy.$$

Define $y_{j^*(T),T}$, $j^*(T) = j^*(T;t,x) \in \mathbf{Z}$ by

$$(5.4) \quad u_{j^*(T),T} - (x - y_{j^*(T),T})^2/2t = \max_j [u_{j,T} - (x - y_{j,T})^2/2t];$$

c.f. (1.9). Also, put

$$R_T(t,x) := \sum_{j \ne j^*(T)} \frac{e^{T^2(u_{j,T} - (x-y_{j,T})^2/2t)} \Delta x_j d_{j,T}(x)}{e^{T^2(u_{j^*(T),T} - (x-y_{j^*(T),T})^2/2t)} \Delta x_{j^*(T)}},$$

$$Q_T(t,x) := \sum_{j \ne j^*(T)} \frac{e^{T^2(u_{j,T} - (x-y_{j,T})^2/2t)} \Delta x_j}{e^{T^2(u_{j^*(T),T} - (x-y_{j^*(T),T})^2/2t)} \Delta x_{j^*(T)}},$$

$$\rho_T(t,x) := d_{j^*(T),T}(t,x) - (x - y_{j^*(T)})/t = \frac{1}{t} \frac{\int_{x_{j^*(T)}/T}^{x_{j^*(T)+1}/T} (y_{j^*(T),T} - y) e^{-T^2(x-y)^2/2t} dy}{\int_{x_{j^*(T)}/T}^{x_{j^*(T)+1}/T} e^{-T^2(x-y)^2/2t} dy}.$$

Then

$$v_T(t,x) = -\frac{(x - y_{j^*(T;t,x),T})/t + R_T(t,x) + \rho_T(t,x)}{1 + Q_T(t,x)}$$

and the theorem follows from

$$\begin{aligned} &(5.5) \qquad & y_{j^*(T;t,x),T} &\Rightarrow y_{j^*(t,x)}, \\ &(5.6) \qquad & R_T(t,x) &\Rightarrow 0, \\ &(5.7) \qquad & Q_T(t,x) &\Rightarrow 0, \\ &(5.8) \qquad & \rho_T(t,x) &\Rightarrow 0. \end{aligned}$$

The proof of (5.5)-(5.8) requires a study of the Poisson convergence of the point process $\nu_T := \{(y_{j,T}, u_{j,T})\}$ in a "parabolic" topology matched to the Hopf-Cole functional, similarly as in [MSW95]. We shall consider the convergence (5.5)-(5.8) for $t = 1, x = 0$ only as the general case requires immaterial changes, see [MSW95], and shall suppress these variables in the subsequent notation.

Let us introduce some terminology. Let $\mathcal{M}(\Gamma)$ be the space of all locally finite point measures on $\Gamma := (G_-, +\infty) \times \mathbf{R}$, with the topology of vague convergence of measures, denoted by $\rightarrow$ (see Kallenberg [Ka86]). Elements $\nu \in \mathcal{M}(\Gamma)$ can be identified with countable sequences $\nu = \{(y_j, u_j)\}_{j \in \mathbf{Z}}$, where $(y_j, u_j) \in \Gamma$.

For $p_1 > 0, p_2 \in \mathbf{R}$, consider the parabola

$$\partial A_{p_1,p_2} = \{(y,u) \in \mathbf{R}^2 : u - p_1 y^2 - p_2 = 0\}, \qquad (5.9)$$

which is the boundary of the "parabolic" set

$$A_{p_1,p_2} = \{(y,u) \in \mathbf{R}^2 : u - p_1 y^2 - p_2 > 0\} \qquad (5.10)$$

above the parabola. Let $\mathcal{A}(\Gamma)$ be the family of all parabolic sets which lie inside Γ; $\mathcal{A}(\Gamma) = \{A_{p_1,p_2} : p_1 > 0, G_- < p_2 < \infty\}$.

Consider the subspace $\mathcal{P}(\Gamma) \subset \mathcal{M}(\Gamma)$ consisting of all measures $\nu \in \mathcal{M}(\Gamma)$ such that $\nu(A) < \infty$ for any $A \in \mathcal{A}(\Gamma)$. Let $\nu_T, \nu \in \mathcal{P}(\Gamma)$. We say that $\nu_T \xrightarrow{\mathcal{P}} \nu$ $(T \to \infty)$ if $\nu_T \to \nu$ (in the vague topology), and, moreover,

$$\nu_T(A) \to \nu(A)$$

for any parabolic set $A \in \mathcal{A}(\Gamma)$ such that $\nu(\partial A) = 0$. The topology in $\mathcal{P}(\Gamma)$ corresponding to the convergence $\xrightarrow{\mathcal{P}}$ will be called the $\mathcal{P}-$topology; it is closely related to the $\mathcal{B}-$ topology introduced in [MSW95]. It is easy to verify that $\nu_T \xrightarrow{\mathcal{P}} \nu$ is equivalent to the statement that for any $A \in \mathcal{A}(\Gamma)$ and any Borel set $B \subset A$ such that $\nu(\partial B) = 0$, the equality $\nu_T(B) = \nu(B)$ holds for all sufficiently large T, and, moreover,

$$\operatorname{dist}(\nu_T \cap B, \nu \cap B) \to 0 \quad (T \to \infty). \qquad (5.11)$$

Denote $\mathbf{P}(\mathcal{M}(\Gamma)), \mathbf{P}(\mathcal{P}(\Gamma))$ the family of probability measures on $\mathcal{M}(\Gamma)$, $\mathcal{P}(\Gamma)$, respectively. Write $\Rightarrow$ and $\stackrel{\mathcal{P}}{\Rightarrow}$ for the weak convergence of probability measures on, or random elements in, $\mathcal{M}(\Gamma)$ and $\mathcal{P}(\Gamma)$, respectively. A characterization of the latter convergence is provided in Lemma 5.1 below, whose proof uses standard probabilistic argument ([Ka86], [MSW95]).

LEMMA 5.1. *Let $P_T, P \in \mathbf{P}(\mathcal{P}(\Gamma))$. Then $P_T \stackrel{\mathcal{P}}{\Rightarrow} P$ if, and only if,*

$$P_T \Rightarrow P, \tag{5.12}$$

and, for any $A \in \mathcal{A}(\Gamma)$,

$$\lim_{k\to\infty} \sup_T P_T(\nu(A) > k) = 0. \tag{5.13}$$

LEMMA 5.2. *For any $T > 0$, the point process $\nu_T = \{(y_{j,T}, u_{j,T})\}$ and the Poisson process $\nu = \{(y_j, u_j)\}$ of Theorem 1.2 belong to $\mathcal{P}(\Gamma)$ a.s. and, as $T \to \infty$,*

$$\nu_T \stackrel{\mathcal{P}}{\Rightarrow} \nu. \tag{5.14}$$

Proof. The relation $\nu \in \mathcal{P}$ a.s. follows from $E\nu(A_{p_1,p_2}) = \int_{\mathbf{R}} G(p_1 y^2 + p_1)dy < \infty$ for any $A_{p_1,p_2} \in \mathcal{A}(\Gamma)$, see (1.10) - (1.12). In a similar way, one can verify the relation $\nu_T \in \mathcal{P}$, $T > 0$.

To prove the convergence (5.14), it suffices to verify (5.12) and (5.13) of Lemma 5.2. The proof of (5.12) is rather standard, making use of (1.10) together with the independence of $\xi_j, \Delta x_j$, and the law of large numbers of (4.6) (see also [LLR83], Th. 5.7.2).

Consider the tightness criterion (5.13). Similarly as in the proof of Lemma 3.1, for any $\epsilon > 0$ one can find $j_0 > 0$ such that $P[|x_j/j| > 1/2\,\forall |j| > j_0] \le 1 - \epsilon$. Let $A = A_{p_1,p_2} \in \mathcal{A}(\Gamma)$, $A^+ := A \cap \{(y,u) : u > 0\}$, $A^- := A \cap \{(y,u) : u \le 0\}$. Then

$$\begin{aligned} P[\nu_T(A) > k] &\le \epsilon + P[\nu_T(A^-) > k/2] \\ &\quad + P[\nu_T(A^+) > k/2, |x_j/j| > 1/2\,\forall |j| > j_0], \end{aligned} \tag{5.15}$$

where

$$\lim_{k\to\infty} \sup_T P[\nu_T(A^-) > k/2] = 0,$$

due to the convergence $\nu_T \Rightarrow \nu$, in the vague topology, and the fact that $A^- \subset \Gamma$ is compact.

Consider the last probability in (5.15), which we denote by $q(T,k)$. As

$$E[\nu_T(A^+)|\{x_j\}] = \sum_j H(A_T + (p_1 y_{j,T}^2 + p_2)_+/B_T),$$

$x_+ := \max(x, 0)$, from (1.11)-(1.12) we obtain

$$\begin{aligned} q(T,k) &= E[P[\nu_T(A^-) > k/2|\{x_j\}]; |x_j/j| > 1/2, |j| > j_0] \\ &\le (2/k)T^{-1}\sum_j \bar{G}_0((\frac{p_1}{4}(j/T)^2 + p_2)_+) + O(k^{-1}) \\ &\le \text{const.}\, k^{-1}\int_{\mathbf{R}} \bar{G}_0((y^2+p_2)_+)dy + O(k^{-1}) \end{aligned}$$

uniformly in $T > 0$ as $k \to \infty$. The last integral being finite, this proves (5.13) and Lemma 5.2, too.

As an easy corollary of Lemma 5.2, see [MSW95], Prop. 4.4, we obtain the convergence

$$(5.16) \qquad (y_{j^*(T),T}, u_{j^*(T),T}) \Rightarrow (y_{j^*}, u_{j^*}),$$

in particular, the convergence (5.5) and

$$(5.17) \qquad u_{j^*(T),T} - y^2_{j^*(T),T} \Rightarrow u_{j^*} - y^2_{j^*}.$$

To prove the theorem, it remains to show (5.6)-(5.8). Denote $A^K_{p_1,p_2} = A_{p_1,p_2}\cap\{(y,u) : |y| \le K\}$, $A^{K^c}_{p_1,p_2} = A_{p_1,p_2}\cap\{(y,u) : |y| > K\}$ $(K < \infty)$. Let $\Omega_{p_2,p_1,K,\delta,T}$ be the set of all $\omega \in \Omega$ satisfying the following four conditions:

$$(5.18) \qquad (y_{j^*(T),T}, u_{j^*(T),T}) \in A^K_{p_1,p_2},$$

$$(5.19) \qquad \{(y_{j,T}, u_{j,T})\} \cap A^{K^c}_{p_1/2,p_2} = \emptyset,$$

$$(5.20) \qquad u_{j,T} - u^2_{j,T} < u^2_{j^*(T),T} - y^2_{j^*(T),T} - 2\delta, \quad \text{for every} \quad j \ne j^*(T),$$

$$(5.21) \qquad \Delta x_{j^*(T)} > e^{-\delta T^2}.$$

We claim that for any $\epsilon > 0$ one can successively find $p_2 > G_-, p_1 \in (0, 1/2), K > 1, \delta \in (0, p_1/2)$ and $T_0 > 0$ such that, for all $T > T_0$,

$$(5.22) \qquad P[\Omega_{p_2,p_1,K,\delta,T}] > 1 - \epsilon.$$

Here, (5.18)-(5.20) follow from Lemma 5.2 (with (5.11) in mind), (5.16), (5.17), and from the corresponding properties of the limit Poisson process $\nu \equiv \{(y_j, u_j)\}$. (5.21) follows from condition (iii) of the theorem; indeed,

$$\begin{aligned} P[\Delta x_{j^*(T)} \le e^{-\delta T^2}] &\le P[\min_{|j|\le 2KT} \Delta x_j \le e^{-\delta T^2}] \\ &+ P[|y_{j^*(T),T})| > K \text{ or } \#\{j : |x_j| \le KT\} > 2KT] \\ &\le 2KTP[\Delta x_0 < e^{-\delta T^2}] + o(1) \\ &\le (2K/\delta T)E\log_+(1/\Delta x_0) + o(1) = o(1) \end{aligned}$$

provided one first chooses K large enough, and then $T \to \infty$.

By (5.22), it suffices to prove (5.6), (5.7) with R_T and Q_T replaced by $R'_T = R_T \mathbf{1}(\Omega_{p_2,p_1,K,\delta,T})$ and $Q'_T = Q_T \mathbf{1}(\Omega_{p_2,p_1,K,\delta,T})$, respectively. By (5.18)-(5.21), one obtains

$$\begin{aligned} Q'_T &\leq e^{-\delta T^2} \sum_{|x_j| \leq KT} \Delta x_j + e^{\delta T^2} \sum_{|x_j|>KT} e^{(\frac{p_1}{2}-1)x_j^2-(p_1-1)x_{j^*(T)}^2} \Delta x_j \\ &\leq O(Te^{-\delta T^2}) + \sum_{|x_j|>KT} e^{-qx_j^2} \Delta x_j, \end{aligned}$$

with $q = p_1/2 - \delta > 0$, and we have used the inequality $|x_j| > KT > |x_{j^*(T)}|$ in the second sum; see (5.18). The last sum in (5.23) vanishes a.s. as $T \to \infty$ in view of (4.6). The proof of (5.6) is similar.

Finally, (5.8) follows from $|\rho_T| \leq \Delta x_{j^*(T)}/T \Rightarrow 0$; indeed, for any $\delta > 0$,

$$P[\Delta x_{j^*(T)} > \delta T] < P[|j^*(T)| > KT] + P[\max_{|j| \leq KT} \Delta x_j > \delta T].$$

The first probability on the right hand side can be made arbitrary small by an appropriate choice of K, see above, while the second vanishes as $T \to \infty$ in view of the law of large numbers of (4.6). Theorem 1.2 is proved. □

REFERENCES

[AMS94] S. Albeverio, S.A. Molchanov, D. Surgailis, *Stratified structure of the Universe and Burgers' equation - a probabilistic approach*, Probab. Theory Rel. Fields, **100** 1994, pp. 457–484.

[AvE93] M. Avellaneda, *Statistical properties of shocks in Burgers' turbulence*, preprint, 1993.

[BGT87] N.H. Bingham, C.M. Goldie, J.L. Teugels, Regular Variation, Cambridge Univ. Press, Cambridge, 1987.

[BuM91] A.V. Bulinskii, S.A. Molchanov, *Asymptotical normality of a solution of Burgers' equation with random initial data*, Theory Probab. Appl., **36** 1991, pp. 217–235.

[DVJ89] D.J. Daley, D. Vere-Jones, An Introduction to the Theory of Point Processes, Springer-Verlag, New York, 1988.

[Fe66] W. Feller, An Introduction to Probability Theory and its Applications, vol. 2. Wiley & sons, New York, 1966.

[FF83] J.-D. Fournier, U. Frisch, *L'équation de Burgers déterministe et statistique*, J. Mec. Theor. Appl., **2** 1983, pp. 699–750.

[FSW95] T. Funaki, D. Surgailis, W.A. Woyczynski, *Gibbs-Cox random fields and Burgers' turbulence*, Ann. Appl. Probab., **5** 1995, pp. 461–492.

[GV64] I.M. Gelfand, N.Ja. Vilenkin, Generalized Functions vol. 4. Applications of Harmonic Analysis, Academic Press, New York, 1964.

[GMaSa91] S.N. Gurbatov, A.N. Malakhov, A.I. Saichev, Nonlinear Random Waves and Turbulence in Nondispersive Media: Waves, Rays and Particles, Manchester Univ. Press, Manchester-New York, 1991.

[IL71] I.A. Ibragimov, Yu.V. Linnik, *Independent and stationary sequences of random variables*, Walters-Noordhoff, Groningen, 1971.

[Ka86] O. Kallenberg, *Random measures*, Akademie-Verlag and Academic Press, Berlin-London, 1986.

[KwW92] S. Kwapien, W.A. Woyczynski, Random Series and Stochastic Integrals: Single and Multiple, Birkhäuser, Boston, 1992.

[LLR83] M.R. Leadbetter, G. Lindgren, H. Rootzén, *Extremes and Related Properties of Random Sequences and Processes.* Springer-Verlag, New York, 1983.

[MSW95] S.A. Molchanov, D. Surgailis, W.A. Woyczynski, *Hyperbolic asymptotics in Burgers' turbulence and extremal processes*, Commun. Math. Phys., **168** 1995, pp. 209–226.

[Si92] Ya.G. Sinai, *Statistics of shocks in solutions of inviscid Burgers' equation*, Commun. Math. Phys., **148** 1992, pp. 601–621.

[SW93] D. Surgailis, W.A. Woyczynski, *Long range prediction and scaling limit for statistical solutions of the Burgers' equation*, In: N. Fitzmaurice et al. (eds.), Nonlinear Waves and Weak Turbulence, with Applications to Oceanography and Condensed Matter Physics, Birkhäuser, Boston 1993, pp. 313–338.

MODELING THE SPATIOTEMPORAL DYNAMICS OF EARTHQUAKES WITH A CONSERVATIVE RANDOM POTENTIAL AND A VISCOUS FORCE

P.L. TAYLOR* AND B. LIN*†

Abstract. A new earthquake model of spatiotemporal dynamics is introduced. In contrast to many other earthquake models, this one does not contain a velocity-weakening frictional force, and is thus an alternative to the Burridge-Knopoff earthquake model. Dissipation in our model occurs only through viscous forces acting in the presence of a non-dissipative random potential. Both small localized and large delocalized events are observed. The scaling behavior of the event probability distribution is found to be non-universal and distinct from that found in other earthquake models. The system loses instability as the strength of the pulling spring becomes large enough. It also shows transitions from behavior exhibiting a wide range of magnitudes of slipping events to showing a narrow range scale in which only large events occur for a certain range of parameters. Effects of varying system size, boundary conditions and pulling speed were investigated.

1. Introduction. Since Burridge and Knopoff's block-and-spring earthquake model [1] was first suggested, there has been much discussion of its many variants [2,3,4,5] and of other earthquake models [6,7]. All these models display spatiotemporal dynamics of stick-slip motion in which slipping events have been observed to have event sizes spanning a wide range of magnitudes. In contrast to the situation in equilibrium statistical mechanics, special values of parameters are not necessary in order to observe wide-range critical fluctuations. This phenomenon of self-organized criticality [8], in which a large-range scale of slipping events occurs without special parameter values is an essential feature of the spatiotemporal dynamics of earthquakes. The stuck state corresponds to the quiet energy-accumulation period, and the slip state to earthquakes in which this energy is released. Scaling properties of the stick-slip dynamics of a version of the Burridge-Knopoff model have been studied extensively by Carlson, Langer and coworkers [2,3], who found numerically a universal power-law distribution for a wide range of magnitudes of slipping events. Stick-slip dynamics have also been observed in many other systems, including granite on granite at large normal forces [9], sandpaper on carpet [10], and rubber sheet on a glass rod [11]. The same large-range-scale critical fluctuations also exist in lattice and cellular automata models [8,12,13], and sandpile systems [14,15].

The most important feature responsible for the stick-slip dynamics in many mechanical models of blocks and springs, such as the Burridge-Knopoff model and its variants [1,2,3,4,5], is the presence of a frictional

* Department of Physics, Case Western Reserve University, Cleveland, OH 44106.

† Present Address: Department of Physics, University of Minnesota, Minneapolis, MN 55455.

force $F(\dot{x})$ whose strength becomes weaker as the velocity increases [2],

$$F(\dot{x}) = \frac{-F_0 \,\mathrm{Sgn}(\dot{x})}{1 + |\dot{x}|} \tag{1}$$

where $\dot{x}$ is the velocity of a block and F_0 is a constant. This non-linear friction function is the source of the stick-slip instability. Numerical calculations in Refs. [2] and [3] indicate that the event probability distributions obey certain scaling laws. For large dissipation, the probability $P(\mu)$ of an event of magnitude μ occurring has the form

$$P(\mu) \simeq \begin{cases} Ae^{-b\mu} & \mu < \mu_0 \\ A'e^{-b'\mu} & \mu > \mu_0 \end{cases} \tag{2}$$

with $b \approx 0.95$ and $b' \approx -1$, where A is a constant and μ_0 the crossover magnitude dividing the regions of small and large events. For small enough dissipation, the event probability roughly shows a single scaling

$$P(\mu) \simeq Be^{-b''\mu} \tag{3}$$

with $b'' \approx 0.35$ and B a constant.

Knopoff and coworkers [5] recently proposed and studied another type of block-spring earthquake model. In their model the velocity-weakening aspect of the frictional force was introduced through the presence of separate sticking and sliding coefficients of friction. The coefficient of sticking friction varied randomly from block to block, and a viscous force was also used.

We propose in this paper a new one-dimensional earthquake model and describe some theoretical and computer-simulation studies designed to demonstrate that velocity-weakening friction is not an essential ingredient of any mechanism describing earthquakes. We also study the scaling properties of magnitude distribution in earthquakes. The block-spring model studied in this paper is similar in some respects to Burridge-Knopoff and other previous earthquake models, but differs from them in its lack of any frictional force term in the equations of motion. Instead, a conservative random Gaussian potential acts on the blocks and mimics the hills and valleys of a rough surface. This non-dissipative force is supplemented by a viscous force that adds the necessary dissipation to the model. While the model has in common with Ref. [4] the presence of a stochastic element, it differs from it and from most previous models in the absence of any velocity-weakening frictional force. The viscous force, which is the only non-conservative element in our model, is velocity strengthening. We will calculate the displacement of blocks, their velocities, and the time distribution of events in order to study spatial and temporal dynamics, and study the event probability distribution in order to examine the scaling law. We will also analyze the dependence of the dynamics on system size, minimum event size, and pulling speed.

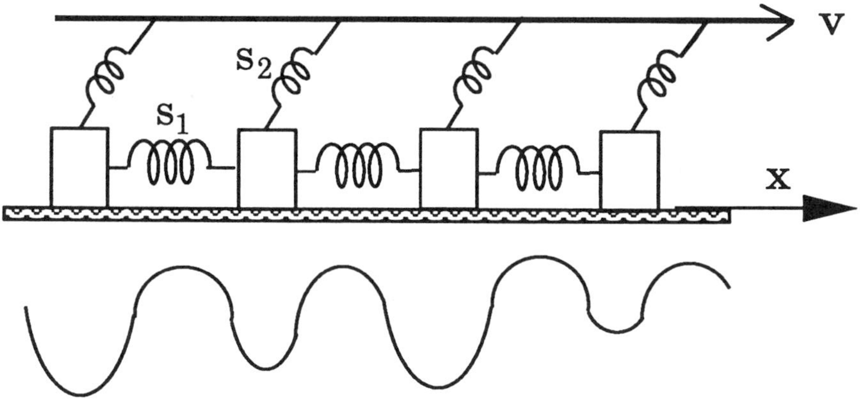

FIG. 1. *Illustration of the one-dimensional block-spring model.*

2. Description of the model. We study the one-dimensional model depicted schematically in Fig. 1. The N blocks of mass m are connected in a chain by Hookean springs of force constant s_1. Each block is attached to a rigid body moving at a constant velocity v (the pulling speed) via Hookean springs of force constant s_2. The blocks are pulled on a fixed rough surface whose interaction with the blocks is represented by a random Gaussian potential $G(x)$,

$$G(x) = -\frac{k}{2d}\sum_{l} R_l\, e^{-d(x-l)^2}, \tag{4}$$

where the l are integers, k is the depth parameter of the potential, and d is the width parameter, and the R_l random numbers between 0 and 1. We have taken the equilibrium spacing between two blocks to be unity, and chose d sufficiently large to make the results independent of its actual magnitude. The frictional force that was described in terms of a macroscopic coefficient of friction in other models is now replaced by the effects of the random potential $G(x)$, and so does not appear in our model. All energy dissipation then occurs solely through a viscous-force term of the same form as was present in the model of Burridge and Knopoff [1].

For this model system the equations of motion are

$$\begin{aligned} m\frac{d^2x_i}{dt^2} &= s_1(x_{i+1} - 2x_i + x_{i-1}) + s_2\Big[vt - (x_i - i)\Big] \\ &\quad -k\sum_{l=-\infty}^{+\infty} R_l(x_i - l)e^{-d(x_i-l)^2} - \gamma\frac{dx_i}{dt}, \qquad i = 1, 2, \cdots, N \end{aligned} \tag{5}$$

where γ describes the strength of the viscous force. The mass m will be taken as unity in our discussion throughout this paper. We investigate the effects of the five independent parameters in Eq. (5): s_1, s_2, v, k and γ. We shall assume an initial configuration of blocks of either a uniform distribution

$$x_i = i, \tag{6}$$

or a spatially non-uniform configuration with each block deviating randomly by a small amount form its equilibrium position, so that

$$x_i = i + (R_i' - 0.5)\delta. \tag{7}$$

Here R_i' is a random number in (0,1) and δ a small constant, typically 0.1.

The model defined by Eqs. (5) differs from that used by Carlson, Langer *et al.* [2,3] in that a random element is introduced into the equations of motion themselves, and does not arise in a translationally invariant system as a consequence of velocity-weakening friction. The random potential (4) provides an instability mechanism more closely in accord with microscopic tribological considerations: a detachment process occurs when some blocks overcome the potential barriers and start sliding. The linear velocity-dependent term $-\gamma\dot{x}_i$ allows energy dissipation of the kinetic energy that would accumulate as work is done by the pulling springs. A similar process has been conjectured to occur on a microscopic scale in the fracture of polymer welds[16].

3. Calculational procedure. Numerical solutions of the coupled differential equations (5) were obtained by use of the standard Runge-Kutta procedure. The system size studied was typically $N = 50$, although some larger variants with N up to 150 were also examined to verify that no qualitative change in behavior would be observed.

The system was always started completely at rest from the positions described by Eq. (6) or (7). Although the stick-slip motion is in general not periodic in time, big slipping events were found to be approximately periodic in time with period $1/v$, as might be expected from the translational invariance of the chain of blocks. The time steps used in the numerical procedure were 0.0002 of this fundamental time unit. We discarded the first 5×10^5 time steps (about 100 cycles of big events) in order to let the system reach a steady state, then collected data for about 10^7 time steps, or about 2200 loading cycles.

It is necessary to set a criterion for the smallest slipping that can be counted as an event. Since the event size is measured by block displacement, a minimum displacement was defined as the threshold for an event. The size $\Delta x_{\min}$ of this minimum displacement was set arbitrarily as the distance traveled at twice the pulling speed v in one time step Δt,

$$\Delta x_{\min} = 2v\Delta t. \tag{8}$$

As a check on the appropriateness of this criterion, two other Δx_{min} were also studied to see their effect on the scaling behavior of the event probability distribution.

Various event-related quantities will be calculated: event distributions in time, space, and size, and velocities of blocks in different events. An event can be defined in a two-dimensional lattice of position and time: if a block at a given (discrete) time has a displacement greater than Δx_{min}, that point is marked in the position-time space to have a slipping point. An event is then a cluster of all connected slipping points. The moment M of an event is defined by the relation [2],

$$M = \sum_i \delta x_i, \tag{9}$$

where i is summed over all blocks contained in the event. The magnitude μ of the event is then defined as

$$\mu = \ln M. \tag{10}$$

The event probability distribution $P(\mu)$ is the frequency of events per unit length of μ. We will normalize the number of events per unit μ interval by dividing by the total number of events in the entire run used to obtain $P(\mu)$.

4. Results of the numerical calculations.

A. Characterizing the spatiotemporal dynamics. Slipping events of various sizes and distributed over space and time were observed. Figure 2(a) shows the block displacements $x_i(t)$. The parameters used in this figure are $s_1 = 30$, $s_2 = 20$, $k = 350$, $d = 50$, $v = 0.015$ and $\gamma = 4.0$. We see both big events (large slipping involving a large number of blocks) and small events (events with small slipping and a small number of blocks), and the time spent in slipping is much less than the time spent in loading. One time unit in the figure is 67 time steps.

Figure 2(b) shows another view of the same spatiotemporal dynamics. Here we plot equal-time contour lines on the $\hat{x}_i(t)$-$x_i(t)$ plane, where the cumulative displacement $\hat{x}_i(t)$ is as defined in Ref. [11],

$$\hat{x}_i(t) = vt + x_i(t). \tag{11}$$

The areas where the lines are dense represent sticking, while the white areas where the lines are widely separated portray slipping events. Only some of the 50 blocks are shown in Fig. 2(a), as we have magnified part of the system to show more clearly the small slipping events.

The stick-slip dynamics of various event sizes can also be seen by plotting the velocities of the blocks in position and time for this same simulation, as shown in Fig 3. Relatively small velocities with a small number of

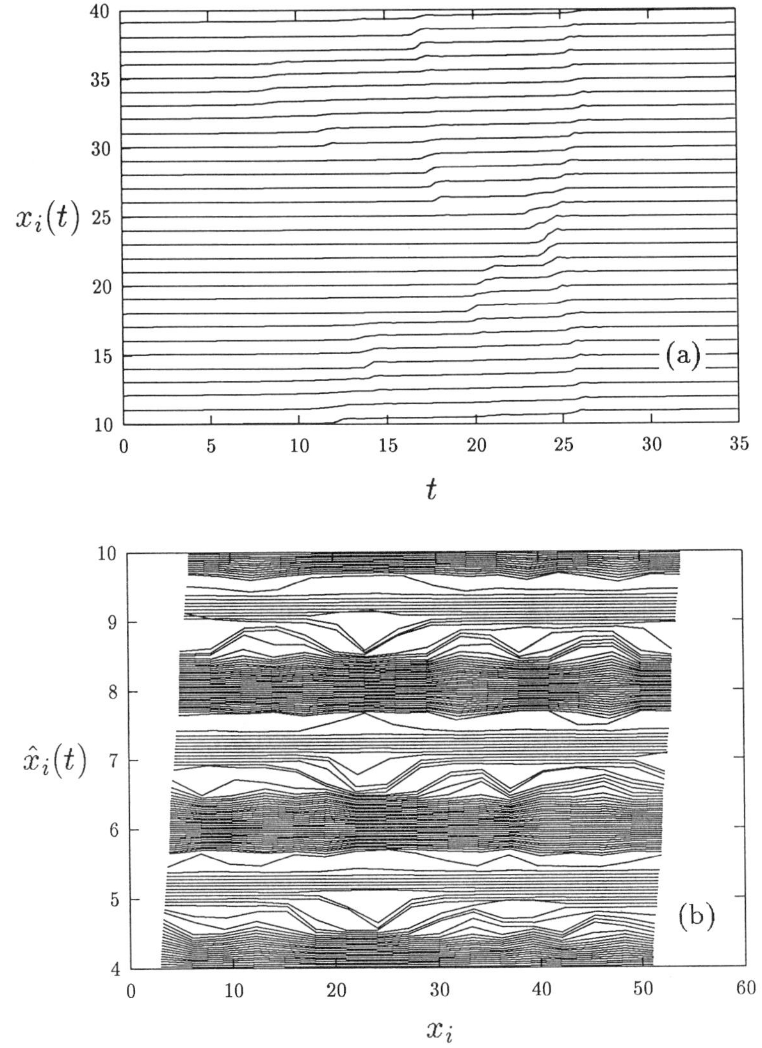

FIG. 2. *The model shows a wide range of slipping events. Shown here is a typical system with parameters* $s_1 = 30$, $s_2 = 20$, $k = 350$, $d = 50$, $v = 0.015$ *and* $\gamma = 4.0$. *(a) Displacement of blocks as a function of time, (b) Equal-time contour map showing the cumulative displacement* $\hat{x}_i(t)$ *as a function of position and time.*

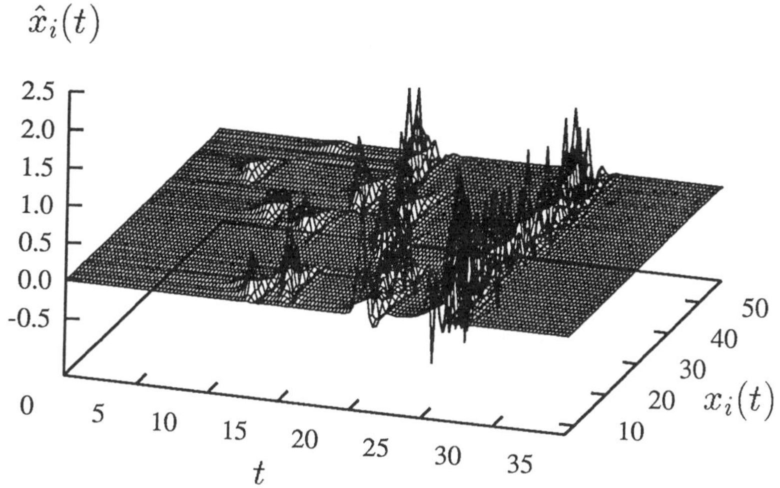

FIG. 3. *In this replotting of the data of Fig. 2, the velocities of blocks are shown as a function of position and time. Both localized events with small velocities and big events with large velocities are seen.*

blocks involved in each peak represent small localized slipping events. We note that the pulling speed is $v = 0.015$, and that small events have velocities of the order of v. Also apparent in Fig. 3 are big events involving a large number of blocks and having velocities greater than the pulling speed v by up to two orders of magnitude.

We can clearly see the time distribution of events in Fig. 4, which displays the size (magnitude μ) of events as a function of time. Even though we impose a minimum displacement for an event to be noted, the range of moments of slipping events span more than three orders of magnitude. Big events repeat roughly every time interval $1/v$, although the sizes are not the same in each cycle. Small events are less regularly distributed in time and magnitude in each cycle, but tend to group together in time in each cycle. There is a relatively long quiet period of time after big events, and this is shown as a gap between two clusters of events. Each cycle thus begins with quiet creeping motion, and then small events occur, quickly triggering big events.

B. Scaling behavior of the magnitude distribution. A large number of numerical runs were completed in order to study the scaling behavior of the magnitude probability distribution $P(\mu)$ of all events. A typical plot of $\ln P(\mu)$ vs. μ is shown in Fig. 5(a). The curve marked with filled circles in Fig. 5(a) corresponds to a system with relatively small dissipation, $\gamma = 3.0$. Other parameters for this distribution are $s_1 = 60$, $s_2 = 20$, $k = 250$, $d = 50$, $v = 0.015$. We refer this parameter set as the standard parameter set. For small events the distribution follows the

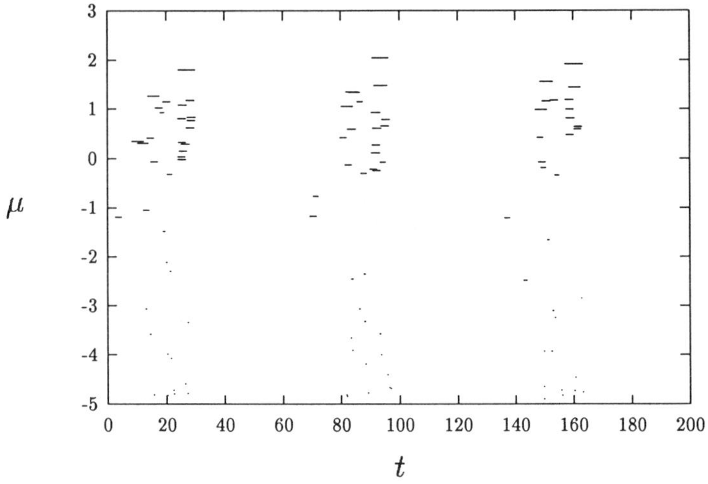

FIG. 4. *Time distribution of events. Slipping events form clusters in time. There is a long quiet creeping motion after big events and before the next event cluster.*

Gutenberg and Richter law [17]

$$P(\mu) = Ae^{-b\mu}, \tag{12}$$

where A is a constant and the scaling exponent b in this case is about 1, which agrees with the predictions of Refs. [2,3] and with seismological data [17]. When, however, the potential depth and the dissipation is increased to $k = 320$ and $\gamma = 5.0$, but with the other parameters unchanged, the probability distribution $P(\mu)$ yields a smaller exponent $b \approx 0.65$, as shown by the curve marked with open circles in Fig. 5(a). This contradicts the prediction of Ref. [3] that weaker dissipation decreases the exponent b. The effect of strong potential and large dissipation on slipping dynamics can also be seen in Fig. 6, the cumulative displacement, which shows large stuck (black) and slipping (white) areas and little small-slipping region. The flatter probability distribution in our model in the small-event regime as potential and dissipation are increased indicates that scattered small slipping events are depressed and less visible and the system's motion is dominated by larger, more concentrated slippings. We argue that our model's prediction of event magnitude distribution better agrees with earthquake data. Collected data from many earthquakes [18,19,20,21] have shown that the scaling exponent b of the surface magnitude distribution is about one, while this exponent decreases to lower values as the depth is increased at which the earthquake magnitude is measured. This decrease in the exponent has been attributed [18] to a decrease in degree of fracturing and an increase in residual stress. The potential-well depth parameter k in our model is responsible for the stress between blocks, and thus corresponds to the stress at earthquake faults. A larger k causes greater local

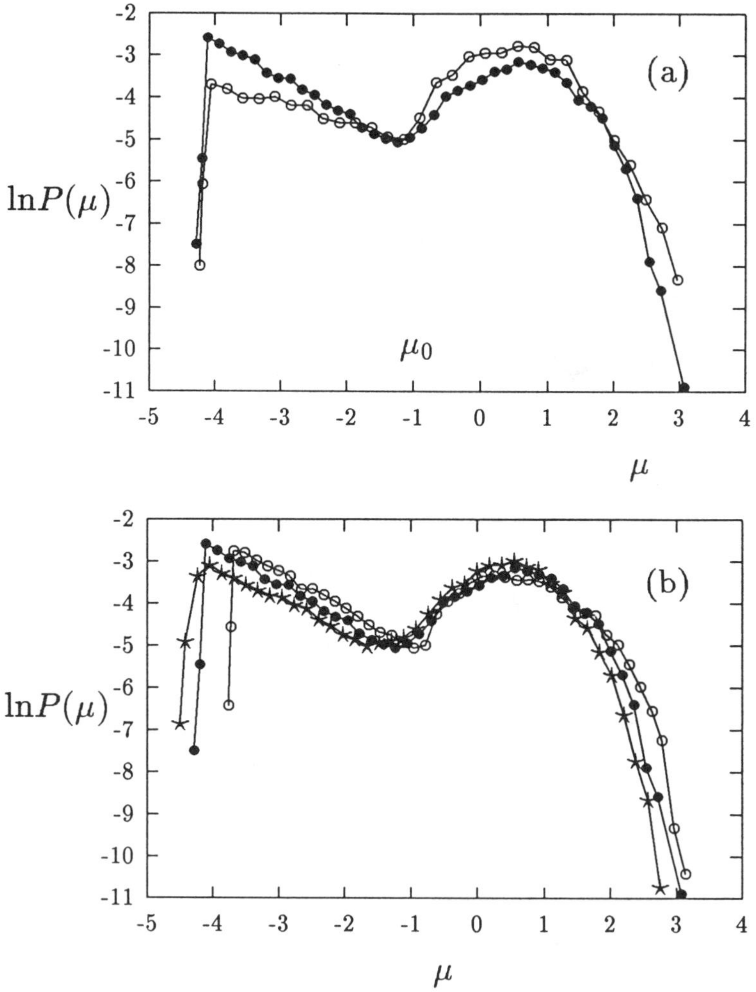

FIG. 5. *(a) Logarithm of event probability distribution as a function of magnitude μ. The stronger dissipation (curve with open circles) flattens the distribution in the small-event region. Compared are the system with standard parameter set, $s_1 = 60$, $s_2 = 20$, $k = 250$, $d = 50$, $v = 0.015$ and $\gamma = 3.0$ (filled circles), and the system with $k = 320$ and $\gamma = 5.0$ (open circles). (b) Scaling of distribution is not changed as the interblock spring parameter s_1 is varied. Unlike the case for the all other parameters, however, different s_1 changes crossover magnitude μ_0. All three curves have the same standard parameters except that $s_1 = 30$ for stars, $s_1 = 60$ for filled circles and $s_1 = 90$ for open circles.*

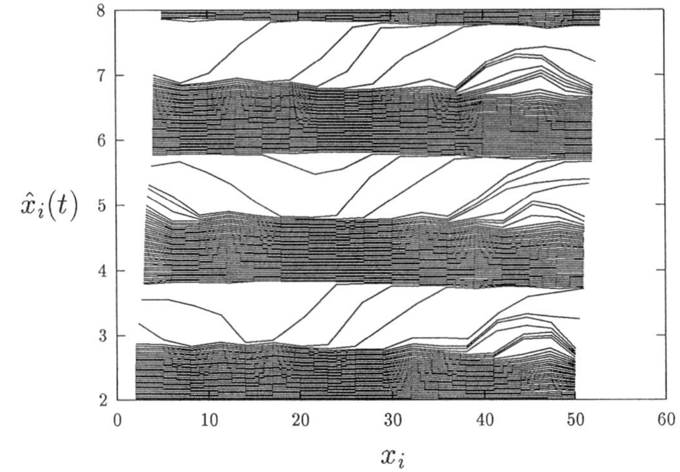

FIG. 6. *Cumulative displacement for a system with strong dissipation. Parameters are the same as for the flatter curve in Fig. 5(a). Few small events are visible, and the system is predominantly in either the stuck or the large-slipping state.*

and global stress and energy buildup before slipping events (earthquakes) occur. The parameter γ in our model describes the degree of dissipation in the system. As dissipation increases, fracturing of the system should decrease because there will be less energy transmitted.

We do observe in our calculations some scaling behavior that agrees with that found in other earthquake models. The scaling exponent b seems invariant under variation of the parameter s_1 that describes the interblock interaction. This seems to indicate that the magnitude-frequency distribution should obey the same exponential law for earthquakes occurring in different areas. This agrees with the seismological data from 1983-1984 earthquakes of Campi Flegrei, Italy [18]. Figure 5(b) shows the probability distributions for three different values of s_1, and they all appear to have the same scaling exponent b. Another common feature in scaling behavior is the crossover magnitude μ_0 (Eq. (2), Fig. 5(a)) which marks the crossover from the small-event region scaling to the large-event region scaling, and remains unchanged under variations of all parameters except s_1, in the regions we have studied. The probability distributions in Figs. 5(a), 9 and 10 have different parameters, but show the same crossover magnitude $\mu_0 \approx -1$. When the interblock interaction parameter s_1 is changed, however, μ_0 changes. As shown in Fig. 5(b), the probability distribution with small s_1 has a smaller crossover magnitude than that with larger s_1, and the distribution curve is shifted to the small-μ side when s_1 is decreased. A smaller s_1 thus has more small events but fewer large events than does a larger s_1 because the stiffer interblock springs eliminate smaller slipping events.

As a test of size effects in our model, we have also run calculations on systems with size $N = 100$ and $N = 150$. We obtained the same scaling exponent b for the parameters used in the smaller system with $N = 50$. Except for a greater total event number, no difference was found between the large and small systems. This indicates that the self-organized criticality in our model is not an artifact of the finite size. In particular, the rapid decrease in the probability distribution for events of magnitude $\mu > 3$ does not reflect any limitation imposed by the size of the system studied. For example, an event in which each block in a chain of 150 blocks moves by only one lattice spacing would have an event magnitude of $\mu > 5$. The almost complete absence of such events in our simulation is a further indication that size effects were not important.

C. Suppression of instabilities. In our model, the pulling springs s_2 are an abstraction of the effects of the shear modulus. Increasing s_2 reduces the relative effectiveness of cooperative motion in the chain. As a consequence of a more rapid response of the contact region to shear, a large enough s_2 will suppress the interblock interactions, and the system will be seen as individual blocks being pulled by separate springs. The solution to the system's equations of motion would then be simply $\{x_i\} = vt$, a stable creeping motion. A loss of instability was observed in our model calculations when the parameter s_2 is substantially increased. Figure 7 shows one such system. All the parameters are the same as those in Fig. 5(a) (i.e., the standard set of parameters) except that $s_2 = 100$. In this case, big events disappear, leaving the event distribution in a narrow range of event size (Figs. 7(a) and 7(b)). The prominent high-velocity large slips are replaced by a quiet creeping motion with a velocity of the order of the pulling speed (Fig. 7(c)).

Changing some of the other parameters of our model allows us to show a transition of dynamical motion to one without small events. This loss of small events can be induced by increasing either the potential or the viscosity. Large static friction (large k) prevents some small events from occurring, since the local stress for these blocks is not large enough to overcome the high potential barrier. A stronger dissipation (large γ) reduces the magnitude of small events and makes them undetectable (smaller than the minimum magnitude for an event). Figure 8 shows the event probability distribution $P(\mu)$ for two cases inside this regime where no small events occur. For large k (normal γ) or large γ (normal k) the distribution shows that only large events are formed. If both k and γ are increased the transition occurs at smaller values than needed for either k or γ alone since the two effects are cumulative.

A large-event cutoff can be seen in the event probability distribution $P(\mu)$. The cutoff value is mainly determined by the parameter k. A larger k results in a larger maximum event size, since more energy is then stored in the system before it can be released. The effect of k on the maximum

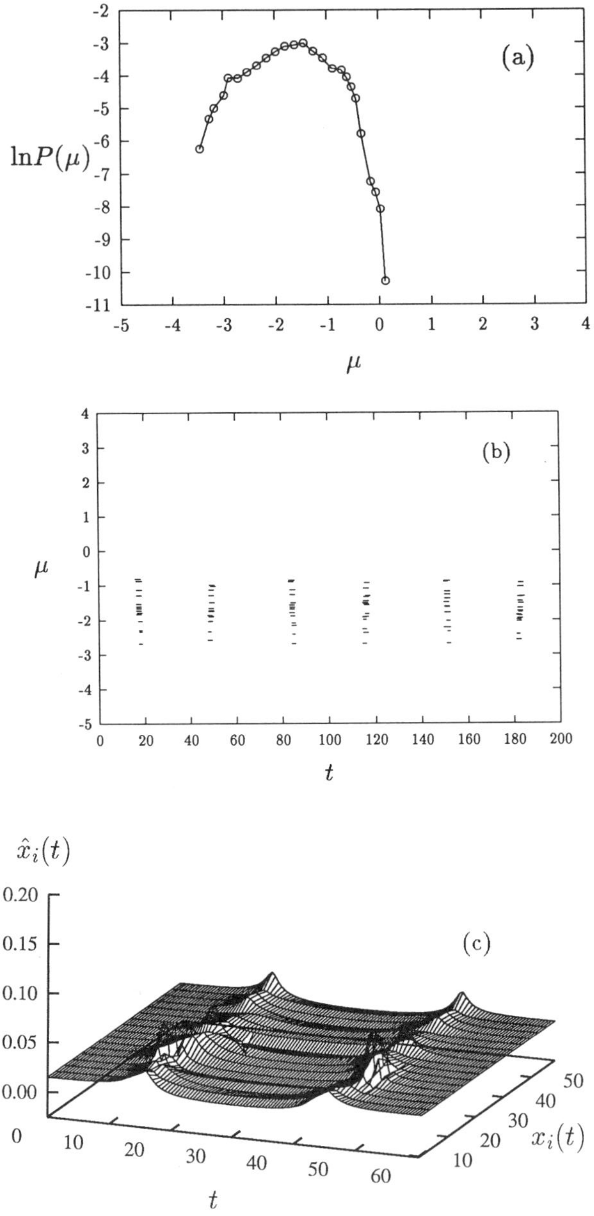

FIG. 7. *System loses instability when the parameter s_2 is sufficiently large. Essentially no events having $\mu > 0$ occur and the motion is dominated by small creeping events with velocities of the order of the pulling speed. Standard parameters except that $s_2 = 100$. (a) Probability distribution, (b) time distribution of events and (c) velocity as a function of position and time.*

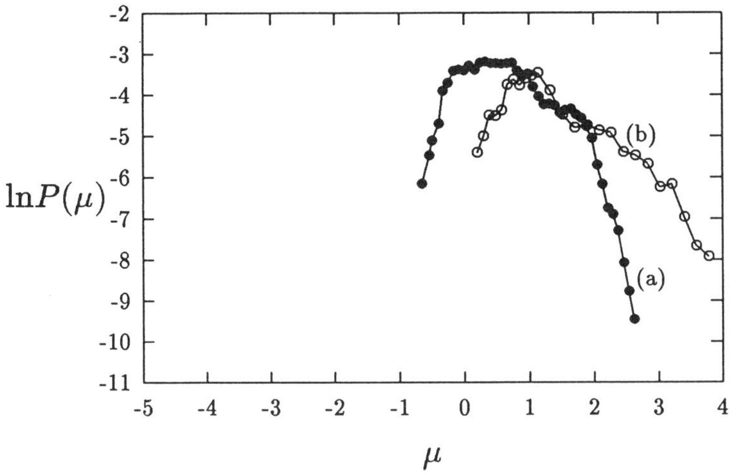

FIG. 8. *Transition of dynamical motion to the regime where only large-scale events are seen. This transition occurs when either (a) dissipation parameter γ or (b) potential depth parameter k is increased substantially. For curve (a) $\gamma = 7.0$ and (b) $k = 800$. Other parameters are standard.*

event size can be seen in Fig. 8. The maximum size is also affected by γ and by s_1, as can be seen in Figs. 8 and 5, because these parameters govern the total number of blocks involved in the largest events. As was discussed in section IV B, this cutoff is not an effect of the finite size of the system.

D. Effects of changing the minimum displacement and pulling speed. It is important in measuring slipping events to exclude measurement noise from recorded events, so we set the criterion that a minimum displacement $\Delta x_{\min}$ would constitute a measured event. Any displacement (in a given time interval) smaller than $\Delta x_{\min}$ was ignored. In all the preceding calculations $\Delta x_{\min}$ was as defined in Eq. (8). Any decrease in $\Delta x_{\min}$ is expected to extend the probability distribution curve to the small-magnitude end, since smaller events are included. To verify that variation of $\Delta x_{\min}$ will not change the essential features of the event magnitude distribution we plot the distributions corresponding to three different $\Delta x_{\min}$ in Fig. 9. All three distributions were calculated using the same standard parameter set but with $\Delta x_{\min} = 1.5v\Delta t$, $2.0v\Delta t$ and $2.5v\Delta t$ respectively. A smaller $\Delta x_{\min}$ threshold, as expected, does give a probability function having an extended distribution in the small-event region. The slight decrease in the whole distribution curve for a small $\Delta x_{\min}$ is due to the renormalization of the probability function $P(\mu)$ following the increase in the total number of events. The scaling behavior and the structure of $P(\mu)$ are not changed by the change of threshold value $\Delta x_{\min}$. The crossover magnitude μ_0 is also invariant under this change. These results confirm

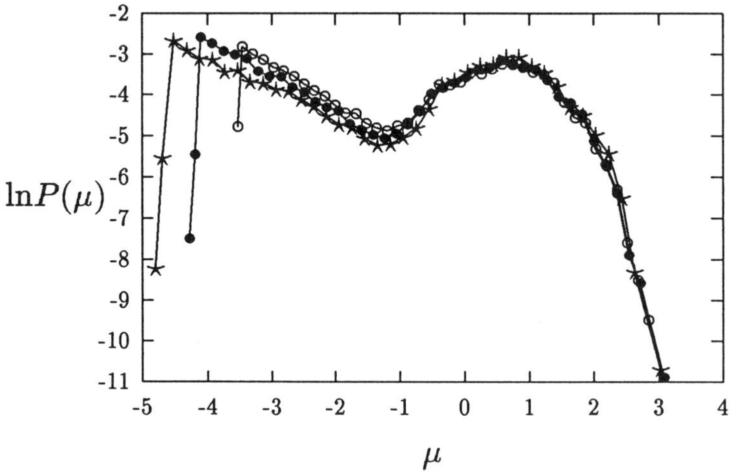

FIG. 9. *Varying the minimum displacement* $\Delta x_{\min}$ *does not change the system's scaling behavior. All distributions have the standard parameters, and* $\Delta x_{\min} = 1.5v\Delta t$ *for open-circle curve,* $2.0v\Delta t$ *for filled-circle curve and* $2.5v\Delta t$ *for star curve.*

that the threshold $\Delta x_{\min}$ is acting as a measurement control quantity, and not as a system parameter.

Changing the pulling speed has a very similar effect to that of changing $\Delta x_{\min}$. Figure 10 compares probability distributions for three different pulling speeds with all other parameters being standard. They have the same distribution pattern, scaling behavior and crossover magnitude as we have seen in the case of different thresholds $\Delta x_{\min}$. At lower pulling speeds, we see the distribution curve extend further into the small-magnitude end and a reduction in the small-event region. These reflect the fact that when the pulling speed is decreased smaller events are recorded, but also indicate that intermediate-size events increase in probability as the pulling speed is lowered. The crossover magnitude is again unchanged for different pulling speeds, as we can see from Fig. 10. While the pulling speed does not change the scaling behavior of magnitude-frequency distribution, it does change the length of the time period between major slipping events. Since the loading period is proportional to $1/v$, larger pulling speeds increase the frequency of major earthquake sequences.

E. Insensitivity to changes in boundary conditions and initial configuration. In systems in which long-range correlations occur it is important to verify that surface effects are not playing an unsuspected role in determining the system dynamics. In our model we used a chain of blocks and springs in which the ends were free. To see if these boundary conditions were causing any effects we performed calculations on systems with periodic boundary conditions for both $N = 50$ and $N = 100$. In contrast to the report in Ref. [2] that periodic boundary conditions tend

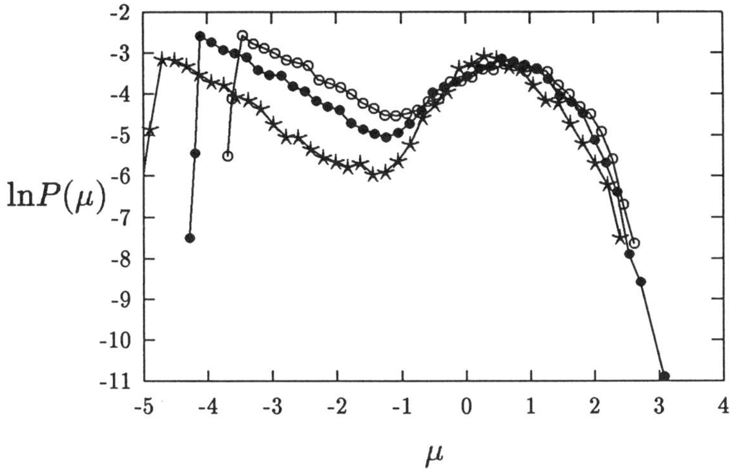

FIG. 10. *The effect of changing the pulling speed v is similar to that of changing the threshold $\Delta x_{\min}$. While a lower speed creates additional smaller events, the dynamics of the system are not changed. For the curve with stars $v = 0.005$, the curve with filled circles $v = 0.015$ and for the curve with open circles $v = 0.03$. All other parameters are standard.*

to eliminate irregular motion and produce smooth, periodic solutions, we found almost identical results from the two kinds of boundary condition.

For the earthquake models with translational invariance, an irregular initial configuration like Eq. (7) is needed to produce stochastic slipping events exhibiting a wide range of magnitudes; otherwise a uniform initial configuration with every block in its equilibrium position will result in periodic solutions that only slowly diverge into irregular motion. This is not the case in our model. The random-depth potential causes uneven motion among blocks, with blocks at weaker potentials moving while others remain trapped in their positions. The independence of the results on the initial configuration was confirmed by numerical calculations with uniform initial configurations as in Eq. (6). The insensitivity of our model to boundary conditions and initial configurations makes it a robust model for studying stick-slip phenomena in earthquakes.

5. Conclusions. In this paper we have proposed and studied a one-dimensional mechanical model for the spatiotemporal dynamics of earthquakes. In contrast with some block-spring earthquake models, the friction between the blocks and the surface is not of the velocity-weakening type. Instead a conservative random Gaussian potential and velocity-strengthening dissipation term are used to describe the static friction and viscosity. The model displays various slipping events with a wide range of magnitudes. These dynamics characteristic of self-organized criticality have been observed in our calculations for a wide range of model parameters.

Although the instability mechanism of our model is totally different from the conventional velocity-weakening friction, the scaling behavior in our model agrees qualitatively with seismological data. This suggests that the dynamic origin of earthquakes could be other than velocity-weakening friction. Our model thus proposes an alternative way of studying earthquakes and self-organized spatiotemporal dynamics. A loss of instability in the system is predicted by the model when the shear modulus is large enough. The model also seems to be a robust system for self-organized slipping in that the characteristic spatiotemporal dynamics do not depend on the size of system, boundary conditions, initial configurations and pulling speed, nor on the threshold that defines a minimal event.

6. Acknowledgments. This work was supported by the National Science Foundation, and was funded by Grant No. DMR91-22227.

REFERENCES

[1] R. Burridge and L. Knopoff, *Bull. Seis. Soc. Am.* **57**, 341 (1967).
[2] J.M. Carlson and J.S. Langer, *Phys. Rev. Lett.* **62**, 2632 (1989); J.M. Carlson and J.S. Langer, *Phys. Rev. A* **40**, 6470 (1989).
[3] J.M. Carlson, J.S. Langer, B.E. Shaw and C. Tang, *Phys. Rev. A* **44**, 884 (1991).
[4] G.L. Vasconcelos, M. de Sousa Vieira and S.R. Nagel, *Physica A* **191**, 69 (1992).
[5] L. Knopoff, J.A. Landoni and M.S. Abinante, *Phys. Rev. A* **46**, 7445 (1992).
[6] H. Takayasu and M. Matsuzaki, *Phys. Lett. A* **131**, 244 (1988); M. Matsuzaki and H. Takayasu, *J. Geophys. Res.* **96**, 19925 (1991).
[7] K. Christensen and Z. Olami, *J. Geophys. Res.* **97**, 8729 (1992).
[8] P. Bak, C. Tang and K. Wiesenfeld, *Phys. Rev. Lett.* **59**, 381 (1987).
[9] T.E. Tullis and J.D. Weeks, *Pure Appl. Geophys.* **124**, 383 (1986).
[10] J. Feder and H.J.S. Feder, *Phys. Rev. Lett.* **66**, 2669 (1991).
[11] D.P. Vallette and J.P. Gollub, *Phys. Rev. E* **47**, 820 (1993).
[12] J. Lomnitz-Adler, L. Knopoff and G. Martinez-Mekler, *Phys. Rev. A* **45**, 2211 (1992).
[13] K. Chen, P. Bak and S.P. Obukhov, *Phys. Rev. A* **43**, 625 (1991).
[14] L. Kadanoff, S.R. Nagel, L. Wu and S. Zhou, *Phys. Rev. A* **39**, 6524 (1989); H.M. Jaeger, C. Liu and S.R. Nagel, *Phys. Rev. Lett.* **62**, 40 (1989).
[15] G.A. Held, D.H. Solina, D.T. Keane, W.J. Haag, P.M. Horn and G. Grinstein, *Phys. Rev. Lett.* **65**, 1120 (1990).
[16] B. Lin and P. L. Taylor, *Macromolecules* **27**, 4212 (1994), *Phys. Rev. E* **49**, 3940 (1994).
[17] B. Guterberg and C.F. Richter, *Ann. Geofis* **9**, 1 (1956).
[18] G. Vilardo, G. Alessio and G. Luongo, *J. Volcanol. Geotherm. Res.* **48**, 115 (1991).
[19] V. Karnik, *Some Characteristics of Seismic Activity* in "Seismicity of the European Area/2", ed. V. Karnik, Reidel Publishing Company (1971), pp. 123.
[20] S.J. Gibowicz, *Bull. Seismol. Soc. Am.* **63**, 517 (1973); *Tectonophysics* **23**, 283 (1974).
[21] M. Bath, *Tectonophysics* **95**, 233 (1983).

MASS TRANSPORT BY BROWNIAN FLOWS

CRAIG L. ZIRBEL* AND ERHAN ÇINLAR†

Abstract. We consider the motion of a mass distribution in a random velocity field which is "δ-correlated" in time but which has arbitrary spatial correlations. We discuss the rigorous formulation of this problem in terms of Brownian flows and stochastic calculus. We use numerical simulations to illustrate the effect of the flow on the mass. We present results concerning the evolution of the mass distribution and, in particular, the long-time asymptotics of the center of mass and relative dispersion.

Key words. Stochastic flows; Brownian flows; Mass transport

AMS(MOS) subject classifications. 60H10, 60G57, 76F05

1. Introduction. In statistical fluid mechanics, an important problem is the behavior of a passive tracer carried by a random velocity field. The problem is generally difficult, because of the statistical dependence of particle motions on each other. In this paper we consider this problem with velocity fields that are "δ-correlated" in time but have arbitrary spatial correlations. The spatial correlations make joint particle motions non-trivial, while the short temporal memory makes possible various explicit calculations.

We begin with a general description of the mass transport problem, introducing our notation along the way. Let v be a time-dependent random velocity field on $\mathbb{R}^d$. Let $F_{st}(x)$ be the position at time t of a particle that was at x at an earlier time s. Then the motion of the particle in the velocity field is described by

$$F_{ss}(x) = x; \qquad \frac{d}{dt}F_{st}(x) = v(F_{st}(x), t), \qquad t > s \tag{1.1}$$

For $s \leq t$ fixed, F_{st} is a random transformation from $\mathbb{R}^d$ into $\mathbb{R}^d$. The family F of transformations F_{st}, $0 \leq s \leq t < \infty$, is called a *random flow*, since F satisfies the flow equations

$$F_{ss} = \text{identity} \qquad \text{for each } s \geq 0, \tag{1.2}$$

$$F_{st} \circ F_{rs} = F_{rt} \qquad \text{for } 0 \leq r \leq s \leq t. \tag{1.3}$$

Mass transport is concerned with the evolution of a mass distribution carried by the flow. The mass distribution may be continuous or discrete (a

* Department of Mathematics and Statistics, University of Massachusetts at Amherst, Amherst, MA 01003-4515, or zirbel@math.umass.edu. The work of the first author was conducted at the Institute for Mathematics and its Applications.

† Department of Civil Engineering and Operations Research, Princeton University, Princeton, NJ 08544. The second author would like to acknowledge the support of the Office of Naval Research through their Grant No. N00014-92-J-1088 P00001.

collection of individual particles). We use the unified formalism of measures to deal with both cases. Let $M_t(R)$ be the amount of mass in region $R \subseteq \mathbb{R}^d$ at time t. In terms of M_0, the equation for conservation of mass is

$$M_t(R) = M_0(\{x \in \mathbb{R}^d : F_{0t}(x) \in R\}) \tag{1.4}$$

For t fixed, M_t is a *random measure* on $\mathbb{R}^d$, that is, for each realization of the velocity field (or, equivalently, of the flow F), it defines a measure on $\mathbb{R}^d$. If M_0 has a smooth density $\rho_0(x)$, then M_t admits a random density $\rho_t(x)$, and (1.4) is equivalent to the familiar conservation equation

$$\frac{\partial \rho}{\partial t} + \nabla \cdot (\rho v) = 0 \tag{1.5}$$

The mass transport problem is to describe the evolution of M_t in time, using the probability law of the velocity field v.

An important, much-studied quantity is the *mean measure* of M_t, namely the measure μ_t defined by $\mu_t(R) = \mathbf{E}M_t(R)$, where $\mathbf{E}$ denotes expectation (average over the probability space). Commonly, one describes μ_t in terms of its density m_t with respect to the Lebesgue measure on $\mathbb{R}^d$. The density $m_t(x)$ is the mean tracer concentration. There has been and continues to be much interest in obtaining partial differential equations for m_t.

A second order concern is the statistical dependence between concentrations at different points in space, for this begins to describe the spatial structure of M_t. Assuming that M_t has the (random) density $\rho_t(x)$ at x, one is interested in the joint distribution of $\rho_t(x)$ and $\rho_t(y)$ for $x \neq y$, or with less ambition, their covariance. Of course, $\mathbf{E}\rho_t(x) = m_t(x)$.

For many applications, it is important to have descriptive measures of where the mass is concentrated and how spread out it is likely to be. As we shall see in Sections 4 and 5, mean concentration provides little information in this regard. Instead, we will emphasize random descriptors like the *centroid* $C_t = (C_t^i)$ and *dispersion matrix* $D_t = (D_t^{ij})$. These are defined by

$$C_t^i = \frac{1}{M_0(\mathbb{R}^d)} \int_{\mathbb{R}^d} M_t(dx) x^i, \tag{1.6}$$

$$D_t^{ij} = \frac{1}{M_0(\mathbb{R}^d)} \int_{\mathbb{R}^d} M_t(dx)(x^i - C_t^i)(x^j - C_t^j), \tag{1.7}$$

(in the case of continuous mass, $M_t(dx)$ may be replaced by $\rho_t(x)dx$).

Our aim in this paper is to present our results regarding M_t in the case where the random velocity field v is given by

$$v(x,t) = u(x,t) + \dot{U}_0(x,t) \tag{1.8}$$

where u is a deterministic velocity field and $\dot{U}_0$ is the formal time derivative of a Brownian motion with a spatial parameter. Formally, $\dot{U}_0$ is a Gaussian random vector field with mean zero and covariance given by

$$\mathrm{Cov}(\dot{U}_0^i(x,s), \dot{U}_0^j(y,t)) = a^{ij}(x,y)\delta(t-s), \tag{1.9}$$

where δ is Dirac's delta function at 0, and a is a spatial covariance tensor. The corresponding flow is then called a *Brownian flow.*

In the next section, we put the equation (1.1) of motion in rigorous form in the case of Brownian velocity fields (1.8). Also in that section, we give a brief introduction to isotropic Brownian flows, which figure prominently in the rest of the paper.

In Section 3 we discuss numerical simulations of Brownian flows, which are complicated by the presence of stochastic integrals. We prove an error estimate justifying the use of the Euler method for simulating trajectories. We use this method to generate pictures of the evolution of mass distributions in compressible and incompressible isotropic Brownian flows.

In Section 4 we list some results concerning the evolution of M_t, including the stochastic differential equation it satisfies, the partial differential equations for the mean and covariance of the density ρ, and some formulas for the mean and covariance of the centroid and dispersion matrix.

Finally, in Section 5 we specialize to the case of isotropic Brownian flows. We report on results we have obtained elsewhere concerning the behavior of C_t and D_t as $t \to \infty$.

2. Brownian flows. A *random flow* on $\mathbb{R}^d$ is a collection $F = \{F_{st};\ 0 \le s \le t < \infty\}$ of random transformations from $\mathbb{R}^d$ into $\mathbb{R}^d$ satisfying, almost surely, the flow equations (1.2) and (1.3). These F_{st} are defined on a probability space $(\Omega, \mathcal{H}, \mathbf{P})$ in such a way that the mapping $(\omega, s, t, x) \mapsto F_{st}(\omega, x)$ is jointly measurable with respect to the product σ-algebra on $\Omega \times \mathbb{R}_+ \times \mathbb{R}_+ \times \mathbb{R}^d$. We call F a *flow of homeomorphisms* if, almost surely, all the mappings $F_{st} : x \mapsto F_{st}(x)$ are homeomorphisms.

The phrase 'almost surely' means 'for all ω in a set $\Omega_0 \in \mathcal{H}$ such that $\mathbf{P}(\Omega_0) = 1$.' One may also substitute 'with probability one.' The realization ω is usually suppressed as an argument of random variables; for example, $F_{st}(\omega, x)$ will always be written $F_{st}(x)$.

In this section we will be concerned with flows which solve stochastic differential equations. First, however, we must introduce the notion of a Brownian motion with a spatial parameter and define the stochastic integral with respect to such a Brownian motion. For a comprehensive introduction to this theory, we refer the reader to KUNITA (1990).

2.1. Brownian motion with a spatial parameter. Let $u = \{u^i(x,t);\ i = 1, \ldots, d,\ x \in \mathbb{R}^d,\ t \in \mathbb{R}_+\}$ be a deterministic vector field and let $a = \{a^{ij}(x,y);\ i,j = 1, \ldots, d,\ x, y \in \mathbb{R}^d\}$ be the covariance tensor of a vector-valued random field on $\mathbb{R}^d$. Let U be a Gaussian random vector field on $\mathbb{R}^d$ with mean and covariance given by

$$\mathbf{E}U^i(x,t) = \int_0^t dr\, u^i(x,r) \tag{2.1}$$

$$\mathrm{Cov}(U^i(x,s), U^j(y,t)) = a^{ij}(x,y)(s \wedge t) \tag{2.2}$$

where $s \wedge t$ is the smaller of s and t. If u and a satisfy conditions (2.8) through (2.11) below, then there exists a version of U which is continuous in (x,t). The field U is called a *Brownian motion with a spatial parameter* because it is continuous and has independent increments: if $s \le t$, the random field

$$U(\cdot,t) - U(\cdot,s) \equiv \{U(x,t) - U(x,s);\ x \in \mathbb{R}^d\}$$

is independent of the history of U before s. The field u is called the *drift* of U and a is called its *covariance*. We could have allowed the covariance to depend on t also, but this generality is not particularly important to us here. We will often decompose U as

$$U(x,t) = \int_0^t dr\, u(x,r) + U_0(x,t), \tag{2.3}$$

where U_0 is a Brownian motion with zero drift and covariance a.

Let X_t, $t \ge 0$ be a process in $\mathbb{R}^d$ which is predictable with respect to the natural filtration of U, and such that $\int_0^T dr\, a^{ii}(X_r, X_r) < +\infty$ almost surely. The integral of U along X, denoted $\int_s^t U(X_r, dr)$, is defined to be the limit in probability of

$$\sum_{k=0}^{N-1} U(X_{t_k\wedge t}, t_{k+1} \wedge t) - U(X_{t_k \wedge t}, t_k \wedge t)$$

as the width of the partition $s = t_0 \le t_1 \le \cdots \le t_N = T$ tends to zero. Here $T > t$ is some fixed terminal time. This is similar to the usual stochastic integral, except that all the action takes place in the spatial argument of U.

For practical purposes there are only a few things we need to know concerning the integral of U along X. First, using the decomposition (2.3), the integral can be written as

$$\int_s^t U(X_r, dr) = \int_s^t dr\, u(X_r, r) + \int_s^t U_0(X_r, dr) \tag{2.4}$$

The first integral on the right side is the usual Lebesgue-Stieltjes integral. Second, the process $\int_s^t U_0(X_r, dr)$, $t \ge s$ is a continuous mean-zero martingale, and if Y is another predictable process, then the joint quadratic variation of the integrals of U along X and along Y satisfies

$$(2.5)\qquad \left\langle \int_0^t U_0^i(X_r, dr), \int_0^t U_0^j(Y_r, dr) \right\rangle = \int_0^t dr\, a^{ij}(X_r, Y_r)$$

These facts will be used repeatedly in the rest of the paper.

2.2. Stochastic differential equation for F. Now we are able to write a stochastic differential equation for a random flow F based on the Brownian motion U:

$$(2.6)\qquad F_{st}(x) = x + \int_s^t U(F_{sr}, dr), \qquad t \geq s$$

Using (2.4) this can also be written as

$$(2.7)\quad F_{st}(x) = x + \int_s^t dr\, u(F_{sr}(x), r) + \int_s^t U_0(F_{sr}, dr), \qquad t \geq s$$

Suppose that u and a satisfy the conditions

$$(2.8)\qquad |u(x,t)| \leq K_1(1 + |x|)$$

$$(2.9)\qquad |u(x,t) - u(y,t)| \leq K_2|x - y|$$

$$(2.10)\qquad |a^{ij}(x,y)| \leq K_3(1 + |x|)(1 + |y|)$$

$$(2.11)\ |a^{ij}(x,y) - a^{ij}(x',y) - a^{ij}(x,y') + a^{ij}(x',y')| \leq K_4|x - x'||y - y'|$$

for all x, x', y, y' in $\mathbb{R}^d$ and t in $\mathbb{R}_+$, where K_1, K_2, K_3, K_4 are finite constants. Then by Theorem 4.2.5 of KUNITA (1990), there exists a flow $F = \{F_{st};\ 0 \leq s \leq t < \infty\}$ of homeomorphisms satisfying (2.6) and such that

(2.12) (*a*) the mapping $(x,t) \mapsto F_t x$ is continuous almost surely
(*b*) $F_{t_1 t_2}, \ldots, F_{t_{n-1} t_n}$ are independent when $t_1 \leq t_2 \leq \ldots \leq t_n$

Because of (b), the independence of multiplicative increments, F is called a *Brownian flow*.

As a consequence of (2.12), for fixed $x_1, \ldots, x_n$ in $\mathbb{R}^d$, the n-point motion

$$\{F_{0t}(x_1), \ldots, F_{0t}(x_n);\ t \geq 0\}$$

is a continuous, strong Markov process, or diffusion, in $\mathbb{R}^{d \times n}$. The drift of each component is u, and the joint quadratic variation satisfies

$$(2.13)\qquad \langle F_{0t}^i(x), F_{0t}^j(y) \rangle = \int_0^t dr\, a^{ij}(F_{0r}(x), F_{0r}(y))$$

which can be seen from (2.4), (2.5), and (2.6). In particular, the generator A of the one-point motion $F_{0t}(x)$, $t \geq 0$ satisfies

$$(2.14)\quad A_t f(x) = \tfrac{1}{2} \sum_{i,j=1}^{d} a_d^{ij}(x) \partial_i \partial_j f(x) + \sum_{i=1}^{d} u^i(x,t) \partial_i f(x), \qquad t \geq 0$$

where $a_d(x) = a(x,x)$ is the 'diagonal' of a.

We can interpret (2.13) in this way: if two particles are near each other, they will move together in concert, since the joint quadratic variation increases rapidly when $F_{0t}(x)$ is near $F_{0t}(y)$. If they are far apart, the joint quadratic variation increases slowly, indicating that their motions are nearly independent.

2.3. Alternative formulations. We now describe three alternative ways of thinking about a Brownian flow F satisfying (2.6). First, under conditions (2.10) and (2.11), we may write U_0 as

$$(2.15)\qquad U_0(x,t) = \sum_{k=1}^{\infty} u_k(x) W_k(t),$$

where $u_1, u_2, \ldots$ are deterministic Lipschitz vector fields on $\mathbb{R}^d$ and W_1, $W_2, \ldots$ are independent one-dimensional Wiener processes. The covariance a of U_0 is related to the u_k through

$$(2.16)\qquad a^{ij}(x,y) = \sum_{k=1}^{\infty} u_k^i(x) u_k^j(y)$$

Equation (2.7) for F becomes, in differential notation,

$$(2.17)\quad F_{ss}(x) = x; \quad dF_{st}(x) = u(F_{st}(x),t)dt + \sum_{k=1}^{\infty} u_k(F_{st}(x)) W_k(dt), \ t > s$$

This makes clear the relationship of (2.6) and (2.7) to standard stochastic differential equations. The difference is that in this equation there are an infinite number of noise terms (otherwise we would limit the class of covariances obtainable via (2.16)) and we are interested in solving the equation for all x in $\mathbb{R}^d$ simultaneously.

Second, if we write (2.7) in differential notation, formally divide through by dt, and write $\dot{U}_0(x,t)$ in place of $U_0(x,dt)/dt$, we obtain

$$(2.18)\quad F_{ss}(x) = x; \quad \frac{d}{dt} F_{st}(x) = u(F_{st}(x),t) + \dot{U}_0(F_{st}(x),t), \quad t > s$$

This is of the form (1.1) for velocity fields of the form (1.8). Now $\dot{U}_0$ is what is sometimes called a 'generalized process,' so its covariance must be inferred from integrals of $\dot{U}_0$ against test functions. For convenience we choose these to be of the form $1_{[s,t]}$:

$$\begin{aligned} \mathbf{E}\int_s^t dr\, \dot{U}_0(x,r)\int_{s'}^{t'} dr'\, \dot{U}_0(y,r')^T \\ &= \mathbf{E}(U_0(x,t)-U_0(x,s))(U_0(y,t')-U_0(y,s')) \\ &= a(x,y)\mathrm{Leb}([s,t]\cap[s',t']) \\ &= \int_s^t dr \int_{s'}^{t'} dr'\, a(x,y)\delta(r-r') \end{aligned}$$

This justifies the following heuristic for equation (2.18): the flow F is driven by a deterministic velocity field u plus noise that is δ-correlated in time but spatially correlated with covariance tensor a.

Finally, suppose that v is a mean-zero random velocity field on $\mathbb{R}^d$ whose law is stationary in time. For each $\varepsilon > 0$, define a flow G^ε via the classical equation

$$G^\varepsilon_{ss}(x) = x; \quad \frac{d}{dt}G^\varepsilon_{st}(x) = \varepsilon v(G^\varepsilon_{st}(x),t), \quad t > s \tag{2.19}$$

Define another flow F^ε by $F^\varepsilon_{st} = G^\varepsilon_{s/\varepsilon^2,t/\varepsilon^2}$ and consider the limit of F^ε as $\varepsilon \to 0$. Under reasonable mixing conditions on v, plus some draconian restrictions on the moments of its derivatives, the flows F^ε will converge in law to a Brownian flow F having drift u given by

$$u^i(x) = \sum_{j=1}^d \int_0^\infty dr\, \mathbf{E}v^j(x,0)\frac{\partial v^i}{\partial x^j}(x,r) \tag{2.20}$$

and covariance a given by $a(x,y) = \bar{a}(x,y) + \bar{a}(y,x)^T$, where

$$\bar{a}^{ij}(x,y) = \int_0^\infty dr\, \mathbf{E}v^i(x,0)v^j(y,r) \tag{2.21}$$

This result is shown in Section 5.6 of KUNITA (1990), along with precise conditions on v.

2.4. Isotropic Brownian flows. Isotropic Brownian flows are the Brownian versions of random isotropic flows that have been of interest in statistical turbulence ever since the seminal work of KOLMOGOROV (1941). See MONIN and YAGLOM (1971) for an account of the classical case where F is obtained from (1.1) with v an isotropic random vector field.

Isotropic Brownian flows were characterized and studied by LE JAN (1985) and BAXENDALE and HARRIS (1986) simultaneously. The following brief review is taken from their work. The reader may also wish to consult the review article by DARLING (1989).

A Brownian flow F based on a Brownian motion U with drift u and covariance a is isotropic if and only if

$$u \equiv 0; \qquad a(x,y) = b(x-y), \qquad x, y \in \mathbb{R}^d, \tag{2.22}$$

where b is an *isotropic* covariance tensor, that is, $O^T b(Oz)O = b(z)$ for every orthogonal matrix O and every z in $\mathbb{R}^d$. Under these conditions, for every rigid motion R on $\mathbb{R}^d$, that is, a combination of translation, rotation, and/or reflection, the flow G defined by $G_{st} = R \circ F_{st} \circ R^{-1}$ has the same probability law as F. In other words, the law of F is invariant under all rigid motions of $\mathbb{R}^d$.

When $d = 1$, isotropy requires only that $u = 0$ and $b(z) = b(-z)$. The following condition guarantees the existence of a non-trivial isotropic flow F satisfying (2.6) and such that, almost surely, all the maps $F_{st} : x \mapsto F_{st}(x)$ are C^1-diffeomorphisms. See KUNITA (1990), Theorem 4.6.5.

CONDITION 2.1. *For $d = 1$, both b and b'' are bounded and continuous, and b is not identically constant.*

In two or more dimensions, isotropy forces b to have a special form, due to YAGLOM (1957); see also YAGLOM (1987):

$$b^{ij}(z) = b_N(|z|)\delta_{ij} + \frac{z^i z^j}{|z|^2}(b_L(|z|) - b_N(|z|)) \tag{2.23}$$

where b_L and b_N are real-valued functions on $\mathbb{R}_+$ defined in terms of two finite measures, Φ_P and Φ_S, on $\mathbb{R}_+$:

$$b_N(r) = A\int_0^\infty \Phi_P(ds)L_m(rs) + A\int_0^\infty \Phi_S(ds)[L_{m-1}(rs) - L_m(rs)] \tag{2.24}$$

$$b_L(r) = A\int_0^\infty \Phi_P(ds)[L_m(rs)-(rs)^2 L_{m+1}(rs)] + A(d-1)\int_0^\infty \Phi_S(ds)L_m(rs) \tag{2.25}$$

Here $m = d/2$, $A = 2^{m-1}\Gamma(m)$, and $L_m(r) = J_m(r)/r^m$, with Γ denoting the gamma function and J_m denoting the Bessel function of the first kind of order m. The measure Φ_P gives rise to the potential (irrotational) part of the flow and Φ_S to the solenoidal (incompressible) part. Indeed, when $\Phi_P = 0$, the flow F is incompressible, that is, it preserves Lebesgue measure.

CONDITION 2.2. *For $d \geq 2$, the measures Φ_P and Φ_S have finite fourth moments and put no mass on the set $\{0\}$.*

Under this condition, b_L and b_N are C^4-functions decaying to 0 at $+\infty$, there exists an isotropic flow F satisfying (2.6), and, almost surely, all the maps $F_{st} : x \mapsto F_{st}(x)$ are a C^1-diffeomorphisms. See BAXENDALE and HARRIS (1986).

The behavior of b_L and b_N near 0 is quadratic:

$$b_L(r) = b_0 - \tfrac{1}{2}\beta_L r^2 + O(r^4), \qquad r \to 0, \tag{2.26}$$

$$b_N(r) = b_0 - \tfrac{1}{2}\beta_N r^2 + O(r^4), \qquad r \to 0, \tag{2.27}$$

where b_0, β_L, β_N are strictly positive constants given by

$$\begin{aligned}
(2.28)\quad b_0 &= \frac{1}{d}\Phi_P(0,\infty) + \frac{d-1}{d}\Phi_S(0,\infty) \\
(2.29)\quad \beta_L &= \frac{3}{d(d+2)}\int_0^\infty \Phi_P(ds)s^2 + \frac{d-1}{d(d+2)}\int_0^\infty \Phi_S(ds)s^2 \\
(2.30)\quad \beta_N &= \frac{1}{d(d+2)}\int_0^\infty \Phi_P(ds)s^2 + \frac{d+1}{d(d+2)}\int_0^\infty \Phi_S(ds)s^2
\end{aligned}$$

Moreover, the maxima of b_L and b_N occur at 0.

Let us set $b_0 = b(0)$ for $d = 1$. Then for all dimensions, the generator A of the one-point motion as defined in (2.14) becomes

$$Af(x) = \tfrac{1}{2}\sum_{i,j=1}^{d} b^{ij}(0)\partial_i\partial_j f(x) = \tfrac{1}{2}b_0\Delta f(x), \tag{2.31}$$

where Δ is the Laplacian operator. We recognize this as the generator of a Brownian motion in $\mathbb{R}^d$ with zero drift and covariance matrix $b_0 I$.

2.4.1. Lyapunov exponents. The Lyapunov exponents $\lambda_1 > \lambda_2 > \ldots > \lambda_d$ of the flow are the values taken by the limits

$$\lim_{t\to\infty}\frac{1}{t}\log \mid DF_{0t}(x)\xi \mid$$

as the vector ξ varies over the unit sphere in $\mathbb{R}^d$; here $DF_{0t}(x)$ is the Jacobian matrix of the map F_{0t} evaluated at x. The Lyapunov exponents are deterministic in an isotropic Brownian flow. For $d = 1$ and b satisfying 2.1, we have $\lambda_1 = \frac{1}{2}b''(0) \leq 0$. For $d \geq 2$, they are given by:

$$\lambda_i = \frac{d-i}{2}\beta_N - \frac{i}{2}\beta_L\,, \qquad i = 1,\ldots,d. \tag{2.32}$$

When F is incompressible, we have $\Phi_P = 0$ which gives $(d+1)\beta_L = (d-1)\beta_N$, and hence the top Lyapunov exponent is

$$\lambda_1 = \frac{d}{2}\beta_L > 0. \tag{2.33}$$

When F is potential, $\Phi_S = 0$, we have $\beta_L = 3\beta_N$, and

$$\lambda_1 = \frac{d-4}{6}\beta_L, \tag{2.34}$$

which is negative for $d = 2$ and $d = 3$.

2.4.2. Separation process. For fixed x and y in $\mathbb{R}^d$, the distance at time t between two particles started from x and y is

$$Z_t = |F_t x - F_t y|. \tag{2.35}$$

The process $Z = \{Z_t;\ t \geq 0\}$ is called the *separation process* (or pair distance process) with initial value $|x - y|$. When F is isotropic, Z is a one-dimensional diffusion on $\mathbb{R}_+$. Under conditions 2.1 and 2.2, Z satisfies the stochastic differential equation

$$dZ_t = \sqrt{2(b_0 - b_L(Z_t))}dW_t + (d-1)\frac{b_0 - b_N(Z_t)}{Z_t}dt \tag{2.36}$$

The boundary point 0 is absorbing and inaccessible from $(0, \infty)$. When $d = 1$, the drift part vanishes, we write b in place of b_L, and Z becomes a martingale diffusion.

The following proposition lists the asymptotic behavior of the process Z as $t \to \infty$; see LE JAN (1985) and BAXENDALE and HARRIS (1986) for the proof. Conditions 2.1 and 2.2 are in force.

PROPOSITION 2.1. *The process Z is transient on $(0, \infty)$ unless $d = 2$ and $\lambda_1 \geq 0$. More specifically, if $Z_0 > 0$,*

(i) *if $d \geq 4$ or if $d = 3$ and $\lambda_1 \geq 0$, then $Z_t \to +\infty$ almost surely;*
(ii) *if $d = 3$ and $\lambda_1 < 0$, then Z_t converges to either 0 or $+\infty$, each with strictly positive probability;*
(iii) *if $d = 2$ and $\lambda_1 > 0$, then Z is null-recurrent and converges to $+\infty$ in probability;*
(iv) *if $d = 2$ and $\lambda_1 = 0$, then Z is null-recurrent;*
(v) *if $d = 2$ and $\lambda_1 < 0$, or if $d = 1$, then $Z_t \to 0$ almost surely*

It follows from this proposition and (2.33) that $\mathbf{P}\{Z_t \to 0\} = 0$ for incompressible flows and in general whenever $d \geq 4$. For $d \leq 3$, $\mathbf{P}\{Z_t \to 0\}$ becomes strictly positive if the potential part is large enough.

3. Numerical simulations of Brownian flows. Our purpose in this section is to describe and implement a procedure for the numerical simulation of Brownian flows. Our emphasis is on gaining insight into the nature of Brownian flows through an adequate simulation method, rather than on developing high-order schemes or making extremely accurate numerical estimates.

Let F be a Brownian flow satisfying the equation

$$F_{st}(x) = x + \int_s^t U(F_{sr}(x), dr), \qquad t \geq s \tag{3.1}$$

where U is a Brownian motion with drift u and covariance a satisfying conditions (2.8) to (2.11). In this section we will always take $s = 0$; for brevity we will write F_t for the map F_{0t} and $F_t x$ in place of $F_{0t}(x)$.

3.1. Euler approximation method. We will use the Euler approximation to define a collection $\{G_t;\ 0 \le t \le T\}$ of maps approximating F on the interval $[0,T]$. First, fix a time-step $\delta > 0$ for the approximation. Define a partition $\{t_k\}$ of $[0,T]$ by $t_k = k\delta,\ k = 0, 1, \ldots$. The definition of G is recursive: for each x in $\mathbb{R}^d$,

$$G_0 x = x \tag{3.2}$$

$$G_t x = G_{t_k} x + U(G_{t_k} x, t) - U(G_{t_k} x, t_k), \qquad t \in (t_k, t_{k+1}] \tag{3.3}$$

Note that the map $(x,t) \mapsto G_t x$ is continuous.

For the purpose of visualizing the action of the flow, an important issue is how closely the n-point trajectories under G approximate those under F when both are based on the same Brownian motion U. The following estimate shows that the approximate solution G is accurate to order $\sqrt{\delta}$.

THEOREM 3.1. *Let $T > 0$. Then there exists a constant C depending on T such that, for all n in $\mathbb{N}$, $x_1, \ldots, x_n$ in $\mathbb{R}^d$, and $\delta < T$, we have*

$$\mathbf{E} \sup_{t \le T} \sum_{m=1}^{n} |F_t x_m - G_t x_m|^2 \ \le\ C\delta\Big(n + \sum_{m=1}^{n} |x_m|^2\Big) \tag{3.4}$$

Proof. First observe that

$$\mathbf{E} \sup_{t \le T} \sum_{m=1}^{n} |F_t x_m - G_t x_m|^2 \ \le\ \sum_{m=1}^{n} \mathbf{E} \sup_{t \le T} |F_t x_m - G_t x_m|^2,$$

so we need only consider $\mathbf{E}\sup_{t \le T} |F_t x - G_t x|^2$ for fixed x in $\mathbb{R}^d$. We will show that this is less than $C\delta(1 + |x|^2)$, which will complete the proof.

To simplify notation, let $X_t = F_t x$ and $Y_t = G_t x$. We need to consider piecewise constant approximations to X and Y, defined by

$$\bar{X}_t = X_{t_k}, \qquad \bar{Y}_t = Y_{t_k}, \qquad t \in (t_k, t_{k+1}]$$

We can represent Y compactly in terms of $\bar{Y}$, for note that by (3.3),

$$Y_t = Y_{t_k} + \int_{t_k}^{t} U(\bar{Y}_r, dr)$$

Writing the same expression for $Y_{t_k}, Y_{t_{k-1}}, \ldots$ yields

$$Y_t = x + \int_0^t U(\bar{Y}_r, dr)$$

¿From (3.1) we have $X_t = x + \int_0^t U(X_r, dr)$, so we can apply the Lemma following this proof to obtain:

$$\mathbf{E} \sup_{r \le t} |X_r - Y_r|^2 \le C_1(T)\mathbf{E} \int_0^t dr\, |X_r - \bar{Y}_r|^2, \tag{3.5}$$

for all $t \le T$. In order to estimate $\mathbf{E}|X_t - \bar{Y}_t|^2$, we add and subtract $\bar{X}_t$:

$$\mathbf{E}|X_t - \bar{Y}_t|^2 \le 2\mathbf{E}|X_t - \bar{X}_t|^2 + 2\mathbf{E}|\bar{X}_t - \bar{Y}_t|^2 \tag{3.6}$$

For the second term, it will suffice to use the inequality $\mathbf{E}|\bar{X}_t - \bar{Y}_t|^2 \le \mathbf{E}\sup_{r\le t}|X_r - Y_r|^2$.

Next, we obtain an upper bound for $\mathbf{E}|X_t - \bar{X}_t|^2$. First, note that by the definitions of X and $\bar{X}$, for t in $(t_k, t_{k+1}]$ we have

$$X_t - \bar{X}_t = \int_{t_k}^t U(X_r, dr)$$

Thus, using the decomposition (2.4),

$$\mathbf{E}|X_t - \bar{X}_t|^2 \le 2\mathbf{E}\left|\int_{t_k}^t dr\ u(X_r, r)\right|^2 + 2\mathbf{E}\left|\int_{t_k}^t U_0(X_r, dr)\right|^2 \tag{3.7}$$

We treat these two terms separately, writing each as a sum over components $1, \ldots, d$.

First, by Jensen's inequality,

$$\begin{aligned}\mathbf{E}\left|\int_{t_k}^t dr\ u^i(X_r, r)\right|^2 &\le (t - t_k)\mathbf{E}\int_{t_k}^t dr\ (u^i(X_r, r))^2 \\ &\le \delta K_1 \int_{t_k}^t dr\ \mathbf{E}(1 + |X_r|)^2,\end{aligned}$$

the last inequality by condition (2.8) on u. Second, since $\int_0^t U_0^i(X_r, dr)$ is a mean-zero martingale,

$$\begin{aligned}\mathbf{E}\left|\int_{t_k}^t U_0^i(X_r, dr)\right|^2 &= \mathbf{E}\left\langle\int_{t_k}^t U_0^i(X_r, dr)\right\rangle \\ &= \mathbf{E}\int_{t_k}^t dr\ a^{ii}(X_r, X_r) \\ &\le K_3 \int_{t_k}^t dr\ \mathbf{E}(1 + |X_r|)^2,\end{aligned}$$

by condition (2.10) on a. Finally, by Lemma 4.5.3 of KUNITA (1990), there exists a constant $C_2 > 0$ such that

$$\mathbf{E}(1 + |X_r|)^2 \le 2e^{C_2 T}(1 + |x|^2)$$

for all $r \le T$. We have shown that

$$\begin{aligned}\mathbf{E}|X_t - \bar{X}_t|^2 &\le 2d(\delta K_1 + K_3)\int_{t_k}^t dr\ 2e^{C_2 T}(1 + |x|^2) \\ &\le C_3(T)\delta(1 + |x|^2)\end{aligned} \tag{3.8}$$

for some $C_3(T) > 0$.

Now let $f(t) = \mathbf{E}\sup_{r\leq t}|X_r - Y_r|^2$. Equations (3.5) through (3.8) show that for $t \leq T$,

$$\begin{aligned} f(t) &\leq 2C_1(T)\left[\int_0^t dr\ C_3(T)\delta(1+|x|^2) + \int_0^t dr\ f(r)\right] \\ &\leq C_4(T)\delta(1+|x|^2) + 2C_1(T)\int_0^t dr\ f(r) \end{aligned}$$

By Gronwall's Lemma,

$$\begin{aligned} f(t) &\leq C_4(T)\delta(1+|x|^2)\exp(2C_1(T)t) \\ &\leq C_5(T)\delta(1+|x|^2), \end{aligned}$$

which completes the proof. □

LEMMA 3.1. *Let U be a Brownian motion with drift u and covariance a satisfying (2.9) and (2.11), respectively. Let X and Y be predictable processes such that $\int_0^T dr\ a^{ii}(X_r, X_r) < \infty$ almost surely; similarly for Y. Then for each $T \geq 0$, there is a positive constant $C(T)$, not depending on X and Y, such that for all $t \leq T$,*

$$(3.9)\ \mathbf{E}\sup_{s\leq t}\left|\int_0^s U(X_r, dr) - \int_0^s U(Y_r, dr)\right|^2 \leq C(T)\mathbf{E}\int_0^t dr\ |X_r - Y_r|^2$$

The proof is found in the proof of Lemma 3.4.2 of KUNITA (1990). Our proof of Theorem 3.1 was suggested by the proof of Theorem 10.2.2 of KLOEDEN and PLATEN (1992).

We conclude from Theorem 3.1 that the second moment of the error between the n-point trajectories of F and G converges to 0 as $\delta \to 0$. This is good enough for the purpose of illustration, but if we were interested in more complicated functionals of the trajectories we would want to choose a higher order method. See TALAY (1990).

For numerical simulations, we want to generate the values of the trajectories at discrete times $t_1, t_2, \ldots$ using formula (3.3), which reads

$$(3.10)\qquad G_{t_{k+1}}x = G_{t_k}x + U(G_{t_k}x, t_{k+1}) - U(G_{t_k}x, t_k)$$

Consider the random field $U(x, t_{k+1}) - U(x, t_k)$ appearing in this equation. This is a Gaussian random field whose mean and covariance are $\int_{t_k}^{t_{k+1}} dr\ u(x, r)$ and δa. Thus, the random field V_k defined implicitly by

$$(3.11)\qquad U(x, t_{k+1}) - U(x, t_k) = \int_{t_k}^{t_{k+1}} dr\ u(x, r) + \sqrt{\delta}V_k(x)$$

is Gaussian with mean 0 and covariance a. Moreover, the collection $\{V_k, k = 0, 1, \ldots\}$ is independent and identically distributed.

Our outline of the numerical simulation procedure will be complete once we describe how to generate the fields V_k. For the simulation of n-point motion, the field V_k only needs to be generated at the points $G_{t_k}x_1, \ldots, G_{t_k}x_n$, and this can be done in general if need be. A particularly simple case occurs when a is isotropic, so that $a(x,y) = b(x-y)$ for some isotropic covariance b. We now discuss this case in detail.

3.2. Spectral method for isotropic vector fields. The *spectrum* of a covariance matrix $(b^{j\ell})$ is a matrix-valued measure f on $\mathbb{R}^d$ such that

$$b^{j\ell}(z) = \int_{\mathbb{R}^d} e^{iz\cdot k}\, \mathrm{f}^{j\ell}(dk) \tag{3.12}$$

(YAGLOM (1987) 22.4.68). The spectrum of an isotropic covariance can be expressed in terms of the measures Φ_P and Φ_S which define b via (2.23), (2.24), and (2.25). We have

$$\mathrm{f}^{j\ell}(dk) = \frac{\sigma_k(ds)}{\Sigma_d |k|^{d-1}} \left[\tfrac{k^j k^\ell}{|k|^2} \Phi_P(d|k|) + (\delta_{jl} - \tfrac{k^j k^\ell}{|k|^2}) \Phi_S(d|k|) \right] \tag{3.13}$$

where $\sigma_k(ds)d|k|$ is the volume element in spherical coordinates, $\sigma_k(ds)$ is the area element at k on the sphere of radius $|k|$, and Σ_d is the area of the unit sphere in $\mathbb{R}^d$, $\Sigma_d = 2\pi^{d/2}\Gamma(d/2)^{-1}$ (YAGLOM (1987) 22.4.174).

To generate a realization of V, first fix the number N of modes desired and let ξ_n^m, ζ_n^m, $n = 1, \ldots, N$, $m = 1, \ldots, d$ be independent standard Gaussian variables. Let k_n, $n = 1, \ldots, N$ be independent random vectors in $\mathbb{R}^d$ with the following common distribution: $\frac{k}{|k|}$ is uniform on the unit sphere and $|k|$ has distribution μ, where μ is some probability measure on $(0, \infty)$ such that Φ_S and Φ_P are absolutely continuous with respect to μ. We may choose μ. Define functions ψ_S and ψ_P on $\mathbb{R}_+$ in terms of the Radon-Nikodym derivatives of Φ_S and Φ_P with respect to μ:

$$\psi_S^2 = \frac{d\Phi_S}{d\mu}, \qquad \psi_P^2 = \frac{d\Phi_P}{d\mu} \tag{3.14}$$

Now define the random field V by setting $V^j(x)$ equal to

$$\frac{1}{\sqrt{N}} \sum_{m,n} (\xi_n^m \cos k_n \cdot x + \zeta_n^m \sin k_n \cdot x)(\tfrac{k_n^j k_n^m}{|k_n|^2} \psi_P(|k_n|) + (\delta_{jm} - \tfrac{k_n^j k_n^m}{|k_n|^2}) \psi_S(|k_n|))$$

The following proposition is the justification of this method.

PROPOSITION 3.1. *For each N in $\mathbb{N}$, V has mean 0 and isotropic covariance b based on Φ_P and Φ_S as in Section 2. Moreover, as $N \to \infty$, the finite dimensional distributions of V converge to those of a Gaussian random field.*

Proof. Consider $\mathbf{E}V(x)$. Condition on k_n. The variables ξ and ζ have mean 0, so $\mathbf{E}V(x) = 0$. Next consider $\mathbf{E}V^j(x)V^\ell(y)$, writing $V^j(x)$ as a sum over n, m, and $V^\ell(y)$ as a sum over n', m'. Condition on k_n and $k_{n'}$ to bring out the factor:

$$\mathbf{E}(\xi_n^m \cos k_n \cdot x + \zeta_n^m \sin k_n \cdot x)(\xi_{n'}^{m'} \cos k_{n'} \cdot x + \zeta_{n'}^{m'} \sin k_{n'} \cdot x)$$

$$= \delta_{mm'}\delta_{nn'} \cos k_n(x-y)$$

The equality follows by independence of the ξ and ζ and by the angle addition formula. Each of the remaining N terms are identical, so we drop $\frac{1}{N}\sum_{n=1}^N$ and the index n. A sum over m remains.

Multiplying the factors containing ψ_P and ψ_S and summing over m yields, after some algebra,

$$\begin{aligned}
&\mathbf{E}V^j(x)V^\ell(y)\\
&= \mathbf{E}(\tfrac{k^jk^\ell}{|k|^2}\psi_P(|k|)^2 + (\delta_{j\ell} - \tfrac{k^jk^\ell}{|k|^2})\psi_S(|k|)^2) \cos k \cdot (x-y)\\
&= \int \mu(d|k|)\frac{\sigma_k(ds)}{\Sigma_d |k|^{d-1}}[\tfrac{k^jk^\ell}{|k|^2}\frac{d\Phi_P}{d\mu}(|k|) + (\delta_{j\ell} - \tfrac{k^jk^\ell}{|k|^2})\frac{d\Phi_S}{d\mu}(|k|)] \cos k \cdot (x-y)\\
&= \int_{\mathbb{R}^d} e^{ik\cdot(x-y)} \mathfrak{f}^{j\ell}(dk)\\
&= b^{j\ell}(x-y)
\end{aligned}$$

The integral against $i \sin k \cdot (x-y)$ is zero because the factor inside the brackets is even under $k \to -k$. By the central limit theorem the finite dimensional distributions of V converge to Gaussians as $N \to \infty$. □

3.3. Simulations of an isotropic flow. We now describe the particulars of our simulations of a two-dimensional isotropic Brownian flow. We begin by choosing the measures Φ_P and Φ_S which define the isotropic covariance b via (2.24) and (2.25):

$$\Phi_P(d\alpha) = 4(1-\eta)p^2\alpha^3 e^{-p\alpha^2} d\alpha \tag{3.15}$$

$$\Phi_S(d\alpha) = 4\eta s^2\alpha^3 e^{-s\alpha^2} d\alpha, \tag{3.16}$$

where η, p, and s are positive constants. We will use η to vary the relative strengths of the potential and solenoidal components. In particular, $\eta = 1$ corresponds to a solenoidal (incompressible) flow and $\eta = 0$ to a potential flow.

In two dimensions, formulas (2.24) and (2.25) become

$$b_N(r) = \int_0^\infty \frac{J_1(r\alpha)}{r\alpha}\Phi_P(d\alpha) + \int_0^\infty J_1'(r\alpha)\Phi_S(d\alpha) \tag{3.17}$$

$$b_L(r) = \int_0^\infty J_1'(r\alpha)\Phi_P(d\alpha) + \int_0^\infty \frac{J_1(r\alpha)}{r\alpha}\Phi_S(d\alpha) \tag{3.18}$$

Making use of a table of integrals allows us to show that

$$b_N(r) = \eta(1-\frac{r^2}{2s})\exp(-\frac{r^2}{4s}) + (1-\eta)\exp(-\frac{r^2}{4p}) \tag{3.19}$$

$$b_L(r) = \eta\exp(-\frac{r^2}{4s}) + (1-\eta)(1-\frac{r^2}{2p})\exp(-\frac{r^2}{4p}) \tag{3.20}$$

$$b_0 = 1 \tag{3.21}$$

$$\lambda_1 = \eta(\frac{1}{2s}+\frac{1}{2p}) - \frac{1}{2p} \tag{3.22}$$

In particular, $\lambda_1 = 0$ when $\eta = \frac{s}{s+p}$.

We turn to generating isotropic random fields using the spectral method described above. We give, for general d, a procedure which is very easy to implement. First, let $k^1, \ldots, k^d$ be independent Gaussian random variables with mean zero and variance γ. Then the vector $\frac{k}{|k|}$ is uniform on the unit sphere, and $|k|^2 = (k^1)^2 + \cdots + (k^d)^2$ has the gamma distribution with shape index $\frac{d}{2}$ and scale parameter $\frac{1}{2\gamma}$. Writing the gamma density explicitly and changing variables shows that the distribution μ of $|k|$ is given by

$$\mu(dr) = \frac{2}{(2\gamma)^{\frac{d}{2}}\Gamma(\frac{d}{2})} r^{d-1}\exp(-\frac{r^2}{2\gamma})dr \tag{3.23}$$

Now one calculates ψ_S and ψ_P from (3.14).

Here we are interested in $d = 2$. The functions ψ_S and ψ_P are given by

$$\psi_S^2(r) = \frac{d\Phi_S}{d\mu}(r) = 4\eta\gamma s^2 r^2 \exp(r^2(\frac{1}{2\gamma} - s)) \tag{3.24}$$

$$\psi_P^2(r) = \frac{d\Phi_P}{d\mu}(r) = 4(1-\eta)\gamma p^2 r^2 \exp(r^2(\frac{1}{2\gamma} - p)) \tag{3.25}$$

We should choose γ to make the exponential factor close to 1.

In what follows, we chose $p = 1$, $s = 2$, $\gamma = \frac{3}{4}$, $\delta = 0.005$, and $N = 64$.

It is important to keep in mind the temporal and spatial scales of the flow. As noted above, for this flow $b_0 = 1$, so that the one-point motion $F_t x$, $t \geq 0$ is Brownian with zero drift and variance given by

$$\mathrm{Var}(F_t^i x) = t, \qquad i = 1, 2,\ t \geq 0$$

The coordinate processes $F_t^1 x$, $t \geq 0$ and $F_t^2 x$, $t \geq 0$ are independent.

The covariance b defines another spatial scale. The functions $b_L(r)$ and $b_N(r)$ are positive for r less than about 1.5, and they are essentially zero for r larger than about 7. Recalling the discussion following equation (2.13), we see that if two points are within about one unit distance of each other, their motions will be somewhat coherent and they will be likely to stay together for some time. Points further than about 7 units apart diffuse almost independently.

3.3.1. The density of a mass distribution. The purpose of the next two figures is to illustrate the effect compressibility has on the density of a mass distribution carried by a flow. We follow the motion of 2500 points. Initially, they are evenly spaced on a rectangular lattice with 50 rows and 50 columns. The x and y coordinates both range from -10 to 10. This approximates a uniform continuous mass distribution.

Figure 1 shows the motion of these points in an incompressible flow ($\eta = 1$) while Figure 2 shows them in a pure potential flow ($\eta = 0$). In the incompressible case, the number of atoms per unit area stays roughly constant, as it should. The potential case is strikingly different. By time 1.0, there are clearly demarcated cells containing little or no mass, the mass being concentrated on the cell boundaries.

Two comments on the presentation: The axes are adjusted in each frame to keep all the points in view. This makes the mass distribution look less spread out than it really is. Second, due to limited resolution, closely-spaced points are plotted as though they overlap.

3.3.2. The boundary of a mass distribution. We have seen the effect of a flow on the density *inside* a mass distribution. Now we consider how the *boundary* of a mass distribution is deformed as the mass spreads into the surrounding space.

Recall that for fixed t, the map F_t is a homeomorphism from $\mathbb{R}^d$ to $\mathbb{R}^d$, so the boundary will be a connected, non-self-intersecting curve for all time, provided that it was so at time 0. We model the boundary with a large number of points initially evenly spaced around a circle. We draw it at each time by connecting the points with lines. The parameters of the simulation are the same as above.

We begin by presenting simulations of the deformation of a circle of radius 1, made up of 2500 evenly-spaced points. The incompressible case is shown first (Figure 3). Recall that the Lebesgue measure of the set enclosed by the curve remains constant in time. Note how elongated the distribution becomes by time 7.

The potential case (Figure 4) shows much more elongation combined with an apparent decrease in the area enclosed by the image of the circle. Note that between time 4 and time 7, the points are compressed from a scale of about 3.5 by 4 to a scale of 0.2 by 0.2.

In Figure 5, we increase the radius of the initial circle to 10 and the number of points to 10,000. This results in an apparent increase in spatial

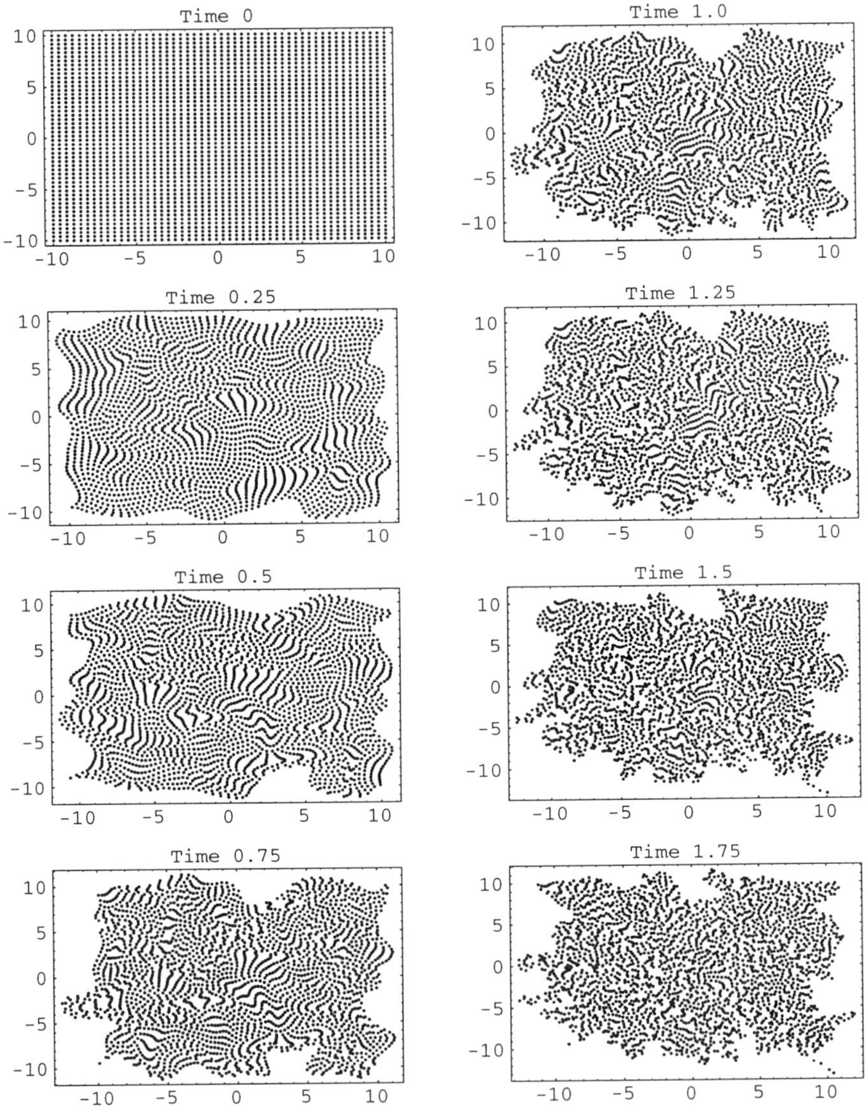

FIG. 1. *Incompressible case, with* $\eta = 1$ *and* $\lambda_1 > 0$. *Note that the mass density remains fairly constant in time.*

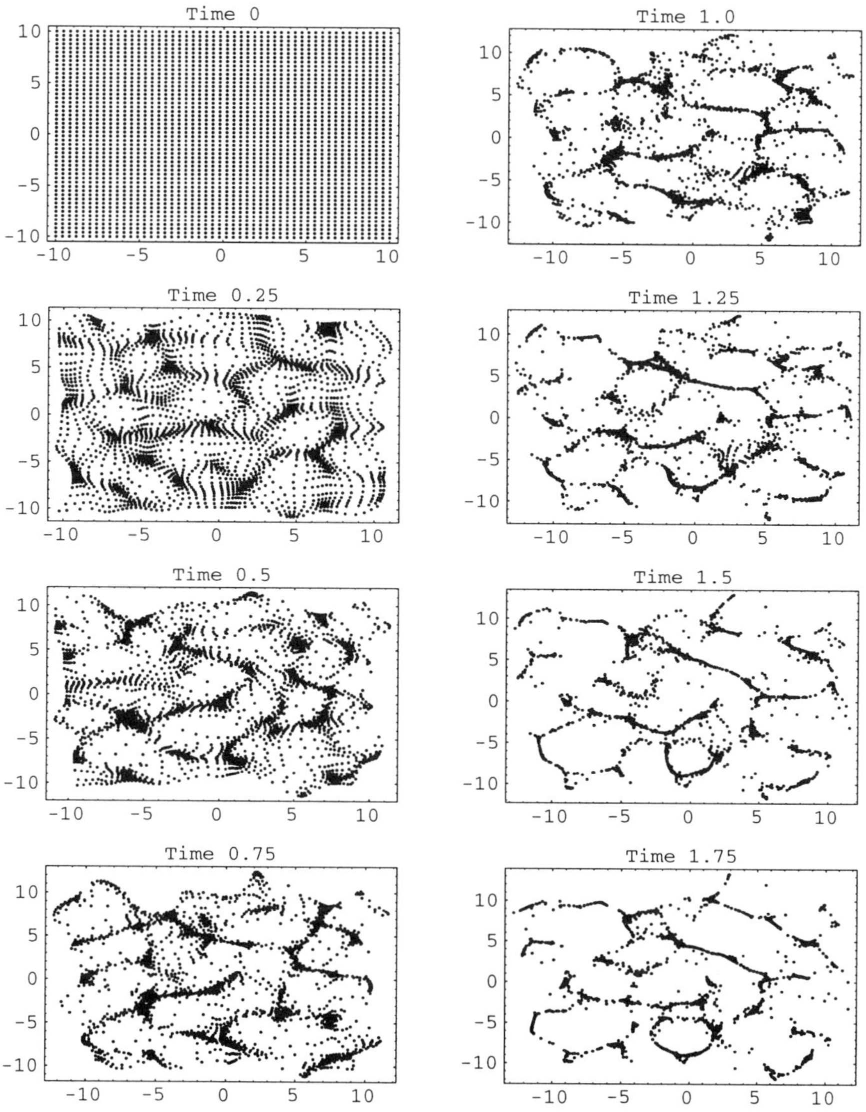

FIG. 2. *Potential case, with $\eta = 0$ and $\lambda_1 < 0$. Note how well-defined the cell boundaries are at times 1.5 and 1.75.*

structure. This structure is resolved reasonably well in the incompressible case – the points we followed did not separate too far, except in a few cases, by time 7. In the potential case, however, some neighboring points are very rapidly separated from one another, so that we quickly see a few small clumps of points separated by long distances. We have omitted the picture since it is not very informative.

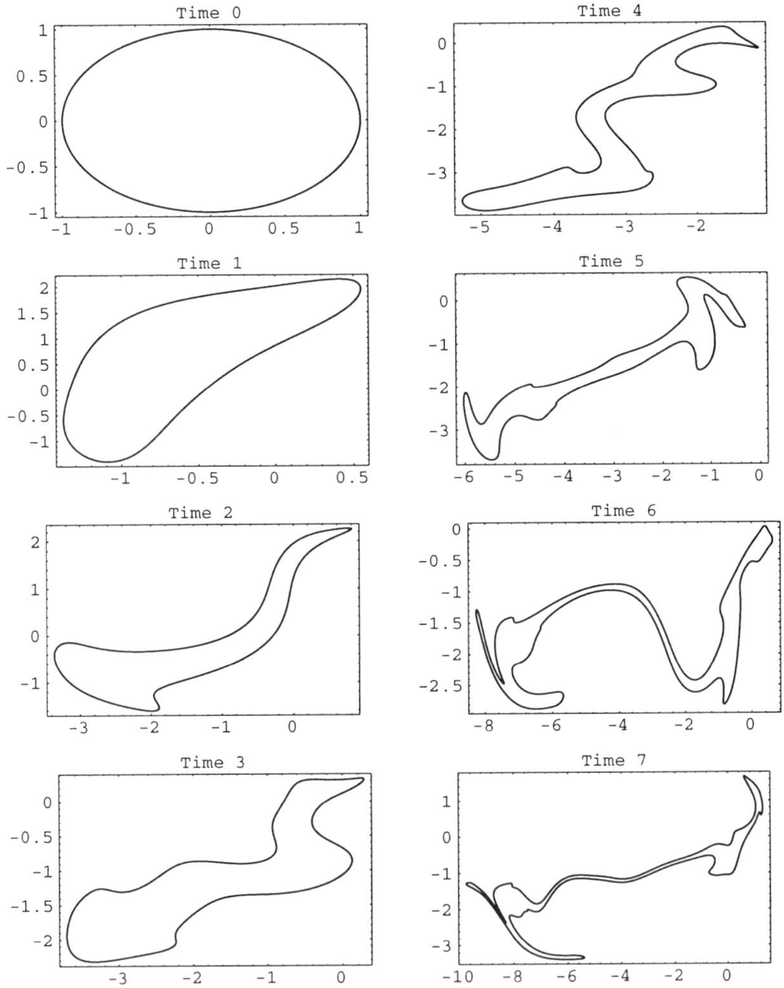

FIG. 3. *Incompressible case, with* $\eta = 1$ *and* $\lambda_1 > 0$. *Initial radius = 1. Note how elongated the curve is by time 7.*

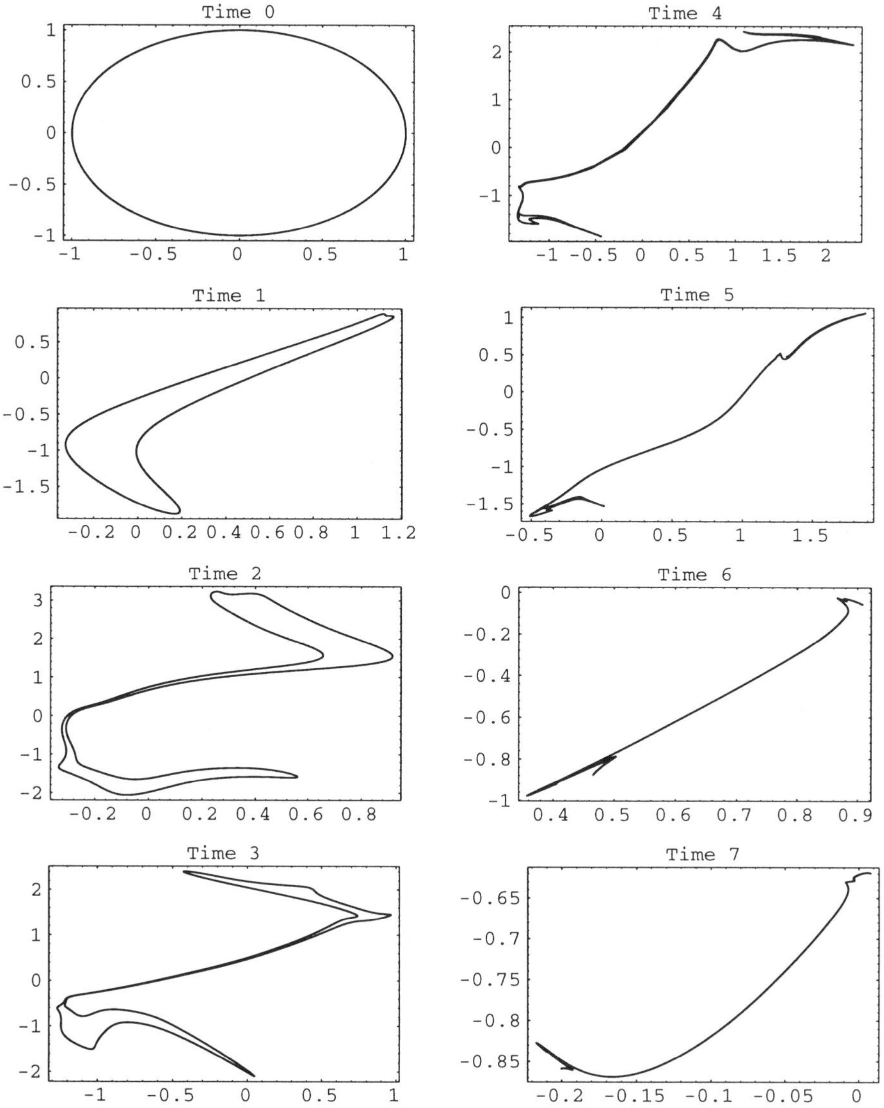

FIG. 4. *Potential case, with* $\eta = 0$ *and* $\lambda_1 < 0$. *Initial radius = 1. Note the rapid compression between times 4 and 7.*

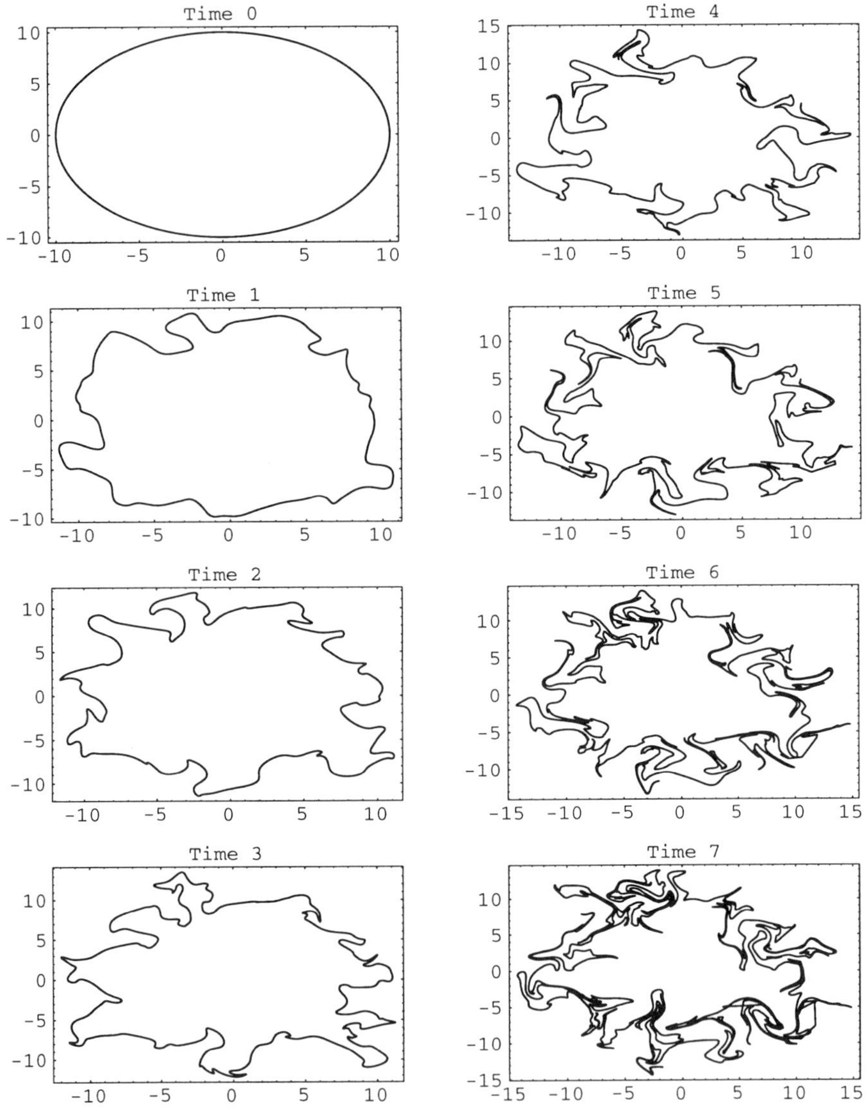

FIG. 5. *Incompressible case, with* $\eta = 1$ *and* $\lambda_1 > 0$. *Initial radius = 10.*

4. Formulas for the evolution of M_t. In this section we present a number of rigorous results and formulas describing the evolution of a mass distribution in a Brownian flow. Recall from Section 1 that M_0 is a finite, deterministic measure on $\mathbb{R}^d$ describing a mass distribution at time 0, and at time t the mass is described by the random measure M_t given by

$$M_t(R) = M_0(\{x \in \mathbb{R}^d \ : \ F_{0t}(x) \in R\}) \tag{4.1}$$

Throughout this section we will assume, without further mention, that F is a Brownian flow on $\mathbb{R}^d$ satisfying equation (2.6) where U is a Brownian motion with drift u and covariance a which satisfy (2.8)-(2.11). We will also write $F_t x$ as a shorthand for $F_{0t}(x)$.

4.1. Limiting behavior. To get our bearings, we first list a result on the limiting behavior of M_t as $t \to \infty$. See KUNITA (1990), Theorems 4.3.9 and 4.3.10 for the proof.

THEOREM 4.1. *a) Suppose that for all x in $\mathbb{R}^d$, the matrix $a(x,x)$ is strictly positive definite:*

$$\sum_{i,j=1}^{d} a^{ij}(x,x)\xi^i\xi^j > 0, \qquad \xi \in \mathbb{R}^d,\ \xi \neq 0 \tag{4.2}$$

Moreover, suppose that the one-point motion under F has an invariant distribution π with $\pi(\mathbb{R}^d) = 1$. Then for each Borel subset R of $\mathbb{R}^d$,

$$\lim_{t\to\infty} \frac{1}{t}\int_0^t ds\ M_s(R) = \pi(R) \quad \textit{almost surely} \tag{4.3}$$

b) Suppose the one-point motion does not have a finite invariant measure. Then for each bounded Borel subset R of $\mathbb{R}^d$,

$$\lim_{t\to\infty} \mathbf{E}M_t(R) = 0 \tag{4.4}$$

□

4.2. Stochastic calculus of M_t. To understand the evolution of the random measure M_t, it is useful to consider the process $\int M_t(dx)f(x), t \geq 0$, for various choices of f. For convenience we will write $M_t f$ in place of $\int M_t(dx)f(x)$. Note that by the definition of M_t, we have

$$M_t f = \int M_t(dx)f(x) = \int M_0(dx)f(F_t x) \tag{4.5}$$

For bounded f, the integral $M_t f$ will always exist and be finite. To deal with unbounded f, we introduce the following condition:

CONDITION 4.2. *There exist constants $p > 2$ and $C < +\infty$ such that $\int M_0(dx)|x|^p < +\infty$ and both $|f|$ and $|Af|$ are bounded by the function $C(1+|x|)^p$. Moreover, $f \in C^2(\mathbb{R}^d, \mathbb{R})$ and for each $i = 1, \ldots, d$, the function $\partial_i f$ is of polynomial growth.*

If this condition is satisfied, then $M_t f,\ t \geq 0$ is well defined almost surely, and the expectation of $M_t f$ is finite. The same is true of $\int_0^t ds\, M_s f$. Moreover, if f is continuous, then $M_t f,\ t \geq 0$ is also continuous. These facts and the following result on the stochastic calculus of $M_t f$ appear in ZIRBEL (1995a).

THEOREM 4.3. *Suppose that M_0 and f satisfy Condition 4.2.*

a) The process $M_t f$ satisfies the following equation:

$$M_t f = M_0 f + \int_0^t ds\, M_s(Af) + L_t, \tag{4.6}$$

where L is a continuous martingale with mean 0, given by

$$L_t = \sum_{i=1}^d \int M_0(dx) \int_0^t \partial_i f(F_s x) U_0^i(F_s x, ds). \tag{4.7}$$

The quadratic variation of L can be computed using part (b).

b) Let N_0 be a finite deterministic measure on $\mathbb{R}^d$ and define N_t by $N_t = N_0 \circ F_t^{-1}$. Suppose that N_0 and a function g satisfy Condition 4.2 with $\bar{p}$ in place of p, and in addition that for $x, y \in \mathbb{R}^d, i, j = 1, \ldots, d$,

$$|\partial_i f(x) \partial_j g(y) a^{ij}(x, y)| \leq C(1 + |x|)^p (1 + |y|)^{\bar{p}} \tag{4.8}$$

Then the joint quadratic variation of $M_t f$ and $N_t g$ satisfies

$$\langle M_t f, N_t g \rangle = \int_0^t ds \int M_s(dx) \int N_s(dy) \sum_{i,j=1}^d \partial_i f(x) \partial_j g(y) a^{ij}(x, y). \tag{4.9}$$

4.3. Mean measure, spatial covariance, Laplace functional. It is quite natural to inquire into the behavior of such simple quantities as the mean, covariance, and Laplace transform of random variables. Here we discuss the analogous quantities for the random measure M_t.

The *mean measure* μ_t of M_t is a deterministic measure defined for each $t \geq 0$ by

$$\mu_t(R) = \mathbf{E} M_t(R), \qquad \text{Borel } R \subseteq \mathbb{R}^d \tag{4.10}$$

Often, μ_t will have a density m_t with respect to Lebesgue measure. Such a density is called the *mean concentration.*

The *spatial covariance measure* η_t is a deterministic measure on $\mathbb{R}^d \times \mathbb{R}^d$ defined for each $t \geq 0$ by

$$\eta_t(R_1 \times R_2) = \mathbf{E} M_t(R_1) M_t(R_2), \qquad \text{Borel } R_1, R_2 \subseteq \mathbb{R}^d \tag{4.11}$$

It is clear that this definition can be extended to any number of factors, yielding higher moments of M_t. If M_t has the (random) density $\rho_t(x)$, then η_t will have a density $c_t(x, y)$ which satisfies

$$c_t(x, y) = \mathbf{E} \rho_t(x) \rho_t(y) \tag{4.12}$$

The covariance of $\rho_t(x)$ and $\rho_t(y)$ is then equal to $c_t(x,y) - \mathbf{E}\rho_t(x)\mathbf{E}\rho_t(y)$.

The *Laplace functional* $\mathcal{L}_t$ of M_t is defined for each Borel map f from $\mathbb{R}^d$ to $\mathbb{R}_+$ by

$$\mathcal{L}_t(f) = \mathbf{E}e^{-M_t f} \tag{4.13}$$

This is the analogue of the Laplace transform. The Laplace functional uniquely characterizes the law of M_t.

The following result concerns the evolution of μ_t, η_t, and $\mathcal{L}_t$ as t increases. Recall that A denotes the generator of the one-point motion $F_t x$, $t \geq 0$; see (2.14). Recall the notation $a_d(x) = a(x,x)$. The adjoint A^* of A is the operator defined implicitly in equation (4.14) below.

We will need the following condition in connection with the density of the mean measure.

CONDITION 4.4. *The function a_d is uniformly elliptic: For some $\delta > 0$,*

$$\sum_{i,j=1}^{d} a_d^{ij}(x)\xi^i\xi^j \geq \delta|\xi|^2, \qquad x,\xi \in \mathbb{R}^d,\ t \in [0,T]$$

Also, the functions a_d^{ij}, $\frac{\partial a_d^{ij}}{\partial x^i}$, $\frac{\partial^2 a_d^{ij}}{\partial x^i \partial x^j}$, u^i, $\frac{\partial u^i}{\partial x^i}$ exist, are bounded, and are uniformly Holder continuous in $\mathbb{R}^d \times [0,T]$. □

For the spatial covariance measure, we will need:

CONDITION 4.5. *For some $\delta > 0$,*

$$\sum_{p,q=1}^{2}\sum_{i,j=1}^{d} a^{ij}(x_p,x_q)\xi_p^i\xi_q^j \geq \delta(|\xi|_1^2 + |\xi|_2^2), \qquad x_1,x_2,\xi_1,\xi_2 \in \mathbb{R}^d,\ t \in [0,T]$$

Also, the functions a^{ij}, $\frac{\partial a^{ij}}{\partial x_p^i}$, $\frac{\partial^2 a^{ij}}{\partial x_p^i \partial x_q^j}$, u^i, $\frac{\partial u^i}{\partial x^i}$ exist, are bounded, and are uniformly Holder continuous in $\mathbb{R}^d \times \mathbb{R}^d \times [0,T]$ and $\mathbb{R}^d \times [0,T]$. □

PROPOSITION 4.1. *a) Suppose that M_0 has compact support and Condition 4.4 holds. Then μ_t has a density m_t in $C^{2,1}(\mathbb{R}^d \times (0,\infty))$ which satisfies the* advection-diffusion *equation:*

$$\frac{\partial m_t}{\partial t} = A^* m_t = \tfrac{1}{2}\sum_{i,j=1}^{d} \frac{\partial^2}{\partial x^i \partial x^j}(a_d^{ij} m_t) - \sum_{i=1}^{d} \frac{\partial}{\partial x^i}(u^i m_t) \tag{4.14}$$

b) Suppose that M_0 has compact support and Condition 4.5 holds. Then η_t has a density c_t in $C^{2,2,1}(\mathbb{R}^d \times \mathbb{R}^d \times (0,\infty))$ which satisfies

$$\frac{\partial c_t}{\partial t} = A_x^* c_t + A_y^* c_t + \tfrac{1}{2}\sum_{i,j=1}^{d} \frac{\partial^2}{\partial x^i \partial y^j}(c_t a^{ij}) \tag{4.15}$$

where A_x^* is A^* acting on the first argument of c_t, etc.

c) Suppose that M_0 and f satisfy Condition 4.2 and that f satisfies (4.8) with $g = f$ and $\bar{p} = p$. Then the derivative of $\mathbf{E}\exp(-M_t f)$ with respect to t equals

$$\mathbf{E}\exp(-M_t f)[-M_t(Af) + \tfrac{1}{2}\int M_t(dx)\int M_t(dy)\sum_{i,j=1}^{d}\partial_i f(x)\partial_j f(y)a^{ij}(x,y)]$$

(4.16)

Proof. a) Denote by P_t the transition kernel of the one-point motion under F, that is, $P_t(x, A) = \mathbf{P}(F_t x \in A)$. By Condition 4.4, the measure $P_t(x, \cdot)$ has a density $p_t(x, y)$ which is once continuously differentiable in t and twice in y (KARATZAS and SHREVE (1988), Section 5.7). Moreover, p satisfies,

$$\frac{\partial p}{\partial t} = A^* p, \qquad t > 0$$

where A^* acts on the second argument of p_t and the first argument is held fixed.

By the definition of M_t and μ_t, for Borel $R \subseteq \mathbb{R}^d$,

$$\begin{aligned}\mu_t(R) = \mathbf{E}M_t(R) &= \mathbf{E}\int M_0(dx)1_R(F_t x)\\ &= \int M_0(dx)\mathbf{E}1_R(F_t x)\\ &= \int M_0(dx)\int p_t(x,y)1_R(y)dy,\end{aligned}$$

which shows that μ_t has a density $m_t(y) = \int M_0(dx)p_t(x, y)$. Now M_0 has compact support and the time and space derivatives of p_t are continuous. This justifies the interchanges of integrals with derivatives in the following computation, and also shows that m_t is in $C^{2,1}(\mathbb{R}^d \times (0, \infty))$.

$$\begin{aligned}\frac{\partial m}{\partial t}(y) &= \int M_0(dx)\frac{\partial p}{\partial t}(x,y)\\ &= \int M_0(dx)A^* p_t(x,y)\\ &= A^* m_t(y)\end{aligned}$$

This completes the proof of part (a).

b) This is quite similar to the proof of part (a), except that here we use the two-particle transition kernel $P_t^{(2)}(x_1, x_2; A_1, A_2)$. Condition 4.5 guarantees the existence of a density for this transition kernel. The density satisfies an evolution equation of the form $\frac{\partial p^{(2)}}{\partial t} = A_2^* p^{(2)}$, where A_2^* is the adjoint of the generator of the two-point motion under F. This operator is defined implicitly in (4.15).

c) Let g be a twice differentiable function from $\mathbb{R}$ to $\mathbb{R}$. By Itô's Lemma,

$$\begin{aligned} g(M_t f) &= g(M_0 f) + \int_0^t g'(M_s f) dM_s f + \tfrac{1}{2}\int_0^t g''(M_s f) d\langle M_s f, M_s f\rangle \\ &= g(M_0 f) + \int_0^t g'(M_s f) M_s(Af) ds + \int_0^t g'(M_s f) dL_s \\ &+ \tfrac{1}{2}\int_0^t ds\, g''(M_s f) \int M_s(dx) \int M_s(dy) \sum_{i,j=1}^d \partial_i f(x) \partial_j f(y) a^{ij}(x,y) \end{aligned}$$

The second equality follows from parts (a) and (b) of Theorem 4.3. Now let $g(x) = \exp(-x)$. Take expectations and the time derivative to give

$$\frac{d}{dt}\mathbf{E}\exp(-M_t f) = -\mathbf{E}\exp(-M_t f) M_t(Af)$$

$$+ \tfrac{1}{2}\mathbf{E}\exp(-M_t f) \int M_t(dx) \int M_t(dy) \sum_{i,j=1}^d \partial_i f(x) \partial_j f(y) a^{ij}(x,y)$$

This is the desired result. □

4.4. Centroid and Dispersion Calculations. The centroid C_t and the dispersion matrix D_t were defined in Section 1. We are now able to elucidate their semimartingale structure and provide some formulas for their means and covariances. All of these results are consequences of Theorem 4.3. See ZIRBEL (1995a) for more details.

For simplicity we take $M_0(\mathbb{R}^d) = 1$. Suppose that for some $p > 2$, $\int M_0(dx)|x|^p < +\infty$. Define $f^i(x) = x^i$. Then $C_t^i = M_t f^i$, and by Theorem 4.3,

$$C_t = C_0 + \int_0^t ds\, M_s u + \int M_0(dx) \int_0^t U_0(F_s x, ds), \tag{4.17}$$

the last term of which is a martingale. We also obtain

$$\langle C^i, C^j\rangle_t = \int_0^t ds \int M_s(dx) \int M_s(dy) a^{ij}(x,y) \tag{4.18}$$

Applying Itô's formula to $C_t^i C_t^j$, we get

$$C_t^i C_t^j = C_0^i C_0^j + \int_0^t C_s^i dC_s^j + \int_0^t C_s^j dC_s^i + \langle C^i, C^j\rangle_t$$

Assume that $\int M_0(dx)|x|^4 < +\infty$. Taking expectations drops out the martingale terms and yields

$$\mathbf{E} C_t^i C_t^j = C_0^i C_0^j + \mathbf{E}\int_0^t ds\, C_s^i M_s u^j + \mathbf{E}\int_0^t ds\, C_s^j M_s u^i$$

$$+\mathbf{E}\int_0^t ds \int M_s(dx) \int M_s(dy) a^{ij}(x,y)$$

In other words,

$$\mathbf{E}C_t^i C_t^j = C_0^i C_0^j + \mathbf{E}\int_0^t ds \int M_s(dx) \int M_s(dy) E^{ij}(x,y), \tag{4.19}$$

where $E^{ij}(x,y) = a^{ij}(x,y) + x^i u^j(y) + u^i(x) y^j$. The covariance of C_t^i and C_t^j can now be computed easily.

By the definition of D_t, we have

$$D_t^{ij} = \int M_t(dx)(x^i - C_t^i)(x^j - C_t^j) = \int M_t(dx) x^i x^j - C_t^i C_t^j$$

Thus, if we set $g^{ij}(x) = x^i x^j$, we have $D_t^{ij} = M_t g^{ij} - C_t^i C_t^j$. By Theorem 4.3 and the fact that $Ag^{ij}(x) = E^{ij}(x,x)$, we have

$$M_t g^{ij} = M_0 g^{ij} + \int_0^t ds \int M_s(dx) E^{ij}(x,x) + L_t^{ij},$$

the last term of which is a martingale. Taking the expectation of D_t^{ij} yields

$$\mathbf{E}D_t^{ij} = D_0^{ij} + \mathbf{E}\int_0^t ds \int M_s(dx) \int M_s(dy)(E^{ij}(x,x) - E^{ij}(x,y)). \tag{4.20}$$

The calculation of $\langle D^{ij}, D^{k\ell}\rangle_t$ begins by substituting the expressions above for $M_t g^{ij}$ and $C_t^i C_t^j$ into $\langle D^{ij}, D^{k\ell}\rangle_t$. Standard properties of quadratic variations and part (b) of Theorem 4.3 then yield

$$\langle D^{ij}, D^{k\ell}\rangle_t = \int_0^t ds \int M_s(dx) \int M_s(dy)(x - C_s)^T H(x,y)(y - C_s), \tag{4.21}$$

where the matrix H is defined by

$$H^{mn} = \delta_{im}\delta_{kn} a^{j\ell} + \delta_{jm}\delta_{kn} a^{i\ell} + \delta_{im}\delta_{\ell n} a^{jk} + \delta_{jm}\delta_{\ell n} a^{ik} \tag{4.22}$$

with the argument (x,y) suppressed. Note that H has at most four non-zero entries. One can now calculate $\mathbf{E}D_t^{ij} D_t^{k\ell}$ and then the covariance of D_t^{ij} and $D_t^{k\ell}$, as above with C_t.

5. Centroid and dispersion in isotropic flow. In this section we specialize to the case of isotropic Brownian flows. We will quickly run through the results of Section 4 for this case, then report our results concerning the limiting behavior of C_t and D_t as $t \to \infty$.

Throughout this section, F is an isotropic Brownian flow with covariance b which satisfies conditions 2.1 and 2.2. The initial measure M_0 is deterministic and satisfies $M_0(\mathbb{R}^d) = 1$ and $\int M_0(dx)|x|^p < +\infty$ for some $p > 2$.

5.1. Formulas for M_t. ¿From Theorem 4.1, we see that $\mathbf{E}M_t(R) \to 0$ for all bounded Borel $R \subseteq \mathbb{R}^d$, since the one-point motion is d-dimensional Brownian motion, which has the Lebesgue measure as its invariant measure. Thus, the mass 'moves off to ∞' in some sense.

More precisely, since the transition density of Brownian motion is smooth, we may rework the proof of Proposition 4.1 without further conditions on b and M_0 to yield the evolution equation for the mean tracer concentration m_t:

$$\frac{\partial m_t}{\partial t} = \tfrac{1}{2} b_0 \Delta m_t, \tag{5.1}$$

which is the heat equation. Similarly, the evolution equation for the spatial covariance density c_t becomes

$$\frac{\partial c_t}{\partial t} = \tfrac{1}{2} b_0 \Delta_x c_t + \tfrac{1}{2} b_0 \Delta_y c_t + \tfrac{1}{2} \sum_{i,j=1}^{d} \frac{\partial^2}{\partial x^i \partial y^j} (c_t a^{ij}) \tag{5.2}$$

The formulas for the centroid and dispersion matrix simplify considerably for isotropic flows. The centroid is a martingale satisfying

$$C_t = C_0 + \int M_0(dx) \int_0^t U(F_s x, ds) \tag{5.3}$$

Its covariance is given by

$$\operatorname{Cov}(C_t^i, C_t^j) = \mathbf{E} \int_0^t ds \int M_s(dx) \int M_s(dy) b^{ij}(x-y) \tag{5.4}$$

The expectation of the dispersion matrix can be written compactly as

$$\mathbf{E} D_t^{ij} = D_0^{ij} + b_0 t \delta_{ij} - \operatorname{Cov}(C_t^i, C_t^j). \tag{5.5}$$

5.2. Behavior of C_t and D_t as $t \to \infty$. The centroid and dispersion matrix are meant to be rough descriptors of where the mass distribution is and how spread out it is, respectively. These are very important matters in any real-world mass transport problem.

Consider, for example, a common pollution remediation problem. A certain quantity of an industrial chemical has been spilled and is spreading underground. One would like to know whether the plume it forms is tightly concentrated or widely dispersed and, in either case, how far it has moved from the site of the spill. In the absence of detailed information about the soil layers and groundwater flow patterns, one needs to take a statistical approach to design a soil sampling and remediation strategy. The relevant concerns then are how dispersed the chemical is *likely* to be, and how far it *may* have moved from the spill site.

This is a notoriously difficult problem.

The results we present here are rigorous and exact, but of course are limited to isotropic Brownian flows. Nevertheless, we believe that this case captures many of the essential difficulties of the general problem, and may serve as a useful point of comparison for other models.

It happens that the continuous or discrete nature of M_0 plays an important role in the asymptotics of C_t and D_t. Generally speaking, the presence of a discrete (atomic) component masks more subtle effects which occur with a non-atomic distribution. If M_0 has an atomic component, we let σ denote the sum of the squares of the masses of the atoms of M_0. Otherwise, M_0 is diffuse and we set $\sigma = 0$. In either case, we have $\sigma < 1$, unless M_0 consists of exactly one atom, which is a trivial case we henceforth exclude.

One further point deserves mention. If we fix the dimension d and the variance b_0 of the one-point motion, we completely specify the one-point motion and the evolution equation (5.1) for the mean concentration m_t. Even so, there remains a wide range of possible behaviors for the centroid and dispersion matrix. Thus, it is not enough to know the mean concentration of M_t.

5.2.1. Linear growth rates. The first result is a fairly complete picture of the linear growth rates of $\mathrm{Cov}(C_t^i, C_t^j)$ and $\mathbf{E}D_t$ in the long run. This proposition covers all dimensions $d \geq 1$ and all cases except the case $d = 2$ and $\lambda_1 \geq 0$. The proof appears in ZIRBEL and CINLAR (1996). It is an easy consequence of (5.4) and Proposition 2.1.

PROPOSITION 5.1. *a) If $d \geq 3$ and $\lambda_1 \geq 0$, then*

$$\lim_{t\to\infty} \frac{d}{dt}\mathrm{Cov}(C_t^i, C_t^j) = b_0\sigma\delta_{ij}, \qquad \lim_{t\to\infty} \frac{d}{dt}\mathbf{E}D_t^{ij} = b_0(1-\sigma)\delta_{ij}$$

b) If $d = 3$ and $\lambda_1 < 0$, then

$$\lim_{t\to\infty} \frac{d}{dt}\mathrm{Cov}(C_t^i, C_t^j) = b_0(1 - p(1-\sigma))\delta_{ij}, \quad \lim_{t\to\infty} \frac{d}{dt}\mathbf{E}D_t^{ij} = b_0 p(1-\sigma)\delta_{ij}$$

where $p \in (0,1)$ equals $\int M_0(dx) \int M_0(dy)\mathbf{P}\{\lim_{t\to\infty} |F_t x - F_t y| = +\infty\}$.

c) If $d = 1$, or $d = 2$ and $\lambda_1 < 0$, then

$$\lim_{t\to\infty} \frac{d}{dt}\mathrm{Cov}(C_t^i, C_t^j) = b_0\delta_{ij}, \qquad \lim_{t\to\infty} \frac{d}{dt}\mathbf{E}D_t^{ij} = 0$$

Moreover, all these results hold with $\frac{d}{dt}$ replaced by $\frac{1}{t}$. □

We are interested in how quickly $\mathrm{Var}(C_t^i)$ and $\mathbf{E}D_t^{ii}$ increase, for these tell how far the centroid may be from its initial location and how spread out the mass is, respectively. The gist of this result is that for large d and $\lambda_1 \geq 0$, dispersion increases at the maximum rate possible, and for small d and $\lambda_1 < 0$, dispersion is not so pronounced, but the centroid moves a great deal.

We are able to give more detailed information in two cases, $d \geq 3$ and $\lambda_1 > 0$, and $d = 2$ and $\lambda_1 > 0$.

5.2.2. Three or more dimensions. Next, we consider isotropic Brownian flows in three or more dimensions and with strictly positive Lyapunov exponent. We will give very exact bounds on the deviation of $\mathrm{Cov}(C_t^i, C_t^j)$ and $\mathbf{E}D_t^{ij}$ from their asymptotic linear growth rates. The proof is given in ZIRBEL (1995b).

We need to assume that the covariance b satisfies:

$$\int_0^\infty dz\ z(|b_N(z)| + |b_L(z)|) < +\infty \tag{5.6}$$

Essentially, we need b_N and b_L to decay faster than $\frac{1}{z^2}$ as $z \to \infty$. Also, we require

$$\int_{\mathbb{R}^d} M_0(dx) \int_{\mathbb{R}^d \setminus \{x\}} M_0(dy) \log^+ \frac{1}{|x - y|} < +\infty \tag{5.7}$$

where $\log^+ z = \max(0, \log z)$. This says that the diffuse component of M_0 must not have strong concentrations of mass.

Let s denote the scale function of the separation process Z_t, and let m be the density of the speed measure of Z. Note that $s(\infty) < \infty$ when $\lambda_1 > 0$, since in that case $Z_t \to \infty$ almost surely.

THEOREM 5.1. *Suppose $d \geq 3$, $\lambda_1 > 0$, and the two conditions above hold. Then for all $t \geq 0$, and $i, j = 1 \ldots d$, we have*

$$|\mathrm{Cov}(C_t^i, C_t^j) - b_0 \sigma t \delta_{ij}| \leq K \tag{5.8}$$

$$|\mathbf{E}D_t^{ij} - D_0^{ij} - b_0(1 - \sigma) t \delta_{ij}| \leq K, \tag{5.9}$$

where K is a finite constant equal to

$$\int M_0(dx) \int_{y \neq x} M_0(dy) \int_0^\infty dz\ m(z)(2|b_N(z)| + |b_L(z)|)(s(\infty) - s(|x - y| \wedge z))$$

Thus, $\mathrm{Cov}(C_t^i, C_t^j)$ and $\mathbf{E}D_t^{ij}$ stay extremely close to their linear asymptotes as $t \to \infty$. More remarkable is what occurs when M_0 is diffuse, so the 'linear asymptote' of $\mathrm{Cov}(C_t^i, C_t^j)$ is 0.

COROLLARY 5.1. *Suppose M_0 is diffuse. Then, for all $t \geq 0$ and $i, j = 1 \ldots d$,*

$$|\mathrm{Cov}(C_t^i, C_t^j)| \leq K \tag{5.10}$$

$$|\mathbf{E}D_t^{ij} - D_0^{ij} - b_0 t \delta_{ij}| \leq K \tag{5.11}$$

Moreover, the martingale C_t converges almost surely as $t \to \infty$ to a random vector C_∞ with mean C_0 and variance bounded by K.

Proof. The inequalities are immediate from the Theorem. The variance of C_t^i is bounded, so C^i is a uniformly integrable martingale, which means it converges almost surely. □

This is a remarkable result. Apparently, in this case, dispersion relative to the centroid is strong enough that the centroid, a spatial average over the mass distribution, converges. This is a sort of strong law of large numbers for the spatial average, which indicates that the particles making up the mass distribution have quite weakly dependent motions at long times.

On the other hand, as soon as we have $\lambda_1 < 0$, the variance of C_t grows linearly in time, by Proposition 5.1.

5.2.3. Two dimensions with positive Lyapunov exponent. Finally, we give a limit theorem for the case $d = 2$ and $\lambda_1 > 0$. The proof appears in ZIRBEL (1995b).

We will need two conditions on M_0. For some $\epsilon > 0$,

$$\int M_0(dx)e^{\epsilon|x|} < +\infty \tag{5.12}$$

$$\int_{\mathbb{R}^d} M_0(dx) \int_{\mathbb{R}^d\setminus\{x\}} M_0(dy)\frac{1}{|x-y|^\epsilon} < +\infty \tag{5.13}$$

The first condition requires the mass to be localized in space. The second condition will be satisfied, for example, if the diffuse component of M_0 has bounded density with respect to Lebesgue measure. We also need a condition on the covariance b:

$$\int_0^\infty dz\; z\; |b_L(z) + b_N(z)| < +\infty \tag{5.14}$$

We introduce the notation $\mathrm{Var}(C_t)$ for $\mathrm{Var}(C_t^1) + \mathrm{Var}(C_t^2)$, and B_t for $D_t^{11} + D_t^{22}$.

THEOREM 5.2. *Suppose $d = 2$, $\lambda_1 > 0$, and the conditions above hold. Then,*

$$\lim_{t\to\infty} \frac{\mathrm{Var}(C_t) - 2b_0\sigma t}{\log t} = \frac{1-\sigma}{2b_0}K \int_0^\infty dz\; m(z)(b_L(z) + b_N(z)) \tag{5.15}$$

$$\lim_{t\to\infty} \frac{2b_0(1-\sigma)t - \mathbf{E}B_t}{\log t} = \frac{1-\sigma}{2b_0}K \int_0^\infty dz\; m(z)(b_L(z) + b_N(z)) \tag{5.16}$$

where $K = \lim_{z\to\infty} \frac{z}{m(z)}$. Moreover, if the flow is incompressible and $\lim_{z\to\infty} z^2 b_L(z) = 0$, then the right side equals 0. □

When M_0 is diffuse, we have $\sigma = 0$ and so (5.15) becomes,

$$\lim_{t\to\infty} \frac{\mathrm{Var}(C_t)}{\log t} = \frac{K}{2b_0}\int_0^\infty dz\, m(z)(b_L(z) + b_N(z)) \tag{5.17}$$

In the incompressible case, if $\lim_{z\to\infty} z^2 b_L(z) = 0$, we even get

$$\mathrm{Var}(C_t) \ll \log t \tag{5.18}$$

$$0 \le 2b_0 t - \mathbf{E}B_t \ll \log t \tag{5.19}$$

as $t \to \infty$.

Consider what we know now about two-dimensional isotropic Brownian flows. Beginning with the incompressible case, $\mathrm{Var}(C_t)$ increases much slower than $\log t$. As we add a potential component to the flow, λ_1 decreases, and $\mathrm{Var}(C_t)$ goes as $\log t$. We cross the point $\lambda_1 = 0$ (for which we have no result as yet), and suddenly $\mathrm{Var}(C_t)$ is asymptotically linear. Similar (but opposite) comments apply to relative dispersion by formula (5.5). Thus it is that the degree of compressibility in the flow exerts a huge effect on the behavior of a mass distribution in the flow.

REFERENCES

P.H. BAXENDALE and T.E. HARRIS (1986) *Isotropic stochastic flows.* Ann. Prob. **14**, No. 4, 1155–1179.

R.W.R. DARLING (1989) *Isotropic stochastic flows: a survey.* Diffusion Processes and Related Problems in Analysis, Vol. II: Stochastic Flows. M. Pinsky, Ed.

I. KARATZAS and S.E. SHREVE (1988) *Brownian Motion and Stochastic Calculus.* Springer-Verlag New York.

P.E. KLOEDEN and E. PLATEN (1992) *Numerical Solution of Stochastic Differential Equations.* Springer-Verlag Berlin Heidelberg.

A.N. KOLMOGOROV (1941) *The local structure of turbulence in an incompressible viscous fluid with very large Reynolds numbers.* Dokl. Akad. Nauk SSSR **30**, 301–305.

H. KUNITA (1990) *Stochastic flows and stochastic differential equations.* Cambridge University Press.

Y. LE JAN (1985) *On isotropic Brownian motions.* Z. Wahrsch. verw. Gebiete. **70**, 609–620.

A.S. MONIN and A.M. YAGLOM (1971) *Statistical Fluid Mechanics: Mechanics of Turbulence.* The MIT Press, Cambridge, Massachusetts.

D. TALAY (1990) *Simulation and numerical analysis of stochastic differential systems: a review.* Tech. Report 1313, INRIA.

A.M. YAGLOM (1957) *Some classes of random fields in n-dimensional space, related to stationary random processes.* Theory Probab. Appl. **2**, 273–320.

A.M. YAGLOM (1987) *Correlation theory of stationary and related random functions.* Vol. 1; *Basic Results.* Springer-Verlag New York.

C.L. ZIRBEL (1995a) *Random measures carried by Brownian flows.* Preprint.

C.L. ZIRBEL (1995b) *Translation and dispersion of mass by isotropic Brownian flows.* Preprint.

C.L. ZIRBEL and E. ÇINLAR (1996) *Dispersion of particle systems in Brownian flows.* Adv. Appl. Prob. **28**, 53–74.

IMA SUMMER PROGRAMS

1987 Robotics
1988 Signal Processing
1989 Robustness, Diagnostics, Computing and Graphics in Statistics
1990 Radar and Sonar (June 18 - June 29)
New Directions in Time Series Analysis (July 2 - July 27)
1991 Semiconductors
1992 Environmental Studies: Mathematical, Computational, and Statistical Analysis
1993 Modeling, Mesh Generation, and Adaptive Numerical Methods for Partial Differential Equations
1994 Molecular Biology
1995 Large Scale Optimizations with Applications to Inverse Problems, Optimal Control and Design, and Molecular and Structural Optimization
1996 Emerging Applications of Number Theory
1997 Statistics in Health Sciences.

SPRINGER LECTURE NOTES FROM THE IMA:

The Mathematics and Physics of Disordered Media
Editors: Barry Hughes and Barry Ninham
(Lecture Notes in Math., Volume 1035, 1983)

Orienting Polymers
Editor: J.L. Ericksen
(Lecture Notes in Math., Volume 1063, 1984)

New Perspectives in Thermodynamics
Editor: James Serrin
(Springer-Verlag, 1986)

Models of Economic Dynamics
Editor: Hugo Sonnenschein
(Lecture Notes in Econ., Volume 264, 1986)

The IMA Volumes in Mathematics and its Applications

Current Volumes:

1 **Homogenization and Effective Moduli of Materials and Media** J. Ericksen, D. Kinderlehrer, R. Kohn, and J.-L. Lions (eds.)
2 **Oscillation Theory, Computation, and Methods of Compensated Compactness** C. Dafermos, J. Ericksen, D. Kinderlehrer, and M. Slemrod (eds.)
3 **Metastability and Incompletely Posed Problems** S. Antman, J. Ericksen, D. Kinderlehrer, and I. Muller (eds.)
4 **Dynamical Problems in Continuum Physics** J. Bona, C. Dafermos, J. Ericksen, and D. Kinderlehrer (eds.)
5 **Theory and Applications of Liquid Crystals** J. Ericksen and D. Kinderlehrer (eds.)
6 **Amorphous Polymers and Non-Newtonian Fluids** C. Dafermos, J. Ericksen, and D. Kinderlehrer (eds.)
7 **Random Media** G. Papanicolaou (ed.)
8 **Percolation Theory and Ergodic Theory of Infinite Particle Systems** H. Kesten (ed.)
9 **Hydrodynamic Behavior and Interacting Particle Systems** G. Papanicolaou (ed.)
10 **Stochastic Differential Systems, Stochastic Control Theory, and Applications** W. Fleming and P.-L. Lions (eds.)
11 **Numerical Simulation in Oil Recovery** M.F. Wheeler (ed.)
12 **Computational Fluid Dynamics and Reacting Gas Flows** B. Engquist, M. Luskin, and A. Majda (eds.)
13 **Numerical Algorithms for Parallel Computer Architectures** M.H. Schultz (ed.)
14 **Mathematical Aspects of Scientific Software** J.R. Rice (ed.)
15 **Mathematical Frontiers in Computational Chemical Physics** D. Truhlar (ed.)
16 **Mathematics in Industrial Problems** A. Friedman
17 **Applications of Combinatorics and Graph Theory to the Biological and Social Sciences** F. Roberts (ed.)
18 ***q*-Series and Partitions** D. Stanton (ed.)
19 **Invariant Theory and Tableaux** D. Stanton (ed.)
20 **Coding Theory and Design Theory Part I: Coding Theory** D. Ray-Chaudhuri (ed.)
21 **Coding Theory and Design Theory Part II: Design Theory** D. Ray-Chaudhuri (ed.)
22 **Signal Processing Part I: Signal Processing Theory** L. Auslander, F.A. Grünbaum, J.W. Helton, T. Kailath, P. Khargonekar, and S. Mitter (eds.)

23 **Signal Processing Part II: Control Theory and Applications of Signal Processing** L. Auslander, F.A. Grünbaum, J.W. Helton, T. Kailath, P. Khargonekar, and S. Mitter (eds.)

24 **Mathematics in Industrial Problems, Part 2** A. Friedman

25 **Solitons in Physics, Mathematics, and Nonlinear Optics** P.J. Olver and D.H. Sattinger (eds.)

26 **Two Phase Flows and Waves** D.D. Joseph and D.G. Schaeffer (eds.)

27 **Nonlinear Evolution Equations that Change Type** B.L. Keyfitz and M. Shearer (eds.)

28 **Computer Aided Proofs in Analysis** K. Meyer and D. Schmidt (eds.)

29 **Multidimensional Hyperbolic Problems and Computations** A. Majda and J. Glimm (eds.)

30 **Microlocal Analysis and Nonlinear Waves** M. Beals, R. Melrose, and J. Rauch (eds.)

31 **Mathematics in Industrial Problems, Part 3** A. Friedman

32 **Radar and Sonar, Part I** R. Blahut, W. Miller, Jr., and C. Wilcox

33 **Directions in Robust Statistics and Diagnostics: Part I** W.A. Stahel and S. Weisberg (eds.)

34 **Directions in Robust Statistics and Diagnostics: Part II** W.A. Stahel and S. Weisberg (eds.)

35 **Dynamical Issues in Combustion Theory** P. Fife, A. Liñán, and F.A. Williams (eds.)

36 **Computing and Graphics in Statistics** A. Buja and P. Tukey (eds.)

37 **Patterns and Dynamics in Reactive Media** H. Swinney, G. Aris, and D. Aronson (eds.)

38 **Mathematics in Industrial Problems, Part 4** A. Friedman

39 **Radar and Sonar, Part II** F.A. Grünbaum, M. Bernfeld, and R.E. Blahut (eds.)

40 **Nonlinear Phenomena in Atmospheric and Oceanic Sciences** G.F. Carnevale and R.T. Pierrehumbert (eds.)

41 **Chaotic Processes in the Geological Sciences** D.A. Yuen (ed.)

42 **Partial Differential Equations with Minimal Smoothness and Applications** B. Dahlberg, E. Fabes, R. Fefferman, D. Jerison, C. Kenig, and J. Pipher (eds.)

43 **On the Evolution of Phase Boundaries** M.E. Gurtin and G.B. McFadden

44 **Twist Mappings and Their Applications** R. McGehee and K.R. Meyer (eds.)

45 **New Directions in Time Series Analysis, Part I** D. Brillinger, P. Caines, J. Geweke, E. Parzen, M. Rosenblatt, and M.S. Taqqu (eds.)

46 **New Directions in Time Series Analysis, Part II**
D. Brillinger, P. Caines, J. Geweke, E. Parzen, M. Rosenblatt, and M.S. Taqqu (eds.)

47 **Degenerate Diffusions**
W.-M. Ni, L.A. Peletier, and J.-L. Vazquez (eds.)

48 **Linear Algebra, Markov Chains, and Queueing Models**
C.D. Meyer and R.J. Plemmons (eds.)

49 **Mathematics in Industrial Problems, Part 5** A. Friedman

50 **Combinatorial and Graph-Theoretic Problems in Linear Algebra**
R.A. Brualdi, S. Friedland, and V. Klee (eds.)

51 **Statistical Thermodynamics and Differential Geometry of Microstructured Materials**
H.T. Davis and J.C.C. Nitsche (eds.)

52 **Shock Induced Transitions and Phase Structures in General Media** J.E. Dunn, R. Fosdick, and M. Slemrod (eds.)

53 **Variational and Free Boundary Problems**
A. Friedman and J. Spruck (eds.)

54 **Microstructure and Phase Transitions**
D. Kinderlehrer, R. James, M. Luskin, and J.L. Ericksen (eds.)

55 **Turbulence in Fluid Flows: A Dynamical Systems Approach**
G.R. Sell, C. Foias, and R. Temam (eds.)

56 **Graph Theory and Sparse Matrix Computation**
A. George, J.R. Gilbert, and J.W.H. Liu (eds.)

57 **Mathematics in Industrial Problems, Part 6** A. Friedman

58 **Semiconductors, Part I**
W.M. Coughran, Jr., J. Cole, P. Lloyd, and J. White (eds.)

59 **Semiconductors, Part II**
W.M. Coughran, Jr., J. Cole, P. Lloyd, and J. White (eds.)

60 **Recent Advances in Iterative Methods**
G. Golub, A. Greenbaum, and M. Luskin (eds.)

61 **Free Boundaries in Viscous Flows**
R.A. Brown and S.H. Davis (eds.)

62 **Linear Algebra for Control Theory**
P. Van Dooren and B. Wyman (eds.)

63 **Hamiltonian Dynamical Systems: History, Theory, and Applications**
H.S. Dumas, K.R. Meyer, and D.S. Schmidt (eds.)

64 **Systems and Control Theory for Power Systems**
J.H. Chow, P.V. Kokotovic, R.J. Thomas (eds.)

65 **Mathematical Finance**
M.H.A. Davis, D. Duffie, W.H. Fleming, and S.E. Shreve (eds.)

66 **Robust Control Theory** B.A. Francis and P.P. Khargonekar (eds.)

67 **Mathematics in Industrial Problems, Part 7** A. Friedman

68 **Flow Control** M.D. Gunzburger (ed.)

69 **Linear Algebra for Signal Processing**
A. Bojanczyk and G. Cybenko (eds.)
70 **Control and Optimal Design of Distributed Parameter Systems**
J.E. Lagnese, D.L. Russell, and L.W. White (eds.)
71 **Stochastic Networks** F.P. Kelly and R.J. Williams (eds.)
72 **Discrete Probability and Algorithms**
D. Aldous, P. Diaconis, J. Spencer, and J.M. Steele (eds.)
73 **Discrete Event Systems, Manufacturing Systems, and Communication Networks**
P.R. Kumar and P.P. Varaiya (eds.)
74 **Adaptive Control, Filtering, and Signal Processing**
K.J. Åström, G.C. Goodwin, and P.R. Kumar (eds.)
75 **Modeling, Mesh Generation, and Adaptive Numerical Methods for Partial Differential Equations** I. Babuska, J.E. Flaherty, W.D. Henshaw, J.E. Hopcroft, J.E. Oliger, and T. Tezduyar (eds.)
76 **Random Discrete Structures** D. Aldous and R. Pemantle (eds.)
77 **Nonlinear Stochastic PDEs: Hydrodynamic Limit and Burgers' Turbulence** T. Funaki and W.A. Woyczynski (eds.)
78 **Nonsmooth Analysis and Geometric Methods in Deterministic Optimal Control** B.S. Mordukhovich and H.J. Sussmann (eds.)
79 **Environmental Studies: Mathematical, Computational, and Statistical Analysis** M.F. Wheeler (ed.)
80 **Image Models (and their Speech Model Cousins)**
S.E. Levinson and L. Shepp (eds.)
81 **Genetic Mapping and DNA Sequencing**
T. Speed and M.S. Waterman (eds.)
82 **Mathematical Approaches to Biomolecular Structure and Dynamics**
J.P. Mesirov, K. Schulten, and D. Sumners (eds.)
83 **Mathematics in Industrial Problems, Part 8** A. Friedman
84 **Classical and Modern Branching Processes**
K.B. Athreya and P. Jagers (eds.)
85 **Stochastics Models in Geosystems**
S.A. Molchanov and W.A. Woyczynski (eds.)
86 **Computational Wave Propagation**
B. Engquist and G.A. Kriegsmann (eds.)

FORTHCOMING VOLUMES

1993–1994: *Emerging Applications of Probability*
Mathematical Population Genetics
1994–1995: *Waves and Scattering*
Wavelet, Multigrid and Other Fast Algorithms (Multiple, FFT) and Their Use in Wave Propagation
Waves in Random and Other Complex Media
Inverse Problems in Wave Propagation
Singularities and Oscillations
Quasiclassical Methods
Multiparticle Quantum Scattering with Applications to Nuclear, Atomic, and Molecular Physics
1995 Summer Program: *Large Scale Optimization with Applications to Inverse Problems, Optimal Control and Design, and Molecular and Structural Optimization*
1995–1996: *Mathematical Methods in Materials Science*
Mechanical Response of Materials from Angstroms to Meters
Phase Transformations, Composite Materials, and Microstructure
Disordered Materials
Particulate Flows: Processing and Rheology
Interface and Thin Films
Nonlinear Optical Materials
Numerical Methods for Polymeric Systems
Topology and Geometry in Polymer Science
Mathematics in Industrial Problems, Part 9
1996 Summer Program: *Emerging Applications of Number Theory*
Applications and Theory of Random Sets